Natural Rubber:

Biology, Cultivation and Technology

Developments in Crop Science

Developments in Crop Science 23

Natural Rubber:

Biology, Cultivation and Technology

edited by

M. R. Sethuraj and N. M. Mathew

Rubber Research Institute of India
Kottayam-686009, Kerala, India

ELSEVIER

Amsterdam — London — New York — Tokyo   1992

ELSEVIER SCIENCE PUBLISHERS B.V.
Sara Burgerhartstraat 25
P.O. Box 211, 1000 AE Amsterdam, The Netherlands

Library of Congress Cataloging-in-Publication Data

Natural rubber : biology, cultivation, and technology / edited by M.R.
  Sethuraj and N.M. Mathew.
      p.   cm. -- (Developments in crop science ; 23)
   ISBN 0-444-88329-0 (acid-free paper)
   1. Hevea. 2. Rubber.   I. Sethuraj, M. R. II. Mathew, N. M.
  III. Series.
  SB291.H4N38 1992
  633.8'952--dc20
                                                            92-28340
                                                                 CIP

ISBN 0-444-88329-0

This book is printed on acid-free paper.

Transferred to digital printing 2006

PREFACE

Professor Richard Evans Schultes has rightly stated "no single species of plant has, in the short span of 100 years, so utterly altered life styles around the globe as <u>Hevea brasiliensis</u>", which today accounts for 98% of World's natural rubber production. Neither has any other plant product been used in such a wide range of industrial applications as natural rubber. Although the tree and its product have been known for long in the Amazonian valley, its introduction as a plantation crop in the East in the last quarter of the 19th century paved the way for systematic studies on improvement of the plant and scientific methods of its exploitation and processing of the crop. The discovery of the pneumatic tyre and the subsequent automobile revolution further strengthened the commercial viability of rubber plantations and even in the face of stiff competition from synthetic rubbers, introduced during and after the Second World War, natural rubber has been playing a significant role as an industrial raw material.

In spite of this spectacular development, the number of comprehensive publications on the scientific aspects of natural rubber has been very limited. Even the available books have been mostly confined to either the plantation or the technology aspects only. We consider a single book covering all the important aspects of natural rubber from its history, production, processing to sophisticated engineering applications may not be out of place, and hence this book.

The book follows a monograph style of presentation and consists of twenty five chapters covering a broad spectrum of subjects related to natural rubber. The authors have been drawn from different countries reflecting the truly international contribution to the development of this unique commodity. Most of the authors are active scientists who have considerable experience in scientific research on natural rubber. This is evident from the richness of information and the depth of treatment in each chapter. While we have attempted to maintain a certain amount of uniformity of presentation, we have tried to retain the individual styles of the authors, as far as possible. The author of Chapter 12, Dr. P.D. Abraham expired since contributing the manuscript. As we did not want to make alteration in the manuscript without the consent of the author, we have, more or less, retained Chapter 12 in the original form. Some of the chapters have become longer than expected. Although we have pruned them considerably, certain detailed treatments of practical importance have been deliberately retained. The book contains one chapter each on rubber wood and ancillary products, considering their influence on the commercial viability of rubber plantations. Also, a chapter on guayule has been included as details on this possible alternative source of natural rubber are rarely seen in similar books.

We believe that this could be a useful reference book on natural rubber for the purpose of higher learning and research and also for enlightened planters and manufacturers. We also believe that the detailed treatment and references to the latest work on the different aspects will stimulate further research and development work in this field.

M.R. SETHURAJ
N.M. MATHEW

ACKNOWLEDGEMENT

The editors express their sincere thanks and appreciation to the authors of various chapters for their elegant contributions, which only have made this monograph a useful reference book. As suggested by the author, manuscript of Chapter 3, was referred to Professor Richard E. Schultes of Harward Botanical Museum. The useful comments by Professor Schultes are gratefully acknowledged. Some of the tables and figures are reproduced from earlier publications with permission from the publishers. Such instances have been indicated wherever necessary. We acknowledge with thanks the permission received from the respective publishers. Dr. Mercykutty Joseph and Ms. Jayashree Madhavan have helped us very much at various stages of preparation of the manuscript and in preparing the subject index. The typing of the camera ready manuscript was elegantly done by Mr. R. Babu. Some of the artwork presented in the book were prepared by Mr. K.P. Sreeranganathan. The editors are very much thankful to them for their very valuable assistance. The cooperation extended by Ms. Hetty Verhagen of Elsevier Science Publishers is also gratefully acknowledged.

The editors are very much grateful to the Chairman, Rubber Board of India, for permission to publish this book.

M.R. SETHURAJ

N.M. MATHEW

CONTENTS

CONTRIBUTORS

P.D. ABRAHAM, Formerly of: The Rubber Research Institute of Malaysia, 260 Jalan Ampang, Kuala Lumpur, Malaysia.

P.W. ALLEN, International Rubber Research and Development Board, Chapel Building, Brickendonbury, Hertford SG13 8NL, England.

Y. ANNAMMA VARGHESE, Rubber Research Institute of India, Kottayam-686009, Kerala, India.

BABY KURIAKOSE, Rubber Research Institute of India, Kottayam-686009, Kerala, India.

K. BERGER, Economic and Social Institute, Free University, P.O. Box 7161, 1007 MC Amsterdam, The Netherlands.

A.K. BOWMICK, Rubber Technology Centre, Indian Institute of Technology, Kharagpur-721302, India.

D.S. CAMPBELL, Tun Abdul Razak Laboratory, Malaysian Rubber Producers' Research Association, Brickendonbury, Hertford SG13 8NL, England.

S.K. DE, Rubber Technology Centre, Indian Institute of Technology, Kharagpur-721302, India.

V. HARIDASAN, Rubber Research Institute of India, Kottayam-686009, Kerala, India.

HUANG ZONGDAO, South China Academy of Tropical Crops, Baodao Xincun, Danxian, Hainan, Peoples Republic of China.

J.L. JACOB, Institut de Recherches sur le Caoutchouc, Department Caoutchouc du CIRAD, B.P. 5035, 34032 Montpellier Cedex, France.

K. JAYARATHNAM, RUbber Research Institute of India, Kottayam-686009, Kerala, India.

K.P. JONES, Tun Abdul Razak Laboratory, Malaysian Rubber Producers' Research Association, Brickendonbury, Hertford SG13 8NL, England.

JOSEPH G. MARATTUKALAM, Rubber Research Institute of India, Kottayam-686009, Kerala, India.

A.K. KRISHNAKUMAR, Rubber Board, NRETC, Bhalukiatilla, Kunjaban, P.O., Agartala-799006, India.

C. KURUVILLA JACOB, Rubber Research Institute of India, Kottayam-686009, Kerala India.

A. de S. LIYANAGE, School of Agriculture and Institute for Research, Education and Training in Agriculture, University of South Pacific, Private Bag, Apia, Western Samoa.

LOW FEE CHON, Rubber Research Institute of Malaysia, 260 Jalan Ampang, Kuala Lumpur, Malaysia.

LUKMAN, Balai Penelitian Perkebunan Sungei Putih, P.O. Box 416, Medan, Indonesia.

N.M. MATHEW, Rubber Research Institute of India, Kottayam-686009, Kerala, India.

C. METHERELL, Run Abdul Razak Laboratory, Malaysian Rubber Producers' Research Association, Brickendonbury, Hertford SG13 8NL, England.

F.S. NAKAYAMA, U.S. Water Conservation Lab, 4331 E Broadway Road, Phoenix, AZ85040, United States of America.

NAMITHA ROY CHOUDHURY, Formerly of: Rubber Technology Centre, Indian Institute of Technology, Kharagpur-721302, India.

S.W. PAKIANATHAN, Formerly of: Rubber Research Institute of Malaysia, 260 Jalan Ampang, Kuala Lumpur, Malaysia.

A.O.N. PANIKKAR, Rubber Research Institute of India, Kottayam-686009, Kerala, India.

PAN YANQING, South China Academy of Tropical Crops, Baodao Xincun, Danxian, Hainan, Peoples Republic of China.

S.N. POTTY, Rubber Research Institute of India, Kottayam-686009, Kerala, India.

D. PREMAKUMARI, Rubber Research Institute of India, Kottayam-686009, Kerala, India.

J.C. PREVOT, Institut de Recherches sur le Caoutchouc, Department Caoutchouc du CIRAD, B.P. 5035, 34032 Montpellier Cedex, France.

P. SANJEEVA RAO, Rubber Research Institute of India, Kottayam-686009, Kerala, India.

C.K. SARASWATHY AMMA, Rubber Research Institute of India, Kottayam-686009, Kerala India.

A.C. SEKHAR, Sanjogtha, 146 NE Layout, Seethammadhara, Visakhapatnam-530013, India.

M.R. SETHURAJ, Rubber Research Institute of India, Kottayam-686009, Kerala, India.

H.P. SMIT, Economic and Social Institute, Free University, P.O. Box 7161, 1007 MC Amsterdam, The Netherlands.

A. STEVENSON, Materials Engineering Research Laboratory Ltd., Tamworth Road, Hertford SG13 7DG, England.

S.J. TATA, Formerly of: Rubber Research Institute of Malaysia, 260 Jalan Ampang, Kuala Lumpur, Malaysia.

K.R. VIJAYAKUMAR, Rubber Research Institute of India, Kottayam-686009, Kerala, India.

P.R. WYCHERLEY, Kings Park and Botanical Garden, West Perth, Western Australia 6005, Australia.

CHAPTER 1

HISTORICAL DEVELOPMENT OF THE WORLD RUBBER INDUSTRY

K.P. JONES

Malaysian Rubber Producers' Research Association, Brickendonbury, Hertford SG13 8NL, England

P.W. ALLEN

International Rubber Research and Development Board, Chapel Building, Brickendonbury, Hertford SG13 8NP, England

INTRODUCTiON

The term 'rubber' describes a group of materials which are highly elastic: a strip of rubber can be stretched severalfold without breaking and will return quickly to its original length on releasing the stretching force. Although high elasticity by itself is not the prime consideration in most applications of rubbers, it is associated with various other properties which are more or less common to all types of rubber and which render them very useful for the manufacture of many products, the most familiar being the pneumatic tyre.

This property of elasticity stems from the fact that all rubbers are composed of long, flexible molecules ('linear high polymers'). As prepared, either by nature (natural rubber) or by man (synthetic rubbers), rubbers are not especially strong and not very flexible, but these necessary properties are imparted by contriving that the long molecules become cross-linked by the process of vulcanization. The useful properties of vulcanized rubber (eg. strength, resilience) are determined by the polymeric nature of the raw rubber as modified by the crosslinks, and these properties are further modified by the use of fillers. The art of designing a 'vulcanizate' for a particular application was developed from the middle of the 19th century on empirical grounds; the concept of polymers did not come into being until 1920 (Staudinger, 1920) and indeed did not gain general acceptance until the late 1930s (Allen and Jones, 1988).

"There is no single chemical composition which uniquely corresponds to 'rubber'. Natural rubber has a particular chemical structure, and the various synthetic rubbers have their own, there being in fact only one synthetic rubber which is structurally more or less identical with natural rubber" (Allen, 1972). Among the family of rubbers, natural rubber has a unique position, not just because of its particular chemical structure and the consequent physical properties, but because it has two aspects: it is a rubber, but it is also an agricultural commodity.

As a rubber, it is used in the fabrication of a wide range of industrial and domestic articles and, despite competition with various types of synthetic rubbers, it remains indispensible for the manufacture of many products.

By provenance, it is an agricultural commodity, being a byproduct of the cell metabolism of many species of plants, mostly belonging to the family Euphorbiaceae. Most rubber-containing plants are native to the tropics: Hevea spp., Castilla spp. and Manihot spp. in Tropical America, Funtumia elastica and Landolphia spp. in Africa, Ficus elastica in Asia. There are also some rubber-bearing species of Compositae: Parthenium argentatum (Central America), Taraxacum (USSR), Solidago spp. (USA).

With the exception of Parthenium (guayule), which remains a small-scale producer of rubber in Mexico, the only significant commercial source of natural rubber is Hevea brasiliensis, a forest tree which is indigenous to the tropical rain forests of Central and South America. In this and similar species, the rubber exists within the tree in the form of latex, a colloidal suspension of rubber particles. Despite speculation, the function of latex in the tree remains unknown (Webster and Paardekooper, 1989).

The name for 'rubber' in most Indo-European languages except English is derived from the Amerindian name for rubber trees: cachuchu: 'weeping wood'. This is certainly more apt than the anomalous English name, which is derived from what is a minor attribute of the material: its ability to erase (rub out) pencil marks. The English name is attributed to the great British scientist, Joseph Priestley.

ORIGIN

The Spanish name (caucho) serves to remind us of the ecological origin of the majority of rubber-bearing plants, because Spain was the principal colonial power in Tropical America at the time when rubber started to become known in Europe. Long before that time, rubber was widely known (Schurer, 1957) within the ancient civilizations of Central and South America and was used in rituals. Effigies were made from it and it was burnt as an unpleasant form of incense. The word used to describe rubber was associated with blood: the Mayan word for blood, olli or ulli, also stood for the effusion from Castilla elastica.

The Spanish played games with hollow leather balls, but they were amazed at the behaviour of the rubber balls which they saw being used in Haiti and Yucatan. To the astonishment of the Spaniards (they used phrases like "beyond belief") the much heavier American balls bounced six times higher than leather balls. Columbus returned from America with a rubber ball (Schurer, 1958; Serier, 1988) to illustrate one of the wonders of the New World. The American Maya civilizations were capable of creating huge buildings, but without mortar; knew about the resilience of rubber yet were unaware of the wheel. The earliest reference to rubber in European literature is in a manuscript of 1530 by Peter Martyr, De orbe novo.

By 1615 the Spanish in Central America had discovered the use of latex in waterproofing. As no method was known of preserving the latex, a proofing industry for leather and fabrics grew up around Orizaba in Mexico and the output was exported. Thus there is nothing novel about manufacturing rubber products in the country of origin!

The history of establishment of natural rubber as an exceedingly useful material stretches back into prerecorded history. But there is one particular period of time - from the middle of the 18th century to about the end of the first decade of the 20th century - during which most of the basic ingredients of the industry as it is today, came into being.

At the start of this period, what little knowledge there was concerning the provenance and use of natural rubber, was confined to Tropical America. By the end of this period the fact that the tree Hevea brasiliensis is the most productive source of natural rubber was known throughout the world; large numbers of Hevea trees had been established in regions of the world far removed from its original home; methods for manipulation, shaping and use of the raw material had been devised; there was extensive trade, both in the raw material and in rubber products, including tyres; and there was some understanding, albeit limited, of the scientific attributes of natural rubber. Also, the basic possibilities for synthesizing rubbery polymers were coming to be understood, paving the way for future growth of natural rubber's rivals: the synthetic rubbers.

The historical development of the industry over this period will be separated into two components: (i) identification and exploitation of organized plantings (plantations) of Hevea brasiliensis, and (ii) development and application of the technology of utilization of natural rubber.

18th CENTURY "EUROPEANIZATION" OF NATURAL RUBBER

Despite the ingenuity which the inhabitants of various countries in Tropical America had displayed in making use of natural rubber, further progress awaited the confluence of "a flourishing scientific life and the collaboration of scientifically-minded statesmen in Europe with the explorers on the spot" (Schurer, 1951). Such were the conditions in Europe by the 18th century and, since Hevea existed only in Tropical America, it was those European countries, actively operating in that continent, which were likely to be able to be successful: France, Portugal, Spain. Of these, only France possessed the needed high level of scientific enquiry, and the honours for bringing natural rubber to the attention of the scientific community in Europe undoubtedly belong to two Frenchmen: Fresneau and de la Condamine.

Charles de la Condamine had been sent in 1735 by the French Royal Academy of Science on an expedition to measure a degree of the meridian on the equator, as part of an operation to determine the shape of the globe. He travelled via Peru to Ecuador, and it was in Quito that he became aware of certain primitive uses of rubber, in the form of torches and bottles (Coates, 1987). After completing the purpose of the expedition, de la Condamine crossed the Andes to Belem where he saw more uses of rubber, but without seeing any rubber trees. His paper of 1745 to the Academy attracted the attention of Francois Fresneau who was at that time an engineer working in Cayenne, capital of what is now French Guiana, and it is to Fresneau rather than to de la Condamine that credit must be given for making the first systematic observations on rubber; he was "the first man to have set out on a carefully planned search for the tree ..... and to have given a description of Hevea brasiliensis and of the methods of tapping and preparation of crude rubber" (Schurer, 1951).

French interest in rubber continued throughout the 18th century, a particularly important contribution being made by the botanist F Aublet who published the first taxonomic description of Hevea in 1775. In passing, it may be noted that the taxonomy of the genus has undergone a number of changes since that time (Schultes, 1970). The name itself is a Latinized version of the Ecuadorian Indian name, Hheve, and there was some earlier competition with other possible names (Siphonia, Caoutchoua).

GENESIS OF THE RUBBER MANUFACTURING INDUSTRY

It is significant that rubber technology does not feature in the volume of Singer's monumental History of Technology (1958) devoted to the high period of the Industrial Revolution, namely 1759-1850. This is not surprising since the nineteenth century was twenty years old before a satisfactory solvent had been found for rubber, and Hancock had discovered mastication. Nevertheless, Singer's decision to postpone discussion of rubber to the subsequent fifty year period deprives Macintosh and Hancock of their true position within the manufacturing history; namely as key figures of the Industrial Revolution. If George Stephenson is the father of railways (which he did not invent) then Hancock is the father of the rubber products industry.

Many had sought to find suitable solvents for rubber since its arrival in Europe in the late 18th century. At the beginning of the 19th century Glasgow was a major centre of the chemical industry and in that city Charles Macintosh was an industrial chemist who was eager to exploit the waste products from the relatively new coal gasification industry. In 1818, James Syme, an Edinburgh medical student, discovered (Schurer, 1952) that coal tar naphtha was an efficient solvent for rubber. This work was brought to the attention of Macintosh via Professor Thomas Thomson of Glasgow University. Macintosh's specific skill came in exploiting the naphtha-based rubber solution as a waterproof layer between two fabrics. The double fabric layer lessened the disadvantages inherent in the use of unvulcanized rubber - namely tackiness. Thus the "macintosh" was born.

At about the same time, Thomas Hancock became interested in using rubber and in 1820 patented elastic fastenings (Pickles, 1958) for gloves, shoes and stockings. These garters were made by cutting strips from "bottles" of Para rubber. Inevitably, this led to a considerable amount of waste material. Innovators are rarely natural communicators, thus the rubber industry is particularly fortunate in that Hancock was able to record (Duerden, 1986) the way in which he discovered mastication from observing that rubber "pieces with fresh cut edges ..... would perfectly unite; but the outer surface, which had been exposed, would not unite", but "it occurred to me that if minced up very small, the amount of fresh-cut surface would be greatly increased and by heat and pressure might possibly unite sufficiently for some purposes". Hancock built a small wooden machine to shred his rubber: this consisted of a hollow cylinder studded with teeth within which a studded core could be rotated. The effort required to shred the rubber did not decrease with time, but increased: when the machine was opened a homogeneous roll of rubber was found. Hancock did not protect his invention by patenting, but called his device a "pickle" to confuse potential competitors. Hancock's hand-driven, wooden masticator was rapidly transformed into steam-driven metal machines.

Hancock was able to use his "pickle" to supply the Macintosh factory in Manchester with masticated rubber, thus greatly enhancing productivity. Hancock went to considerable lengths to keep his process secret from out-siders, including this important customer. In 1837 he was forced to patent the masticator, as Macintosh's proofing patent was being challenged (Schurer,

1962) and there was a risk that Hancock's endeavours might be left un-protected.

The post-mastication, pre-vulcanization rubber industry was able to support a wide range of applications: pneumatic cushions, mattresses, pillows and bellows, hose, tubing, solid tyres, shoes, packings and springs. Rubber was widely used in surgical applications, and rubber hose was used in both fire-engines and beer-engines. It was even used in the buffer springs of railway wagons. Sole crepe, an unvulcanized rubber, has continued to be used until today.

Britain with its mild and predominantly wet climate was probably the ideal place to market a macintosh based on unvulcanized rubber. For American entrepreneurs the product was inadequate for the harsh winters and hot summers experienced there: rubber articles alternately became excessively rigid or sticky, and this led to a loss of confidence in the new industry, and the closure of many rubber factories.

This failure came to the attention of Charles Goodyear, a hardware dealer, who thought that rubber was a good material which required to be modified to avoid temperature induced defects. Goodyear attempted to remove the stickiness of uncured rubber (Simmons, 1939) by mixing it with magnesia, then boiling this mixture in lime. Next he attempted to decorate this with bronze powder, but this was not successful. So he attempted to remove the bronze with nitric acid. This led to a mess which was dis-carded. But a few days later he noted that the surface had lost its sticki-ness. The use of nitric acid nearly asphyxiated Goodyear, but it did lead to some recognition (Barker, 1938) and probably enabled him to hire Nathaniel Hayward (who unlike Goodyear had been associated with the American rubber industry). It was Hayward who introduced Goodyear to the idea of using sulphur on, rather than in, rubber: this happend in September 1838. Goodyear encountered many difficulties, both financial and personal, before he decided to try the effect of heat upon a mixture of rubber, sulphur and white lead. An accidental over-heating of one of the specimens produced charring but no melting. When he repeated the process before an open fire, again charring occured in the centre, but along the edges there was a border which was not charred but perfectly cured. Further tests showed that the new substance thus obtained did not harden in the winter cold and was not softened by the summer heat; it was also proof against solvents that dissolved the native gum. He had thus attained the object of his long search.

Obtaining financial and other assistance from William Rider, a New York rubber manufacturer, Goodyear continued his experiments and in 1841 succeeded in making the elastic compound uniformly in continuous sheets, by passing it through a heated cast iron trough. This was the first successful operation of vulcanization as an industrial process. On 6 December 1842, Goodyear had a specification prepared and this was deposited in the Patent Office of the United States as a claim for invention. The application for an English patent was not lodged until 1844, the reasons for the delay being mostly financial.

Public opinion in the United States was still very hostile to rubber, and Goodyear was anxious to interest manufacturers abroad in his discovery. He enlisted the services of Stephen Moulton, an Englishman then resident in the United States, who was about to return to England. Goodyear requested Moulton to take with him samples of his 'improved rubber' to show to appropriate people, especially the Macintosh Company, with the objective of selling the secret of manufacture. Eventually samples reached Thomas Hancock via a mutual friend, William Brockedon (who was to coin the word "vulcanization" for the process).

Hancock immediately recognised the significance of the samples and was able to deduce the presence of sulphur from a yellowish bloom on the surface. He took out a provisional patent on the use of sulphur in rubber and then set out to establish how vulcanization took place, eventually finding out that strips of rubber immersed in molten sulphur changed in character. Hancock applied for a patent for this in November 1843, a matter of weeks before Goodyear belatedly applied for an English patent. Litigation followed which Hancock won.

With the benefit of hindsight Goodyear may appear to have been naive to entrust his samples to Stephen Moulton for transit to England, and Hancock may appear rapacious in his quest for the source of the modification for Goodyear's samples. Hancock's contemporaries tend to support the latter view. Alexander Parkes, the inventor of the cold cure process (using a solution of sulphur chloride in carbon disulphide) marked his own copy of Hancock's Personal Narrative with a note:

> I think it is a sad thing for Mr Thomas Hancock to try to claim the discovery of vulcanization from the fact of the vulcanized rubber being first brought by Goodyear from America and pieces given to Brockedon, Hancock's co-partner and others.....

The American Geer (1922) generously noted that Hancock merely attempted to match his competitor's samples and "those of my readers who are chemists in the rubber business, will recognize this as one of the daily demands made upon them". Duerden (1956) observes that at "the time of Hancock's birth (1786) rubber had no applications of any real consequence although its potential was fully appreciated; by his death (1865) rubber manufacture was established as a major industry with a great potential for growth that was to be realized to the full in the years to come".

E.M. Chaffee of the Roxburg Rubber Company in the United States patented the calender in 1836. This enabled sheets of uniform thickness to be produced and also assisted in frictioning rubber to fabrics. H. Bewley patented an extruder for gutta percha in 1845, and this was soon adopted by the rubber industry. Thus the industry was suitably equipped with processing machinery (Pickles, 1958).

FROM WILD RUBBER TO PLANTATION RUBBER

At the beginning of the 19th century the source of natural rubber remained, as it had for many centuries before, a number of species of trees growing wild in the rain forests of Tropical America. By the end of the century the source had started to shift, with the establishment of organized plantings (plantations) of Hevea brasiliensis in several countries, geographically far removed from the native habitat of Hevea.

This transformation has two distinct elements: (i) the concept of taking a tree which grows naturally (a few per hectare) as part of the forest eco-system, and establishing plantations in which there are some hundreds of trees per hectare, and (ii) the concept of establishing these plantations, not in the vicinity of the native habitat of the tree but in other continents. Granted the evident merits of plantations, interest in a shift from wild rubber is not surprising, but an outside observer might well ask why was this shift not concentrated in Tropical America? The answer is two-fold: there were strong geo-political pressures to centralize production away from South America, and there have been perennial problems (notably the fungal leaf disease, South American Leaf Blight) with plantations of Hevea in the Amazon Basin. Even so, production of rubber in Tropical America was a thriving albeit badly managed business, throughout the 19th century. Great fortunes were created at the expense of appalling conditions for the tappers ('seringuieros' - the name being derived from 'syringe', an early application of rubber).

Pressure from two directions influenced both these actions. Those industrialists who were starting, mainly in Europe, to manufacture rubber products were, like all manufacturers, anxious to widen the source of supply of their raw material. It is interesting to note that a key figure in the industrialization scene, Thomas Hancock, wrote in 1857: "as this substance has now become an article of large and increasing consumption, plantation of these trees may, in a few years, produce a beneficial return" (Coates,

1987).

It was actions by a few British officials during the mid-19th century which was largely instrumental in bringing about the transformation. The story is complex, and can only be summarized here; for comprehensive accounts see Dean (1987) and an earlier account by Wycherley (1968).

By the 1850s, thanks to the work of Richard Spruce, there was full identification of Hevea brasiliensis and some knowledge of the methods used in Tropical America to extract rubber from the trees. During the 1860s, three people in London played decisive roles: James Collins, curator of the museum of the Pharmaceutical Society; Clements Markham, an official at the India Office; Joseph Hooker, director at the Royal Botanic Gardens at Kew. Collins had been interested in Amazonian rubber for some years, and published in 1868 and 1869 reports which attracted the attention of Markham. It was undoubtedly Markham who was the prime mover; he is described by Dean as "a remarkable figure in nineteenth-century British imperialism". He had previously been responsible for the successful transfer of the quinine-bearing plant, Cinchona, from Peru to India, an enterprise which he undertook "as a means of improving the conditions of the mass of the Indian population" (Dean, 1987). He decided that what had been done for Cinchona should be done for Hevea, given that the supply of wild rubber might soon become inadequate.

After commissioning in 1872 another report from Collins on rubber-bearing plants, Markham arranged via Hooker that Kew Gardens would receive seeds, and transport seedlings to the Far East. In 1873 he decided to purchase three collections of Hevea brasiliensis. The first (2000 seeds) was from Charles Farris, who had returned from Brazil with the seeds: a few seedlings were produced at Kew, and six were subsequently sent to Calcutta but appear not to have survived. The second was by Henry Wickham; the third by Robert Cross.

Wickham is a key figure in the story. Even though some people (and he himself) may have over-rated the contribution that he made to the subsequent establishment of the rubber plantation industry, there is no doubt that in 1873 he was the right man, at the right time, and at the right place.

Wickham, who was at that time living and working near Santerem in the Rio Tapajos region of the upper Amazon, was already interested in Hevea, and had published some information on the tree, thus satisfying Markham that he would be a suitable person to make a collection. Markham then arranged that the British consul in Belem should be instructed to obtain seeds of Hevea, with a mention of "a Mr Wickham, at Santerem, who may do the job".

The story of the way in which Wickham organized his famous collection of 1876 has been told many times, often with little reference to the facts. In later years, Wickham himself encouraged the idea that his seeds had been loaded aboard a ship under the nose of a gunboat which "would have blown us out of the water had her commander suspected what we were doing". Dean (1987) expresses the view that it was essential to Wickham's subsequent and largely unsuccessful attempts to achieve fame and fortune that he should foster the notion that his operation was attended by extreme personal danger. It is curious that no comprehensive biography of Wickham has been written, but much information on his life is provided in a series of articles by Lane (1953).

In June 1876 seventy thousand of Wickham's seeds arrived at Kew Gardens; only 2700 seem to have germinated. According to the Kew records (see Baulkwill, 1989), 1900 of the seedlings were sent to the Botanic Gardens at Colombo, where 90 per cent survived; 18 went to the Botanic Gardens at Bogor, Indonesia, where two survived; and 50 went to Singapore where probably none survived.

During the same year, there was another collection of Hevea in Brazil, by Robert Cross, who departed from England after the arrival of the Wickham seeds; he returned in November with 1000 seedlings collected in the Lower Amazon. The subsequent fate of these is a mystery; the general opinion seems to be that none of Cross's material survived, though Baulkwill suggests that "some small admixture of Cross genetic material cannot be entirely ruled out". This matter is of some importance, because the Wickham and Cross collections were made in different regions of the Amazon, so that their materials would certainly have possessed differing genetic compositions. The probability is that the entire Hevea industry has developed from Wickham's 2700 seedlings, collected in the Upper Amazon, a very narrow genetic base.

As just noted, during 1876 seedlings from Kew Gardens were received in Sri Lanka, Singapore (and subsequently Malaysia) and Indonesia. In India, the first were received in 1878, from Sri Lanka. It was in fact Sri Lanka which was the centre of early activity, the Heneratgoda Botanic Gardens in Colombo becoming a major source of rubber seeds, for domestic use and also for export. Much valuable early development work was done in Colombo, for example on latex flow and the use of acetic acid to coagulate latex (Parkin, 1910), and on diseases (Petch, 1911).

The key figure during this period of early development was Henry Ridley, who was Director of the Singapore Botanic Gardens over 1888-1911.

His principal contribution was to develop what is still today the basic method for tapping the tree: the removal at each tapping of a thin layer of bark from the cut end, thus permitting a smooth flow of latex and allowing the bark to regenerate. This was a great improvement on the method which was being used in Brazil, involving slashing the tree, causing great damage to the cambial layer and consequently to the productivity and life of the tree.

This was by no means the sum of his contributions: Ridley became the 'product champion' for natural rubber. As Dean writes, "Ridley was a whirlwind" who devoted much of his energy for 24 years towards the encouragement of rubber planting. Baulkwill (1989) observes that Ridley and his associates were responsible for a number of major contributions to the art: the importance of tapping early in the morning, the effects of daily and alternate daily tapping, the best age for starting tapping, the influence of girth and direction of tapping cut on yield, the value of contour planting, density of planting and use of smoke-houses. By the conclusion of his term of office, the basic methodology for production of rubber from Hevea had been established, and there have been, in a sense, remarkably few changes since then.

PNEUMATIC TYRES

To flourish, the nascent plantation industry needed an expanding market. Coates (1987) probably places undue emphasis on the importance of rubber in electrical insulation. This was an important market, subsequently eroded by thermoplastic materials, notably PVC. Even in Victorian times rubber was subject to competition from textiles, paper, vulcanized bitumen, and gutta percha as cable materials.

Robert William Thomson, a Scottish engineer, invented the pneumatic tyre (Tompkins, 1981) whilst working in London during the 1840s. The tyre itself was of sensible construction (inner tube with valve plus casing), but the means of fitment to the wheel rim was rather complex. Nevertheless, the utility of the invention was clearly demonstrated and the patent (No. 10,990) of 1845 clearly indicates the advantages of the pneumatic tyre:

> The nature of my said invention consists in the application of elastic bearings round the tires (sic) of the wheels of carriages for the purpose of lessening the power required to draw the carriages, rendering their motion easier, and diminishing the noise they make when in motion.

The idea did not become a commercial success, probably because he lacked the obsessive drive of a Goodyear to implement it fully.

The pneumatic tyre was re-invented by another Scot – John Boyd Dunlop – who was a veterinary surgeon practising in the north of Ireland. His young son appealed to his father for a device which would make his tricycle run more smoothly on granite paving. After a number of experiments Dunlop re-invented and patented the pneumatic tyre. The advantages of the invention were quickly appreciated as bicycling was extremely popular at that time – what has remained the definitive bicycle design (the Rover safety cycle) had been introduced in 1885. Thomson was forty years too early: Dunlop, who belongs to the very small class of successful, amateur inventors, fortuitously re-invented something at the right time.

Self-propelled vehicles driven by steam were too heavy to be considered for the pneumatic tyre, but the early motor cars designed in Germany and France were much more likely candidates. Asa Briggs (1983) notes that the British class structure was a major inhibitory factor in the implementation of an automotive industry. Until 1895 self-propelled vehicles had to be preceded by a pedestrian bearing a red flag. In the same year the Michelin brothers successfully competed in the Paris-Bordeaux car race with a vehicle fitted with pneumatic tyres (Jemain, 1982). This gave the French firm a key competitive edge.

Developments followed both in the design and composition of the pneumatic tyre, and in the condition of the roads on which they ran. In the early days of motoring car tyres could cost as much as $100 and yet only run for some 750 km, but by 1920 the cost had fallen to $30 and the expected 'mileage' had extended to 21,000 km (Dickerson, 1969). A life of 50,000 km was being claimed by the late 1930s (Litchfield, 1938).

The earliest aircraft landed on wooden skids, but these were rapidly superseded by bicycle or motor cycle tyres; in 1910 the first specialist aircraft tyres were marketed. Although trucks and buses made a major contribution to military transport during the First World War, most ran on solid tyres and continued to do so for some years after it. The pneumatic truck tyre emerged in the USA around 1917, but some very heavy vehicles continued to use solid tyres until as late as the 1950s. The trucks and buses demobilized at the end of the First World War revolutionized travel patterns in rural Western Europe. Similarly, but on a vastly greater scale, the Model T Ford opened up rural America and began the era of mass motoring which is continuing to grow even in the country of its birth. It would be difficult to postulate whether the rise of the automobile or the increase in the power of telecommunications has had a greater influence on the spread of democracy.

PLANTATION RUBBER: THE EARLY YEARS

During the 40 years from the first arrival of rubber seeds in Asia to the outbreak of the First World War, there was established a substantial volume of commercial production of and trade in natural rubber from plantations in various Asian countries.

This development was not smooth; in the early days there was little interest in growing Hevea in countries such as Malaysia which already had well-established and profitable food crops such as coffee and tea, and no doubt they regarded rubber, an industrial raw material, as being outside their normal terms of reference. Coates (1987) observes that "nearly twenty years were to go by without anyone (in Malaysia) evincing the least commercial interest in the trees", and by 1900 the total planted area was only 2800 ha (Malaysia: 2400, Sri Lanka: 400).

Several factors conspired eventually to force the pace. One of these is a complex story involving the abolition of slavery in Brazil from 1885 onwards and its effects on coffee production and prices (Coates, 1987). This, plus political changes in Brazil and severe disease problems affecting coffee production in S E Asia, especially Malaysia, where it had been a significant crop, led to the position where coffee growers there were starting to express interest in alternative crops. It is ironical that it should be coffee which was involved, because the coffee plant is a native of Africa, not Tropical America, and the dominance of Brazilian coffee production stems from the introduction to Brazil by Francisco Palheta in the early 18th century of the superior arabica variety. Given the frequent though incorrect accusation that Hevea was 'stolen' from Brazil, an observation by Dean is pertinent: "if it is necessary to view the transfer of plant species as theft, then perhaps Palheta may be looked upon as evening the score". Be that as it may, "the upshot was that plantation owners in Asia turned to rubber, leaving the coffee to South America" (Allen, 1972).

There was a more fundamental reason to expand rubber production: "a high and increasing price" (Barlow, 1978), stimulated by the arrival of motoring (i.e. tyres) plus growing consumption of rubber in other products, especially cables. The supply of wild rubber from Tropical America became increasingly unable to cope with this growing demand, and by 1913 the production of plantation-produced rubber had overtaken that of wild rubber. From then onwards, rubber from Tropical America became a minor component of world production.

The price rise from 1900 onwards (£ 275 per tonne in 1900, rising to £ 1400 in April 1910, the highest ever price in actual money terms) conveyed very strongly the message that rubber production on (mainly)

European-owned rubber estates in S E Asia could be a very profitable operation. Production expanded, not only in the then British colonies, who as noted, were the first to receive planting material, but also in Indonesia, then the Dutch East Indies, which had received an early supply of seeds from Sri Lanka as a "friendly gesture" (Coates, 1987). Indeed it was the Dutch researchers in Indonesia who pioneered breeding work with Hevea from 1912 onwards, and the use of breeding to produce high-productivity planting material subsequently became a very important - some would say the most important - contribution to the ability of natural rubber to weather economic storms from various directions (Simmonds, 1989).

Concomitant with rising production and prices, the period from 1900 up to the start of the First World War saw the cultivation of Hevea in Asia become an accepted component of tropical agriculture, and there developed the necessary range of infra-structural facilities: a marketing system which made use of the already-established commodity markets in Singapore, London and New York; 'agency houses' to supervise the plantation companies and to market their output; a system for assessing the 'quality' of the merchandise; shipping arrangements, and so on. Coates (1987) has provided a detailed account of the history of such matters.

At this stage, production was mainly from European-owned estates, though signs were emerging that Hevea would be a useful crop for small-holders, and in Malaysia the proportion of production belonging to locally-owned estates as well as to smallholders started to rise slowly. Also, because of the need to supply labour in increasing numbers, there was a considerable increase in immigration of plantation workers (eg. from India to Malaysia), a feature which continued for decades.

THE INTERWAR YEARS

By the start of the First World War, production of natural rubber on plantations in several Asian countries was firmly established, and no doubt the producers hoped for a steady growth in markets and profitability. This was not to be: the years between the two World Wars brought many problems, and this was perhaps the most unhappy period during the whole history of the industry.

As just noted, natural rubber prices reached a peak in 1910, but the consequent fast growth in output led to falling prices during the First World War and afterwards. In an attempt to arrest the price drift, the British Government instigated the 'Stevenson Scheme' (1922) to restrict output in the then British colonies (McFadyean, 1944). The Scheme was only moderately successful; prices increased over 1924-26, but this had the unwelcome effect of stimulating smallholder production in other countries, notably Indonesia. This, plus the onset of the world Depression in 1929, led to a catastrophic fall in prices which reached their lowest-ever level in 1932. To cope with this, the 'International Rubber Agreement' was brought into being in 1934 by the United Kingdom, India, the Netherlands, France and Thailand. The Agreement laid down production quotas for existing plantings, and new planting was forbidden.

The Agreement was certainly an improvement over the Stevenson Scheme, but like the Scheme it neglected the interests, even the existence, of small-holders, and tended to imbue the producers with a restrictionist outlook. On the credit side, it had two side effects: stimulation of research aimed to increase consumption, and establishment of well-organized statistics via the International Rubber Study Group (Allen and Jones, 1988).

COMPOUNDING

In 1916 Banbury (Killeffer, 1962) invented the internal mixer which greatly increased productivity in the mixing process. At around the same time organic chemistry gradually began to contribute. Firstly, came a number of empirical contributions to compounding (through the development of accelerators and antioxidants) and gradually the increased understanding of the nature of vulcanization, oxidation and the composition of rubber in turn led to further improvements.

Goodyear's original vulcanizing mix consisted of sulphur with lead carbonate. The latter acted as an inorganic accelerator, but this was later supplanted by lead and zinc oxides. Typical 19th century formulations contained high amounts of zinc oxide (Norris, 1939). Lampblack had been introduced shortly after the discovery of vulcanization as a means of reducing light-induced ageing, but the introduction of carbon black as a reinforcing agent did not occur until S.C. Mote (Ruffell, 1952) of the India Rubber, Gutta Percha and Telegraph Works in Silvertown, London exploited it in 1904 to improve wear resistance.

Davis (1951) considered that Thomas Rowley of Manchester should have been recognized for his work (1881) which enabled sulphur to be reduced to only 2 per cent through the addition of ammonia, and commends Oenslager, Marks and Spence (Oenslager, 1933) for their "memorable work" on thiocarbanilide which was introduced as an accelerator during the First World War. At about the same time Hoffman and Gottlob (1915) applied for a patent for methylene bases as accelerators and Boggs (1919) recognized the accelerating and antioxidant properties of beta-naphthylamine and para-phenylenediamine. Molony (1920), working like Goodyear in his kitchen, synthesized tetramethylthiuram disulphide in 1917 and found it to be a powerful accelerator. It is hardly surprising that Naunton is alleged to have called (Leyland, 1971) this a period of "inspired empiricism", although the pioneering academic work at the College de France in Paris on oxidation and antioxidants by Moureau and Dufraise (Dufraisse, 1937) is important for a wide range of materials including rubber.

In 1921 diphenylguanidine was introduced commercially and by 1923 guanidine accelerators were in general use (Alliger and Sjothun, 1964). Improved channel blacks were marketed at about the same time. Mercapto-benzothiazole (MBT) was first marketed in 1925 and this, together with the introduction of powerful antioxidants, produced marked improvements to ageing resistance and wear. The beneficial influence of naturally occurring fatty acids on vulcanization was known at least as early as 1912, but their additive use gradually emerged (Sheppard, 1929) during the 1920s. The earliest antioxidants had all been simple commonly occurring materials: materials like creosote, naphthalene, camphor, etc. From about 1918 organic chemicals (amines, phenols, aldehyde-amines, ketone-amines and nitroso compounds) were gradually introduced (Semon, 1937). Some of these materials (notably alpha-naphthylamine) were found to be carcinogenic (Parkes, 1976) and were withdrawn, but not before one of the best-documented occupational cancers had sadly emerged. The effect of ozone was first fully recognized in the 1920s and waxes were recognized as a means of affording protection. Sulphenamide accelerators were discovered (Hofmann, 1967) in the mid-1930s by Zaucker (1933) and by Harmann (1937): the key cyclohexylbenzthiazyl derivative was first marketed in 1940.

LATEX PROCESSING

Hancock had proposed using latex in 1824. Following the discovery of ammonia preservation by Johnson in 1853, Hancock and Silver (1864) patented several applications, but commercial exploitation did not take place until concentrated latex was marketed in the 1920s. By the late 1920s over one hundred British patents were being granted each year, and this figure was still rising in 1939. Some of the developments may be regarded as adaptations of earlier techniques for handling rubber solutions: these included dipping, and spreading, but many of the outlets required the development of completely new technologies, the most notable of which was foam rubber. The Talalay process introduced sophisticated heat-exchange techniques (the frothed latex is frozen at around -30°C prior to vulcanization), enabling large quantities of foam rubber to be manufactured. By 1953 Murphy estimated that production was approaching 100,000 tonnes, although the introduction of polyurethane foams precluded further growth. Pendle (1989) estimates that some 85,000 tonnes are still produced (mainly within the Soviet Union).

Foam rubber and to a lesser extent rubberized fibres invaded the furniture market from which rubber had previously been almost entirely absent, enabling many changes to be wrought in furniture design. Foam rubber seating found outlets in vehicles and in a wide range of public buildings. Foam rubber mattresses enabled new standards in hygiene to be established in institutions like army barracks, hotels and hospitals.

Latex entered the tyre and general rubber goods industry to a limited extent (as in tyre cord dipping and in fire-hose lining). It generated its own breed of technical personnel, its own technology and even a specialised literature. The liquid nature of latex makes it very versatile especially in its prevulcanized form. In 1938 Kratz noted its use in special paper products, in dental supplies, food containers and in face masks for the theatrical stage.

ARRIVAL OF SYNTHETIC RUBBERS

Rubber had excited the curiosity of a number of chemists during the 19th century. Faraday (1826) had elucidated the basic chemical composition, though not the structure, of natural rubber, and in 1879 Bouchardat (1879) had succeeded in decomposing rubber into its monomeric component, isoprene, and had polymerized this back into rubber. Similarly, Tilden (1884) had synthesized rubber from isoprene made from turpentine. These and similar academic exercises, though, could not form the basis of viable processes for large-scale production of synthetic rubbers. This had to await the evolution of more practical polymerization methods based on starting materials other than isoprene which is a difficult material to manufacture.

What was needed was an external stimulus, and this was provided by the advent of the First World War, during which Germany was denied access to the Asian rubber plantations, and made substantial progress towards viable processes, resulting in a modest production (150 tonnes per month) of a somewhat inadequate synthetic rubber, 'methyl rubber'. Similarly, there was activity in the USSR. By the 1940s Germany was producing synthetic rubbers at over 100,000 tonnes per year.

The low prices of the 1930s undoubtedly caused a lack of interest in the development of economic competitors to natural rubber, though several important speciality synthetic rubbers were developed: Thiokol (1930), polychloroprene (1931), nitrile rubber (1936) and butyl rubber (1940). Polyurethanes were also developed in the same period.

The second, and conclusive, stimulus was the over-running by Japan of the S E Asian rubber plantations from 1941, thus depriving the United States and its Allies of a vital war material. One reaction was the renewed interest in rubber-bearing plants other than Hevea: Funtumia and Landolphia from Africa; Parthenium (guayule) from the southern USA and Mexico (Polhamus, 1962). Production from such sources was stepped up, but the quantities were miniscule (23,000 tonnes from Africa between 1942 and 1945), and this was clearly not the way to solve the problem. What was needed was a large volume of synthetic rubbers with the capability of (more or less) substituting for natural rubber.

It was the USA which succeeded in achieving this, in a remarkably short space of time, and on a prodigious scale: the US emergency programme to develop synthetic rubbers capable of satisfying their war needs was second only to the much more publicized atom bomb project in scale. Ironically, it was based on pre-War German technology which had succeeded in the development of a viable synthetic rubber based, not on isoprene, but on a copolymer of styrene and butadiene (Buna-S in Germany; GR-S in the USA; now SBR). US access to this development came via a commercial agreement (Morton, 1982) made in 1929 between the Standard Oil Company of New Jersey and I.G. Farbenindustrie in Germany.

By 1944 the USA was producing over 600,000 tonnes per year of synthetic rubbers (Phillips, 1963). These were not fully technically-competitive with natural rubber; war-time SBR was designed to meet an emergency, which it did very well, but - had it not been improved further - it would not have been able to compete on technical grounds with natural rubber in normal times.

THE SYNTHETIC RUBBERS SINCE 1945

The pattern of development for the first decade or so after the end of the Second World War was set by two main factors: (i) the achievement of substantial improvements in the technology of SBR ('cold polymerization', oil extension), and (ii) a substantial acceleration in demand for rubber, stimulated by the enormous growth in motor (cars, trucks) transport as the world returned to normality. Together, these created the position where natural rubber was for the first time faced with serious competition. Natural rubber producers did not have the capability to rapidly expand production to meet the growing demand, and the gap between their production and world demand for rubber was readily filled by the improved SBR and by other synthetic rubbers (Allen and Jones, 1988). As a result, natural rubber's share of the world market for rubber steadily declined, from effectively 100 per cent pre-War to about 30 per cent today.

SBR is not a 'better' rubber than natural rubber: in many respects it is technically inferior but it is relatively cheap to produce, is outstandingly good in tyre treads, its main market, and can replace natural rubber in many run-of-the-mill products, though not in those products (eg. high-performance engineering components) which require the special properties of natural rubber. The increasing dominance of SBR in the market-place is, as just noted, primarily the result of the supply problem for natural rubber, whose potential market share is substantially greater than its actual share (Allen et al., 1974).

A significant development with synthetic rubbers took place in 1953-54: invention of means to synthesize the chemical replica of natural rubber - synthetic cis-1,4-polyisoprene (McMillan, 1979). The resultant synthetic rubber (IR) went into commercial production duing the 1960s, but curiously its impact on natural rubber has been less significant than might have been expected. It is technically not quite as good as natural rubber, and the starting monomer, isoprene, is relatively expensive to produce. Only in the USSR, which dislikes using foreign exchange to import natural rubber, has there been any really significant establishment of capacity for production of IR.

Another development has been the growth in production and use of speciality synthetic rubbers. Some such (eg. polychloroprene rubber, nitrile rubber) were in production before 1939, and since then a number of new types have come into being. Most do not compete with natural rubber: they are not used in the manufacture of tyres and other main-stream rubber products, but in products where the need for their special properties (eg. resistance to high temperatures and oils) outweighs their relatively high cost.

RESPONSES BY NATURAL RUBBER PRODUCERS

Over the 40 years since the end of World War II, natural rubber producers have been faced with increasingly heavy competition from synthetic rubbers, fuelled by the very low cost of oil up to the time of the first 'oil shock' of 1973.

For much of this period, natural rubber producers had been on the defensive, and among their various responses to this competitive situation the following three main themes were especially significant.

(1)   Replanting to replace old low-yielding trees with improved clones of much higher productivity, together with sustained effort on all aspects of agronomic efficiency

(2)   Improvements in the 'presentation' of natural rubber to the consumers

(3)   Searches for new markets, via new applications and new forms of rubber.

The need to replant arises from a combination of two features: first that the productivity of <u>Hevea</u> reaches a maximum about 15 years after planting, and thereafter decreases, and second, that there is a continual stream of new, improved clones coming forward as the result of sustained tree breeding.

On this second point, it has been noted earlier that it was the Dutch researchers who undertook the first experiments with breeding, since when there has been a dramatic improvement in productivity. The original Wickham importations would have yielded about 200-300 kg/ha/year: the more recent clones can improve on this by a factor of 10, and it is not surprising that <u>Hevea</u> breeding has been described as "one of the outstanding success stories in the agriculture of this century" (Simmonds, 1985).

Planters have always recognized the need to maintain a proper re-planting schedule, despite the cost. This need became even more clamant in Malaysia in the 1950s ("replant or die") and a special Mission of Enquiry (Mudie) in 1954 made specific recommendations, not only on replanting but on the provision of fiscal incentives, with special reference to the needs of smallholders. Stemming from this, several agencies (eg. FELDA in Malaysia) came into being in a number of countries, to encourage replanting and kindred matters.

One obvious way in which the producers could seek to improve the competitive position was to emulate the superior 'presentation' of the synthetics. These were marketed with technical specifications in contrast to the time-honoured, distinctly archaic system which had been used by natural rubber producers since the start of the century (Allen, 1964), and also were offered in sensible-size small bales which were preferred by consumers over the very large traditional bales of natural rubber. Thus the question was being asked: why cannot natural rubber be presented in a similar fashion? Curiously, the commodity markets seemed uninterested in change (their outlook is traditionally conservative) and the impetus for change came from the research side, who recognized not only the logic of the change but also the fact that the arrival of new ways of converting the raw material (eg. field latex) into the marketable product ('block rubber' as distinct from Ribbed Smoke Sheet etc) made it easier to adopt new presentation methods.

The first step was taken in 1965, by Malaysia, with the launch of the Standard Malaysian Rubber (SMR) scheme, which replaced the former visual grading scheme by a set of technical specifications, the primary purpose of which was to define purity in terms of content of particulate dirt, and to limit the concentration of various undesirable materials (eg. moisture). Other countries soon followed suit, and by 1986, forty two per cent of world production was as Technically Specified Natural Rubber.

As agriculture rather than technology was the genesis of the Hevea plantation industry, there has always been perhaps a tendency for the industry to concentrate its mind on agricultural features such as productivity rather than on the equally important matter of developing new markets. As long ago as 1931 Gallagher pointed out the need to devote more effort to "explore all avenues for the consumption of raw rubber" (Gallagher, 1931), and since that time there has been an input of effort to devise new forms of the raw material (eg. Thermoplastic Natural Rubber, Epoxidized Natural Rubber, Liquid Natural Rubber), and to establish new applications. In this latter respect, the existence of synthetic rubbers has had the beneficial effect of forcing natural rubber producers to consider carefully, exactly what are the particular merits of their material. This led to the important view that natural rubber can compete very successfully with the synthetics in products which are technically-demanding, and especially in high-performance engineering components (springs, bearings). Large and growing markets have been established in this area.

One curious feature is that whilst the synthetic rubber industry has claimed (Salisbury, 1971) to be interested in establishing new uses, its efforts have not been attended by a marked success, despite the ability for many synthetic rubbers to perform in environments traditionally hostile to natural rubber. Certain engineering markets, notably automotive suspensions (Brice and Jones, 1988) remain unconquered; there is also a major potential outlet in energy storage devices.

Natural rubber latex concentrate has also been subject to competition since 1945, before which it was a unique material. In foam (for furniture etc) it has been largely displaced by polyurethane foams, but in other products (dipped goods, threads) it retains strong and growing markets which have recently been stimulated (in the case of dipped goods such as surgical and examination gloves, contraceptives) by the AIDS problem.

All of the features just noted are in a sense internal to the industry, but natural rubber has also had to content with the impact of external features: the 'oil shocks' of the 1970s, the arrival of the radial-ply tyre, ups and downs in the world economy, and so on. Analysis of the impact of all these is a complex matter which has been the subject of two important studies in the past decade: a World Bank Report of 1978 (Grilli, 1978) (which was perhaps over bullish) and the Malaysian 'Task Force of Experts' in 1983 which made a realistic analysis of the situation as seen by Malaysia.

For the producers, price is of course the main consideration, and fluctuations in the price of natural rubber - a consequence of the fact that it is sold in open commodity markets - have frequently been the source of unfavourable comments, by consumers as well as by producers. For many years there has been interest in the merits of trying to smooth out price fluctuations, and perhaps also to raise prices in the long term, by use of a buffer stock. The idea is not new (it has been used for tin and cocoa) and not all experts are agreed as to the usefulness of the concept. Be that as it may, there was growing pressure throughout the 1970s to initiate action, and in 1979 the International Natural Rubber Agreement was brought into being. This should not be compared with the earlier attempts to regulate price (eg. the 1922 Stevenson Scheme; the 1934 Agreement) because, unlike these, it involves consuming as well as producing countries. It is too soon to assess the long-term effects of this Agreement, but many would agree that it has been successful in achieving its immediate objective of stabilizing prices within agreed limits.

CONCLUDING COMMENTS

The history of exploitation of Hevea brasiliensis, and development of consumption of its product, natural rubber, is a microcosm of world history, embodying the change from colonialism to independence, from parochialism to internationalism. It shows how the industry has withstood the many shocks that it has been subjected to by the vicissitudes of the world economy. It demonstrates how well-targetted science (agricultural, technological) can assist a natural, environmentally-desirable material not merely to survive but to prosper in the face of competition from synthetic materials.

One can only conclude by observing that, given its past history, natural rubber and its producers, need have no fear of the future.

REFERENCES

Allen, P.W., 1964. The evolution of market grades. Rubb. Dev., 17: 90-97.
Allen, P.W., 1972. Natural Rubber and the Synthetics. Crosby Lockwood, London. pp. 255.
Allen, P.W., Thomas, P.O. and Sekhar, B.C., 1974. The techno-economic potential of NR in major end uses. MRRDB, Kuala Lumpur.
Allen, P.W. and Jones, K.P., 1988. A historical perspective of the rubber industry. In: A.D. Roberts (Editor), Natural rubber science and technology. Oxford University Press, Oxford, pp.1-34.
Alliger, G. and Sjothun, I.J., 1964. Vulcanization of elastomers. Reinhold, New york. pp. 410.

Barker, P.W., 1938. Charles Goodyear. Rubber Age, 43: 229-230; 294-296; 307-309; 44: 29-31; 88-90.

Barlow, C., 1978. The Natural Rubber Industry: Its Development, Technology, and Economy in Malaysia. Oxford University Press, Kuala Lumpur.

Baulkwill, W.J., 1989. The history of natural rubber production. In: C.C. Webster and W.J. Baulkwill (Editors), Longman Scientific & Technical, Harlow, pp. 156.

Boggs, C.R., 1919. Accelerator for vulcanizing rubber. United States Patent 1,296,469.

Bouchardat, G., 1889. Compt. Ren. 89: 1117.

Brice, R.E. and Jones, K.P., 1988. The evolution of new uses for natural rubber. In: A.D. roberts (Editor), Natural Rubber Science and Technology. Oxford University Press, Oxford, pp. 1038-1077.

Briggs, A., 1983. A Social History of England. Weidenfeld and Nicolson, London.

Coates, A., 1987. The Commerce in Rubber - The First 250 Years. Oxford University Press, Singapore.

Davis, C.C., 1951. Some of the real pioneers of the rubber industry. Ind. Rubb. Wld., 123: 433-440.

Dean, W., 1987. Brazil and the Struggle for Rubber: A Study in Environmental History. Cambridge University Press, Cambridge.

Dickerson, W.N., 1969. Tires at a dollar a mile. Mod. Tire Dealer, 50(12): 62-67.

Duerden, F., 1956. Thomas Hancock - an appreciation. Plast. Rubb. Int., 11(3): 22-26.

Dufraisse, C., 1937. Auto-oxidation and deterioration by oxygen, pro-oxygens and anti-oxygens. In: C.C. Davis and J.T. Blake (Editors), The Chemistry and Technology of Rubber. Reinhold, New York, pp. 440-523.

Faraday, M., 1826. On pure caoutchouc, and the substances by which it is accompanied in the state of sap, or juice. O.J. Sci., 21: 19.

Gallagher, W.J., 1931. The need for balanced research in the rubber industry. Bull. Rubb. Grow. Assn., 13: 326.

Geer, W.C., 1922. The Reign of Rubber, Century, New York.

Grilli, E.R., Agostini, B.B. and T Hooft Welvaars, M., 1978. The World Rubber Economy: Structure, Changes, Prospects. World Bank, Washington.

Hancock, T., 1857. Personal Narrative. Longmans, London.

Hancock, T. and Silver, S.W., 1864. Ornamental moulded articles. British Patent 3094.

Hancock, T. and Silver, S.W., 1864. Rubber products and articles. British Patent 3110.

Harman, M.W., 1937. Vulcanization characteristics of mercaptobenzothiazole derivatives. Rubb. Chem. Technol., 10: 329-335.

Hofmann, F. and Gottlob, K., 1915. Vulcanizing rubber. United States Patent 1,126,469.

Hofmann, W., 1967. Vulcanization and Vulcanizing Agents. Maclaren, London. pp. 371.

Jemain, A., 1982. Michelin: un siecle de secrets. Calmann-Levy, Paris.

Johnson, W., 1853. Preservation of latex, British Patent 467.

Killeffer, D.H., 1962. Banbury: The Master Mixer. Palmerton, New York.

Kratz, G.D., 1938. Latex aids decentralization of American rubber industry. Rubb. Age (NY), 43: 361-362.

Lane, E.V., 1953. The life and work of Sir Henry Wickham. India Rubb. J. 125: 962-965; 1076-1078; 1118-1120; 1149-1151. 1954, 126: 25-27; 65-67; 95-98; 139-142; 177-180.

Lambourn, L.J., 1958. The indispensable pneumatic tyre. Trans. Instn Rubb. Ind., 34: 118.

Leyland, B.N., Pryer, W.R., Sweeney, T. and Thompson, J., 1971. Compounding ingredients: 50 years of progress. J. Instn Rubb. Ind., 5: 51.

Litchfield, P.W., 1938. Benefits of tire improvement - 1929 to 1937. India Rubb. Wld., 98(5): 37.

McFadyean, A., 1944. History of rubber regulation 1934-43. Allen & Unwin, London.

McMillan, F.M., 1979. The Chain Straighteners. Macmillan, London.

Molony, S.B., 1920. Vulcanizing rubber. United States Patent 1,343,224.

Morton, M., 1982. History of synthetic rubber. In: R.B. Seymour (Editor), History of Polymer Science and Technology. Marcel Dekker, New York, pp. 225-238.

Mudie, R.F., Raeburn, I.R. and Marsh, B., 1954. Report of the mission of enquiry into the rubber industry of Malaya. Government Press, Kuala Lumpur.

Murphy, E.A., 1955. Development of latex foam rubber. Trans. Instn Rubb. Ind., 31: 90.

Norris, W., 1939. Compounding in the nineteenth century. India Rubb. Wld., 101(1): 60-62.

Oenslager, G., 1933. Organic accelerators. Ind. Eng. Chem., 25: 232-237.

Parkes, H.G., 1976. The epidemiology of the aromatic amine cancers. In: Searle, C.E., Chemical carcinogens. American Chemical Society, Washington, pp. 462-480.

Parkin, J., 1910. The right use of acetic acid in the coagulation of Hevea latex. Ind. Rubb. J., 40: 752.

Pendle, T.D., 1989. A review of the moulded latex foam industry. Cellular Polym., 8: 1-14.

Petch, T., 1911. The physiology and diseases of Hevea brasiliensis. Dulau, London.

Phillips, C.F., 1963. Competition in the synthetic rubber industry. University of North Carolina, Chapel Hill.

Pickles, S.S., 1958. Production and utilization of rubber. In: Singer, C. and others. A history of technology. v.5. The late nineteenth century, c1850-c1900. Clarendon Press, Oxford, pp. 752-775.

Rowley, T., 1881. Vulcanizing articles or fabrics of India rubber. British Patent 787.

Ruffell, J.F.E., 1952. Compounding ingredients. In: P. Schidrowitz and T.R. Dawson (Editors), History of The Rubber Industry. Heffer, Cambridge.

Salisbury, T.E., 1971. The development of new uses for rubber by both the consumer and the producer. Proceedings of 12th Annual Meeting. International Institute of Synthetic Rubber Producers, New York, pp. 62.

Schultez, R.E., 1970. History of taxonomic studies in Hevea. Bot. Rev., 36(3): 197-276.

Schurer, H., 1951. Bicentenary of the discovery of rubber in Europe. J. Rubb. Res., 20(2): 9-18.

Schurer, H., 1952. The macintosh. India Rubb. J., 122: 632-637.

Schurer, H., 1957. Rubber - a magic substance of Ancient America. Rubb. J., 132: 543-549.

Schurer, H., 1958. The Spanish discovery of rubber. Rubb. J., 135: 269-273.

Schurer, H., 1962. The trials of Thomas Hancock. Part I. The 'macintosh' problem. Rubb. Plast. Wkly., 143: 943-949.

Semon, W.L., 1937. History and use of materials which improve ageing. In: C.C. Davis and J.T. Blake (Editors), The Chemistry and Technology of Rubber. Reinhold, New York, pp. 414-439.

Serier, J.B., 1988. Importance et symbolisme du caoutchouc chez les Precolombiens de l'aire mesoamericaine. Rev. gen. Caout. Plast., (683) 45-49.

Sheppard, J.R., 1929. Stearic acid in litharge-cured rubber compounds. Ind. Eng. Chem., 21: 732-738.

Simmonds, N.W., 1985. The strategy of rubber breeding. Proceedings of the International Rubber Conference, 1985, Kuala Lumpur, Malaysia, 3: 115-126.

Simmonds, N.W., 1989. Rubber breeding. In: C.C. Webster and W.J. Baulk-
    will (Editors), Rubber. Longman Scientific & Technical, Harlow, pp.
    85-124.
Simmons, H.E., 1939. Charles Goodyear, the persistent researcher. India
    Rubb. Wld., 101(1): 48-49.
Singer, C. and others, 1958. A history of technology. v. 4. The industrial
    revolution, c1750-c1850. Clarendon Press, Oxford.
Staudinger, H., 1920. Uber polymerisation. Ber. Dtsch Chem. Ges., 53: 71.
Task Force of Experts. 1983. The Malaysian Natural Rubber Industry
    1983-2300. MRRDB, Kuala Lumpur.
Tompkins, E., 1981. The History of The Pneumatic Tyre. Eastland Press.
Webster, C.C. and Paardekooper, E.C., 1989. The botany of the rubber
    tree. In: C.C. Webster, and W.J. Baulkwill (Editors), Rubber. Longman
    Scientific & Technical, Harlow, pp. 57-84.
Wycherley, P.R., 1968. Introduction of Hevea to the Orient. Planter, 44: 1.
Zaucker, E., 1933. Substituted sulfenamides. German Patent 586,351.

CHAPTER 2

THE OUTLOOK FOR NATURAL RUBBER PRODUCTION AND CONSUMPTION

H.P. SMIT and K. BURGER

Economic and Social Institute, Free University, 1007 MC Amsterdam,
The Netherlands

This chapter describes the possible future of natural rubber (NR),
and is divided into four sections. The first section briefly reviews the
recent developments in natural rubber economy. In the second section the
relationships among various factors, ranging from planting policies to saturation in the passenger car market, are analysed. These relationships form
the basis for forecasting and assessment of policy measures in the third
section. The last section summarises the findings.

A GENERAL REVIEW OF THE NATURAL RUBBER ECONOMY

Production

The world NR production more than doubled between 1955 and 1985
after the turbulent first half of this century (Table 1). Since the end of
the 1950s, the share of Malaysia, the leading producer, increased from 36.4%
in 1955 to 44% in 1975. In 1980, however, it declined to 40.1% and further
to 33.5% and less than 30% in 1985 and 1989, respectively. Production in
Indonesia declined to a minimum in 1960, recovered considerably in the following decade, remained on a plateau between 1970 and 1977, to grow fairly
rapidly again after 1977, roughly keeping pace with the growth in world
NR production in the 1980s. Thailand has shown the fastest growth among
the larger producers, her production increasing to a level of 1.1 m tonnes
in 1989. These three large producers together had a share of 78.3% in 1960,
79.8% in 1980 and 76.8% in 1985 and 1988. Sri Lanka, the fourth largest
producer of NR in 1975, registered a declining trend and the production
fell from 152704 tonnes in 1979 to 133151 tonnes by 1980 and thereafter remained stagnant at that level. Late comers like China and the Philippines
have shown fast growth rates. India had increased her share in world NR
production from 1.2% in 1955 to over 5% in 1988. The remaining countries
together produced only 6.7% in 1985.

An important aspect of the NR plantation industry is the size of the
holdings. Two groups can be distinguished: smallholdings and estates. The

TABLE 1

Production of natural rubber ('000 tonnes).

	1955	1960	1965	1970	1975	1980	1985	1988
Malaysia (a)	705	785	917	1269	1459	1530	1470	1660
Indonesia	749	620	716	815	823	1020	1130	1235
Thailand	132	171	216	290	355	501	726	975
Sri Lanka	95	99	118	159	149	133	138	122
India	23	25	49	90	136	155	198	255
Philippines	(c)	(c)	6	20	52	70	90	85
Vietnam	66	74	61	28	20	49	53	66
Kampuchea	28	35	49	13	10	-	22	27
China	-	-	-	-	25	113	188	240
Other Asia and Oceania	19	15	19	18	22	25	22	27
Liberia	39	48	49	83	83	78	84	108
Nigeria	31	59	69	65	68	47	52	68
Cote d'Ivoire	-	-	3	11	15	23	41	63
Other Africa	31	42	33	53	49	47	36	53
Brazil	22	23	29	25	19	28	40	33
Other Latin America	6	7	7	7	12	20	23	26
Total (b)	1946	2003	2341	2946	3297	3839	4313	5035

(a) Includes Singapore for 1955 and 1960.
(b) Including allowances for statistical discrepancies.
(c) Included in other Asia and Oceania.

Source: Rubber Statistical Bulletin, IRSG.

TABLE 2

Rubber area of estates and smallholdings ('000 hectares).

Territory	Year	Estates	Smallholdings	Total
Peninsular Malaysia	1985	418.1	1244.6	1662.7
Sarawak	1985	2.0	206.0	208.0
Sabah	1985	8.7	69.3	78.0
Malaysia, total	1985	428.8	1519.9	1948.7
Indonesia	1977	465.5	1862.0	2327.5
Thailand	1979	75.9	1442.1	1518.0
Sri Lanka	1975	105.6	122.0	227.6
Philippines	1983	57.7		57.7
Vietnam	1983			115.2
Kampuchea	1984			19.0
Liberia	1973	76.7	43.1	119.8
Nigeria	1982			185.0
Cote d'Ivoire	1982			41.5
Brazil	1965	9.8	10.2	20.0

Source: Rubber Statistical Bulletin.

rationale for this division is the difference between the two groups in owner-
ship, role of wages/salary earner, efficiency and productivity, access to
markets and information. Table 2 gives the area under rubber in different
countries. Dividing production by the corresponding area gives an approximate
average yield per hectare in each country. In comparing average yields
between countries, it must be remembered that discrepancies in the statistics
of registered area may have caused some errors. Nevertheless, the differences
in average yield between countries are so large that only a small part of
it can be explained by possible faulty area registration.

Consumption

World rubber consumption, both natural and synthetic has increased
dramatically in the past three decades, even though the economic recession
in the early eighties severely depressed the rubber market. A broad picture
of the growth in rubber consumption is given in Table 3.

TABLE 3

Total rubber consumption 1965-1985 (in '000 tonnes).

	1965	1970	1975	1980	1985	1988
USA	2088	2517	2630	2565	2726	2868
Canada	141	186	252	280	268	285
Japan	377	779	870	1312	1487	1665
Australia	80	94	100	101	82	98
Germany F.R.	366	559	557	601	614	675
France	277	419	434	530	468	496
United Kingdom	369	473	437	379	327	367
Italy	200	310	338	420	404	452
EEC	1410	2095	2215	2445	2312	2570
Total W. Europe	1585	2360	2515	2800	2760	3030
Eastern Europe and USSR	1175	1535	2475	2975	3142	3340
China	170	290	350	495	660	890
India	85	118	161	217	303	393
Republic of Korea	18	38	104	259	305	343
Taiwan	10	33	61	150	180	233
Brazil	64	122	235	325	333	403
Rest of the World	382	544	760	1066	1061	1576
Grand Total World	6190	8660	10460	12545	13295	15130

Source: World Rubber Statistics Handbook, IRSG.
        Rubber Statistical Bulletin, IRSG.

While for the world as a whole, rubber consumption increased by 150%
during 1965-85, some countries have shown a more moderate though still
substantial increase. Extremely high growth rates have been achieved by
Japan, some developing countries, Eastern Europe and USSR.

Data on rubber consumption, split up between the tyre and non-tyre sectors, are available for a limited number of countries only. The percentage share of the tyre sector is given in Table 4. For the eight countries together, this share is remarkably stable over time until 1980. However, after 1980 a significant decline can be seen.

TABLE 4

Percentage share of the tyre sector in total rubber consumption.

	1965	1970	1975	1980	1985
USA	64	64	62	59	55
Canada	71	78	65	68	55
Brazil	71	65	63	63	67
Germany F.R.	57	51	47	46	44
France	58	63	65	70	62
United Kingdom	52	51	49	48	41
Italy	56	51	48	42	41
Japan	49	51	62	63	60
Sub-total	60	59	59	58	53

Note: derived from Rubber Statistical Bulletin.

Until 1940 NR was the only source of rubber. Large scale production of synthetic rubber (SR) emerged during the second world war, when supply of NR was insufficient, largely due to blocked supply lines. In the fifties and sixties production of SR increased rapidly because, demand, particularly from the automotive sector grew much faster than supply of NR, thus creating a reduction in the share of NR from 75% in 1950 to 30% in 1980. However, the decline in the share of NR was gradually stopped because certain end uses, particularly radial passenger car tyres and commercial vehicle tyres still needed a large share of NR. On an average, about 1/3 of the rubber in a passenger car tyre is NR while in a heavy duty truck tyre about 2/3 is NR. Apart from the preferred proportion of rubbers in end products, certain countries show a preference for NR while others seem to use SR more than that would be required. A major reason for this is domestic production of NR or SR.

Prices

During the last decade a number of developments in the price front have taken place, of which may be mentioned the effect of the International Natural Rubber Organization, which achieved price stabilization. The other major developments are the increased direct trade and the shift of rubber consumption away from Western Europe and the USA to countries in South-East

Asia. Table 5 shows average annual prices for selected years. The last set
of prices refers to the daily market indicator price (DMIP), an average
of prices of various grades in various markets in Malaysian/Singapore currency. The instability in prices is apparent from this table. For this reason,
the United Nations Conference on Trade and Development at its Fourth Session
in Nairobi in May 1976, adopted Resolution No. 93 (IV) containing the integrated programme for commodities.   Pursuant to that resolution, UNCTAD

TABLE 5

Prices of natural rubber, all prices in local currencies per tonne.

		1960	1965	1970	1975	1980	1985	1988
London	RSS 1	289	191	180	288	663	642	710
	RSS 3	281	186	175	277	638	612	693
New York	RSS 1	841	566	463	659	1625	924	1287
	RSS 3	830	556	454	634	1565	897	1180
Kuala Lumpur	RSS 1	2383	1544	1244	1357	3123	1886	3098
	RSS 3	2313	1512	1193	1300	2987	1798	3012
	SMR 20						1735	2779
Singapore	RSS 1	2383	1544	1244	1346	3079	1665	2328
DMIP*	average					281	171	255
	daily high					290	188	332
	daily low					274	160	216

* INRO's Daily Market Indicator Price.

convened a series of meetings to prepare for a natural rubber agreement.
At the end of the Fourth Session on October 6, 1979, the International Natural
Rubber Agreement was established. The agreement came into force provisionally
on October 23, 1980 and definitively on April 15, 1982. It expired five years
later and was extended for two years until October 22, 1987. In March 1987,
a second international agreement was reached in continuation of the previous
agreement, under roughly the same terms.

A graph relating the DMIP and the price ranges is presented
in Fig. 1. The area between the upper two lines is called the 'may-sell'
range, while at the bottom, a similar area is shown as the 'may-buy' range.
The top line shows the 'must-sell' price and the bottom line the 'must-buy'
price. Early in 1980, prices of RSS 1 had peaked at US $1.60 per kg; they
fell to the 'may-buy' level in 1981, stayed there in 1982, rose to the 'may-
sell' level of US $1.15 in 1983, fell again in 1985 and rose afterwards to
the level of around US $1.50 in mid 1988, which is the upper limit in the

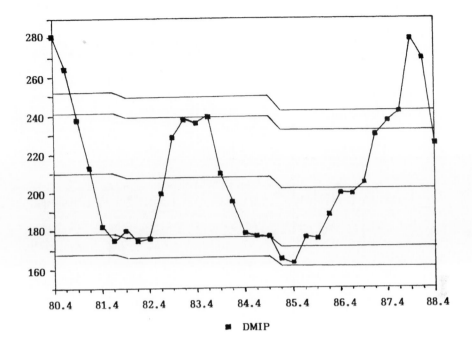

Fig. 1.   DMIP INRA ranges.

Agreement. The last quarter of 1988 showed prices in the 'no-action' range. Buffer stock intervention took place in 1982 and 1985 and in September 1987 the buffer stock contained 362000 tonnes. Later in the year prices rose to the 'may-sell' level and just before the expiry of the first Agreement rubber could be sold from the buffer stock.

ANALYSIS OF THE STRUCTURE OF THE NATURAL RUBBER ECONOMY

The outlook of the natural rubber economy as presented in this chapter is based on an elaborate model. For the purpose of judging the quality of the analysis on which forecasts and policy suggestions are made, the model is briefly described here. The complete model consists of

a) long-term analysis of natural rubber supply,

b) long-term analysis of total rubber demand, and

c) quarterly model which is developed to describe short-term reactions of demand, supply and prices to each other, as well as in relation to buffer stock intervention in the market. Schematically the model is depicted in Fig. 2.

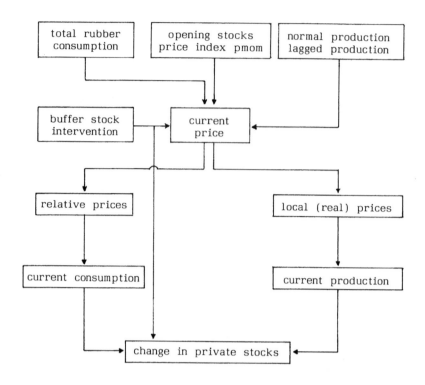

Fig. 2.   Schematic representation of the model.

LONG-TERM SUPPLY TRENDS OF NATURAL RUBBER

Production potential

Rubber trees can reach the age of 60 years or more, but the yield per tree is not constant. Trees can be tapped after about 6 years, with the yield normally increasing to reach its maximum when the trees are approximately 12 years old. As they grow older, yield normally declines and at the age of 50, it decreases to almost nil. The supply of natural rubber, therefore, depends not only on the area planted with rubber, but also on their age-composition. The long-term supply model establishes the link between the area with rubber on the one hand and the 'normal' production of NR on the other.

The long-term supply models are constructed for the three major producing countries. In Malaysia and Indonesia a further distinction is made between estates and smallholdings, as these groups may differ in yield level, technological progress and replanting behaviour. In Thailand, the third largest producer, the estate sector is too small to warrant separate treatment.

Production trends for natural rubber, using a vintage approach

To quantify the concept 'production trends', we use the term 'normal production'. This is defined as that level of production in a certain year, which would have been realized had there been no price influence. The reason for employing such a concept is that for real developments in the market, only the short-term price effect need be superimposed. How normal production can be estimated in relation to other variables is indicated in this section. This is done for smallholdings and estates in Peninsular Malaysia, Indonesian estates and Thai rubber producers.

Estimating normal production

In analysing production of NR the following elements are essential: planted area, new planting, replanting and uprooting, age and the yield profiles, technological progress, other factors influencing normal production and prices. The reason for such an approach is that a hypothetical hectare of rubber plantation will have the following yield profile. Planted in 1955, it provides rubber from the 1960s onwards, declining in the 1970s. In the 1970s and 1980s decisions about replanting must be made, otherwise, production from that hypothetical plantation would decline to very low levels in the years after. If this hectare of plantation is uprooted and not replanted the reduction of NR production potential depends on the yield profile and the age of the plantation at the time of uprooting. If this area is replanted, no production can be expected during the immaturity period. Afterwards production starts again, following a yield profile. However, the yield profile to be followed, will be considerably higher than the old one because of technological progress. Finally, the intensity at which tapping is carried out will depend on price and labour availability.

Estimation of normal production therefore requires answers to the following questions:

(i) what is the composition of the total area of NR according to the year of planting (the vintages);

(ii) what percentage is uprooted or replanted because of age, disease or damage (the discarding system);

(iii) what is the average yield profile for a hectare of rubber during its life (the yield profile);

(iv) how does technological progress in quality of clones affect yield profiles of areas planted in various years;

(v) what is the influence of other exogenous factors influencing normal production, e.g., weather and slaughter tapping.

It will be clear to anybody familiar with NR statistics, that data to do such an analysis are not available. Below will be shown the methodology to include all relevant variables in the analysis and how the data base and the relationships have been developed. For further details see Smit (1984).

Area distribution by vintages and discarding systems

It is clear that an essential element in analysing NR production is total area divided by age groups or vintages. A vintage of year r is the area planted in the year r e.g. r = 1955. As the years pass by, the 1955 vintage will reduce in size owing to uprooting or replanting. Data on area planted per year are reasonably accurate for many years. Other data available include total area under rubber and areas discarded per year.

A serious problem arises in determining area distribution by vintages, i.e. the number of hectares planted in year r and still existing in the year of tapping, year t, which is the year of analysis. Discarding of area, meaning uprooting for other crops or replanting by rubber, can be derived for total area, but cannot be split up according to vintages. In other words, it is not known how much of the 1955 vintage is discarded in 1956, 1957 and so on. However, one may assume that the percentage of a vintage r still remaining in year t and thus having an age k = $t-r$, will be related to k. First, little will be discarded; then, when the trees become less productive, discarding will increase.

If the area distribution by vintage would be known in a certain year $t-1$, e.g. 1959, thus splitting up the area planted 1 year ago, 2 years ago and so on, and if one would know the discarding percentage p_1, p_2, for age 1, 2 etc., then it would be easy to calculate how much would be left for each vintage in the year 1960. Using the same percentages one could then carry on and calculate the area distribution for 1961, 1962 etc. One does not know p_k exactly, but one may assume that the shape of p_k can be approximated by a sigmoid curve. By summing up all discarded areas over the vintages in 1960, one obtains the total discarded area in 1960 and similarly for the years thereafter. The p_k must be chosen in such a way that the calculated discarded area in each year equals the actual discarded area in each year. This implies that for each year a specific point of inflexion might shift the sigmoid curve to the left when there is large scale replanting and uprooting and shift it to the right when discarding is limited.

Yield profiles and technological progress

All Hevea clones follow a general yield curve over time. It is clear that in this study no detailed analysis of the area for the various clones can be undertaken. Because of the large degree of aggregation that is inevitable in the set-up of this study, an average of the various yield profiles has been used. The actual (commercial) yield profile will be lower than the ideal yield profile. The ideal yield profile therefore needs to be multiplied with a certain factor, in order to reduce the ideal yield profile to actual levels. This multiplication factor will be different for different countries, and, within countries, for estates and smallholdings. One of the reasons may be the selection of clones. This selection of clones, of which a certain vintage is composed, is fixed once a vintage is established. However, the composition of the different vintages (hence different planting years) may vary over time, implying that the multiplication factor may need to increase over time in view of the technological progress achieved. The assumption is that each vintage will have an average yield profile, which is a constant fraction of the ideal yield profile: if a yield profile is estimated to be, for example 0.3 times the ideal yield profile, then the profile is suppressed to 30% of the original shape with a value of 330 kg per ha in the second year of tapping and 720 kg per ha in the 14th year of tapping. Of course, later vintages may be composed of better clones, thus increasing the average yield. In the example, the fraction of 0.3 may become 0.4 for the vintage planted a number of years later. The fractions are estimated per vintage by relating area and ideal yield profile to production and then deriving the fractions which create actual yield profiles that are consistent with area composition and production. For example, estimated yield profiles for the 1950 and the 1980 vintages are about 55% and 70% respectively of the ideal yield profile. As a matter of clarification, this means that trees planted in 1950 have a yield profile which is about 55% of the ideal yield profile. This vintage, of course, will reach high production levels only after 1960.

We have discussed the components of normal production above: vintages, discarding systems, yield profiles and technological progress. Add to this, factors such as labour availability and the picture is complete. Of course, no data on normal production are available. The approach is to explain actual production in terms of prices and the various elements of normal production.

Production outlook for other countries

Production by countries other than Peninsular Malaysia, Indonesia and

Thailand could not be analysed in detail for lack of data. For these areas production trends have been estimated using more simple models. Drawing from various sources, such 'trend projections' for different countries have been summarised in Table 6. These trend estimates and projections are used in the quarterly model in interaction with demand and prices.

TABLE 6

Projections of NR production trends.

		1985	1990	1995	2000
Malaysia	Estates	563	463	408	342
	Smallholdings	974	1095	1127	1131
	East Malaysia	39	68	78	80
	Total	1576	1626	1613	1553
Indonesia	Estates	318	349	367	373
	Smallholdings	730	874	1018	1163
	Total	1048	1223	1385	1536
Thailand		732	1067	1332	1569
Sri Lanka		144	166	208	225
India		198	300	400	500
China		188	275	350	400
Other Asia		176	229	316	407
Africa		213	397	530	556
Latin America		63	63	81	99
World total		4338	5346	6215	6845

LONG-TERM DEMAND TRENDS FOR ALL RUBBERS

The automotive sector was the major pushing factor behind the dramatic growth of the rubber industry. Passenger cars and commercial vehicles on the road determine how many tyres are needed for replacement of worn-out tyres. Then there are tyres for other vehicles. Of all rubbers, 40-50% is used for products other than tyres. These products will be grouped into a 'general products sector'. Most attention is paid to the tyre sector, as this is the area where most of the rubber is going and where about 75% of all natural rubber is consumed.

Passenger cars in use

On a global scale, the number of passenger cars in use saw dramatic growth since the early 1950s: the figure in 1985 was about seven times that of 1950, implying a compound annual growth rate of 5.7%. It will be clear that income is one of the major determinants of passenger cars in use. Next to income level, population size is the basic explanation underlying the number of cars in use. For many advanced countries, growth in the last decade has been less than before, caused by economic recession as well as a possible approach towards saturation. Because of differences in demographic, geographic and other characteristics between countries, the saturation level, in principle, will be different for different countries. After establishing the model, saturation levels in cars per 1000 persons can be derived, followed by the number of cars in use. For this, various types of composite S-shaped curves have been used, with income, population and time as explanatory variables. After aggregation, on a global scale, it is found that cars in use will grow from 53, 195 and 370 million respectively in 1950, 1970 and 1985 to about 525 million in 2000.

Commercial vehicles in use

It is interesting to note that, on a global scale, growth in the number of commercial vehicles in use is extremely close to that of passenger cars. The 1985 figure was also about seven times the figure in 1950, indicating a compound annual growth rate of 5.6%, which is only 0.1% less than the growth rate for passenger cars. The important explanatory variables in the analysis of usage of commercial vehicles are: annual road transportation in tonne-kilometre, average capacity of vehicles, average degree of capacity utilization and average annual distance driven. The main problem in empiricizing this relationship is the data base. This leaves the number of commercial vehicles in use to be determined broadly by the volume of road transportation, which in turn, is closely related to GDP. However, as countries reach high levels of development the composition of GDP may become such that the need for transportation may increase less than the increase in GDP. This is included in the analysis by allowing non-linear relationships and even a possible saturation level. The outlook is that the number of commercial vehicles in use in the world will grow from levels of 17, 51 and 112 million in 1950, 1970 and 1985 respectively to 165 million in 2000.

Demand for passenger cars and commercial vehicles

Demand for passenger cars and commercial vehicles is represented by new registrations since this variable is more appropriately covered by statistics than sales of passenger cars and commercial vehicles. Production follows demand and as there are, in general, no real production shortages, new registration determines demand for original equipment tyres. New registrations can be divided into increase in the vehicle park and discards. This implies that it is appropriate to explain new registrations using a definitive equation. Discards are related to past levels of new registrations. The vehicle park has already been analyzed and projected above. No data are available on discards. However, they can be derived by deducting the increase in total vehicles in use from new registrations.

The basic determinants of discarding are average age and the mortality curve. One could argue that average life of vehicles should be different for different vintages **r** (construction years) and for different years of usage **t**. Average life per vintage **r** and per year **t** might therefore be indicated with μ_{tr}. Since no data on μ_{tr} are available, one has to estimate these data on the basis of such variables as new registrations, discards and number of vehicles in use. It has not been feasible to arrive at such a two dimensional variable μ_{tr} for each country. While having to choose between an analysis based on μ_r, a vintage determined average life, or on μ_t, a usage-year fixed average life, it was derived that the quality of the vehicles (technical and age) was less relevant than the quality of the economy and therefore projections could be based on μ_t using a vintage approach. Estimates of μ_t for cars in 1985 range from 9.4 years in Japan through around 11 in many European countries, and 12.6 in the USA to 15.3 in Italy and 15.9 in Australia. New registrations of cars will grow from 32 million in 1985 to between 45 and 50 million in 2000. In the case of commercial vehicles, the number of new registrations are 11 million in 1985 and 15 million in the year 2000.

Tyre demand for passenger cars and commercial vehicles

The past three decades have shown rapid changes in the tyre scene. In the early sixties virtually the only type of tyres were the cross-ply tyres. Then the radial tyre started conquering the market in Europe. First there was the textile belted radial tyre; more recently the steel belted radial has become increasingly important. In the USA, meanwhile, the bias belted tyre had become popular, both with glass and steel belts. In Japan, the move to radial gathered momentum in the seventies as well. The most important advantages of the radial tyre compared to the conventional cross-ply or bias belted tyre is 'tyre distance' defined as the number of kilo-metres over which the tyres can last from the start until replacement. Radial tyre distance will reach 50 to 100% more than that of conventional cross-ply tyres, while bias belted tyres are somewhere in between.

Virtually all passenger cars have four wheels. Add to this the spare tyre, so all new cars are equipped with five tyres. However, newly developed tyres (the so-called safety tyres or run-flat tyres) may reduce the average number of tyres per new passenger car. For commercial vehicles it is extremely difficult to obtain adequate statistical information on the number of tyres per vehicle. The main determinant is the average vehicle size. The estimated number of tyres per commercial vehicle per country ranges from 6.0 for Japan to 8.5 for the United Kingdom. By simply multiplying the number of tyres per vehicle by the number of vehicles newly registered as derived above, it is possible to make projections of the number of tyres for original equipment, divided into conventional and radial tyres.

Demand for replacement tyres is determined by tyre life which can be derived from tyre distance in kilometres divided by driving distance per vehicle in kilometres per year. A complicating factor, however, is the variation in driving distance as well as in tyre distance within a country or a region. This is why, in the analysis, probability distributions have been applied. As data on world tyre production are available for a number of years, the tyre replacement model has first been used to estimate average tyre distance per region. Afterwards projections of average tyre distance and average driving distance and that of demand for replacement tyres are derived. One of the major problems in this analysis is the determination of the average tyre distance, the average number of kilometres achieved with one set of tyres per type. The starting point for the analysis is a series of data on world tyre production, both for passenger cars and commercial vehicles. After deduction of tyres for original equipment, data on tyres for replacement are available.

A prior information on differences in average tyre distance between types and between countries is partly available from statistics and industry sources. On this basis, for each region average tyre distance and its annual increase for each type is assumed to have a fixed ratio to a basic world wide estimate of average tyre distance per type, thus allowing for country specific differences. Radial passenger car tyres, on an average, now-a-days last between 35000 and 60000 km, depending on the country. For commercial vehicle tyres this range is 55000 to 100000 km. The number of new tyres for passenger cars will grow from 495 million in 1985 to 613 million in 2000. For commercial vehicles the numbers are 190 million in 1985 and 237 million in 2000.

Consumption of rubber in the tyre sector

In the previous paragraph the number of new tyres for passenger cars and commercial vehicles was forecast. Also incorporated in the analysis is the number of retreaded tyres. Together with new tyres, they provide the need for rubber, as analysed for the country, where the vehicle is being driven on roads, by straightforward multiplication with the rubber weight per type of tyre. However, no data are available on the average weight of tyre. Using the model, estimated average rubber needs per passenger car radial tyre were derived. In 1985 they ranged from 3.7 kg in Japan to 5.1 kg in the USA. For commercial vehicles, these figures were 10.0 kg in Japan and 24.5 kg in the United Kingdom.

Thus far, rubber demand has been derived and projected for passenger car and commercial vehicle tyres. This leaves untreated, those groups of tyres used for tractors, aircraft, motor cycles, scooters and bicycles as well as inner tubes. Rubber consumption for such other tyres as well as for retreading material was derived using a fixed share of consumption in the tyre sector. This percentage was different for various countries.

Rubber consumption in general products sector

Specific non-tyre end-uses for rubber number in thousands. To mention a few: rubber thread, rubberised cloth, footwear, window strips, engine mountings, conveyor belts, hoses, rubber sheets, roofing sheets, rubber gloves, carpet backing, elastic rubber bands, fishing ropes and soft-balls. Availability of data, manpower and time are barriers in undertaking a detailed analysis of all possible end-uses. Instead, total non-tyre rubber demand has been explained from GDP. Presently used projections also give room to recent developments in this sector where thermoplastic elastomers have taken their share of the potential rubber market.

World rubber consumption

World consumption projections can now be obtained directly by adding the projections of the various component sections. Results for selected years for world total are presented in Table 7 for the three economic growth scenarios.

TABLE 7

Projections of world total rubber consumption: three scenarios ('000 tonnes)

	1985	1990	1995	2000
Scenario G1	13340	15265	16545	17625
Scenario G2	13340	15393	17250	18943
Scenario G3	13340	15521	17996	20407

SHORT-TERM ANALYSIS

Short-term supply response to prices for natural rubber

Supply equations were estimated, based on 'normal' production levels.

These normal production levels described above are derived on a yearly basis and were converted to quarterly figures. Consequently, quarterly supply behaviour refers to tapping only and need not distinguish between utilization and change of production capacity. Such changes would be incorporated in the normal production levels and these are assumed not to depend on prices. Hartley et al. (1987), Hwa (1985) and Tan (1984) also reported on the lack of price responsiveness of production capacity. The ratio of actual to normal production was taken as the dependent variable in the short-term supply equations. A double logarithmic specification related this ratio to seasonal dummies and to the ratio of the Singapore price of NR, converted into local currency and adjusted for export duties in Malaysia and Thailand, and the consumer prices. Quarterly supply elasticities were in the order of 0.13 to 0.18 for smallholdings and for Thailand and 0.10 and 0.06 (0.14 and 0.08 on a longer term) for Malaysian and Indonesian estates respectively. Production of these five groups represents about 75% of the world production. The production of the rest of the world was analysed separately. We used a similar approach, including real prices and normal production levels for the world's sixth largest producer, Sri Lanka, but without success. Amongst the other producers, only the Philippines showed a reliable response to prices, with an elasticity of 0.07 (short-term) and 0.43 (longer term). India and China were treated exogenously and the rest of the world, though analysed at country level, was assumed to follow exogenous trends.

Short-term demand response to prices for natural rubber

Demand for natural rubber is related to total (natural and synthetic) rubber demand. As price of rubber constitutes only a small portion of the cost of end products (mainly tyres), it can be assumed to play no significant role in deciding the total use of rubber. Given a certain total demand for rubber, however, the share of NR in this aggregate depends on NR prices relative to synthetic rubber prices. The price of RSS 1 was taken as the representative price for natural rubber expressed in US dollars in the Singapore market, and unit values of the US SBR exports were used as competitive synthetic rubber prices. As an additional explanatory variable, a trend has been included to account for the decline in the share of NR until 1978, due as this was to an expansion of the non-tyre use of rubber, and to account for the increase in its share since 1978 when the decline was more than compensated for by the increasing use of radial tyre, which has a higher NR content than conventional tyres. In view of the large share that radial tyres now have, the upward trend is not expected to continue.

Prices

Quarterly price determination on world markets is made between buyers (most NR consuming tyre manufacturers) and sellers (mostly stockholders). Hence, a current quarter's production can hardly play a role, as this production cannot physically reach the consumers. The relevant considerations regarding the prices, therefore, are those of the NR consumers and of the stockholders. Production plays a role only as an indication for the stockholders about what future replenishment of their stocks is to be expected. The main determining endogenous factors of the prices are total world rubber consumption (NR and SR), opening stocks and expected production, being represented by the four-quarter average, lagged by two quarters.

Prices are determined in a world market, influenced by changing rates of inflation, of interest and of currency exchange. These monetary influences on price were captured in two additional variables, viz. the price index of minerals, ores and metals, and the US dollar/SDR exchange rates. The inclusion of the price index for minerals, ores and metals (abbreviated to 'pmom') serves two ends. Firstly, the prices pertain to commodities that are in demand by the industry that also uses rubber, and in this way, pmom complements the exogenous variable describing total rubber demand. Secondly, the commodity prices underlying pmom, like copper and aluminium prices, are sensitive to monetary changes. A declining US dollar, declining interest rates or rising inflation rates, induce conversion of money balances into commodities (Frankel, 1984).

It was estimated that changes in total rubber consumption have a bearing on the price of NR in US dollars. This is incorporated by including the consumption variable measured in tonnes and adjusted for seasonal fluctuations. Additionally the US dollar/SDR exchange rate was found to have an effect of its own on the Singapore price of NR in US dollars.

In spite of the large seasonal fluctuations in supply and demand, prices do not show this cyclic pattern. Thus, stockholders appear to well play their role of accommodating these short term fluctuations. In the price equation this is incorporated by de-seasonalising the level of total consumption and the level of private stocks.

Finally, the purchases for the bufferstock that took place in 1982 and 1985, and sales from the buffer stock in 1987 and 1988 were to be included in the equation. Their influence on the prices proved to be modelled best when they were considered as reduction in the private opening stocks of the quarter. Since these purchases are neither consumption nor production, this result is quite plausible. Buffer stock sales in 1988 fell outside the estimation period.

The estimated log-linear equation corroborates the dollar exchange rate effect, displaying an elasticity of about 0.16. Effects of consumption and production were of about equal size, on an average, showing an elasticity of 2, whereas the direct effect of a change in opening stock was 0.5, measured as elasticity. As expected, the price index of minerals, ores and metals had an elasticity of unity.

Stocks

The model is closed by an identity for stock changes, and therefore private stockholders' behaviour is modelled implicitly. Changes in private stocks take place for seasonal reasons, as production and consumption do not match in every quarter; they also occur after buffer stock intervention, as there is hardly any alternative but to buy from or sell to private stockholders. Checks were made to investigate whether the buffer stock should be added to the private stocks, and whether this would have an effect on the prices. This was not so. The private opening stocks, after adjustment for current buffer stock purchases, appeared to be preferable. Buffer stock purchases thus reduce the quantity available for sales in that quarter and in later quarters, but the cumulated quantity of NR, taken out of the market, but still stored somewhere, did not appear to have any further influence on the prices. The estimated price equation can be used to simulate the behaviour of the buffer stock. The equation can be solved for that change in buffer stock that would bring prices to a certain level. Simulations for the period 1982-1988 showed that apparently, buffer stock intervention was to defend a price, laying some t per cent above the minimum ('must-buy') price of the agreement, i.e. around the 'may-buy' level. A full description of the model is given by Burger and Smit (1989).

PROJECTIONS AND POLICY SIMULATIONS

Introduction

In this section, first a standard set of projections for the world natural rubber economy is presented. Then the following policy measures are assessed:
(a) the effect of buffer stock operations under the second International Natural Rubber Agreement;
(b) The effects of concerted effort on the part of major producers to temporarily restrain production by increasing replanting.

All assessments are made against a reference scenario, which will be sketched in the next sub-section and with the assumption that planting policies are more or less as they were in the past, and that throughout the period until the year 2000 the INRO may intervene in the market according to the regulations of INRA-2.

Standard scenarios: outlook for the year 2000

The long-term outlook for normal production levels and the outlook for long-term demand for all rubbers are already given in Tables 6 and 7 respectively. Comparison of the trends in normal production and in total rubber demand shows that in the medium term future, supply tends to outstrip demand. Actual supply differs from normal production levels because of its sensitivity to prices. In order to have an approximate equilibrium of natural rubber demand and supply, the latter should be below normal levels, and the share of NR consumption should increase, both of which can be accomplished by decreasing the relative prices and this is what

TABLE 8

Reference projections of NR supply, demand and prices
(volumes in '000 tonnes, price in S$/tonne)

Year	Production	Consumption	Stocks	Price
1988	5040	5105	1663	2328
1989	5120	5300	1483	1851
1990	5359	5464	1378	1904
1995	6078	6082	1364	1687
2000	6648	6646	1384	1811

is projected. Table 8 shows the projections for actual NR supply and demand and the corresponding levels of NR prices, according to the basic scenario.

In these projections, an inflation rate of 3% per year is incorporated, and the real price decline of NR is, therefore, much more pronounced. The base projection does not lead to market intervention from the buffer stock, as prices will remain above the lower limit of the Agreement. The share of NR in the total rubber consumption is projected to increase less than that in the past 10 years and to reach a level of 35% by the year 2000. A regional breakdown of total and natural rubber consumption is given in Table 9.

In view of the uncertainty about future demand prospects, analyses were made of the effects of lower and higher rubber demand growth (the G1 and G3 scenarios). In both cases, NR prices would fall after a brief rise in 1990. In the low-growth scenario prices would go down to the 'may-buy' level of INRA and trigger mild but prolonged market intervention from

TABLE 9

Total rubber consumption and natural rubber consumption in the year 2000, standard scenario ('000 tonnes).

	Total rubber	Natural rubber
USA	1646	1098
Canada	345	136
Japan	1527	790
Australia	107	74
Germany F.R.	750	276
France	486	242
United Kingdom	388	186
Italy	492	189
EC	2834	1139
Total W. Europe	3424	1295
Eastern Europe and USSR	4283	491
China	1791	758
India	771	386
Rep. of Korea	810	379
Taiwan	516	212
Brazil	581	161
Rest of the world	3142	868
Grand Total	18943	6648

buffer stock. In this scenario there is not much chance for sale from accumulated buffer stock before the turn of the century. In the high-growth scenario prices would hardly go down initially and would rise afterwards,

TABLE 10

Projections of NR production by major producers in the year 2000.

Malaysia, Peninsular

	Estates	334
	Smallholdings	1096
	Total	1430

Indonesia

	Estates	350
	Smallholdings	1155
	Total	1505

Thailand	1457
Rest of the world	2255
World total	6648

reaching 2900 S$/tonne in the year 2000. This rise in prices would just keep real prices on the increase. Although the eventual price would be above the upper limit of INRA, there may not be any buffer stock available to prevent prices from rising.

On the basis of these scenarios a case can be made for some rationalization of supply to prevent real prices from falling in the years to come. This is discussed below.

Evaluation of policy measures

The most obvious policy measure in the NR market is the presence of a buffer stocking mechanism. As indicated above, only in case of low demand growth (or very low inflation rates) will there be a buffer stock intervention in the market and in that case there are only meagre prospects for the disposal of the stock. As indicated in Herrmann, Burger and Smit (1990) the presence of the INRO can hardly be assumed to stabilize prices to such an extent that major investment decisions on plantations are influenced. The above scenarios show likewise that there is only a small chance for buffer stock intervention in the market, unless the price range is adjusted upwards.

Turning to an investment oriented supply management instrument, the following three figures show what would happen in the NR market if the three major producers would agree on an intensive replanting scheme. Replanting subsidies abound in all producing countries and appear to be the policy instrument most easy to use. It is assumed here that subsidies are increased to such an extent that replanting between 1991 and 1993 in Malaysia, Thailand and on the estates in Indonesia will be twice the area that was assumed for the reference scenario. New plantings and area were assumed not to be affected. The additional replanting was translated into new levels for normal production in these countries. These new levels should initially be lower, because productive trees are replaced by immature trees. After 5 to 8 years, normal production levels would increase again, when the new trees come into production. The new level of normal production are fed into the quarterly model to assess their effect on world market prices and actual production.

Figure 3 shows the effect on actual production in the three countries taken together for the medium and high-growth scenarios. For each growth scenario production levels will be slightly reduced by the the rationalization measure until 1998 or 2000 when the levels coincide. It is noteworthy that the effect of demand growth outweighs the rationalization effect, and that high production levels can be reached, when these are called for by the consumers, even in the case of reduced normal production levels. Fig. 4 shows the effect on prices: in the standard growth scenario, prices are

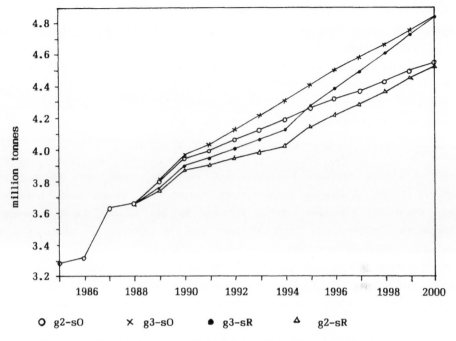

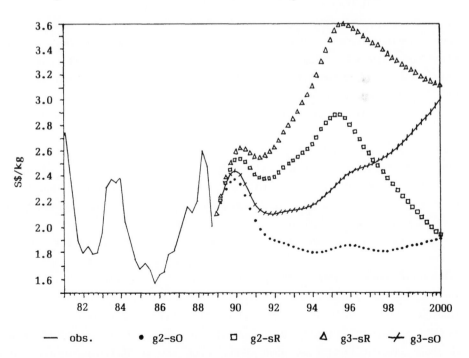

O g2-sO   × g3-sO   ● g3-sR   △ g2-sR

Fig. 3.   NR Production - Indonesia + Malaysia + Thailand.

— obs.   ● g2-sO   □ g2-sR   △ g3-sR   ✝ g3-sO

Fig. 4.   NR prices - double replanting and standards.

shown to remain around their present level in case of rationalization for the period until 1996, after which they will begin to fall, coinciding with the 'unrationalized' level in the year 2000. In the case of high demand growth, prices will initially move upward and will meet the original projection again in the year 2000. The price effect clearly outweighs the volume effect, at least until the end of the century. Other supply rationalization measures are discussed by Smit and Burger (1986).

SUMMARY

A vintage approach was applied to both the demand and the long-term supply sides of natural rubber economy. On the demand front allowance was made for saturation levels of cars in use in the major tyre-using countries. The forecasts, derived from separate analyses for passenger cars and for commercial vehicles, show that the number of vehicles in use will increase from 370 million passenger cars in 1985 to 525 million in the year 2000. For commercial vehicles the corresponding numbers are 112 and 165 million. This involves new registrations increasing from 32 million passenger cars in 1985 to 45-50 million in 2000. For commercial vehicles, the numbers are 11 and 15 million. Derived demand for new tyres, including tyres for replacement will increase from 495 million passenger car tyres in 1985 to 613 million in 2000. For commercial vehicles, the demand for tyres in 1985 was 190 million and in 2000, 237 million tyres is expected. This, combined with the expected increase in demand for rubber for non-tyre use, will lead to a world demand for rubber increasing from 13.3 million tonnes in 1985 to 18.9 million tonnes in 2000, or in the case of high growth, to 20.4 million tonnes. The share of natural rubber will be between 34 and 35%, depending on the price, which also depends on supply. From a detailed analysis of yield and age composition of the rubber area in the major producing countries, it was derived that the normal production of NR is expected to increase from 4.2 million tonnes in 1985 to 6.8 million tonnes in 2000. A quarterly model, including buffer stock intervention and exogenous price trends that are assumed to rise by 3% per year, yielded forecasts of the price of NR, which is expected to decline from a level of 1850 S$/tonne in 1989 to 1700 S$/tonne in 1995 and slightly higher thereafter. With high demand growth, NR prices would go up to S$2900 after initially staying at or slightly above S$2000 per tonne. An alternative supply scenario, involving twice as much replanting during 1991-1993, was expected to lead to higher prices for the whole period. At first, prices would remain at their present level, but after 1996, when the additionally planted trees come into production, prices would fall, reaching the levels that were predicted for the standard case by the year 2000.

Regional distribution of supply appears to tend towards declining share for Malaysia (down to 23%) and increasing shares for Thailand (up to 22%) and Indonesia (up to 23%) and for the rest of the world.

Natural rubber consumption is shifting away from the traditional areas and by the year 2000, the share of Japan, South Korea and Taiwan should have increased from 18 to 21%, whereas the EC and the USA would stabilize around 17%.

REFERENCES

Allen, P.W., Thomas, P.O. and Sekhar, B.C., 1973. The Techno-Economic Potential of NR in Major End-Uses. Malaysian Rubber Research and Development Board, Monograph No. 1, Kuala Lumpur.

Burger, K. and Smit, H.P., 1989. Long-term and short-term analysis of the natural rubber market. Weltwirtschaftliches Archiv. Vol. 125(4).

Carr, J.D., Jampasut, P. and Smit, H.P., 1988. The world rubber economy, changes and challenges. International Rubber Study Group, London.

Frankel, J.A., 1984. Commodity prices and money: lessons from international finance. Ame. J. Agric. Econ., 66(5): 560-566.

Harley, M.J., Nerlove, M. and Peters, R.K., 1987. An analysis of rubber supply in Sri Lanka. Ame. J. Agric. Econ., 69(4): 755-761.

Herrmann, R., Burger, K. and Smit, H.P., 1989. Commodity policy, price stabilization versus financing. Paper presented at the CEPR Conference on "Primary Commodity Prices: Economic Models and Economic Policy". London, 16-17 March 1989.

Hwa, Erh-Cheng, 1989. A model of price and quantity adjustments in primary commodity markets. J. Policy Model, 7(2): 305-338.

International Rubber Study Group (IRSG), Rubber Statistical Bulletin, Various issues, and (1986), World Rubber Statistics Handbook. Volume 3, 1960-1985. London.

Smit, H.P., 1984. Forecasts for the world rubber economy to the year 2000. Globe Industry Report No. 2, MacMillan, London.

Smit, H.P. and Burger, K., 1986. Cooperation in marketing of natural rubber: a simulation analysis of supply alternatives. UNCTAD/ST/ECDC/34, Geneva.

Tan, C. Suan, 1984. World rubber market structure and stabilization. World Bank Staff Commodity Papers, No. 10.

CHAPTER 3

THE GENUS HEVEA - BOTANICAL ASPECTS

P.R. WYCHERLEY

Kings Park and Botanic Garden, West Perth, Western Australia 6005.

INTRODUCTION

The first botanical description of the genus Hevea and of the type species H. guianensis was by Fusee Aublet (1775). The generic name Siphonia was proposed by L.C. Richard in 1779 as the equivalent of Aublet's Hevea and in 1781 H.F. Gmelin suggested Caoutchoua elastica to replace H. guianensis, however these names were not published until 1791. H. brasiliensis (1824), H. pauciflora, H. spruceana and H. rigidifolia (1854) together with some subsequently discarded concepts were originally described under Siphonia, which was however reduced under Hevea by J. Mueller - Argoviensis (1865). The taxonomic treatments were by botanists working in European herbaria on material much of it collected by Richard Spruce during his explorations of the Amazon and the Andes 1849-1864.

The history of infrageneric classification of Siphonia (Bentham 1854 and Baillon, 1858) and of Hevea (Mueller - Argoviensis, 1865; Huber, 1906; Pax, 1910 and Ducke, 1923 to 1935) has been reviewed by Schultes (1977b), whose recognition of two subgenera Hevea and Microphyllae is followed here. Schultes (1987) has also reviewed the variation in the number of taxa recognised in the genus. This rose from 11 species (Mueller - Argoviensis, 1874) to 24 species (Huber, 1906) and then declined through 17 species (Pax, 1910), 12 species (Ducke, 1935) to 8 species (Seibert, 1947) and then rising slightly to the present 10 species, the last being due in part to the discovery of a new species H. camargoana (Murca Pires, 1981).

Previously specific rank was given to what are now considered to be local or transient variants, reduced to varieties or forms if still recognised at all. This modern trend to fewer species has been adopted by Ducke (1935), Baldwin (1947), Seibert (1947), Murca Pires (1971) and Schultes (1987), although Baldwin and Seibert tended to distinguish many putative interspecific hybrids. About 100 binomials and trinomials have been published for naturally occurring Hevea, but relatively few of these taxa have been retained. The number of cultivars selected amounts to many thousands, but the number planted on a large scale is much less.

Thus, after a little more than 200 years, ten species are recognised in the genus Hevea, which are in the order of the first descriptions of the concepts:- H. guianensis, H. brasiliensis, H. pauciflora, H. spruceana, H. rigidifolia, H. benthamiana, H. nitida, H. microphylla, H. camporum and H. camargoana. All are placed in the subgenus Hevea, except for H. microphylla in the subgenus Microphyllae.

DISTRIBUTION OF THE GENUS HEVEA

The genus Hevea occupies the Amazon Basin and parts of the adjacent uplands. The Amazon rises in the Andes and flows eastwards to the Atlantic Ocean. The broad lowlands of the Amazon Basin are bordered to the north by the Guayana Shield and to the south by the Central Brazilian Shield. The tributaries of the Upper Orinoco flow northward from the divide or watershed of the Guayana Shield, which is also the border between Brazil and Venezuela, whence the Rio Negro and its affluents flow southward to the Amazon. The land rises steeply from the lowlands of the Amazon Basin to the divide so that the uplands form a relatively narrow zone on the south side (Fig. 1).

Hevea extends from the lowlands into the foothills on the south side of the divide of the Guayana Shield. But north of the divide, Hevea is limited to the humid forests of the Upper Orinoco in Venezuela and does not extend into the seasonally arid savannahs further north.

South of the Amazon Hevea penetrates into the foothills of the Mato Grosso but not so far as the depression in the middle of the Central Brazilian Shield. Hevea occurs also in the zone at the western edge of the Amazon Basin bordering on the Andean syncline including foothills and uplifts such as that about Iquitos.

The soils of the lowlands of the Amazon Basin are mainly pale yellow (kaolinistic) latosols of low fertility. Groundwater latosols and low humic gley soils occur in locally extensive areas about the major water courses. A variety of soils are found in the uplands.

The natural range of Hevea falls into two main climatic regions, namely an ever humid region north of the Upper Amazon and a monsoonal region covering the rest. The mean monthly temperature does not fall below 18°C in any month throughout virtually the whole range of Hevea. The climatic types of Walter and Lieth in which Hevea occurs are as follows:

12   Constantly humid equatorial.   Precipitation at least twice the evaporation every month. Along the north of the Upper Amazon including the Upper Rio Negro and Uaupes (Vaupes in Colombia).

52

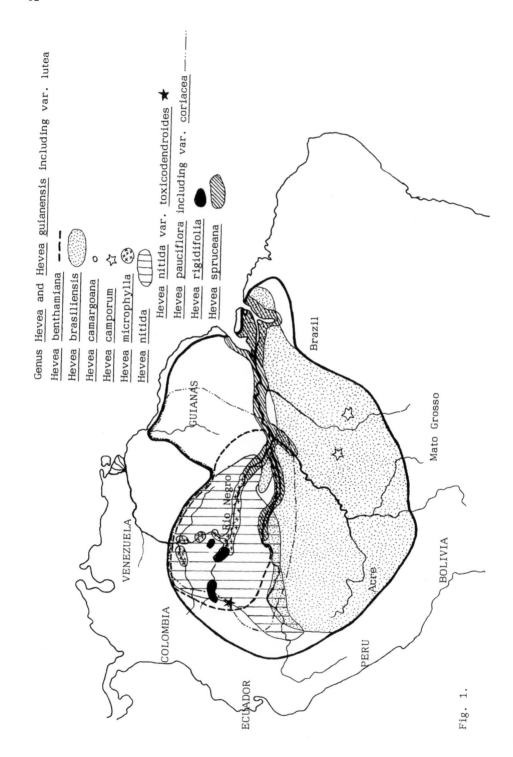

Fig. 1.

13a   Humid equatorial.   Precipitation exceeds evaporation. Brief less wet period. Mainly south of the equator. Lower Rio Negro, Solimoes section of Amazon from Manacapuru to above Tefe and south west of Tefe.

13b   Humid equatorial.   Precipitation exceeds evaporation. Brief less wet period(s). Tendency to bimodality. Mainly north of the equator, includes Brazil, Rio Branco and Para north of the equator, most of the Guianas and the Upper Orinoco in Venezuela.

I(II)b   Transitional from humid equatorial to tropical monsoonal. Annual precipitation exceeds evaporation. Dry spell of one to three months. Monomodal. A broad band south west from the Amazon delta including the lower Amazon below Manaos, and the Rio Madeira to beyond Porto Velho.

II 1   Monomodal tropical monsoon region.   Dry spell of three to five months. Annual precipitation in excess of 1500 mm. This is largely in or about the foothill region between the Amazon Basin and the Andes to the west and the Mato Grosso to the south in Peru, Bolivia and Brazil. The boundary with 12 is at about 6°S.

II 4   Monomodal tropical monsoon region.   Dry spell of three to five months. Annual precipitation less than 1500 mm. This is where Hevea penetrates into the Mato Grosso.

DISTRIBUTION OF THE SPECIES

Hevea occurs in the following countries: Bolivia, Brazil, Colombia, Ecuador, French Guiana, Guyana, Peru, Surinam and Venezuela. All ten species occur in Brazil. Seven species have been found in Colombia. Five occur in Venezuela and four in Peru.

The approximate distribution of the species is given in Figure 1 and in Table 1. The areas indicated in Figure 1 are the approximate limits of the species, some of which are disctontinuous in their distribution within these limits and all of them are restricted to some degree by habitat preferences within their areas.

Hevea guianensis and its variety lutea occur throughout the whole range of the genus. H. guianensis var. marginata is found in the general region of Manaos.

Hevea brasiliensis occupies most of the area south of the Amazon, penetrating north of the river to the west of Manaos.

Hevea pauciflora and its somewhat more common variety coriacea (=H.

confusa) occupies much of the area north of the Amazon from the Guianas to the Amazonas district of northern Peru, but it also occurs south of the Amazon on the lower Madeira and along the Solimoes reaches of the Amazon. The distribution is discontinuous perhaps because of habitat preferences.

Hevea nitida also occurs mainly north of the river, although there are populations south on the lower Madeira and in northern Peru. The distribution is markedly disjunct into almost separate populations within its general area. It is absent from the eastern half and from most of the southern part of the range of the genus. Hevea nitida var. toxicodendroides is restricted to quartzitic soils on sandstone mesas in the Vaupes and Amazonas regions of Colombia.

Hevea benthamiana is found only north of the Amazon and west of Manaos, the Rio Negro - Vaupes system runs through the middle of its range. It does not extend so far west as H. nitida, H. pauciflora, H. guianensis and their varieties because of their different habitat requirements.

Hevea spruceana lines the banks of the Amazon from its confluence with the Rio Ica (known as the Rio Putumayo on the borders of Colombia and Peru) in the Upper Solimoes reaches down to the delta, almost to the sea. H. spruceana occurs also on the Lower Madeira, Rio Negro and the lower reaches of the other major tributaries.

Hevea microphylla is not common and is confined to the middle and upper reaches of the Rio Negro and the Rio Guainia.

Hevea rigidifolia is an endemic, restricted to localities on the Upper Rio Negro and Rio Vaupes in Brazil, Colombia and Venezuela.

Hevea camporum is an endemic restricted to an area far south of the Amazon about the headwaters of the Marmelos and Manicore, tributaries of the Rio Madeira and of the Caruru, a tributary of the Rio Tapajos.

Hevea camargoana is the most recently discovered species (1975). It is found in the north eastern part of the Isla Marajos (Amazon Delta).

CENTRE OF GENETIC DIVERSITY

The following species occur in the Rio Negro region: H. benthamiana, H. guianensis and its variety Lutea, H. microphylla, H. nitida, H. pauciflora and its variety coriacea, H. rigidifolia and H. spruceana, comprising seven species and two varieties of importance. The area where the more abundant species overlap (H. benthamiana, H. guianensis, H. nitida and H. pauciflora) has H. nitida var. toxicodendroides on its western margin, and their common range borders on H. brasiliensis and H. spruceana along the southern boundary (Solimoes and lower Madeira). Only H. camporum and H. camargoana are distant from the Rio Negro and surrounding regions. There can be little

TABLE 1

TAXA OF HEVEA ARRANGED IN APPROXIMATE DESCENDING ORDER OF THE EXTENT OF THEIR NATURAL RANGES AND
ABUNDANCE IN THE WILD.

Taxa, species and varieties.	Range	Habitat	Growth characteristics	Exploitation in the wild and cultivation	Summary
H.guianensis & its var. lutea	Throughout whole range of genus	Well drained but moist soils	Tall trees, "short shoots", suberect mature leaflets.	Exploited in the wild, but not cultivated.	++ *
H.brasiliensis	Ca.southern half south of Amazon	Well drained, seldom flooded	Tall trees, "winter" trunk sometimes swollen in wild.	Exploited in the wild. Extensively cultivated.	-- ***
H.pauciflora & its var. coriacea	Most of northern half, sl.sth, discontinuous	Well drained, often sandy, rocky.	Small to medium trees, "short shoots" tough leaves.	Not exploited due to high proportion of resins.	++ 0
H.benthamiana	North of the Amazon, western quadrant	Swamp forests, often flooded	Medium trees, "winter", trunk often swollen in wild.	Exploitation in wild. Some culti-vation of hybrids.	+ **
H.nitida	Also north-west quadrant, discontinuous	Usually swamps, occ. moist rocks	Medium trees, "short shoots', vivid green folded leaflets.	Not exploited. Resins etc. give poor latex, rubber.	+ 0
H.spruceana	Amazon (Ica to delta), Negro & L.Madeira Rivers.	Muddy banks and islands often flooded.	Medium trees, "short shoots", trunk sometimes swollen.	Not exploited due to high proportion of resins.	+ 0
H.microphylla	Middle and upper reaches Rio Negro.	Sandy soils, flooded regularly, riverine.	Small trees, swollen trunk, slender crown	Not exploited, probably due to resins, etc.	+ 0
H.rigidifolia	Restricted Upper Rio Negro and Rio Vaupes	Dry, well drained, rocky soils.	Small trees, "short shoots", very tough leaves.	Not exploited due to resins and variety.	+ 0
H.nitida var. toxicodendroides	Restricted Upper A paporis, Colombia.	Quartzitic soils sand-stone mesas	Shrubs	Not exploited. Too small and rare.	- 0
H.camporum	Restricted far south Amazon tributaries	Dry savannah	Shrubs	Not exploited. Too small and rare.	@ 0
H.carmargoana	Restricted Marajo Island, Amazon delta	Transition savannah seasonal muddy swamps	Small - medium trees	Not exploited. Only recently discovered	@ 0

+   = present in Rio Negro region centre of genetic diversity of genus Hevea.
++ = ditto, variety also.
-   = absent from Rio Negro region, but occurs outside near boundary of
      that area.
-- = ditto, range extends to parts remote from that area.
@   = absent and distant from Rio Negro region.

*   = exploited in the wild but not cultivated.
** = exploited in the wild, limited use of hybrids in cultivation.
*** = exploited extensively in the wild and widely cultivated.
O   = not exploited.

---

doubt that the Rio Negro is the centre of genetic diversity in Hevea.

Geomorphologically the centre of diversity is near the northern margin of the Amazon Basin where the uplands of the Guayana Shield rise. The soils about the centre of diversity are lowland latosols and riverine gleys, but the region surrounding the centre of diversity includes varied upland soils.

Climatically the centre of diversity itself is entirely within the constantly Humid Equatorial Region 12. A relatively small part of the immediately surrounding zone is in the Humid Equatorial Regions 13a and 13b.

Thus Hevea appears to have evolved under an almost constantly humid climate where edaphic and topographic conditions, however, give rise to some variety of soils and ecological habitats. The constant humidity means that new foliage will almost certainly be borne under conditions favouring attack by fungal diseases and some degree of resistance would seem essential for survival, although only a few selections from this centre of diversity have shown very high resistance to South American Leaf Blight. The constant humidity may also lead to absence of any strong influence acting as a coordinating stimulus of such phenological responses as defoliation, flushing of new foliage or flowering. If so, this may reduce opportunities for cross pollination between species in the wild, which with the short flight range of most pollen vectors and spatial separation due to habitat preferences, may account for the relatively few putative hybrids found in nature. Some hybrids have been reported from disturbed areas, for example on the margins of cultivation.

The species extending beyond the region surrounding the centre of diversity have adapted to progressively more seasonal conditions and longer dry periods each year. Seasonality coordinates change of foliage and flowering. H. brasiliensis has its range entirely outside the primary centre of diversity, although it does occur in the secondary centre of diversity about Borba on the lower Rio Madeira where five of the widespread species occur.

HABITAT PREFERENCES

Hevea microphylla occurs on sandy or lateritic soils subject to heavy flooding for at least four months of the year usually on river banks or islands.

Hevea spruceana is found on muddy soils of islands and river banks subject to periodic heavy inundation.

Hevea benthamiana grows in swamp forest or, in Manaos, on well drained sandy soil often in association with Mauritia palm on alluvial flood plains and islands flooded every year.

Hevea nitida grows in periodically heavily flooded swamps but also on rocky hillsides near falls and rapids and in highland forest sands above the annual flood level.

Hevea brasiliensis is found mainly on well drained sites, but it can grow where brief or light inundations occur periodically, such as high river banks. In the south eastern Amazon region of Brazil and Peru, it grows well above the annual flood.

Hevea guianensis and its variety lutea grow on well drained but moist soils such as high river banks seldom flooded or near small streams or in the uplands.

Hevea pauciflora and its variety coriacea occur on well drained soils such as high river banks, sandy soils and rocky hillsides.

Hevea rigidifolia grows on dry, well drained, sandy "caatingas".

Hevea nitida var. toxicodendroides has been found on quartzitic soils on top of sterile sandstone mesas.

Hevea camporum occurs in dry savannahs.

Hevea camargoana occurs in savannah and locally in woodlands flanking seasonally flooded swamps.

GROWTH FORM

There is considerable growth variation within every species. Comparative data are often lacking.

Under conditions of optimum development in nature H. brasiliensis is one of the tallest species in the genus growing upto 40 m in height. H.guianensis is also a tall tree in its native habitat, being 30 to 35 m in height. Hevea benthamiana, typical H. nitida and H. spruceana are reported as medium sized trees in nature upto about 27 m tall. H. pauciflora var. coriacea is somewhat smaller and often rather low (15 m), but the typical form of H. pauciflora shares with H. microphylla and H. rigidifolia being described as small trees upto 18 m. Trees of H. camargoana range from 2 to 12 m in the wild. Under cultivation it may grow much taller, perhaps as much as 25 m. Hevea nitida var. toxicodendroides and H. camporum are both diminutive shrubs, usually little more than 2 m tall.

In general the different species and varieties have preserved their relative growth forms in cultivation. It has been opined that if these variants were used as rootstocks, they would induce similar growth forms in the scions. There is hardly any evidence for this. Admittedly the small shrubby forms have never been tried as rootstocks.

H. microphylla is swollen in the trunk or bottle-butted with a slender stem and sparse crown. To a lesser degree H. benthamiana sometimes displays a swollen trunk and a narrow crown. The trunks of H. brasiliensis and H. spruceana are sometimes swollen. All these seem to be responses to periodic flooding and do not persist in cultivation. H. pauciflora is the only species, sometimes reported as being corpulent, in cultivation and naturally on drained soils.

The desired form in Hevea has been described as dumpy rather than dwarf. The evidence to-date is that this is not induced by rootstocks in Hevea and that where this growth form occurs in nature it is in response to environmental conditions such as flooding. Xeromorphic characters such as a lignotuber have been reported in plants growing under savannah conditions.

PATTERN OF GROWTH

Seibert (1947) drew attention to the occurrence of short shoot or spur type growth in Hevea in which the internodes were strongly compressed and marked by scale leaves. This feature is not conspicuously developed in the best known species H. brasiliensis and consequently escaped mention previously.

It appears that generally in those species in which short shoots are transient, poorly developed and inconspicuous, the trees 'winter'. All the foliage is shed and the trees are bare for a brief period before new shoots extend bearing young foliage distally and scale leaves proximally, the latter often bearing flowers in their axils. Whereas in those species in which short shoots are well developed and obvious, the new shoots emerge bearing flowers and foliage before the old leaves fall. However, sometimes the old foliage falls before the new shoots emerge, even in some trees displaying conspicuous short shoot development.

Three species, H. brasiliensis, H. benthamiana and H. microphylla have inconspicuous short shoots and 'winter' strongly. Lim and Rao (1976) found that strong wintering, complete defoliation prior to refoliation, reduced the severity of attack by Oidium Secondary Leaf Fall Diseases. This may be due to less inoculum and a modified microclimate in the canopy during the most susceptible phase of new leaf growth. Such 'disease-escape' may be of consequence for other leaf diseases such as South American Leaf Blight. Aerial spraying of defoliants has been used to induce leaf-fall, stimulated heavy wintering and disease-escape.

Five species, H. guianensis, H. nitida, H. pauciflora, H. rigidifolia and H. spruceana have conspicuous well-developed short shoots, at least in the higher orders of branching. This may be the case in H. camporum and H. camargoana also. If so, it can be regarded as the typical growth pattern of the genus Hevea. The immature period before a tree flowers and sets seed is not known for each species. H. brasiliensis and the few other species already used by plant breeders normally flower in about five years. However, H. camargoana has been noted as exceedingly precocious, flowering within one year of planting the germinated seed.

FOLIAGE

All species of Hevea have trifoliate leaves which are folded back at emergence. Subsequently the leaves assume various positions from the bent down, reclinate or declinate through the more or less horizontal to the erect.

Only in H. guianensis and its variety lutea are the mature leaflets erect or distinctly inclined above the horizontal. This may indicate possibly a different balance in natural growth substances compared with other species. The main distinction between H. guianensis and its variety lutea is that the latter has markedly obovate leaves.

The other species and varieties of Hevea have more or less elliptical leaflets borne in more or less horizontal or deflexed positions. Only exceptional features will be considered further.

Hevea benthamiana has a reddish-golden-brown indument or pubescence on the undersides of the leaflets.

Hevea camargoana has pale green undersides to the glabrous leaves.

Hevea nitida has strongly reclinate leaflets which are moreover markedly folded upward from the midrib. The upper surface is very shiny and the under surface is devoid of scales, so that both surfaces are bright glossy green.

Hevea pauciflora has leaflets glabrous beneath which become leathery and each is marked by a blunt tip with a subterminal gland beneath.

Hevea rigidifolia has glabrous glossy, rigid, thick, leathery leaflets with inrolled margins beneath.

FLOWERS

In all species of Hevea there are separate male and female flowers borne in the same inflorescence, the latter terminating the main branches of the panicle.

The unopened buds, open flowers and central axes of the two sexes are together distinctive of each species. In both sexes of all species there are five perianth lobes. The central axis in male flowers is occupied by the staminal column and in female flowers the pistil is in this position. A disk is found at the base of the central axis. The disk may be strongly or weakly developed. The comparative situation may be judged from the diagrams in Figure 2. Only special features are mentioned in the following account.

The relatively broad flowers of H. spruceana are purplish in colour, unique in the genus, except for a rose-red colouration at the bases of the whitish flowers of H. camargoana and often a reddish colouration at the bases of the flowers in H. pauciflora var. coriacea.

The largest female flowers are those of H. microphylla, although their most distinctive feature is the greatly swollen torus or bulge at the base of the female flowers. Large flowers are also borne by H. rigidifolia.

The tips of the calyx lobes are blunt or calloused in H. brasiliensis, H. nitida (including var. toxicodendroides) and H. pauciflora (including var. coriacea). The calyx lobes are strongly twisted in H. rigidifolia.

The disks of the female flowers are large, obvious and markedly lacerated in H. rigidifolia, H. nitida, H. pauciflora, H. camporum and H. microphylla, although somewhat obscured by the large torus in the last. The disks of the male flowers are prominent in these five species and also in H. benthamiana. A torn or toothed basal disk is evident in the male flower of H. camargoana.

There is one whorl of three to five anthers in H. camargoana. There are five anthers only (seldom more) in H. guianensis and its variety lutea. These are arranged in one regular whorl in H. guianensis but in an irregular whorl or in two whorls in var. lutea. H. rigidifolia usually has six anthers in two irregular wholrs, but sometimes there are seven to ten anthers. In H. benthamiana and H. spruceana there are usually seven to nine anthers, rarely ten, in two irregular whorls. In H. brasiliensis, H. microphylla, H. nitida and H. pauciflora there are normally ten anthers in two whorls,

Fig. 2.   (a) Male Buds, (b) Male Flowers, (c) Male Axes,
          (d) Female Buds, (e) Female Flowers, (f) Female Axes

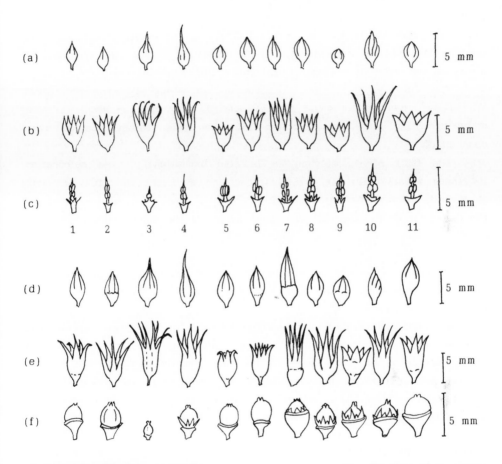

1 Hevea benthamiana Muell-Arg.
2 Hevea brasiliensis Muell-Arg.
3 Hevea camargoana Pires
4 Hevea camporum Ducke
5 Hevea guianensis Aubl.
6 Hevea guianensis var. Lutea Ducke & R.E.Schultes
8 Hevea nitida Mart. ex Muell-Arg.
9 Hevea pauciflora Muell-Arg.
10 Hevea rigidifolia Muell-Arg.
11 Hevea spruceana Muell-Arg.

The flowers of H. nitida
var. toxicodendroides
R.E. Schultes and of
H. pauciflora var.
coriacea (Ducke) are
somewhat smaller than
those of H. nitidia and
of H. pauciflora
respectively.

These were copied from Seibert (1948) except for H. camargoana and H. camporum copies from Pires (1981) and Schultes (1970) respectively. Some errors in scale may have occurred as a result.

which may be regular or irregular. H. camporum bears the anthers in two whorls.

FRUIT

The fruit in all species is a tricolour capsule usually containing three seeds. The fruit is explosively dehiscent in all except two species. The woody valves twist on drying out and throw the seeds far out.

The first exception is H. spruceana in which although the valves are woody, they do not twist strongly enough to throw the seeds far. The other exception is H. microphylla which has thin leathery valves which open slowly.

The fruit of H. microphylla is also exceptional in its colouration. It ripens yellow with six green stripes and a bright red tip. The fruit of H. nitida, H. pauciflora and H. rigidifolia also have reddish tips but the remainder of the capsules are green as in all the other species.

SEEDS

In Table 2 the shape of the transverse cross section of the seed, the approximate average dimensions and the approximate average fresh weight if known, are given.

H. camporum has the smallest seeds and H. spruceana the largest. The seeds of H. guianensis and its var. lutea are similar. The seeds of H. nitida are larger than those of its var. toxicodendroides, the contrast is even greater between the large seeds of H. pauciflora and the small seeds of its var. coriacea. The seed of all species are short lived unless stored under cool or other special conditions.

LATEX AND RUBBER

Commercially acceptable latex and rubber have been obtained from H. brasiliensis, H. benthamiana and H. guianensis in descending order of quality. The latex and rubber from H. nitida var. toxicodendroides are reported to be satisfactory although difficult to obtain in quantity, whereas H. nitida itself yields anti-coagulant latex and rubber of poor quality. Rubbers from H. spruceana and H. pauciflora have a poor reputation, mainly because of the high proportion of resins. H. microphylla and H. rigidifolia also have bad reputations locally in the field. H. camporum does not seem to have ever been tapped.

Considerable variability was reported in the yields of wild trees of H. brasiliensis. The 'preta' or black barked trees were reported on more favourably than the 'vermelha' or red and the 'branca' or white, the last

TABLE 2

Shape, approximate average dimensions and fresh weight of seed of Hevea species

Species	Shape TXS*	Dimensions (mm)	Fresh Weight (g)
Hevea benthamiana Muell-Arg.	Rounded	19 x 14	3
Hevea brasiliensis Muell-Arg.	Rounded	28 x 20	5
Hevea camargoana Pires	Rounded	15 x 10	1½ ?
Hevea camporum Ducke	Rounded	11 x 6	1 ?
Hevea guianensis Aubl. and Hevea guianensis var. lutea Ducke & Schultes	Kite - shaped	21 x 18	2
Hevea microphylla Ule	Triangular - ovate	25 x 14	4
Hevea nitida Mart. ex Muell-Arg.	Kite at one ed	20 x 13	2
Hevea nitida var. toxicodendroides Schultes	Rounded other	14 x 10	1½
Hevea pauciflora Muell-Arg.	Hexagonal	24 x 18	4 ?
Hevea pauciflora var. coriacea Ducke	Rounded - Hexagonal	13 x 10	1½ ?
Hevea rigidifolia Muell-Arg.	Hexagonal	27 x 20	4½ ?
Hevea spruceana Muell-Arg.	Lozenge	38 x 18	7 ?

*TXS = Transverse Cross Section

being the category of the Wickham collection. Wickham had collected from the Rio Tapajos upstream of its confluence with the Amazon apparently to avoid contamination by H. spruceana. However, the reputation for quality of rubber from trees growing much further up river was even better, in particular that from the Acre Territory of Brazil and the bordering Madre de Dios region in Peru. Trees of these provenances were reputed to grow larger and to yield more than those from the lower Amazon. Some of these materials have now been introduced into comparative trials to elucidate the interactions of genotype and environment for these characters, and to broaden the very narrow genetic base used hitherto in Hevea breeding programs.

HYBRIDISATION AND INTRAGENERIC RELATIONSHIPS

Thrips, midges and parasitic Hymenoptera were reported as possible pollinators of Hevea in Malaysia by Sripathi Rao (1961), who reviewed observations elsewhere including several other insect visitors and putative pollen vectors. Apart from artificial or hand pollination, Hevea appears to be obligately insect pollinated; there is no evidence of wind pollination. However, the wide range of insects visiting Hevea and the relatively unspecialised nature of the flowers, suggests that lack of pollination is not a barrier to cross fertilisation in the wild, unless it be due to non-coincidence of flowering or separation by habitat preferences of trees of different taxa as discussed under Distribution of the Species with respect to the relative scarcity of hybrids in the wild.

Reviewing Hevea cytology, Majumder (1964) concluded that there was no chromosome number other than 2n = 36. A few putative hybrids between several different species of Hevea have been collected in the wild. Increasing numbers of interspecific hybrids have been produced in breeding programs. In brief, there is no reason to suppose that there are cytogenetic barriers to interspecific hybridisation in Hevea, or at least none of greater magnitude than in intraspecific crosses. On the other hand lower fertility has been reported in some selfings and inbreeding programs. Hevea trees may be regarded as typically outcrossing.

On such morphological characters as the much swollen torus at the base of the female flower and the pyramidal fruit triangular in cross section, which dehisces slowly, Schultes (1977b) has separated Hevea microphylla into the subgenus Microphyllae, the other species remaining in the subgenus Hevea. If it were not inconvenient to do so, one might, as Baldwin (1947) suggested, regard Hevea or at least the subgenus Hevea as monospecific.

OTHER GENERA

There are thirteen species of Micrandra, one of which M. minor has been tapped as a wild source of rubber. The range of the genus Micrandra includes and extends beyond the range of the genus Hevea. Micrandra may be closely related to Hevea. Vaupesia is very closely related to Micrandra (Schultes, 1955) and may also be considered as a potential extension of the genetic pool. Johannesia princeps is another lactiferous tree of the Euphorbiaceae, which is in the same tribe as Hevea. Annesijoa novaguiensis is believed by some botanists to be closely related to Johannesia and Hevea, but this is questioned by others, because it occurs in Western New Guinea isolated from the other members of the tribe. It grows on Biak Island in Irian Barat. No inter-generic hybrid has been reported in this group, and none has been attempted experimentally.

ACKNOWLEDGEMENT

It is a pleasure to thank Prof. R.E. Schultes for his careful reading of the manuscript and many constructive and helpful amendments and suggestions.

BIBLIOGRAPHY

Aublet, J.B.C.F. 1975.   Histoire des plantes de la Guiane Francaise, 2: 871-873, t 335. P.-F. Didot, j. London, Paris.
Baillon, H.E. 1858.   Etude generale du groupe des Euphorbiacees, 324-327. Victor Masson, Paris.
Baldwin, J.T. Jn. 1947.   Hevea, a first interpretation. J. Hered. 38, 54-64.
Bentham, G. 1854.   On the north Brazilian Euphorbiaceae in the collection of Mr. Spruce. Hooker Journ. Bot. 6: 368-371.
Ducke, A. 1935.   Revision of the Genus Hevea Aubl. mainly the Brazilian species. Arch. Inst. Biol. Veget. Rio de Janeiro 2: 217-346.
Gmelin, H.F. 1791.   Syst. II: 677.
Huber, J. 1902.   Observacoes sobre as arvores de borracha da regiao ama-zonica. Bol. Mus. Paraense Hist. Nat. 3: 345-369.
Huber, J. 1906.   Ensaio d'uma sunopse des especies do genero Hevea sob os pontos de vista systematico e geographico. Bol. Mus. Paraense Hist. Nat. 4: 620-651.
Lim, T.M. and Sripathi Rao, B. 1976.   An epidemiological approach to the control of Oidium secondary leaf fall of Hevea. Proc. Int. Rubb. Conf. 1975 Kuala Lumpur. 3: 293-311.
Majumder, S.K. 1964.   Chromosome studies of some species of Hevea. J. Rubb. Res. Inst. Malaya 18: 269-275.
Mueller-Argoviensis, J. 1865.   Euphorbiaceae. Linnaea 34: 203-204.
Mueller-Argoviensis, J. 1874.   Euphorbiaceae. In: C.F.P. von Martius, (ed.) Flora Brasiliensis 11(2): 297-304.
Murca Pires, J. 1971.   O genero Hevea: descricao das especies e distribucao geografica. Anexo 7, Plano Nacional da Borracha, Ministerio de Industria e do Comercio, Brasilia. (mimeogr.)

66

Murca Pires, J. 1981. Notas de Herbario I. Bol. Mus. Paraense Emilo Goeldi 52: 1-11.

Pax, F. 1910. In: A. Engler, (ed.) Pflanzenreich iv (147): 117-128.

Richard, L.C. 1791. Siphonia. Schreber Gen. II: 656.

Schultes, R.E. 1950. Studies in the genus Hevea III. On the use of the name Hevea brasiliensis. Bot. Mus. Leafl. 14: 79-86.

Schultes, R.E. 1955. A new generic concept in the Euphorbiaceae. Bot. Mus. Leafl., Harvard Univ. 17: 27-36.

Schultes, R.E. 1956. The Amazon Indian and evolution in Hevea and related genera. J. Arn. Arb. 37: 123-148.

Schultes, R.E. 1970. The history of taxonomic studies in Hevea. Bot. Rev. 36: 197-276.

Schultes, R.E. 1977a. Wild Hevea: An untapped source of germplasm. J. Rubb. Res. Inst. Sri Lanka 54: 227-257.

Schultes, R.E. 1977b. A new infrageneric classification of Hevea. Bot. Mus. Leafl., Harvard Univ. 25: 243-257.

Schultes, R.E. 1987. Studies in the Genus Hevea. VIII. Notes on infraspecific variants of Hevea brasiliensis (Euphorbiaceae). Economic Bot. 41(2): 125-147.

Seibert, R.J. 1947. A study of Hevea (with its economic aspects) in the Republic of Peru. Ann. Missouri Bot. Gard. 34: 261-352.

Sripathi Rao, B. 1961. Pollination of Hevea in Malaya. J. Rubb. Res. Inst. Malaya, 17(1): 14-18.

Walter, H. and Lieth, H. 1960. Klimadiagramm Weltatlas. VEB Gustav Fisher Verlag, Jena.

CHAPTER 4

ANATOMY AND ULTRACYTOLOGY OF LATEX VESSELS

D. PREMAKUMARI and A.O.N. PANIKKAR

Rubber Research Institute of India, Kottayam-686009, Kerala, India.

Laticifers are present in a very large number of species belonging to about twenty different families (Metcalfe, 1966), mostly dicotyledons. A few monocotyledons like Allium cepa (Hoffmann, 1933) and the genus Regenellidium of Marsiliaceae (Pteridophyta) are also reported to have laticifers (Gomez, 1982). The taxonomic importance of these structures is doubtful, although it has been a matter of fundamental interest.

The occurrence of laticifers in Hevea and their structure have been studied during the nineteenth century by Scott (1886) and Calvert (1887). Detailed investigations on the bark structure and the laticiferous tissue of the para rubber tree were carried out during the early twentieth century (Bryce and Campbell, 1917; Keuchenius, 1918, Arisz, 1919, 1921; La Rue, 1921; Heusser, 1921; Vischer and Tas, 1922; Bobilioff, 1919, 1920, 1923; Steinmann, 1923).

Nature of laticifers

The type of laticifers is characteristic of the plant species. The complex types of laticiferous systems in plants are classified as articulated and non-articulated according to the mode of origin (De Bary, 1877). Both are tubular structures and are described as latex vessels.

Articulated laticifers are compound in origin comprising a series of cells, either remaining blunt or becoming continuous tubular structures due to partial or complete dissolution of endwalls. Depending on the presence or absence of lateral connections they are further categorised as anastomosing and non-anastomosing respectively. The non-articulated laticifers are more simple in structure. They may remain as single unbranched cells or branched structures extending throughout the shoot and root system as in Euphorbia species. The groups of laticifers in plant kingdom can be classified as indicated below:

Laticifers: Specialised cells or tissues which contain latex.

Non-articulated: Develop from single cells, which elongate and their tips keep pace with growth of the cells of the surrounding meristem penetrating among the new cells.

- Unbranched: More or less straight tubes developing into long structure
  - Cannabis
  - Urtica
  - Catharanthus

- Branched: Each laticifer cell branches repeatedly forming an immense system of tubes.
  - Ascle pias
  - Cryptostegia
  - Euphorbia
  - Ficus
  - Nerium

Articulated: Originate from rows of cells by the partial or complete absorption of the separating walls in early ontogeny.

- Non-anastomosing: Long, compound tubes not connected with each other laterally.
  - Achras
  - Chelidonium
  - Convolvulus
  - Ipomoea

- Anastomosing: Forms anastomoses laterally with cells or tubes of similar nature, all combine forming a reticulum.
  - Argemone
  - Carica
  - Cichorium
  - Hevea
  - Manihot
  - Taraxacum

Classification of laticifers - A schematic representation
(Adapted from Panikkar, 1974)

In Hevea, the principal type of laticifers exploited commercially for its latex is the secondary laticifers, differentiated by the activity of vascular cambium as in the case of vessel elements and hence the term 'latex vessels' is appropriate. They are articulated and belong to the anastomosing, coenocytic type.

Aseptate thread-like laticifers, however, occur in the pith and leaves and also in young branches of Hevea tree. Presence of ray cells containing latex has also been reported (Bobilioff, 1923). Recently Xiuqian (1987) made detailed investigations on the nature of laticifers in Hevea brasiliensis and observed that the primary laticifers found in the seed (cotyledon), leaf, flower, root and young stem are non-articulated which show intrusive growth into the intercellular spaces of the primary phloem and cortex in contrast to the articulated anastomosing secondary laticifers.

Ontogeny

The entire laticifer system in a species may initiate from a few initials present in the embryo, as in certain <u>Euphorbia</u> sp. (Scharffstein, 1932; Rosowski, 1968), which in the course of development, are found on the circumference of the central cylinder. In certain others the primary laticifers originate in the phloem or pericycle as in <u>Taraxacum</u> <u>kok-saghyz</u> or in the hypodermal region. In <u>Allium</u>, the laticifer initials are found in the third layer of leaf mesophyll or in the third layer below the abnormal epidermis of the bulb scale.

Ontogeny of laticifers in <u>Hevea</u> was outlined by Scott (1882) based on the observations on germinating seeds. He could identify the latex vessels as small elongated cells with characteristic granular contents devoid of the aleurone grains. According to him dissolution of cross walls takes place at a stage of root growth approaching 3-4 mm length in the seedlings. He considered the hypodermal system of laticifers as more advanced than the vascular laticifer system. Calvert (1887) identified three systems of laticifers in the stem of <u>Hevea</u>: hypodermal, principal and medullar. But Milanez (1946, 1948, 1951) later found that the primary laticiferous system is differentiated from the procambium in the vicinity of the phloem. More elaborate studies on the development of latex vessels, as reviewed by Gomez (1982), confirm that the principal system of primary laticifers is that observed in the procambial region belonging to the primary phloem proper.

Bobilioff (1919) described two modes of laticifer ontogeny in <u>Hevea</u>, one by the dissolution of cross walls of a row of cells and the other by extension of growth of certain cells. The first type of development in the formation of secondary laticifers was confirmed later (Panikkar, 1974; Xiuqian, 1987). In the course of development of the articulated laticifers in <u>Hevea</u>, before the dissolution of end walls of laticifer initials, small projections are formed on the lateral walls and the tubular projections of the adjacent laticifers come in contact and fuse to form the anastomoses.

Distribution

In <u>Hevea</u> laticifers are present in all organs except wood although occasional presence of plugged vessels having latex has been reported (Bobilioff, 1921, 1923). Non-articulated laticifers of primary origin are usually present in young organs in the primary state of growth. They are also observed in young leaves, flowers, cotyledons and pith.

Secondary laticifers are distributed in the bark, and the commercially exploited part of the tree is the main trunk. A comprehensive description of the structure of mature bark in <u>Hevea</u> (Fig. 1) was made by Bryce and Campbell (1917). In the bark, in addition to the outermost protective tissues consisting of layers of cork cells, there are two more distinguishable zones,

70

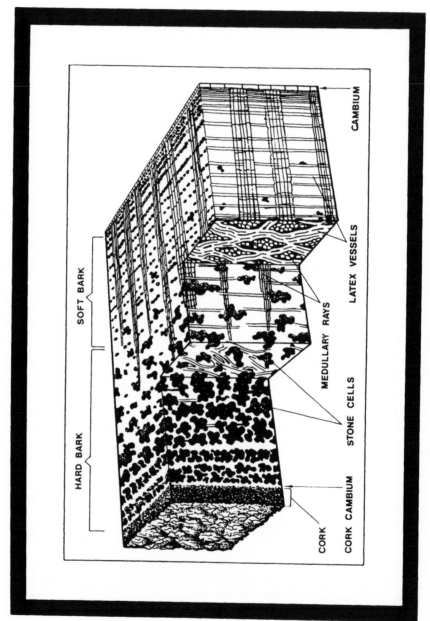

Fig. 1. Mature bark – a three dimensional view.

an inner soft zone and an outer hard zone. Hardness of the outer zone is due to the occurrence of sclerified cells in masses, known as stone cells. There are tannin cells also in the bark. The laticifer differentiation from the cambial derivatives is a rhythmic process and a ring of laticifers are produced each time. Hence the latex vessels in Hevea form concentric rings almost alternating with layers of other phloem tissue. The general structure and organization is the same in the main trunk and branches, though the bark thickness and number of latex vessel rows vary.

In a cross sectional view, latex vessels have a more or less circular shape. The concentric rows of latex vessels are seen almost parallel to the cambium, and are separated by layers of phloem elements.

In a longitudinal section in radial plane the latex vessels have tubular shape, and the vessels representing different rows look like straight tubes interrupted at the location of phloem rays. In the soft bast region the latex vessels are continuous, while most of them in the hard bast region are discontinuous and hence non-functional. In the hard bast region the latex vessels become crushed and broken due to the push and pull exerted by the surrounding tissue, especially the stone cells, during radial growth. The latex vessels between rows are not usually connected although rare instances of anastomosing are reported.

In a tangential longitudinal section of the bark through a ring of laticifers, the system resembles a meshwork. The cross sectional areas of phloic rays represent the sieves of the net. Weaving the phloic rays the tubular latex vessels run up, in a slightly inclined path to the long axis of the tree, in anti-clockwise direction. The degree of inclination varies from 2-7° depending on clone. The inclination of latex vessels was well known to earlier researchers (Petch, 1911; De Jonge, 1916, 1919; Gomez and Chen, 1967) and this trait was taken into account in deciding the slope of tapping cut.

The course of latex vessels from the base upwards interweaving the ray groups leads to a zig-zag pattern. The extent of waviness depends on the size and distribution of the phloic rays and hence is a clonal character (Premakumari et al., 1985). The degree of waviness of latex vessels within a tree varies according to the sampling distance from the cambium.

Premakumari et al. (1985) recorded a range of 6-9 connections, within a distance of 0.22 mm of anastomosing vessels among ten clones and observed that the intensity of anastomosing is a significant clonal character. Xiuqian (1987) observed 1-5 connections per laticifer cell arranged in a regular pattern.

QUANTITATIVE ASPECTS

The quantity of laticiferous tissue in a tree is determined by various factors such as the number of latex vessel rows, density of latex vessels within a ring, distance between vessel rings, distribution pattern of latex vessel rings, size of laticifers and the girth of the tree.

Number of latex vessel rings

The number of latex vessel rings is a clonal character (Bobilioff, 1923; Vischer, 1921, 1922; Sanderson and Sutcliffe, 1921) and the frequency of laticifer differentiation is genetically controlled. The number of latex vessel rows varies depending on the clone, age of the tree, growth rate and seasonal factors. During the active growth period, the rate of laticifer differentiation would be much higher than that during the rest of the year. Significant variation of cambial activity and number of vessel rings between different periods of the year have been noted in clone Gl 1 (Premakumari et al., 1981). Higher number of latex vessel rows was associated with higher rate of cambial activity. Gomez et al. (1972) studied the influence of age of the tree on number of latex vessel rings and observed a linear relation-ship upto about 15 years.

In seedling trees, the number of latex vessel rows decreases with increasing height of the trunk owing to the positive relationship between latex vessel rows and bark thickness. Bark thickness decreases with increasing height due to the coniform nature of stems, compared to the cylindrical trunk in clones. This factor is given due consideration in determining the height and slope of tapping cut.

Gomez et al. (1972) studied 112 clones, aged eight and half, and recorded a mean of 25.6 rings. Gomez (1982) suggested that the increase in the number of laticifer rings, as seen from the reports of various studies over a period of fifty years, can be attributed to genetic improvement.

Concentration of latex vessel rings in the bark at positions distal to the cambium varies among clones. Nearly 10-35% of the vessel rings were observed in the second and third millimeter and the number diminished to zero over a distance of five to eight mm. Clonal differences in the distance between two consecutive latex vessel rings became prominent only beyond the third millimeter from the cambium.

The quantum of laticifers in the virgin bark is influenced by the age of the tree. Gomez et al. (1982) reported that in trees below five years, the laticifer rings were concentrated in the first 4-5 mm; only 40% being in the second. Between five and ten years, however, laticifer rings were more concentrated near the cambium tailing away to zero near to the eighth millimeter and by the 25th year about 75% of the latex vessel rings were oriented at the innermost five mm of the bark.

Density of latex vessels within rings

The number of laticifers per unit circumference of the tree, usually referred to as density of latex vessels, indicates the quantity of laticiferous tissue in terms of laticifer area in the cross section of a tree. Gomez (1982) mentioned a difference in latex vessel density between two distant positions from the cambium. He observed an apparent clonal difference also. Highly significant clonal differences in this trait have been demonstrated later by Premakumari et al. (1985). Among ten clones they recorded a range of 24-31 latex vessels per row per 1.25 mm circumference. The density of latex vessels is also related to the width of phloic rays (Premakumari et al., 1984).

Diameter of latex vessels

The laticifer diameter is a known clonal character. Ashplant (1928) recorded high correlations (0.76) between yield and latex vessel diameter. Frey-Wyssling (1930) and Riches and Gooding (1952) related the diameter of latex vessels to rate of flow of latex, compared the relationship with with Poisseuille's equation for viscous flow in a capillary and demonstrated that the volume of flow is proportional to the fourth power of the radius of the capillary.

Gomez et al. (1972) reported a variation of latex vessel diameter from 21.6 μm to 29.7 μm in eight RRIM clones. In a recent study Prema-kumari et al. (1985) recorded a range of 16.67 μm to 26.87 μm in nine year old plants of ten clones.

Laticifer area index

Laticifer area index is a tentative index proposed by Gomez et al. (1972) to approximate the quantity of laticiferous tissue of a tree in terms of cross sectional area. This index is believed to account all the main quantitative factors which are involved in latex production. The total number of laticifers in a tree is approximately assessed as nfG, where, n is the number of latex vessel rings, f, the density of latex vessels per ring and G, the girth in cm. Hence the total cross sectional area of latex vessel would be nfG (πr^2), where r is the radius of the latex vessel.

Since the latex vessel rings are not interconnected, opening of more latex vessel rings on tapping gives proportionately higher yied. It is conventional to leave approximately one mm bark undisturbed near the cambium to avoid injury to the cambium. About 40% of the latex vessel rings are distributed within the first millimeter from the wood and hence 'n' should be corrected as 0.6 n and again for half spiral cut it should be corrected as 0.3 n. For practical purposes the number of latex vessels cut on each tapping in half spiral cut should be calculated as 0.3 nfG and the cross sectional area as 0.3 nfG (πr^2).

Variations at higher ploidy levels

Only limited studies have been made on the quantitative variations of laticiferous tissue of <u>Hevea</u> at different ploidy levels. A study on the comparative bark anatomy of the mature trees of induced polyploids of GT 1, Tjir 1 and RRII 105 showed considerable reduction in bark thickness, number of latex vessel rows and density of laticifers leading to reduced laticifer area for the tetraploids as compared to their respective diploids (Premakumari et al., 1988b). The effect was not uniform in clones studied but a tendency for higher size and reduced number of cells for the polyploids was indicated.

LATEX VESSELS AND YIELD

Correlations between structural features and yield in <u>Hevea</u> have been studied in detail by several researchers. Bobilioff (1920), La Rue (1921) and Taylor (1926) established highly significant correlations between yield and number of latex vessel rows in seedling materials. Later studies yielded still higher correlation of this structural trait with yield in clones and positive linear relationship of the number of latex vessel rows and bark thickness with yield of <u>Hevea</u> clones in 33 month old nursery plants was obtained (Narayanan et al., 1973). Elaborate studies and yield component analysis (Narayanan et al., 1973; Narayanan et al, 1974; Narayanan and Ho, 1973; Ho et al., 1973) proved that the number of latex vessel rows is the most important single factor related to yield. This character, combined with plugging index and girth, could account for 75% of the yield variations in young plants. The accountability was reduced to 40% at the mature phase which indicated a predominant role of genotype environment interaction. Sethuraj et al. (1974) demonstrated a significant positive correlation between number of latex vessel rows and initial rate of flow. Premakumari et al. (1984) proposed a negative relationship of phloic ray width and yield, mediated through the inverse relationship between laticifer density and phloic ray width.

In addition to the number of latex vessel rings, the density of latex vessels/ring, laticifer diameter and also the distance between laticifer rows are factors determining the quantity of laticiferous tissue expressed in terms of laticifer area in cross section of the bark. Premakumari et al. (1988a) suggested laticifer area and orientation of laticifers as very important factors influencing yield of Hevea clones.

REGENERATION OF LATICIFERS

Hevea trees are exploited for a period of 20-23 years from the commencement of tapping. After the consumption of virgin bark, regenerated bark is exploited and hence regeneration of bark is of great significance. In tapping, only a thin slice of bark, 1.0-1.5 mm thick is shaved off to cut open the latex vessels. The cambium is not injured in this process. Moreover a layer of soft bast which is left intact along with the cambium during tapping gives protection to the cambial layer. Bark is regenerated due to continued activity of the vascular cambium, during which process new phloem tissues are produced and the normal process of laticifer differentiation continues. The protective tissue lost along with the bark is replaced by the formation and activity of a new phellogen below the cut surface (Bobilioff, 1923; Panikkar, 1974). Thus the process of bark regeneration is the function of two cambia, the continued activity of the vascular cambium and the formation and functioning of a new cork cambium.

The immediate effect of tapping in Hevea is a shrinkage of cells at the cut surface, which undergo necrosis. Immediately below the necrotic layer, cells enlarge and divide in irregular planes which later become oriented parallel to the cut surface. A wound phellogen differentiates at the hypodermal, or, occasionally the sub hypodermal, layer which cuts off phellem towards the periphery and phelloderm towards the inner side (Panikkar, 1974).

The extent of bark regeneration and differentiation of laticifer rows depends on the clone. Bobilioff (1923) found that the total number of laticifer rows, after one year of regeneration of bark will become equal to the number which was present in the virgin bark. It is also reported that the rate of regeneration of laticifer rows is determined by the inherent characteristic of the tree for laticifer differentiation in the virgin bark.

The extent of regeneration is controlled by various factors like depth of tapping and age of the tree. Gomez et al. (1972) observed that clonal differences appear to be more important only in the young trees. For very old trees the proportion of uncut vessel rings recorded were only 8-13% for the clones studied, whereas, it was 30-45% in younger trees. The effect of deeper tapping therefore would show only smaller response as the tree grows older.

Regeneration of laticifers in stimulated bark

Significant increase in bark thickness was recorded for stimulated trees by De Jonge (1955, 1957) in experiments using 2,4-D, and 2,4,5-T applied on renewing bark. However, no effect on the number of latex vessel rows was evident. Reduction in the number of latex vessel rows associated with increased periderm formation for oil/petroleum based treatment, has been reported by Gomez (1964) while no characteristic effect on the number of latex vessel rows could be noticed for Ethephon stimulation (Gomez, 1982). However, an increase of periderm tissue due to stimulant application was apparent.

Regeneration of laticifers after microtapping

Delayed formation of latex vessel ring just after microtapping was noticed in normal puncture tapped trees (Hamzah and Gomez, 1981). However, normal activity of the vascular cambium was not permanently affected as it continued normal production of latex vessels. The scar tissue or 'scab' formed on the surface of the bark under the flaky external bark, at the punctured portion, consists of the 'wound zone' and the periderm tissue formed after puncturing. The wound zone is a part of the bark at the punctured portion where the tissues show discolouration due to the presence of lignin, suberin, tannin, etc., which are supposed to aid protection of protoplast against desiccation and decay.

EFFECT OF MINERAL DEFICIENCIES ON LATEX VESSEL FORMATION

The relative importance of soil type on laticifer differentiation was studied by Keuchenius (1920) who recorded an annual average increment of 3.14 latex vessel rings in good soil, 2.14 rings in average soil and 1.74 rings in poor soil. Later studies (Mass, 1923; Hamzah et al., 1975) confirmed the effect of mineral deficiencies on laticifer formation. Deficiencies of nitrogen, phosphorus, potassium and magnesium have major detrimental effect on laticifer differentiation. Likewise calcium and sulphur deficiency also reduced the number of latex vessels. In addition to the major elements, certain minor elements such as manganese, iron and zinc were also found to influence formation of latex vessels.

LATEX VESSELS OF BROWN BAST AFFECTED TREES

Brown bast is considered as a physiological disorder very often related to over exploitation. The disorder mainly originates in the latex vessels and spreads along the vessels. However this does not usually spread from

one regenerated to another and from virgin bark to regenerated bark
(Paranjothy et al., 1975). Brown bast often produces cancerous growths
in the bark tissue leading to malformations of the tree trunk, and detailed
investigations have been made on the bark anatomy of brown bast affected
trees (Rands, 1919, 1921,a,b.c; Rhodes, 1930; Sanderson and Sutcliffe, 1921).
Structural characteristics associated with brown bast incidence mainly involve
depletion of starch in tissues, occurrence of a brown substance similar to
tannin in abundance, frequent occurrence of oil globules and production of
a large number of stone cells. Formation of a type of wound gum, resistant
to acids and alkalies but reacting with tannins and lignin was observed
in brown bast affected trees by Rands (1919, 1921a,b,c) and Rhodes (1930).
Gomez (1982) observed formation of tyloses inside the latex vessels of brown
bast affected tissue. The tyloses later became filled with tannins and their
walls got lignified. The vessels, partially blocked with tyloses, check
latex flow, leading to initiation of dryness. De Fay and Hebant (1980) have
confirmed some of these findings. However, these histological reasons may
only be secondary effects leading to reduced latex flow. The effect of
diminished permeability of the latex vessels due to intensive tapping, as
described by Bealing and Chua (1972), followed by coagulation of latex
may be the major cause of critical reduction in latex flow.

ULTRACYTOLOGY

The anatomy and histology of Hevea and the ontogeny of latex vessels
have been documented well. Hevea latex vessels being a difficult material
for light microscopy due to the out flow of latex on sectioning and latex
dissolution/coagulation during processing, only very little information could
be gathered on the cytology of the latex vessel and its contents. According
to Bobilioff (1923) the very early contention was that latex vessels are
intraxylary spaces between tissues filled with sap which diffused in from
neighbouring cells. The cytoplasmic nature of latex had been first proposed
in 1886 by Berthhold. However, this was still a debatable subject and the
concept that latex is the cell sap bounded by a tonoplast gained more
acceptance when published in standard works of Bobilioff (1923, 1930) and
Frey-Wyssling (1952). Later high speed centrifugation of latex and its
separation into four main fractions by Cook and Sekhar (1953) and subsequent
research by Moir (1959) revived the cytoplasmic theory of latex vessels.
Application of vital stains like neutral red improved the stainability of
different zones and rendered support to earlier reports on the occurrence
of lutoids and its identification as vacuoles (Wiersum, 1957). By this time
there was considerable improvement in electron microscopy. This technique

provided clear micrographs of latex vessels in section which proved, beyond doubt, that latex in Hevea is a specialised type of cytoplasm (Fig. 2).

Cell wall

As in the case of sieve tubes the cell walls of laticifers are mostly made of cellulose. Photomicrographs (Gomez, 1976) reveal higher wall thickness for older vessels. A bulge is often characteristic of the laticifer wall, which can be attributed to the remnants of cross walls. Presence of partially dissolved cross walls at the initial stages of latex vessel ontogeny has been suggested by Dickenson (1969), who has also found the occurrence of microtubules apparently at the periphery of the walls with plasmodesmata between latex vessels as well as between latex vessels and neighbouring cells of very young vessels. In contrast to this, d'Auzac (1988) reported that laticifer cells do not have any pit fields connecting the neighbouring parenchyma. He found perforations connecting a laticifer with other laticifers.

Nucleus

Bobilioff (1923, 1930) reported coenocytic nature of latex vessels in Hevea. According to him the nucleus of a laticifer initial retains central position and during the course of ontogeny the cytoplasm forms a peripheral layer where numerous nuclei are embeded in mature latex vessels. Later observations with the aid of electron photomicrographs by Dickenson (1965) confirmed the coenocytic nature of the latex vessels and the parietal position of nuclei.

Mitochondria

The occurrence of mitochondria in young latex vessels of the tender parts of the tree was found by Dickenson (1965) and later by Gomez (1976). However, mitochondria were rare or absent in tapped latex which may be attributed to its parietal position after tapping. However, more detailed observations are needed to prove the nature and functions of such organs in latex vessels.

Endoplasmic reticulum, ribosomes and proplastids

Occurrence of cytoplasmic organelles like endoplasmic reticulum and ribosomes in young laticifers has been reported (Archer et al., 1963; Dickenson, 1965), and later confirmed by Gomez (1974, 1976). The presence of ribosomes in tapped latex has been reported by Coupe and d'Auzac (1972) after biochemical studies on latex. Occasional presence of plastids and

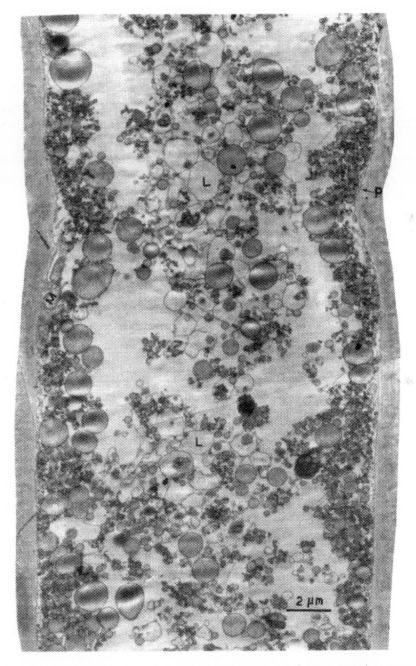

Fig. 2. Ultrastructure of a mature latex vessel from a tapping tree.

pro-plastids has also been noted (Gomez and Moir, 1979).

MAJOR CONSTITUENTS OF LATEX

Fresh Hevea latex is a polydisperse system in which negatively charged particles of various types are suspended in an ambient serum (C-serum). The two main particulate phases contained in Hevea latex are rubber particles constituting 30-45% and lutoid particles, 10-20%. The third type on a quantum basis is the Frey-Wyssling complexes.

Rubber particles

The rubber particles usually have a size ranging from 50 A° to about 30,000 A° (3 μm), although extreme cases having 5 or 6 μm are also found. They are spherical bodies in young trees and potted plants but in mature trees the particles are large, often having a pear shape. The shape in certain cases seems to be a clonal character. Pear shape is reported to be very frequent in clones such as Tjir 1 and PR 107 (Southorn, 1961).

Particle size of greater proportion is beyond the limit of the resolution power of light microscope and reliable information in this regard could be obtained only with electron microscopic observations. Tempel (1952) recorded 1000 A° size at maximum frequency. This was confirmed later by Gomez and Moir (1979). Schoon and van der Bie (1955), observed a multi-modal distribution of rubber particles in latex of mature Hevea trees and proposed that larger particles are formed by the association of smaller particles. Gomez (1966) found a multimodal distribution in latices from young potted plants. In laticifers of very young plants small osmiophylic particles are seen freely in the cytoplasm. The structure and colloidal properties of Hevea latex has been well studied (Cockbain and Philpott, 1963; Ho et al., 1976).

A rubber particle of average size, about 1000 A°, contains hundreds of molecules of the hydrocarbon and is surrounded by a surface film of proteins and lipids. The rubber particles are also associated with tri-glycerides, sterols, sterol esters, tocotrienols and other lipids. Dupont et al. (1976) have confirmed the presence of phosphatidylcholine and small amounts of phosphatidyl ethanolamine in the lipids associated with rubber particles. The protein envelope of rubber particles is visible in sections of osmium stained rubber particles and is approximately 100 A° thick (Andrews and Dickenson, 1961). The envelope carries a negative charge and confers colloidal stability to the rubber particles.

According to Dickenson (1969) there are rubber particles with variously stained regions. An inner osmiophilic region surrounded by a weakly stained periphery is attributed to lack of uniformity when rubber particles are deposited on existing particles during biosynthesis. He has also suggested that the inner particulate inclusion, having 50-80 A° thickness, might be molecules of rubber of molecular weight about 100,000 but further investigations are needed to confirm this.

Lutoids

Lutoids form the next major component of Hevea latex. They are membrane-bound bodies and are mostly larger in size than the rubber particles. They are 2-5 μm in diameter bounded by a unit membrane of about 80 A° thickness. It was Wiersum (1957) who first suggested that the lutoids behave like vacuoles due to stainability with neutral red. Though controversy existed in this regard, the work of Ribaillier et al. (1971) provided evidence for the vacuolar properties of lutoids.

The content of lutoids (B-serum) has a very rapid flocculating action on aqueous suspension of rubber particles in latex, resulting in the formation of microfloccs (Southern and Edwin, 1968). This activity is apparently moderated by the ambient C-serum and is much reduced if B-serum is boiled. Southorn and Yip (1968) demonstrated that this fast initial flocculating action of B-serum is an electrostatic one involving the interaction between the cationic contents of B-serum and the anionic rubber particle surface.

By phase contrast microscopy and application of suitable staining procedures the structure of lutoid particles have been studied in detail. Mainly two types of fibrillar structures have been described. The first type, known as microfibrils, are characteristic of latex vessels in young tissues (Dickenson, 1965, 1969; Audley, 1965, 1966). As seen by phase contrast microscopy of tapped latex from young tissue, the microfibrils are freely suspended in the fluid content of the lutoid B-serum. The microfibrils are seen usually as grouped together in bundles. Each bundle has a diameter of 450-500 A°. Individual microfibrils are several micron long and 70-80 A° in diameter. The microfibrils can be isolated from the sediments of latex from young tissues which on negative staining with phosphotungstic acid shows further details. Each microfibril is a tightly coiled continuous helix with a hollow axis. The diameter of the helix is about 125 A° and that of the hollow axis 30 A°. The microfibrils consist of an acidic protein while nucleic acid seems to be absent. Microfibrils however are not present in tissue or latex collected from the mature bark. It is believed that they disintegrate as the particles mature or else the young lutoids containing microfibrils themselves disintegrate as the tissue ages and are replaced by a population of lutoids without microfibrils. However, the microfibrils do not seem to have vital role in rubber bio-synthesis.

The second type of fibrillar structures, observed in lutoids of latex from mature bark of stimulated trees, are known as 'microhelices', so named (Gomez and Yip, 1975) because of their spring like shape. These structures were first observed by Dickenson (1965, 1969). They are occasionally found in unstimulated trees and their number increases on dilution. However, microhelices are reported to be more frequent in lutoids of tapped latex than in situ latex (Southorn and Edwin, 1968; Gomez and Yip, 1975) and are occasionally observed in latex collected from young tissue also.

As reviewed by Gomez and Moir (1979) the microhelices are approximately 1 μm in length with a diameter of 200 A°, having a fibre width of about 50 A° and an open hollow helix having a 300 A° wide pitch. Dickenson (1965) suggested the formation of microhelices from microfibrils but this has been questioned by Gomez and Yip (1975).

A third type of lutoid inclusion - minute spherical particles in Brownian movement - was observed by Schoon and Phoa (1956). Later Southorn (1960, 1961) found such particles in large numbers in the bottom fraction of ultra-centrifuged latex of long rested trees and this was confirmed by Dickenson (1969). The role of such particles in latex is unknown.

Frey-Wyssling complexes

Yellow globules, in clusters in tapped latex were first noted by Frey-Wyssling (1929). The existence of such particles in groups, associated with a vacuolar body was observed by Southorn (1969) in phase contrast micro-scopy and he found that the individual particles are covered by a membrane: this was confirmed by electron microscopy. Dickenson (1969) named these particles, enclosed as a single structure, as Frey-Wyssling complexes.

The Frey-Wyssling complexes are more or less spherical in shape in a size range of 3-6 μm (diameter) and are bounded by a double membrane. Within the membrane there are two types of particles - large osmiophilic globules in variable numbers and a system of rope-like tubules of about 750 A° diameter, usually embeded in a membrane bound matrix of osmiophilic nature. The complex structure of Frey-Wyssling complexes has been elaborated by Dickenson (1969) who described a series of concentric lamellae of the double unit membrane and the system of tubules and also highly folded invaginations of the inner membrane.

The Frey-Wyssling complexes are considered to have vital role in metabolic activities. Though Dickenson (1969) opined that these structures may be possible sites of rubber biosynthesis, the double membrane and presence of carotene and polyphenoloxidase in the Frey-Wyssling complex led to a tentative suggestion that it is a type of plastid.

TYPES OF LATICIFERS

Gomez (1976) made a comparative study of latex vessels collected at different stages of development or from different positions of a tree and identified mainly five types. The first type was an embryonic vessel in leaf petiole at a stage prior to the fusion of laticifer initials. This resembled a normal living parenchyma cell in cell contents, except for the presence of numerous osmiophilic rubber particles.

A typical latex vessel from the secondary phloem of green stem had osmiophilic rubber particles, ranging from 100 A° to 5000 A° in diameter. Lutoids were prominent and they contained microfibrils. This type of vessel contained mitochondria and occasionally Frey-Wyssling complexes, golgi bodies and chloroplasts.

The third type was the latex vessel collected from the secondary phloem of the tree trunk at mature age, from the inner portion of the bark, i.e., within 1 mm from the wood. This contained numerous osmiophilic rubber particles of smaller size (50 A°-2 µm) in diameter. Lutoids were present but were devoid of microfibrils. Mitochondria were also present.

The fourth type was latex vessel under tapping and had rubber particles in very large numbers. Lutoids and Frey-Wyssling complexes were common and occasionally mitochondria and endoplasmic reticulum (at the periphery) were present. Rarely, nuclei were also detected.

The last type represented senescent vessels in the outer bark. In this type, the rubber particles were comparatively larger in size and the other organelles obscured.

REFERENCES

Andrews, E.H. and Dickenson, P.B. 1961. Preliminary electron microscope observations on the ultrastructure of the latex vessel and its contents in young tissues of Hevea brasiliensis. Proc. Nat. Rubb. Res. Conf. 1960, Kuala Lumpur: The Rubber Research Institute Malaya, 756 pp.

Archer, B.L., Barnard, D., Cockbain, E.G., Dickenson, P.B. and McMullen, A.I., 1963. Structure, Composition and Biochemistry of Hevea latex. In: L. Bateman (Ed.), The Chemistry and Physics of Rubber-like Substances. Maclaren & Sons Ltd., London, 43 pp.

Arisz, W.H., 1919. The structure of the laticiferous vessel system of Hevea. Arch. v.d. Rubberc., 3: 139 pp.

Arisz, W.H., 1921. De waarde van het bartouderzock by Hevea voor de praktyk. Arch. v.d. Rubberc., 5: 81pp.

Arisz, W.H. and Schweizer, J., 1921. Over the looistof van Hevea brasiliensis (on the tannin in Hevea brasiliensis). Arch. v.d. Rubberc., 5: 334 pp.

Ashplant, H., 1928. Yield variability in Hevea brasiliensis. Nature, 121: 1018 pp.

Audley, B.G., 1965. Studies of an organelle in Hevea latex containing helical protein microfibrils, Proc. Nat. Rubb. Prod. Res. Ass. Jubilee Conf., Cambridge, 1964. L. Mullins (Ed.), Maclaren & Sons Ltd., London, 67 pp.

Audley, B.G., 1966. The isolation and composition of helical protein microfibrils of Hevea brasiliensis latex. Biochem. J., 98: 335 pp.

Bealing, F.J. and Chua, S.E., 1972. Output, composition and metabolic activity of Hevea latex in relation to tapping intensity and the onset of brown bast. J. Rubb. Res. Inst. Malaya, 23: 204.

Bobilioff, W., 1919. Onderzoekingen over het onstaan van latexvaten enlatex bij Hevea brasiliensis. (Investigations concerning the origin of latex vessels and latex in Hevea brasiliensis). Arch. v.d. Rubberc., 3: 43.

Bobilioff, W., 1920. Correlation between yield and number of latex vessel rows of Hevea brasiliensis. Arch. v.d. Rubberc. 4: 391 pp.

Bobilioff, W., 1921. Over het voorkomen van gecoaguleerd melksap in de houtvaten van Hevea brasiliensis (on the occurrence of coagulated latex in the wood bundles of Hevea brasiliensis). Arch. v.d. Rubberc., 5: 169 pp.

Bobilioff, W., 1923. Anatomy and Physiology of Hevea brasiliensis. Drukkerijen Ruygrok. Batavia.

Bryce, G. and Campbell, L.E., 1917. On the mode of occurrence of latex vessels in Hevea brasiliensis, Dept. Agric. Ceylon Bull., 30: 1-22.

Bryce, G. and Gadd, C.H., 1923. Yield and growth in Hevea brasiliensis. Dept. Agric. Ceylon Bull., 68:

Calvert, A., 1887. The laticiferous tissues in the stem of Hevea brasiliensis. Ann. Bot., 1: 75-77.

Cockbain, E.G. and Philpott, M.W., 1963. Colloidal properties of latex. In: L. Bateman (Ed.), The Chemistry and Physics of Rubber-like Substances. Maclaren & Sons Ltd., London, p. 73.

Cook, A.S. and Sekhar, B.C., 1953. Fractions from Hevea brasiliensis latex centrifuged at 59000 g. J. Rubb. Res. Inst. Malaya, 14: 163 pp.

Coupe, M. and d'Auzac, J., 1972. Misc en evidence de polysomes functionnels dans le latex d' Hevea brasiliensis (Kunth) Muell. Arg. C.R. Accad. Sci. Paris, 274: 1031 pp.

d'Auzac, J., 1988. Transmembrane transport mechanisms. Application to the laticiferous system. Compte-Rendu du Colloque Exploitation-Physiologie et Amelioration del l'Hevea - 73-89.

De Bary, A., 1877. Vergleichende Anatomie der vegetation-sorgane der Phanerogamen and Farne. W. Engelmann, Leizig.

De Fay Elizabeth and Hebant, C., 1980. Etude histologique des ecorces d' Hevea brasiliensis atteint de la maladie des encoches seches. C.R. Acad. Sci. Paris, 291: 865-868.

De Jonge, A.W.K., 1916. Wetenschappelijke tapproevenbij Hevea brasiliensis. Meded. Agric. Chem. Lab. Buitenz. 14: 1 pp.

DeJonge, A.W.K., 1919. Tapproen bij Hevea brasiliensis (Tapping experiments on Hevea brasiliensis). Arch. v.d. Rubberc., 3: 1 pp.

De Jonge, P., 1955. Stimulation of yield in Hevea brasiliensis. III. Further observations on the effects of yield stimulante. J. Rubb. Res. Inst. Malaya, 14: 383 pp.

De Jonge, P., 1957. Stimulation of bark renewal in Hevea and its eifect on yield of latex. J. Rubb. Res. Inst. Malaya, 15: 53 pp.

Dickenson, P.B., 1965. The ultrastructure of the latex vessel of Hevea brasiliensis. Proc. Nat. Rubb. Prod. Res. Ass. Jubilee Conf., Cambridge, 1964. L. Mullins (Ed.), Maclaren & Sons Ltd., London, 52 pp.

Dickenson, P.B., 1969. Electron microscopical studies of latex vessel system of Hevea brasiliensis. J. Rubb. Res. Inst. Malaya, 21: 543 pp.

Du Pont, J., Moreau, F., Lance, C. and Jacob, J.L., 1976. Phospholipid composition of the membrane of lutoids from Hevea brasiliensis latex. Phytoche., 15: 1215 pp.

Frey-Wyssling, A., 1929. Microscopic investigations on the occurrence of resins in Hevea latex. Arch. v.d. Rubberc., 13: 392.

Frey-Wyssling, A., 1930. Investigations into the relation between the diameter of the latex tubes and the rubber production of Hevea brasiliensis. Arch. v.d. Rubberc., 14: 135.

Frey-Wyssling, A., 1952. Latex flow. In: Frey-Wyssling, A. (Ed.), Deformation and flow in biological systems. North-Holland Publ., Amsterdam, pp. 322-349.

Gomez, J.B., 1964. Anatomical observation on Hevea bark treated with yield stimulants containing 2,4-D. J. Rubb. Res. Inst. Malaya, 18: 226 pp.

Gomez, J.B., 1966. Electron microscopic studies on the development of latex vessels in Hevea brasiliensis Muell. Arg. Ph.D. Thesis, University of Leeds. Thesis submitted for the degree of Doctor of Philosophy University of Leeds.

Gomez, J.B., Narayanan, R. and Chen, K.T., 1972. Some structural factors affecting the productivity of Hevea brasiliensis. I. Quantitative determination of the laticiferous tissue. J. Rubb. Res. Inst. Malaya, 23: 193 pp.

Gomez, J.B., 1974. Ultrastructure of mature latex vessels in Hevea brasiliensis. Proc. 8th Int. Congr. Electron Microsc., Canberra, 1974. Australian Academic of Science, Canberra, 2: 616 pp.

Gomez, J.B. and Yip, E., 1975. Microhelices in Hevea latex. J. Ultrastruct. Res., 52: 76 pp.

Gomez, J.B., 1976. Comparative ultracytology of young and mature latex vessels in Hevea brasiliensis. Proc. Int. Rubber Conf., Kuala Lumpur, 1975. Rubber Research Institute of Malaysia, Kuala Lumpur, 2: 143 pp.

Gomez, J.B., 1982. Anatomy of Hevea and its Influence on Latex Production. Malaysian Rubber Research and Development Board, Publication No. 15, 54 pp.

Gomez, J.B. and Chen, K.T., 1967. Alignment of anatomical elements in the stem of Hevea brasiliensis. J. Rubb. Res. Inst. Malaya, 20: 91 pp.

Gomez, J.B. and Moir, G.K.J., 1979. The ultracytology of latex vessels in Hevea brasiliensis. Monograph No. 4. Malaysian Rubber Research and Development Board, Kuala Lumpur.

Hamzah Samsidar, S., Mahmood, A., Sivanandyan, K. and Gomez, J.B., 1975. Effects of mineral deficiencies on the structure of Hevea brasiliensis Muell. Arg. I. Bark Anatomy. Proc. Int. Rubb. Conf., 1975, Kuala Lumpur.

Hamzah Samsidar, S. and Gomez, J.B., 1981. Anatomy of bark renewal in normal puncture tapped trees. J. Rubb. Res. Inst. Malaya, 29: 86 pp.

Heusser, C., 1921. Tapping tests and bark investigations in Hevea plantation from selected seed. Arch. v.d. Rubberc., 5: 303 pp.

Ho, C.Y., Narayanan, R. and Chen, K.T., 1973. Clonal nursery studies in Hevea. I. Nursery yields and associated structural characteristics and their variations. J. Rubb. Res. Inst. Malaya, 23(4): 305-316.

Ho, C.Y., Subramaniam, A. and Yong, W.M., 1976. The lipids associated with the particles in Hevea latex. Proc. Int. Rubb. Conf. 1975, Kuala Lumpur. Rubber Research Institute of Malaysia, Kuala Lumpur, 2: 441pp.

Hoffman, C.A., 1933. Developmental morphology of Allium cepa. Bot. Gaz., 95: 279-299.

Keuchenius, P.E., 1918. On the structure, the degeneration and the regeneration of latex rings with Hevea trees. Arch. v.d. Rubberc., 2: 837 pp.

Keuchenius, P.E., 1920.   Onderzoekingen over de bast anatomie van Hevea.
    Arch. v.d. Rubberc., 4: 5 pp.
La Rue, C.D., 1921.   Structure and yield in Hevea brasiliensis. Arch. v.d.
    Rubberc., 5: 574 pp.
Mass, J.G.J.A., 1923.   The effect of soil difference and manuring on the
    formation of latex vessels. Arch. v.d. Rubberc., 7: 392 pp.
Metcalfe, C.R., 1966.   Distribution of Latex in the Plant Kingdom. Notes
    from Jodrell Laboratory III. Royal Botanical Gardens, Kew.
Milanez, F.R., 1946.   Nota Previa sobre os laticiferos de Hevea brasiliensis.
    Arquivos do Servico Florestal do Brasil, 2: 39 pp.
Milanez, F.R., 1948.   Segunda nota sobre laticiferos. Lilloa, 16: 193 pp.
Milanez, F.R., 1951.   Galactoplastas de Hevea brasiliensis Muell. Arg.
    Arquivos do Jardim. Botanico de Rio de Janeiro, 11: 39 pp.
Moir, G.F.J., 1959.   Ultracentrifugation and staining of Hevea latex. Nature,
    184: 1626 pp.
Narayanan, R., Gomez, J.B. and Chen, K.T., 1973.   Some structural factors
    affecting productivity of Hevea brasiliensis. II. Correlation studies
    between structural factors and yield. J. Rubb. Res. Inst. Malaya,
    23: 285 pp.
Narayanan, R. and Ho, C.Y., 1973.   Clonal nursery studies in Hevea. II.
    Relationship between yield and girth. J. Rubb. Res. Inst. Malaya,
    23(5): 332 pp.
Narayanan, R., Ho, C.Y. and Chen, K.T., 1974.   Clonal nursery studies
    in Hevea. III. Correlations between yield, structural characters, latex
    constituents and plugging index. J. Rubb. Res. Inst. Malaya, 24: 1 pp.
Panikkar, A.O.N., 1974.   Anatomical studies in Hevea brasiliensis. Ph.D.
    Thesis, Birla Institute of Technology and Science, Pilani, India.
Paranjothy, K., Gomez, J.B. and Yeang, H.Y., 1975.   Physiological aspects
    of brown bast development. Proc. Int. Rubb. Conf., 1975. Rubber
    Research Institute of Malaysia, Kuala Lumpur, 2: 181-202.
Petch, T., 1911.   Tapping Experiments and Their Teachings. In: The Physi-
    ology and Diseases of Hevea brasiliensis. Dulau & Co. Ltd., London.
Premakumari, D., Panikkar, A.O.N., Annamma, Y. and Leelamma, K.P., 1981.
    Studies on the cambial activity in Hevea brasilinesis, Muell. Arg.
    Proceedings of PLACROSYM IV, 1981, Mysore. pp. 425-430.
Premakumari, D., Joseph, G.M. and Panikkar, A.O.N., 1985. Structure of
    the bark and clonal variability in Hevea brasilinesis Muell. Arg. (Willd.
    ex A. Juss.) Ann. Bot., 56: 117-125.
Premakumari, D., Joseph G.M. and Panikkar, A.O.N. 1988a.   Influence
    of the orientation of laticifers and quantity of laticiferous tissue on
    yield in Hevea brasilinesis, Muell. Arg.   J. Plant. Crops, 16(1):
    12-18.
Premakumari, D., Panikkar, A.O.N. and Sobhana Shankar, 1988b. Comparative
    bark anatomy of three induced polyploids of Hevea brasiliensis (Willd.
    ex. A. Juss.) Muell. Arg. and their diploids. Proceedings of PLACRO-
    SYM VIII, Cochin, 1988, India.
Premakumari, D., Sasikumar, B. and Panikkar, A.O.N., 1984. Variability
    and association of certain bark anatomical traits in Hevea brasiliensis
    Muell. Arg.   Proceedings of PLACROSYM VI, 1984, Kottayam, Oxford IBH,
    New Delhi, pp 49-54.
Rands, R.D., 1919.   De bruine binnenbastziekte van Hevea brasiliensis-
    voorloopige mededeeling (Brown bast disease of Hevea brasiliensis -
    Preliminary account). Arch. v.d. Rubberc., 3: 158 pp.
Rands, R.D., 1921a.   Brown Bast Disease of Plantation Rubber. Its Causes
    and Prevention. Medd. v.h. Inst. v. Plantenziekten, No. 47.
Rands, R.D., 1921b.   Histological Studies on the Brown Bast Disease of
    Plantation Rubber, Medd. v.h. Inst. v. Plantenziekten, No. 49.
Rands, R.D., 1921c.   Brown Bast Disease of Plantation Rubber, Its Cause
    and Prevention. Arch. v.d. Rubberc., 5: 223 pp.

Rhodes, E., 1930. Brown bast: Some considerations as to its nature. Quar. J. Rubb. Res. Inst. Malaya, 2: 1 pp.

Ribaillier, D., Jacob, J.L. and d'Auzac, J., 1971. Sur certains characters vacuolaires des lutoides du latex d' Hevea brasiliensis. Muell. Arg. Physiol. Veg., 9: 423 pp.

Riches, J.P. and Goodding, E.G.B., 1952. Studies in the physiology of latex. I. Latex flow on tapping - threoretical considerations. New Phytol., 51: 1 pp.

Rosowski, J.R., 1968. Laticifer morphology in the mature stem and leaf of Euphorbia supena. Bot. Gaz., 129: 113-20.

Sanderson, A.R. and Sutcliffe, H., 1921. Brown bast. Rubber Grower's Association, London.

Scharffstein, G., 1932. Untersuchungen an ungeglieder-ten Milchrohren. Bein. bot. Zbl., 49(1): 197-220.

Schoon, Th.G.F. and Phoa, K.L., 1956. Morphology of the rubber particles in natural latices. Arch. v.d. Rubberc., 33: 195 pp.

Schoon, Th.G.F. and van der Bie, G.J., 1955. Particle size distribution in brominated Hevea latices. J. Polymer Sci., 16: 63 pp.

Scott, D.H., 1882. The development of articulated laticiferous vessels. Quar. J. Micr. Sci., 22: 136 pp.

Scott, D.H., 1886. On the occurrence of articulated laticiferous vessels in Hevea. J. Linn. Soc (Bot), 21: 566-573.

Sethuraj, M.R., Sulochanamma, S. and George, M.J., 1974. Influence of initial flow rate, rows of latex vessels and plugging index on the yield of the progenies of Hevea brasiliensis Muell. Arg. derived from crosses involving Tjir 1 as the female parent. Indian J. Agric. Sci., 44: 354-356.

Southorn, W.A., 1960. Compex particles in Hevea latex. Nature, 188: 165 pp.

Southorn, W.A., 1961. Microscopy of Hevea latex. Proc. Nat. Rubb. Res. Conf. Kuala Lumpur 1960. Rubber Research Institute of Malaya, Kuala Lumpur, 766 pp.

Southorn, W.A., 1969. Physiology of Hevea (latex flow). J. Rubb. Res. Inst. Malaya, 21: 494 pp.

Southorn, W.A. and Edwin, E.E., 1968. Latex flow studies. II. Influence of lutoids on the stability and flow of Hevea latex. J. Rubb. Res. Inst. Malaya, 20(4): 187 pp.

Southorn, S.A. and Yip, E., 1968. Latex flow studies. III. Electrostatic considerations in the colloidal stability of fresh latex from Hevea brasiliensis. J. Rubb. Res. Inst. Malaya, 20(4): 201 pp.

Steinmann, A., 1923. Vereenvoudigde methode voor bastonderzock. Arch. v.d. Rubberc., 7: 199 pp.

Taylor, R.A., 1926. The interrelationship of yield and the various vegetative charactors in Hevea brasiliensis. Rubb. Res. Sch. Ceylon Bull., 43.

van den Tempel, M., 1952. Electron microscopy of rubber globules in Hevea latex. Trans I.R.I., 28: 303.

Vischer, W., 1921. Over de resultaten verkregen met het oculeeren van Hevea brasiliensis op de onderneming pasir waringen (Results obtained with budded trees of Hevea brasiliensis on Pasir Waringen estate). Arch. v.d. Rubberc., 5: 17.

Vischer, W., 1922. Resultaten bereikt met oculaties van Hevea brasiliensis op de onderneming pasir waringen, een bijdrage tot de vraag, in ho-everre aantal latex-vaten en productiviteit raskenmerken zijn (Results obtained with buddings on the Pasir Warigen estate). Arch. v.d. Rubberc., 6: 426 pp.

Vischer, W. and Tas, L., 1922. Results obtained with buddings on the estate. Arch. v.d. Rubberc., 6: 416 pp.

Wiresum, L.K. 1957. Enkele latexproblemen. Vakbl. Biol., 3: 17 pp.

Xiuqian Zhao, S., 1987. The significance of the structure of laticifer with relation to the exudation of latex in Hevea brasiliensis. J. Nat. Rubb. Res., 2: 94-98.

CHAPTER 5

GERMPLASM RESOURCES AND GENETIC IMPROVEMENT

Y. ANNAMMA VARGHESE

Rubber Research Institute of India, Kottayam-686009, Kerala, India.

Hevea brasiliensis (Willd. ex Adr. de Juss.) Muell. Arg. is perhaps the youngest of the major domesticated crops in the world. The genetic base of Hevea in the East is very narrow, limited to a few seedlings collected from a minuscule of the genetic range of Hevea brasiliensis in Boim, near the Tapajos river in Brazil (Wycherley, 1968, Schultes, 1977; Allen, 1984). It is from this small genetic foundation that spectacular yield improvement of about ten times has been achieved. However, the genetic advance gained in the early breeding phases seems to have slowed down in the more recent phases of breeding (Tan, 1987; Ong and Tan, 1987; Simmonds, 1989), for which different reasons have been attributed.

THE GENETIC BASE

The para rubber tree cultivated in South East Asia belongs to the original collection of Sir Henry Wickham in 1876 (see Chapter 1. Eds.). The number of Wickham seedlings which have contributed to the oriental plantation stock is believed to be only around 22, although around 2000 seeds had been sent to Sri Lanka, Singapore, Perak and Java (Wycherley, 1977). It is pertinent to note that all the eastern clones have originated from a relatively few trees surviving from Wickham's original collection, referred to as the 'Wickham base' (Simmonds, 1989).

FACTORS LEADING TO DECREASE IN GENETIC DIVERSITY

The commercially accepted practice of clonal propagation, directional selection for yield and a cyclical assortative breeding pattern have contributed to a further decrease in genetic diversity.

Propagation by budgrafting has become the established commercial practice in H. brasiliensis. With the development of high yielding clones, extensive areas are being planted with a limited number of modern clones of high production potential. With the narrow genetic base and a breeding method based on additivity of gene control (Wycherley, 1969; Simmonds,

1969), clonal propagation has reduced the genetic base still further.

In the ortet selection programme, aimed at development of clones, the mother trees were selected mainly based on yield considerations (Dijkman, 1951). Efforts for hybridisation started during the late twenties, and again the primary concern was productivity improvement. The earlier hybrid clones recorded a fivefold increase in yield over the original un-selected seedlings. However, consequent to such a directional selection for yield, genetic variability in regard to many secondary characters influencing overall performance was ignored (Wycherley, 1969) leading to a certain amount of genetic erosion.

One of the major disadvantages of the conventional breeding system is the cyclical 'generation-wise assortative mating' (GAM), where the best genotypes in one generation were used as parents for the next cycle of breeding. This resulted in a limited number of high yielding clones, majority of which originated from a few prominent parents (Simmonds, 1986, 1989). Most of the clones bred to date in Malaysia can be traced back to about seven early clones viz., Tjir 1, Pil A 44, Pil B 84, PB 24, PB 49, PB 56 and PB 86 (Subramaniam, 1980; Tan, 1987). Similarly, in India the present day popular clones originated from about 20 clones (Fig. 1). The situation is not different in other rubber producing countries in Asia and Africa.

The early selections recorded substantial yield increase over the Wickham material and some earlier clones (GT 1) are still under commercial cultivation and some others (PB 86, PR 107 etc.) popular in certain areas. From an yield level of 450 kg ha^{-1} yr^{-1} in the unselected materials, the early primary (Pil B 84) and hybrid clones (RRIM 501 and RRIM 600) attained an yield of 1000, 1300 and 1550 kg ha^{-1} yr^{-1} respectively (Tan, 1987). How-ever, the yield improvement in further generations did not keep pace with the spectacular achievement in the early phases (e.g., RRIM 700 and 800 series). Thus a slowing down of genetic advance in the more recent breeding phases, in contrast to the remarkable early progress, is evident (Simmonds, 1986; Tan, 1987). Genetic studies, in general, also indicate inbreeding depression and unpredictable interaction when related parents are used in breeding (Simmonds, 1969; Gilbert et al., 1973; Nga and Subramaniam, 1974). So far, a phenotypic selection of parents for high yield was practiced in hybridization programmes. Such a selection, however, did not, in general, take into account the combining ability of the parents in regard to yield components. Parents with complementary yield components produce higher yields than both the parents (Jaijian and Jingxian, 1990).

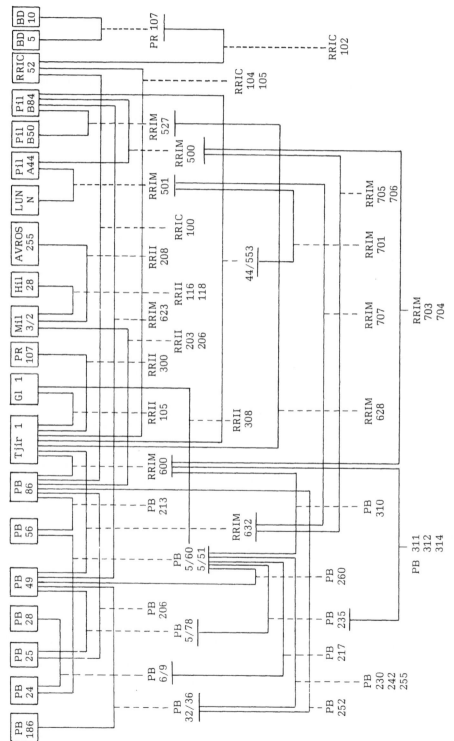

Fig. 1. Pedigree of present day popular clones in India.

EFFORTS TO INTRODUCE NEW GERMPLASM

After the Wickham material, a few introductions, including H. spruceana and H. guianensis, were made in Indonesia from Brazil and Surinam in 1896, 1898 and between 1913 and 1916 (Dijkman, 1951). In Malaysia, seedlings belonging to H. brasiliensis, H. benthamiana, H. guianensis, H. spruceana and H. pauciflora as well as hybrid seeds from different provenances in Brazil were imported during 1951-52 (Brookson, 1956; Tan, 1987). Of the 106 IAN clones imported by Malaysia from Brazil via Sri Lanka in 1952 (Baptiste, 1961), IAN 873 and IAN 717 recorded comparatively good yield, tolerance to Colletotrichum and to some physiological races of Microcyclus (Ong and Tan, 1987). Other introductions to Malaysia were 25 SALB resistant clones from Brazil during 1953-54 (Brookson, 1956) and 821 seeds belonging to seven Hevea spp. from Schultes' museum, Belem, Brazil in 1966 (IBPGR, 1984). Most of the introduced materials have been exchanged between Sri Lanka and Malaysia.

A Franco-Brazilian prospection in 1974 collected budwood from selected genotypes from the forests of Acre, Rondonia and Peru (Compagnon, 1977). Nigeria imported Hevea clones from Malaysia and Sri Lanka in the sixties (Olapade, 1988) which led to the replacement of the low yielding genotypes and these were used as the new parental breeding stock. In India, the main source of genetic material has been the Wickham gene pool. A total of 114 clones introduced during different periods constituted the exotic component of the present gene pool. These exotic materials include two other species (H. benthamiana and H. spruceana), one inter specific hybrid (Fx 516) and two IAN clones (IAN 717 and IAN 873). In Brazil in 1972 collections were made from Acre and Rondonia and breeding was based on mother trees in jungle (Goncalves, 1982).

In general, these later introductions were of relatively small sample size and of low yield potential and hence have contributed less to the plantation industry (Simmonds, 1989).

'GERMPLASM 1981'

Recognizing the fact that international action was imperative to enrich the available genetic variability of Hevea in the orient, the International Rubber Research and Development Board (IRRDB) organized a major collection expedition to the Amazon rain forests in 1981. This effort can be considered as one of the most significant events in the history of rubber germplasm collection, with a view to ensuring a healthy future for the plantation industry.

This joint expedition of the IRRDB and the Brazilian Government (National Centre for Genetic Resources-CENARGEN under EMBRAPA - National Agricultural Research System) collected a total of 64736 seeds (Ong et al., 1983; Mohd. Noor and Ibrahim, 1986), from the states of Acre, Matto Grosso and the territory of Rondonia. These three Western States of Brazil were selected for the expedition as it was known that Rondonia and Acre had vigorous, high yielding trees and that Acre genotypes produced superior quality rubber (Wycherley, 1977; Schultes, 1977, 1987). Ecological differences between these states offered chances of selection of materials suitable for diverse situations. Accessibility to the wild trees in these areas was another consideration. The team also collected budwood from 194 presumably high yielding trees, checking visually the absence of leaf blight and Phytophthora leaf fall (Ong et al., 1983).

The materials collected were first despatched to the National Centre of Rubber and Oil Palm Research (CNPSD) in Manaus, Brazil. After stringent phytosanitary measures at this primary centre, 50% of the seeds and clones were retained at Manaus in compliance with the international code of plant collection. The remaining 50% seeds were sent to two germplasm centres; Malaysia and Ivory Coast. Malaysia received 75% of the collections. The seeds were sent through an intermediate quarantine station at Tun Abdul Razak Laboratory, Brickendonbury, U.K. Budwood of 162 clones were multiplied in the primary nursery at Manaus. The intermediate quarantine nursery in Guadeloupe, Brazil, received budwood materials from this primary nursery.

The transfer of these genotypes to the member institutes in Asian countries from the Malaysian Centre and to the African countries from the Ivory Coast Centre has been initiated and a good number of genotypes have already been introduced.

BASE BROADENING FOR RESISTANCE TO DISEASES

South American Leaf Blight (SALB) caused by the fungus, Microcyclus ulei, the greatest single threat to the world supply of natural rubber, has caused abandoning of ambitious programmes of extensive rubber culti-vation in South American humid tropics. Screening tests for resistance to SALB have shown that none of the oriental clones has resistance to this devastating disease (Baptiste, 1961; Wijewantha, 1965). This may be either due to lack of resistance genes in the Wickham collection or to subsequent erosion of such genes in breeding and selection process. There are also indications of gene erosion to Oidium and Gloeosporium in the original Wickham material (Wycherley, 1977). Many recent reports indicate instances of less serious diseases becoming more severe. Corynespora leaf disease affecting clones RRIC 103, KRS 21 and RRIM 725 in Sri Lanka, a new anthracnose caused by Fusicoccum reported during 1987 in Malaysia, a minor disease caused by Guignardia observed intermittently in Malaysian estates since 1982 affecting clones like PB 235, PB 260 and PB 217 (IRRDB, 1987) and the identification of target leaf spot (Thanathephorus cucumeris) in Malaysia (IBPGR, 1984) are some of the examples. For introgression of disease resistance in high yielding oriental clones it is vital that all available sources of resistant genes, irrespective of their origin, should be utilised in breeding programmes. Each country should therefore conserve indigenous, exotic and wild genetic resources.

GERMPLASM RESOURCES

The different sources of Hevea germplasm are depicted in Fig. 2 (see Chapter 3 for origin and centre of diversity of genus Hevea. Eds.).

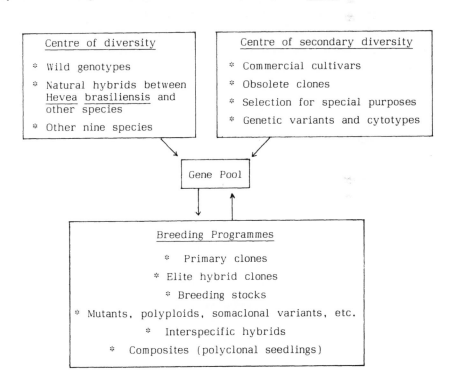

Fig. 2. Spectrum of germplasm of Hevea brasiliensis.

The centres of cultivation include countries like Indonesia, Malaysia, Thailand, China, India, Sri Lanka, Liberia, Nigeria, Zaire, Ivory Coast, Cameroon, Philippines and Burma.

CONSERVATION, EVALUATION AND UTILIZATION OF GENETIC RESOURCES

Conservation of genetic resources of Hevea in perpetuity is an urgent need of the time. IBPGR has included Rubber as one of the highest priority crops for conservation of the entire genepool (Annual Report, IBPGR, 1984). A possible danger of restricting cultivation to a few high yielding strains of any agricultural crop in any geographical area is the narrowing down of genetic variability. This in course of time may lead to disease and pest epidemics. On the other hand, natural diversity of the genetic material is a useful source of individual alleles that confer adaptive advantages such as resistance to diseases and pests, drought, etc. Wild genotypes may be useful in direct introgression of certain oligogenic as well as polygenic characters and also in facilitating plant breeding programmes (Sneep, 1979).

Motives for genetic conservation by breeders of Hevea are classified as specific, general and innovative (Wycherley, 1977). The base collection should include all known variants, even those which are of no immediate use, in order that the widest possible range of material is available for innovative purposes. Even related genera of Hevea, like Micrandra, Johannesia, which yield latex may also be collected and conserved (Wycherley, 1977). Thus germplasm collections should reflect the requirements of the present and should be reservoirs of genetic diversity for future.

Conservation of genetic resources of Hevea is indeed elaborate, expensive and difficult. The germplasm collections, maintained in the field as 'active collections' are exposed to natural calamities as well as pest and disease outbreaks. Efforts to establish 'base collections' for long term maintenance and evaluation of the genetic materials should receive priority in conservation programmes. From this, 'core collections' or a condensed assembly of germplasm should be identified and established for efficient conservation as well as for eliminating redundency in collections. Finally, a 'working collection' to suit short term needs of individual breeders and breeding programmes can be established for current use.

In addition to conventional methods of conservation, in vitro techniques have demonstrated abundant potential in some crop plants. Cryopreservation or freeze preservation of cultures in liquid nitrogen at $-196°C$ offers scope for preservation of rare, elite and desirable germplasm of vegetatively propagated plants with recalcitrant seeds (Balaj, 1986). In the germplasm conservation of a perennial plantation crop like Hevea, cryobiology offers much potential.

The sizeable number of wild genotypes of the 1981 germplasm should be evaluated and documented for valuable wild desirable genes and a strategy developed for utilizing them in breeding programmes. Non-availability of sufficient area for field evaluation and the long life span of the crop limit the scope for detailed evaluations of all the collections. Therefore, it becomes imperative to devise minimum descriptor lists for basic characterisation within the minimum time. A standard set of descriptors are necessary for ensuring uniformity of evaluation of data from among the different research institutes. Objectives in further characterisation and evaluation should be to test the introduced material, choose the right parents in making crosses, test the progenies derived from a heterogenous population and evaluate elite lines for specific attributes.

Preliminary evaluation of the wild genotypes is underway at the different institutes where they have been introduced and established. Electrophoretic studies on these materials as well as of Schultes' collection and Wickham material, using izozyme markers (Chevallier, 1988; Chevallier et al., 1988a) reveal that the genotypes display high variability. Similarly studies on agronomic variability (Demange, 1988) and leaf morphology also indicate a distinction between 1981 germplasm and clones evolved from the Wickham base.

Preliminary evaluation of some of the 1981 germplasm in India (Annamma et al., 1988, 1989a) also reveal wide variability with respect to growth parameters like plant height, girth, number of leaves, number of flushes, bark thickness, number of latex vessel rings and juvenile yield. Studies in Malaysia also indicate considerable variations in growth vigour and disease resistance among and within the genotypes from the different states in Brazil (Ong and Tan, 1987). Prominent leaf scar, very close and distant leaf flushes, dome shaped flushes, close leaves and prominent pulvinus and numerous small narrow crinkled leaves (Fig. 3) are some of the morphological variations observed. Variations in leaf colour, shape, size and number, petiolar and leaflet orientation, dwarf and semidwarf stature were also noted (Annamma et al., 1989a). Wide variations observed among the germplasm genotypes in growth vigour, bark structure, juvenile yield and morphological characters were in accordance with the general expectation that wild and primitive forms from the centre of origin would exhibit such variability. While chances for any direct selection for high productivity appears to be limited (Annamma et al., 1989a), the potential value of the 1981 introduction of wild germplasm from the provenances of Rondonia, Acre and Matto Grosso in Brazil, in providing many valuable genes for incorporation

Fig. 3. Morphological variations in 1981 germplasm accessions.
a. distant leaf flushes (control on left), b. very close leaf flushes,
c. prominent pulvinus, d. numerous small, narrow, crinkled leaves.

into the breeding pool needs no emphasis.

The course of evaluation and utilisation for the three main sources of germplasm viz., indigenous collections, exotic collections and wild materials are outlined in Fig. 4.

With the growing genetic resources available to plant breeders it has become important that the information system should be a cohesive and comprehensive data reporting, processing and communication system for all germplasm users. Computer based information systems have been in use for the last two decades for storage and retrieval of data on germplasm. An information storage and retrieval programme called EXIR has been developed by the scientists of Information Services and Genetic Resources Programme at the University of Colarado, USA to meet specific needs of scientists involved with data management problems.

STRATEGIES FOR GENETIC IMPROVEMENT

Breeding objectives

The ultimate objective of Hevea breeding is to synthesise ideal clones with high production potential combined with desirable secondary attributes. Initial vigour, smooth thick bark with a good latex vessel system, good bark renewal, high growth rate after opening, tolerance to major diseases and wind are considered to be good secondary characters. In addition, low incidence of tapping panel dryness (brown bast) can be a selection criterion. Good response to stimulation and low frequency tapping are attributes aimed at in recent times. When discounted cash flow is considered clones with early attainability of tapping girth and high initial yields are to be preferred over clones with higher yields in later phases of exploitation (Lim et al., 1973).

Specific objectives may however vary depending on agro-climatic and socio-economic requirements. In countries where labour is relatively cheap, clones suitable for high intensity tapping are preferred. On the other hand, clones responding to low frequency tapping should be bred in countries with labour shortage. Similarly in countries with shortage of suitable land, rubber cultivation is being extended to marginal lands, necessitating evolution of clones capable of withstanding stress conditions (drought, cold, high elevation, etc.). Genotypes suitable for high density planting and poor soil fertility could also be aimed at. Even clones with higher timber output could become a future breeding objective.

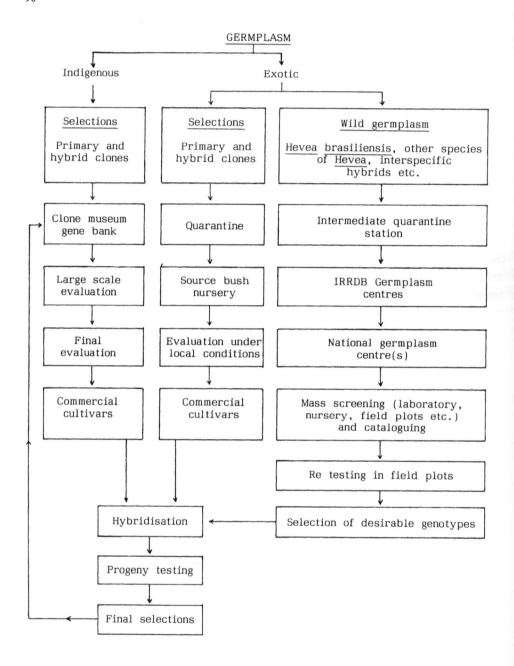

Fig. 4. Flow chart outlining utilisation of Hevea germplasm.

Yield improvement

The conventional breeding methods adopted in Hevea viz., ortet selection and hybridization have contributed to substantial increases in productivity. Two significant developments in the second decade of this century were (i) the realisation of the large variability in yield in the seedling population observed by the Dutch workers in Java and Sumatra (Whitby, 1919) and (ii) the success of budgrafting by Van Helton in 1918. These led to the development of early primary clones through ortet selection or mother tree selection. This oldest breeding method is aimed at systematically screening for outstanding seedling genotypes resultant of genetic recombination in nature.

Rapid progress with mother tree selection was achieved in Indonesia between 1919 and 1926. Screening of very large seedling areas resulted in a number of popular clones such as Tjir 1, PR 107, GT 1, Tjir 16, BD 5, BD 10, AVROS 49, AVROS 255, AVROS 352 and LCB 1320. In Malaysia, outstanding mother trees resulted in clones like PB 86, PB 5/60, PB 5/139, PB 6/9, PB 6/50, PB 28/59, PB 28/83, Pil B 84, Pil D 65, Gl 1, Ch 4, Ch 30, CHM 3, S. Reko, Lun N, etc. At RRIM, large areas of advanced generation seedlings have been established in commercial planting. Further systematic screening programme of seedling trees of PBIG/GG series was initiated in 1972 and a total of 57 clones from two phases of selection have been established in promotion plots (RRIM, 1982). The primary clones developed in Sri Lanka include Mil 3/2, Hil 28, Hil 55, Wagga 6278, Warring 4, RRIC 52, etc. In India, the earlier mother tree selections include 46 clones; RRII 1, RRII 2, RRII 3, RRII 4, RRII 5, RRII 6, RRII 33, RRII 43, RRII 44, etc. which are under large scale evaluation (Marattukalam et al., 1980). In recent years, over 150 preliminary selections have been established in small scale trials. Since the seedling areas are increasingly being replaced with modern clones, further extensive screening for yield, resistance to disease, drought, etc. should be given priority.

With a view to raising seedling planting materials, special polyclonal seed gardens with elite clones as component clones are to be laid out. The choice of parents should be based on General Combing Ability (GCA) estimates (Simmonds, 1989). The evaluation of such polycross population and selection of promising recombinants can be considered as selective breeding. Such multiparent, first generation synthetic varieties (SYN-1-polycrosses of rubber - Simmonds, 1986) have been economically successful for many decades predominantly due to additive genetic control of vigour, yield and high GCA (Tan, 1987; Swaminathan, 1977; Simmonds, 1986). These seedlings are less expensive and are in general more manageable and vigorous. Progenies of good polyclonal seed gardens have agricultural merits, though they are in general behind the best clones in productivity. Hence, there is a continuous need for high quality seed gardens to serve as reservoirs for continued ortet selection. At RRII, studies have indicated scope for identific- ation of likely prepotents on the basis of a performance index, computed for the seedling progenies of clones (Kavitha et al., 1990a).

As a result of hybridisation and selection, a good number of hybrid clones of commercial significance have been evolved. The evaluation of progenies of hybridization series is elaborate and expensive and requires much time. The conventional methods involve preliminary evaluation in small scale clone trials (SSCT) for rough sorting of numerous entries, second stage of evaluation viz., large scale clone trial (LSCT) of selected entries for more accurate statistical trials and final blockwise evaluation of promising selections from small scale and large scale trials for commercial evaluation. Simmonds (1989) suggested generalized lattice designs and rectangular plots to be more efficient than balanced lattices with two replications for small scale clone trials.

The early clones were used as parents of the first hybridisation series and resulted in early hybrid clones of commercial significance (e.g., RRIM 500 and 600 series). The yield levels of these selections were, in general, superior to those of primary clones. The parents of the later hybridization series were the best selections of the earlier series. This sort of cyclical generation-wise assortative mating (GAM) was followed in different Rubber Research Institutes over the past years. In Malaysia, the Rubber Research Institute (established in 1925), developed and released RRIM 500, 600, 700, 800 and 900 series of clones and the Prang Besar Institute developed a series of PB clones. The Indonesian Research Institute for estate crops in Java and Sumatra (BPPM) evolved PR, AVROS, BPM, LCB, PPN and RR clones. The RRIC clones originate from Rubber Research Institute of Sri Lanka, KRS clones from Rubber Research Centre at Hat Yai, Thailand and Dafeng, Haiken, YRITC (Yunnan Research Institute and Tropical Crops) and SCATC (South China Academy of Tropical Crops) clones from China.

In India, commercial planting in a 200 ha plantation in 1902 marked the beginning of rubber plantation industry (Nair et al., 1976). Crop improvement programmes were initiated in 1954 with the inception of RRII. So far over 1,60,000 controlled pollinations have been attempted, around 5700 hybrid seedlings produced and about 1500 clones were developed (Annamma et al., 1990a), of which RRII 100 series (Nair and Panikkar,

1966; Nair and George, 1969; George et al., 1980, Nazeer et al., 1986), 200 series (Saraswathy Amma et al., 1980a), and 300 series (Premakumari et al., 1984) are of commercial importance. The progenies of later hybridization series are under various stages of experimental evaluation.

In West Africa, breeding of clones has been accomplished at Institut de Recherches sur le Caoutchouc (IRCA at Abidjan, Ivory Coast) and at the Rubber Research Institute, Nigeria (RRIN). In Brazil, many earlier attempts to plant rubber failed due to the incidence of SALB. Lack of suitable infrastructure and appropriate technological inputs also contributed to this situation. However, there is a recent resurgence of interest in Brazil on rubber research, the leading agency being the National Agricultural Research System, EMBRAPA, which works on behalf of the national rubber authority, SUDHEVEA.

Yield gap

A wide gap exists between the average commercial productivity and the yield recorded in experimental plots. The national average during 1989 in Malaysia, India, Thailand, Sri Lanka, Indonesia and China (in 1985) were 1062, 988, 812, 766, 650 and 750 kg ha^{-1} yr^{-1} respectively. In general, tne national averages of productivity tend to be only about half of the productivity in good estates. The present yield levels of certain clones in experimental plots and good estates range from 2500-4000 kg ha^{-1} yr^{-1}. The factors responsible for the gap between experimental yield and actual commercial productivity are to be identified and catalogued for each level of management. The occurrence of significant genotype environment (GE) interaction (Paardekooper, 1964; Jayasekera et al., 1977) suggests that trials should be conducted in diverse environments for proper evaluation of the performance of planting materials. This naturally demands extensive areas and higher costs. However, location specific planting recommendations (Ho et al., 1974) would help to maximise productivity.

Hybridization involving divergent genotypes can generate greater range of variability and therefore, it can be expected that by choosing clones for specific situations the present yield level of Hevea brasiliensis can be increased. In addition, for further break-through in yield, breeders will have to work hand in hand with anatomists, physiologists, biochemists and biotechnologists. Sethuraj (1981 and 1987) identified initial flow rate, plugging index, dry rubber content and length of the tapping cut as major yield components in Hevea. Clones may be characterised and catalogued based on yield component analyses and the heritability of physiological characters established (Sethuraj, 1983). An improved breeding system aimed at combining specific yield components and a modified selection programme, utilizing both Wickham clones and fresh germplasm could result in higher yield levels (Naijian and Jingxiang, 1990).

CREATION OF GENETIC VARIABILITY

Polyploidy and mutations

Attempts for broadening the genetic base in _Hevea_ through artificial induction of polyploidy (Shepherd, 1969; Ong and Subramaniam, 1973; Markose et al., 1974; Ling et al., 1988; Saraswathy Amma et al., 1980b; Goncalves et al., 1983) and mutations (Markose et al., 1977; Saraswathy Amma et al., 1983) have been made in limited scale in different countries. The Indian workers succeeded in evolving a tetraploid (Saraswathy Amma et al., 1984) and in synthesising a triploid (Saraswathy Amma et al., 1980b) using clone RRII 105). A spontaneous triploid has also been identified (Nazeer and Saraswathy Amma, 1987). Fruits being the primary source of inoculam for abnormal leaf fall disease, lack of fruit set in triploids may be of significance. A natural genetic variant showing dwarf stature has also been identified, which may be of use in evolving compact forms for high density planting (Markose et al., 1981).

In vitro culture techniques

In _vitro_ techniques in rubber offer various possibilities for propagation and creation of genetic variability for crop improvement programmes and is an adjunct to conventional breeding methods. This approach includes micro-propagation, production of haploids, exploitation of gametoclonal and soma-clonal variability and selection of cell lines to evolve clones resistant to drought, cold, diseases, etc.

Reports on micropropagation studies (Sachithanthavale, 1974; Paranjothy and Ghandhimathi, 1975; Sinha et al., 1985) and anther culture (Chen et al., 1982; Hu Han, 1984) reveal varying degrees of success. In the RRII, scientists have succeeded in perfecting a tissue culture system utilising shoot tips of some of the popular clones and the plants have already been established in a field trial (Ashokan et al., 1988). In addition to mass multiplication, somaclonal variants occurring in the populations raised through tissue culture (Larkin and Scowcroft, 1981) can add to genetic variability in _Hevea_.

The crop being highly heterozygous and vegetatively propagated, lethal genes are likely to survive and be transmitted from generation to generation. This may affect yield and other quantitative characters. Now that it is

possible to develop hemizygous, haploid plants through pollen culture, these
lethal genes are effectively selected against, so that the surviving haploids
carry no lethal genes. Pollen derived plants are reported to be grown in
China (Chen et al., 1982; Hu Han, 1984, Shijie et al., 1990). Chen et al.
(1982) described regeneration of rubber plants from embryoids derived
from the culture of anthers and reported mixture of haploid, diploid and
aneuploid plants among the regenerated plants. These achievements give
optimism for further advancements.

Constraints in breeding

The main constraints in Hevea breeding are lack of reliable early
selection methods, long breeding and selection cycle, low fruit set etc.
Systematic efforts are essential to overcome these.

Early selection methods

Any reliable method for early prediction of yield and yield contributing
factors would result in quck release of cultivars as well as savings on time,
money and land. Realising the significance of juvenile selection, early
workers studied a number of parameters at the young stage in relation to
yield in mature plants. The parameters like girth, height, bark thickness,
latex vessel number, latex vessel and sieve tube diameters and rubber
hydrocarbon in bark and petiole showed very poor and inconsistant relationships (Tan, 1987). Cramer (1938) developed the 'testatex' method using
a special knife consisting of four 'V' shaped blades set one below the other
with which incisions could be made on one to two year old plants for making
qualitative assessment of the latex ozzing out. However, this method was
found useful only for culling (Cramer, 1938; Dijkman, 1951). Other pre-
selection methods, like use of perforating wheel (Meyer, 1950), a single
half spiral sloping cut (Ferrand, 1939) and needle prick test (Waidyanatha
and Fernando, 1972), also have limitations. The modified Hamaker Morris
Mann method (Tan and Subramaniam, 1976), the one widely adopted by many,
consists of successive tappings of two to three year old plants and quantify-
ing the latex yield. Premakumari et al (1989) found that the mean of yield
recorded in different seasons was more reliable than one recording alone.
Annamma et al. (1989b) suggested an incision method for prediction of
juvenile yield at the age of one year. Girth, bark thickness, number of
latex vessel rings, latex vessel density and first to third year yield are
used by breeders for preliminary selection in small scale trials with a
view to determining latex yield at a still younger age.

Fernando and de Silva (1971) using seeds of different clones reported an inverse relationship between oil content of cotyledons with growth and latex yield in some clones and suggested determination of oil content as a possible method for early selection of Hevea seedlings for growth and yield. NPK content (Ho, 1976), number of stomata (Senanayake and Samaranayake, 1970) and gas exchange parameters (Nugawela and Aluthewage, 1985) have also been suggested as early detection parameters. Huang et al. (1981) reported significant association between latex vessel, number of lateral vein and petioles of young clones with mature yield. Zhou et al. (1982) suggested petiolule latex method and lateral vein latex method for predicting rubber yield at the nursery stage. However, the ratio of the petiolule rubber content to dry weight of the middle leaflet suggests that this might be a parameter for early selection, but not a perfect one (Samsuddin and Mohd. Noor, 1988).

Studies on correlation between nursery yield and mature characters revealed only low to moderate correlations (Ho et al., 1985). Premakumari et al. (1989) observed a correlation coefficient of $r = 0.55$ ($P < 0.01$) between immature yield and yield of first year regular tapping, with a gradual fall between yields of later years, in a seedling population. Highly significant association of nursery yield and plugging index with mature yield have also been reported (Ho, 1976).

Nearly 80% of the variation in yield between clones at the nursery stage was accountable by bark thickness, number of latex vessel rings, girth increment and plugging index (Narayanan et al., 1974). In a study on association of characters in hand pollinated progenies, juvenile plant height, girth, latex vessel rows and bark thickness could explain 43% of the yield variability in the nursery stage (Licy and Premakumari, 1988). However, with the available early prediction methods, nursery yield can be considered as a fair predictor of mature yields (Ong et al., 1986; Licy et al., 1990). The present strategy adopted at RRII is to exercise only mild selection (around 20%) of the hybrid progenies based on juvenile characters for further testing, so that minimum loss of the potential high yielders is ensured (Annamma et al., 1990b).

Some attempts have also been made for early prediction of desirable secondary attributes. Nursery screening techniques for Gloeosporium disease (Wastie, 1973), a rapid laboratory method of assessing susceptibility of clones to Oidium (Lim, 1973), in vitro screening for resistance to Phytophthora leaf fall (Chee, 1969), petiolar stomatal characters for Phytophthora leaf fall (Premakumari et al., 1979, 1988), evaluation of clones to black stripe etc., are some of the examples. Quantity of epicuticular waxes and optical properties of leaves were suggested to be useful by Rao et al. (1988) for predicting drought resistance. Further multidisciplinary efforts are desirable to evolve useful early prediction parameters for yield and secondary attributes.

Breeding and selection cycle

With a view to shortening the breeding cycle, Subramaniam (1980) suggested an accelerated method of evaluation. In this trial system, the best one per cent of the hand pollinated progeny is selected based on juvenile yield, during the fifth year, followed by regional clone trials (promotion plot trials) of 50 plants per clone in two replications, in subsequent years. Early results reveal that a fair proportion of the tested clones are promising (Ong et al., 1986). In this method, though identification of a few promising clones at an early age is likely, the main reservation is that only a very small proportion is selected for evaluation. Through a drastic reduction in population size based on nursery yield, many potential yielders as well as best recombinants for secondary characters are not evaluated in a systematic manner. However, such new approaches are useful in identification of some promising clones at an early stage and hence may be considered as an adjunct to conventional breeding.

Markose and Panikkar (1984) suggested establishment of replicated field trials during the third year after hand pollination and task wise trials, if possible in different locations, during the 12th year. This would enable planting recommendation in 24-25 years and simultaneous with the selection process, progeny testing and genetic analysis for combining ability, heritability, genetic advance etc., can be done from this replicated trial.

Low fruit set

Other constraints in Hevea breeding are seasonal flowering pattern, lack of synchronization of flowering in different clones and low fruit set following hand pollinations. Hybridization programmes are carried out during the normal flowering season of January-March in India. Pollen storage could be explored as a means to solve the problem of non-synchronous flowering. An average fruit set of 3-5 per cent is met with instances of success rate ranging from 0-12 per cent within combinations involving the same female parent, indicating considerable influence of the male parent, have been reported (Kavitha et al., 1990b). Severe Oidium infection during the flowering season followed by fruit drop caused by Phytophythora, other factors like clonal difference in female fertility and injury caused to the female flowers, contribute to low fruit set. Application of extra nitrogen to increase fruit set was suggested by Sivanandan and Ghandhimathi (1986). Ghandhimathi and Yeang (1984) suggested a new method of hand pollination and reported insufficiency of pollen as a cause of low fruit set. Kavitha et al. (1989) compared different pollination treatments and found that the treatment using butter paper cover recorded significantly higher fruit set in comparison to other treatments. Even under these methods the extent of final fruit set is too inadequate and hence, further investigations are necessary. With a view to ensuring success of controlled pollination programmes, Kavitha et al. (1990b) suggested establishment of breeding orchards, proper management of trees with optimum doses of fertilizers, induction of off-season flowering to advance the flowering season and synchronise flowering among clones, pollination of 6-8 flowers per inflorescence to reduce fruit load and competition for assimilates and bagging of pollinated flowers with a butter paper cover to minimise injury to floral parts.

Breeding for disease resistance

In the different rubber producing countries, where any particular disease poses serious problems, some efforts in breeding for resistance have been made with varying degrees of success. In general, three leaf diseases viz., abnormal leaf fall (Phytophthora spp.), powdery mildew (Oidium heveae) and secondary leaf spot (Colletotrichum gloeosporoides) cause serious problems in Hevea. In India, abnormal leaf fall and powdery mildew results in varying degrees of yield drop. In Sri Lanka also Phytophthora and Oidium cause severe damage (Baptiste, 1961), while in Indonesia, Oidium is a major problem. Birds eye spot (Drechslera heveae) is reported from Ghana. White root disease is a serious problem in Nigeria. In RRIM nursery also, incidence of root disease has been reported. In Brazil, the hot spot of SALB, damages are also due to target leaf spot (Pellicularia filamentosa - Goncalves, 1968). Breeding for Colletotrichum and Oidium resistance are priorities in Malaysia and Indonesia respectively whereas, in India more emphasis is given for Phytophthora and Oidium resistance in the recent hybridization programmes.

Breeding for SALB (Microcyclus ulei) resistance should be given priority not only in Brazil, but also in other rubber producing countries. Fortunately for the plantation rubber industry of the old world, producing over 90% of the world's natural rubber requirement, this disease is still confined to the tropical Americas. However, in spite of the stringent quarantine measures adopted, the chances of the spread of this disease to the eastern hemisophere cannot be completely ruled out. Promising clones

(Fx 25, Fx 3899 and Fx 3164) were bred by Ford Motor Company in Brazil and by Fire Stone Rubber Company (MDF and MDx series) in Liberia and Guatemala (Edathil, 1986). Subsequently a large number of resistant clones were bred in Brazil, of which only six (Fx 25, 3810, 3899, 3925, IAN 710 and IAN 717) seem to be commercially acceptable. However, in many cases, the resistant clones selected were pathotype specific. Break down of resistance by occurrence of physiologic races (Langdon, 1965) is a serious problem and breeding for race nonspecific resistance should be aimed at. Simmonds (1989) stressed the need of breeding for horizontal resistance in contrast to vertical resistance to SALB. Similarly quantitative or partial resistance rather than qualitative or high level or absolute resistance offer durable resistance to fungal diseases in any crop species (Annamma, 1985; Annamma and Robbelen, 1984). In Hevea available results indicate polygenic inheritance, implying chances of obtaining horizontal resistance, which is more stable and durable. However, extensive studies on the genetic basis of resistance to SALB and other diseases are necessary so as to provide a sound basis for breeding. Simultaneously valuable source of wild genes rendering resistance to different diseases, available in the fresh wild germplasm should be identified, conserved and incorporated in breeding programmes.

Genetic studies

In Hevea the characters of economic importance are, in general, polygenically controlled. In order to ascertain the influence of genes and various nongenic factors in the expression of characters, detailed biometric analysis is required. This, however, is relatively scanty. Varying levels of heritability for yield have been reported.

Heritability (h^2) is hereditory or genotypic variance expressed as percentage of total variance and can be estimated as h^2 = VG/VP (broad sense) or VA/VP (narrow sense) where VG = total genetic variance, VP = phenotypic variance, VA = additive genetic variances. h^2 decreases with the increase in environmental component of variance for the character under observation. It estimates the degree of resemblance between offsprings and parents. Given an estimate of h^2, genetic advance under selection is predicted by R = $ih^2 \sigma p$ where σp is a dimensionless statistical parameter defined by the intensity of selection (Simmonds, 1989). Varying heritability estimates for yield, viz. low (0.21, Alika, 1982), medium (0.42, Liang et al., 1980) and hihg (0.82, Markose, 1984) have been reported.

Predominantly additive inheritance has been established for several seedling characters observed in the nursery (Tan and Subramaniam, 1976).

Statistically, combining abilities are more robust than heritability (Simmonds, 1989). Combining ability studies have revealed that yield and girth variation can be largely accounted for by additive genetic variation (Gilbert et al., 1973; Nga and Subramaniam, 1974). This suggests that phenotypic selection of parents would be generally effective, but selection based on general combining ability will be more precise and reliable. Simmonds (1969) reported that variation in GCA can account for most of the differences between family yields. Tan (1978) reported that GCA estimates on the basis of nursery seedlings can help early identification of promising parents.

Simmonds (1989) identified four correlations as important for planning breeding programmes. These are (1) between test tapping of young budded plants and performance in subsequence clone trials (r = 0.73, rising to 0.85, Ho, 1976), (2) between test tapping of young nursery seedlings and yield in small scale clone trials about 0.3 (Tan, 1987), (3) between yield in small scale clone trials and in subsequent large scale clone trials, and (4) between GCA estimates for seedlings and for the same genotype as clones after five years of tapping. Significant positive correlations of plant height and girth with juvenile yield of seedlings have been reported (Annamma et al., 1989a; Kavitha et al., 1990a).

Correlation coefficients worked out from breeding data have revealed that yield and girth are related to each other and in general positively correlated (Narayanan and Ho, 1973; Liu, 1980). Significant positive correlation of dry rubber yield with volume yield, latex vessel rows and virgin bark thickness have been reported (Narayanan et al., 1974; Wycherley, 1969). Markose (1984) also observed similar associations (Table 1). A negative association of girth and girth increment with rubber yield has been observed (Narayanan, 1973) for trees under tapping where plant assimilates are partitioned in favour of latex formation rather than growth, particularly in the case of high yielding clones. Path coefficient analysis for seven different characters by Markose (1984) revealed that volume of latex had the highest positive direct effect on dry rubber yield.

With the introduction of new amazonian germplasm (1981 collection) and another expedition to the centre of origin contemplated, the general problem regarding the narrow genetic base has been taken care of. Basic genetic studies and breeding strategies based on yield components, with the objective of evolving clones for different agro-climatic situations should prove to be rewarding. The emerging international co-operation in this area will further strengthen the efforts of Plant Breeders.

Table 1. Correlation coefficients between nine characters in twenty different clones (Markose, 1984).

Characters	1	2	3	4	5	6	7	8	9
1. Dry rubber yield	1.0000	0.9466*	0.5404*	0.5925*	0.1027	-0.2114	-0.2573	-0.2540	0.1118
2. Volume of latex/tree/tap		1.0000	0.4396*	0.4566*	0.0835	-0.3186	-0.3800	-0.3739	0.0798
3. Bark thickness			1.0000	0.7300*	0.2746	0.2061	0.1097	-0.1609	0.1605
4. Latex vessel rows				1.0000	0.1976	0.0577	-0.0143	-0.1625	0.3552
5. Dry rubber content					1.0000	0.1813	0.2049	0.1731	-0.2670
6. Girth (1981)						1.0000	0.9577*	0.4712*	0.0129
7. Girth (1982)							1.0000	0.7047*	-0.0202
8. Girth increment								1.0000	0.0762
9. Branching height									1.0000

* Significant at 5% level.

REFERENCES

Alika, J.E. 1982. Preliminary assessment of some hand pollinated progenies of rubber in Nigeria. Indian J. Agric. Sci., 52(6): 367-369.

Allen, P.W. 1984. Fresh germplasm for natural rubber. Span, 27(1): 7-8.

Annamma Varghese, Y. 1985. Selection of mutants showing partial resistance to powdery mildew (Erisiphe graminis, DC. f. sp. hordei Marchal) in barley (Hordeum vulgare L.) after Sodium azide mutagenesis. Indian J. Genet., 45(1): 57-66.

Annamma Varghese, Y. and Robbelen, G. 1984. Sectorial chimeras as a proof of the de novo origin of mutants showing partial resistance to powdery mildew in barley. Z. Pflanzenzuchtg. 92: 265-280.

Annamma, Y., Marattukalam, J.G., Premakumari, D. and Panikkar, A.O.N. 1988. Nursery evaluation of 100 Brasilian genotypes of Hevea in India. Colloque Hevea 88, IRRDB, Paris. 353-364.

Annamma, Y., Marattukalam, J.G., George, P.J. and Panikkar, A.O.N. 1989a. Nursery evaluation of some exotic genotypes of Hevea brasiliensis Muell. Arg. J. Plant. Crops, 16 (suppliment). 335-342.

Annamma, Y., Licy, J., Alice John and Panikkar, A.O.N. 1989b. An incision method for early selection of Hevea seedlings. Indian J. Nat. Rubb. Res., 2(2): 112-117.

Annamma, Y., Joseph G Marattukalam, Premakumari, D., Saraswathy Amma, C.K., Licy, J. and Panikkar, A.O.N. 1990a. Promising rubber planting materials with special reference to Indian clones. Planters' Conference, India, 1990. 62-70.

Annamma, Y., Licy, J. and Panikkar, A.O.N. 1990b. Genetic improvement in Hevea: Achievements, problems and perspectives. National Symposium on New Trends in Crop Improvement of Perennial Species, Kottayam, India.

Asokan, M.P., Sobhana, P., Sushamakumari, S. and Sethuraj, M.R. 1988. Tissue culture propagation of rubber (Hevea brasiliensis) (Willd. ex. Adr. de Juss.) Muell. Arg. clone GT (Gondang Tapen) 1. Indian J. Nat. Rubb. Res. 1(2): 1-9.

Balaj, Y.P.S. 1986. In vitro preservation of genetic resources; Techniques and problems. In: Nuclear techniques and in vitro culture for plant improvement. Proc. IAEA/FAO Symp. Vienna, 1985: 49-57.

Baptiste, E.D.C. 1961. Breeding for high yield and disease resistance in Hevea. Proc. Nat. Rubb. Res. Conf., Kuala Lumpur 1960, 430-445.

Brookson, E.V., 1956. Importation and development of new strains of Hevea brasiliensis by the Rubber Research Institute of Malaya. J. Rubb. Res. Inst. Malaya, 14: 423-448.

Chee, K.H. 1969. Phytophthora leaf disease in Malaysia. J. Rubb. Res. Inst. Malaya, 21: 79-87.

Chen, Z., Quian, L., Quin, M., Xu, X. and Xiao, Y. 1982. Recent advances in anther culture of Hevea brasiliensis (Muell. Arg.) Theor. Appl. Genet. 62: 103-108.

Chevallier, M.H. 1988. Genetic variability of Hevea brasiliensis germplasm using isozyme markers. J. Nat. Rubb. Res., 3: 42-53.

Chevallier, M.H., Lebrun, P. and Normand, F. 1988. Approach to the genetic variability of germplasm using enzymatic markers. Colloque Hevea 1988, IRRDB, Paris, 365-376.

Compagnon, P. 1977. Note on Brasilien clones obtained from Brazil. Workshop on international collaboration in Hevea breeding and the collection and establishment of materials from neo-tropics, Kuala Lumpur, Malaysia, 1977.

Cramer, P.J.S. 1938. Grading young rubber plants with the 'Testatex' knife. Proc. Rubb. Technol. Conf., London. 10-16.

Demange, C.A. 1988. Agronomic variability of germplasm in Cote d' Ivoire. Colloque Hevea 1988, IRRDB, Paris, 403-421.

Dijkman, M.J. 1951. Hevea, Thirty years of research in the Far East. University of Miami Press, Florida.

Edathil, T.T. 1986. South American Leaf Blight - a potential thread to the natural rubber industry in Asia and Africa. Tropical Pest Management, 32(4): 296-303.

Fernando, D.M. and de Silva, M.S.C. 1971. A new basis for the selection of Hevea seedlings. Q. J. Rubb. Res. Inst. Ceylon, 48: 19-30.

Ferrand, M. 1939. Nursery selection in Hevea. Ann. Rep. INEAC, 2: 99-113.

Ghandimathi, H. and Yeang, H.Y. 1984. The low fruit set that follows conventional hand pollination in Hevea brasiliensis: insufficiency of pollen as a cause. J. Rubb. Res. Inst. Malaysia, 32: 20-29.

George, P.J., Bhaskaran Nair, V.K. and Panikkar, A.O.N. 1980. Yield and secondary characters of RRII 105 in trial plantings. IRC, India, 1980.

Gilbert, N.E., Dodds, K.S. and Subramaniam, S., 1973. Progress of breeding investigations with Hevea brasiliensis V. Analysis of data from earlier crosses. J. Rubb. Res. Inst. Malaya, 23: 365-380.

Gonvalves, J.R.C. 1968. The resistance of Fx and IAN rubber clones to leaf diseases in Brazil. Trop. Agriculture, Trin., 45: 331-336.

Goncalves, P. de. 1982. Collection of Hevea materials from Rondonia Territory in Brazil: A preliminary study. Pesq. agropec. bras. Brasilia, 17(4): 575-582.

Goncalves, P. de., Valois, A.C. and Paiva, J.R. de. 1983. Induction and investigation of polyploidy in IAN 717 rubber tree clone. A preliminary study. pesq. agropec. bras. Brasilia, 18(1): 789-796.

Ho, C.Y., Chan, H.Y. and Lim, T.M. 1974. Enviromax planting recommendation - a new concept in choice of clones. proc. Rubb. Res. Inst. Malaysia Plrs' Conf., Kuala Lumpur, 293-320.

Ho, C.Y. 1976. Clonal characters determining the yield of Hevea brasiliensis. Proc. Int. Rubb. Conf., Kuala Lumpur, (2): 17-38.

Huang, X., Wei, L., Zhan, S., Chen, C., Zhou, Z., Yuen, X., Guo, Q. and Lin, J. 1981. A preliminary study of relations between latex vessel system of rubber leaf blade and yield prediction at nursery. Chinese J. Trop. Crops, 2(1): 16-20.

Hu Han, 1984. Crop improvement by anther culture. In: Genetics: New Frontiers. Proc. XV Int. Congress of GEnetics, Vol. IV (Eds. Chopra, V.L., Joshi, B.C., Shrma, R.P. and Bansal, H.C.). 77-81.

IBPGR, 1984. Genetic resources of Hevea. Report of an IBPGR working group. pp.14.

IBPGR, 1984. Annual Report. p. 73.

IRRDB, 1987. Disease warnings. Supplement to IRRDB News Letter, No. 4.

Jayasekera, N.E., Samaranayake, P. and Karunasekara, K.B. 1977. Initial studies on the nature of genotype environment interactions in some Hevea cultivars. J. Rubb. Res. Inst., Sri Lanka, 54: 33-42.

Kavitha K. Mydin, Nazeer, M.A., Licy, J., Annamma, Y. and Panikkar, A.O.N. 1989. Studies on improving fruit set following hand pollination in Hevea brasiliensis (Willd. ex. Adr. de Juss.) Muell. Arg., Indian J. Nat. Rubb. Res., 2(1): 61-67.

Kavitha K. Mydin, Gopinathan Nair, V., Panikkar, A.O.N., Saraswathy, P. and Sethuraj, M.R. 1990a. Prepotency in Rubber. 1. Early estimation through juvenile traits. National Symposium on New Trends in Crop Improvement of Perennial Species, Kottayam, India.

Kavitha K. Mydin, Annamma, Y., Nazeer, M.A., Premakumari, D., Saraswathy Amma, C.K. and Panikkar, A.O.N. 1990b. Controlled pollinations in Hevea - problems and perspectives, IRRDB Breeding Symposium, China.

Langdon, K.R. 1965. Relative resistance or susceptibility of several clones of Hevea brasiliensis and Hevea brasiliensis x Hevea benthamiana to two races of Dothidella ulei, Plant Disease Reporter, 49: 12-14.

Larkin, P.J. and Scow croft, W.R. 1981. Somaclonal variation - a noval source of variability from cell culture for plant improvement. Theor. Appl. Genetc., 60: 197-214.

Liang, M., Yuntong, W., Dongquiong, H., Dehe, Z., Deshum, L., Sengxian, W., Zhnocai, C. and Jialin, F. 1980. Preliminary analysis of genetic parameters of some characters of Hevea seedling. Chinese J. Trop. Crops. 1: 50-52.

Licy, J. and Premakumari, D., 1988. Association of characters in Hand Pollinated Progenies of Hevea brasiliensis (Willd. ex. Adr. de Juss.) Muell. Arg. Indian J. Nat. Rubb. Res., 1(1): 18-21.

Licy, J., Nazeer, M.A., Annamma, Y., Panikkar, A.O.N. 1990. Correlation studies in a hybrid population of Hevea brasiliensis (Willd. ex. Adr. de Juss.) Muell. Arg. National Symposium on New Trends in Crop Improvement of Perennial Species, Kottayam, India.

Lim, T.M. 1973. A rapid laboratory method of assessing susceptibility of Hevea clones to Oidium heveae. Exptl. AGric., 9: 275-279.

Lim, S.C., Ho, C.Y. and Yoon, P.K. 1973. Economics of maximising early yields and shorter immaturity. proc. RRIM Planters' Conference, Kuala Lumpur. 1-16.

Ling, X., Chen, G. and Yang, T. 1988. A method to increase polyploid induction rate in Hevea brasiliensis. Chinese J. of Tropical Crops, 9(2): 9-16.

Liu, N. 1980. A review of quantitative genetic studies on Hevea by RRIM. Chinese J. Trop. Crops. 1: 41-42 (Abs.).

Marattukalam, J.G., Saraswathy Amma, C.K. and George, P.J. 1980. Crop improvement through ortet selection in India, IRC, India (Abs.).

Markose, V.C. 1984. Biometric analysis of yield and certain yield attributes in the para rubber tree; Hevea brasiliensis Muell. Arg. Ph.D. Thesis, Kerala Agricultural University, Vellayani, India.

Markose, V.C. and Panikkar, A.O.N. 1984. Breeding strategies for Hevea improvement. Compte-Rendu du Colloque Exploitation Physiologie et Amelioration de l' Hevea, Paris, 367-373.

Markose, V.C., Sulochanamma, S. and Bhaskaran Nair, V.K. 1974. Mutation and polyploidy in Hevea brasiliensis Muell. Arg. IRRDB Sci. Symp. Cochin, India. 107-116.

Markose, V.C., Panikkar, A.O.N., Annamma, Y. and Bhaskaran Nair, V.K. 1977. Effect of gamma rays on rubber seed germination, seedling growth and morphology. J. Rubb. Res. Inst. Sri Lanka, 54: 50-64.

Markose, V.C., Saraswathy Amma, C.K., Licy, J. and George, P.J. 1981. Studies on the progenies of a Hevea mutant. Proc. PLACROSYM IV. Mysore, India, 58-61.

Meyer, W.H. 1950. Hevea selection in the nursery by means of the perforation method. Bergcultures, 19: 71-79.

Mohd. Noor, A.G. and Ibrahim, N. 1986. Characterization and evaluation of Hevea germplasm. SABRAO Symposium, Malaysia, 1986.

Naijian, L. and Jingxiang, O., 1990. An improved breeding system of Hevea brasiliensis. IRRDB Breeding Symposium, Kunming, China.

Nair, V.K.B. and George, P.J. 1969. The Indian clones: RRII 100 series, Rubber Board Bulletin, 10(3): 115-139.

Nair, V.K.B. and Panikkar, A.O.N. 1966. Progress of investigation on the improvement of rubber (Hevea brasiliensis Muell. Arg.) in India. Rubber Board Bulletin, 8: 201-210.

Nair, V.K.B., George, P.J. and Saraswathy Amma, C.K. 1976. Breeding improved Hevea clones in India. Proc. Int. Rubb. Conf., Kuala Lumpur, V. 4: 45-54.

Narayanan, R. and Ho, C.Y. 1973. Clonal nursery studies in Hevea II. Relationship between yield and girth. J. Rubb. Res. Inst. Malaysia, 23: 332-338.

Narayanan, R., Gomez, J.B. and Chen, K.T. 1973. Some structural factors affecting the productivity of Hevea brasiliensis. J. Rubb. Res. Inst. Malaysia, 23(4): 285-297.

Narayanan, R., Ho, C.Y. and Chen, K.T. 1974. Clonal nursery studies in Hevea. III. Correlation between yield, structural characters, latex constituents and plugging index. J. Rubb. Res. Inst. Malaysia, 24: 1-14.

Nga, B.H. and Subramaniam, S. 1974. Variation in Hevea brasiliensis. I. Yield and girth data of the 1937 hand pollinated seedlings. J. Rubb. Res. Inst., Malaysia, 24: 69-74.

Nazeer, M.A., Markose, V.C., George, P.J. and Panikkar, A.O.N. 1986. Performance of a few Hevea clones from RRII 100 series in large scale trial. J. Plant. Crops, 14(2): 99-104.

Nazeer, M.A. and Saraswathy Amma, C.K. 1987. Spontaneous Triploidy in Hevea brasiliensis (Willd ex Adr. de Juss.) Muell. Arg. J. Plant. Crops. 15(1): 69-71.

Nugawela, A. and Aluthewage, 1985. Gas exchange parameters for early selection of Hevea brasiliensis Muell. Arg. J. Rubb. Res. Inst., Sri Lanka, 64: 13-20.

Olapade, E.O. 1988. General and specific combining abilities for latex yield in Hevea brasiliensis. Colloque Hevea 1988, IRRDB, Paris, 403-421.

Ong, S.H., Mohd. Noor, A.G., Tan, A.M. and Tan, H. 1983. New Hevea germplasm - its introduction and potential. Proc. RRIM Plrs' Conf., Kuala Lumpur, 3-17.

Ong, S.H. and Subramaniam, S. 1973. Mutation breeding in Hevea brasiliensis Muell. Arg. In: Induced Mutations in Vegetatively Propagated Plants, IAEA, Vienna, 117-127.

Ong, S.HY. and Tan, H. 1987. Utilisation of Hevea genetic resources in the RRIM. Mal. Appl. Biol., 16: 145-155.

Ong, S.H., Tan, H., Khoo, S.K. and Sultan, M.O. 1986. Selection of promising clones through accelerated evaluation of Hevea. Proc. IRC 1985, Kuala Lumpur, 157-174.

Paardekooper, E.C. 1964. Report on the RRIM 600 series distributed block trials. (Group A1 trials). Rubb. Res. Inst. Malaya, Res. ARch. Doc., No. 35.

Paranjothy, K. and Ghandimathi, H. 1975. Tissue and Organ Culture of Hevea. Proc. IRC 1975, Kuala Lumpur, 59-84.

Premakumari, D., Annamma, Y. and Bhaskaran Nair, V.K. 1979. Clonal variability for stomatal characters and its application in Hevea breeding and selection. Indian J. agric. Sci., 49: 411-413.

Premakumari, D., George, P.J., Panikkar, A.O.N., Bhaskaran Nair, V.K. and Sulochanamma, S. 1984. Performance of RRII 300 series clones in the small scale trial. Proc. PLACROSYM-V, 148-157.

Premakumari, D., George, P.J. and Panikkar, A.O.N. 1989. An attempt to improve test tapping in Hevea seedlings. J. Plant. Crops. 16 (supplement), 383-387.

Premakumari, D., Panikkar, A.O.N. and Sethuraj, M.R. 1988. Correlations of the characters of petiolar stomata with leaf retention after the incidence of Phytophthora leaf fall disease in Hevea brasiliensis (Willd ex Adr. de Juss.) Muell. Arg. Indian J. Nat. Rubb. Res., 1(1): 22-26.

Rao, G.G.R., Devakumar, A.S., Rajagopal, R., Annamma, Y., Vijayakumar, K.R. and Sethuraj, M.R. 1988. Clonal variation in leaf epicuticular waxes and reflectance: Possible role in drought tolerance in Hevea. Indian J. Nat. Rubb. Res., 1(2): 84-87.

RRIM, 1982. Mother tree selection. Plrs'. Bulletin, 171: 31-32.

Samsuddin, Z. and Mohd. Noor, A.G. 1988. A preliminary study on the ratio of the petiolule rubber content to the leaflet's dry weight and its relationship to nursery yield of Hevea. Colloque Hevea 1988, IRRDB, 469-478.

Satchuthananthavale, R. 1974. Hormonal control of organ formation in explants of Hevea stem sections. IRRDB Sym. Part I, Cochin, India.

Saraswathy Amma, C.K., George, P.J., Nair, V.K.B. and Panikkar, A.O.N. 1980a. RRII 200 series clones. IRC, India, 1980.

Saraswathy Amma, C.K., Licy, J. and Markose, V.C. 1983. Effect of gamma rays on Hevea seeds and the resultant seedlings. Int. Genet. Congress. New Delhi.

Saraswathy Amma, C.K., Markose, V.C., Licy, J., Annamma, Y., Panikkar, A.O.N. and Nair, V.K.B. 1980b. Triploidy in Hevea brasiliensis Muell. Arg. IRC, India, 1980.

Saraswathy Amma, C.K., Markose, V.C., Licy, J. and Panikkar, A.O.N. 1984. Cytomorphological studies in an induced polyploid of Hevea brasiliensis Muell. Arg. Cytologia, 49: 725-729.

Schultes, R.E. 1977. Wild Hevea - an untapped source of germplasm. J. Rubb. Res. Inst., Sri Lanka, 54: 1-31.

Schultes, R.E. 1987. Studies on the genus Hevea. VIII. Notes on intraspecific variants of Hevea brasiliensis (Euphorbiaceae). Economic Botany, 41(2): 125-147.

Senanayake, Y.D.A. and Samaranayake, P. 1970. Intraspecific variation of stomatal density in Hevea brasiliensis Muell. ARg. Q. J. Rubb. Res. Inst., Ceylon, 46: 61-68.

Sethuraj, M.R. 1981. Yield components in Hevea brasiliensis. Theoretical considerations. Plant Cell and Environ., 4: 81-83.

Sethuraj, M.R. 1983. Yield component concept in improving productivity in Hevea brasiliensis. IRRDB Sci. Symp. Beijing, China, 1983.

Sethuraj, M.R. 1987. Rubber. In: Tree Crop Physiology (Eds. M.R. Sethuraj and A.S. Raghavendra). Elsevier Science Publishers Ltd. Netherlands, 193-223.

Shepherd, R. 1969. Aspects of Hevea breeding and plant selection investigation undertaken on Prang Besar estate. Plrs' Bull. Rubb. Res. Inst., Malaya, 104: 207-219.

Shijie, Z., Sherghua, C. and Xueng, X. 1990. A summary report on anther culture for haploid plants of Hevea brasiliensis. IRRDB Breeding Symposium, Kunming, China.

Simmonds, N.W. 1969. Genetical basis of plant breeding. J. Rubb. Res. Inst. Malaya, 21: 1-10.

Simmonds, N.W. 1986. The strategy of rubber tree breeding. Proc. IRC, Kuala Lumpur, 1985. 3: 115-126.

Simmonds, N.W. 1989. Rubber breeding. In: Rubber (Eds. Webster, C.C. and Baulkwill, W.J.). Longman Singapore Publishers. 85-124.

Sinha, R.R., Sobhana, P. and Sethuraj, M.R. 1985. Auxilary buds of some high yielding clones of Hevea in culture. IRRDB Workshop on Tissue Culture, 1985, Kuala Lumpur, malaysia.

Sivanandan, K. and Ghandimathi, H. 1986. Mineral nutrition and reproduction in Hevea: effects of nitrogen fertilization on flowering and fruiting in immature trees of some clones. J. Nat. Rubb. Res. 1, 155-166.

Sneep, J. 1979. In: Proceedings, Broadening the Genetic Base of Crops, Wageningen. 343-344.

Subramaniam, S. 1980. Developments in Hevea breeding research and their future. Seminario Nacional da Seringueira, 3. Manaus.

Swaminathan, M.S. 1977. Recent trends in plant breeding with special reference to the improvement of the yield potential of rubber. J. Rubb. Res. Inst. Sri Lanka (1977) 54: 11-16.

Tan, H. 1978. Estimates of parental combining abilities in rubber based on young seedling progeny, Euphytica 27: 817-823.

Tan, H. 1987. Strategies in rubber tree breeding. Improving vegetatively propagated crops (Es. Abbot and Atkin). Academic Press, London, 27-62.

Tan, H. and Subramaniam, S. 1976. Combining ability analysis of certain characters of young Hevea seedlings. Int. Rubb. Conf. 1975. Kuala Lumpur II: 13-26.

Van Helten, W.M. 1918. Het oculeeren van Hevea. Archis. Rubb. Cult. Ned-Ind. 2, 187-194.

Waidyanatha, U.P. de S. and Fernando, D.M. 1972. Studies on a technique of micro-tapping for the estimation of yields in nursery seedlings of Hevea brasiliensis. Q. J. Rubb. Res. Inst., Ceylon, 49: 6-12.

Wastie, R.L. 1973. Nursery screening of Hevea for resistance to Gloeosporium leaf disease. J. Rubb. Res. Inst., Malaysia, 23: 339-350.

Whitby, 1919. Variation in Hevea brasiliensis. Annals of Botany, 33: 313-321.

Wijewantha, R.T. 1965. Some breeding problems in Hevea brasiliensis. Q. J. Rubb. Res. Inst., Ceylon, 41(1-2): 12-22.

Wyucherley, P.R. 1968. Introduction of Hevea to the Orient. Plrs' Bull. RRIM, 44: 127-137.

Wycherley, P.R. 1969. Breeding of Hevea. J. Rubb. Res. Inst., Malaya, 21: 38-55.

Wycherley, P.R. 1977. Motivation of Hevea germplasm collection and conservation. Workshop on international collaboration in Hevea breeding and the collection and establishment of materials from the neo-tropics, Kuala Lumpur.

Zhou, Z., Yuan, X., Guo, Q. and Huang, X. 1982. Studies on the method for predicting rubber yield at the nursery stage and its theoretical basis. Chinese J. Trop. Crops. 3(2): 1-18.

CHAPTER 6

METABOLISM OF THE LATICIFEROUS SYSTEM AND ITS BIOCHEMICAL REGULATION

J.L. JACOB and J.C. PREVOT

Institut de Recherches sur le Caoutchouc, Avenue du Val de Montferrand,
BP 5035, 34032 Montpellier cedex, France.

LATEX, THE CYTOPLASM OF THE LATICIFERS

Biochemically, latex is true cytoplasm. Latex contains most of the subcellular elements and its metabolic activities can be studied in vitro. The subcellular particles include, besides rubber particles, lutoids, an important vacuolysosomal compartment (Pujarniscle, 1968; Ribaillier et al., 1971), plastids, the Fre-Wyssling particles whose role is not clearly understood (Gomez, 1979; Hebant, 1981) and ribosomes (Coupe et al., 1976). However, neither the nuclei nor the mitochondria, which are visible in situ under the microscope, are expelled during tapping, probably because of their parietal position (Dickenson, 1965) and because of this, biochemical studies concerning nuclear and energy metabolism is difficult in vitro.

THE METABOLISM OF THE LATICIFEROUS SYSTEM AND ITS IMPORTANCE IN PRODUCTION

Tapping causes loss of cell constituents from the laticifers. When flow stops as a result of the complex phenomena which lead to the coagulation of rubber particles and plugging of the wound (Southorn, 1969), regeneration of the latex lost becomes necessary. This involves intense metabolic activity. If there is a sufficiently long interval between two tappings, this regeneration can be complete and the intralaticiferous metabolism then slows down (Jacob et al., 1988a). Although metabolism plays a major role in production through the reconstitution of latex in the laticiferous tissue, factors regulating the latex flow also determines the amount of latex loss and the subsequent rate of catabolic activities (Sethuraj and Raghavendra, 1987). Transport of water and solutes to the laticiferous system also requires biochemical energy (Jacob et al., 1988b).

Isoprenic production, the "royal pathway" of the laticiferous metabolism

The biochemical composition of latex (Compagnon, 1986) shows very clearly that the "royal metabolic pathway" of the laticiferous system is the synthesis of rubber, which forms 35-45 per cent of fresh weight and over 90% of dry weight of latex. All the enzymatic processes are thus coordinated and arranged to result in the biosynthesis of rubber.

Schematic representation of rubber production

Hevea rubber is a macromolecule formed by chains of 5-carbon isoprenic units (Bouchardat, 1875). These units are the precursor of numerous other natural isoprenic substances (sterols, carotenoids, etc.). There may be as many as 10,000 of these units in Hevea rubber (Audley and Archer, 1988). A close study of its structure has shown that the isoprenic bonds are mainly of the cis form; less than 0.2% are in the trans form and these make the first "geranyl geranyl" links in the polyisoprene chain (Archer et al., 1982; Audley and Archer, 1988). According to Kekwick (1988), the number average molecular weight ($\bar{M}_n$) is between 200,000 and 800,000. The weight average molecular weight ($\bar{M}_w$) is frequently bimodal and can be as high as 1,800,000 to 2,000,000. The initial link is produced in situ in the biochemically active isopentenyl pyrophosphate (IPP) (Lynen, 1963).

Numerous investigations have shown that acetate is the most simple initial precurser of IPP (Bandurski and Teas, 1957; d'Auzac, 1965). In Hevea, this acetate molecule is produced mainly by catabolism of sugars (Jacob, 1970; Tupy, 1973). Although Bealing (1975) put forward the hypothesis that cyclitols, which are found in fairly large quantities in latex, might also be the source, it is certain that the sucrose produced by photosynthesis and transported by phloem pathways to the laticifers is the main precursor of acetate and hence of rubber (Lynen, 1969).

The synthesis of cis-polyisoprene can thus be divided schematically into two distinct phases (Fig. 1):

- the first is the conversion of sugars into acetate. It simultaneously provides energy in the form of ATP and reducing power in the form of NAD(P)H;
- the second is isoprenic synthesis itself. It requires ATP, reducing power in the form of NAD(P)H and acetate or acetyl CoA to enable the building of IPP.

Glycolysis

Glucidic catabolism in latex is essentially glycolysis (Jacob, 1970).

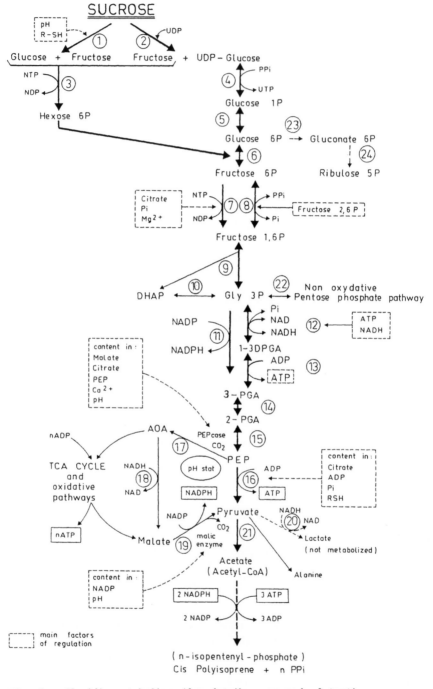

Fig. 1. Glucidic catabolism (for details, see end of text).

It not only generates the energy required for the process of synthesis (particularly that of rubber) but also produces NADH or NAD(P)H (Fig. 1), the latter cofactor is required for isoprenic biogenesis (Lynen, 1969). The sequence is entirely cytosolic and certain stages control it activity. This is the case of invertase (Tupy, 1973; Jacob, 1970). The enzymes involved in the synthesis of sucrose are: pyrophosphate; fructose-6-phosphate 1-phosphotransferase (PP-PFK), UDPG pyrophosphorylase and sucrose synthetase (Tupy and Primot, 1982). These enzymes can also effectively slow down glycolytic activity (Tupy and Primot, 1982; Tupy, 1988a).

Pyruvate-acetate or acetyl CoA transformation is essential between glucidic catabolism in latex and isoprenic synthesis. However, other metabolic pathways are connected with glycolysis and can lead to pathways other than that of rubber production. These directions are nevertheless very partial and severely controlled; the diversion by PEPase to the synthesis of organic acids and of the energy metabolism is one such example (Jacob et al., 1988). The oxidative part of the pentose cycle from glucose-6-phosphate (G6P) potentially exists but is probably not functional in situ owing to insufficient NADPH. In contrast, there is no doubt that its non-oxidative part from dihydroxyacetone phosphate (DHAP), capable of producing phosphate riboses linked to the nucleic acid metabolism, is able to function (Arreguin and Rock, 1967; Jacob et al., 1988a).

Isoprenic anabolism

Isoprenic anabolism is the major synthetic process in latex since over 90% of dry matter is cis-polyisoprene. The general pattern (Fig 2) of this synthesis is now known (Lynen, 1969; Kekwick, 1988); it consists of the linking of the initial monomers, isopentenyl pyrophosphate (IPP) and dimethylallyl pyrophosphate (DMPP) from the acetate molecule produced by glycolysis.

The first stage is the activation of the acetate molecule into acetyl CoA by an acetyl thiokinase present in serum C (Jacob et al., 1988a).

The second stage is the condensation of two molecules of acetyl CoA into acetoacetyl CoA (Lynen, 1969); the reaction equilibrium is orientated mainly towards the splitting of acetoacetyl CoA.

The third stage is the production of β-hydroxyl-β-methylglutaryl Coenzyme A (HMG CoA) by means of a HMG-CoA synthetase combining a molecule of acetoacetyl CoA and a molecule of acetyl CoA. This reaction is probably irreversible (Lynen, 1969), which would explain the possible balance of the previous reaction towards systhesis of acetoacetyl CoA in spite of its unfavourable reaction equilibrium.

Conversion of HMG-CoA into mevalonic acid (MVA) is probably very important in the regulation of isoprenic anabolism. It corresponds to a reduction which requires the specific presence of two molecules of NADPH by a HMG-CoA reductase (Hepper and Audley, 1969; Sipat, 1982; Wititsuwanna-kul, 1986). The enzyme appears to be located on membrane structures.

Mevalonic acid is then activated as phosphomevalonic acid by a cytosolic mevalonate kinase (Skilleter et al., 1966). The next stage consists of fresh activation of phosphomevalonate into pyrophosphomevalonate (PPMVA) by a phosphomevalonate kinase (Williamson and Kekwick, 1968).

Production of the rubber monomer, isopentyl pyrophosphate, takes place from PPMVA by means of a decarboxylase and is accompanied by dehydration of the molecule. This has been studied in Hevea by Chesterton and Kekwick (1968).

The formations of polyisoprene chain has been reviewed recently by Audley and Archer (1988). Its initiation cannot take place without the formation of an electronically stable initial isomer molecule; dimethylallyl pyrophosphate (PPDMA) or a similar allylic compound which can be complexed with a glyco or lipoprotein. The isomerase which is responsible for this reaction is found at the surface of rubber particles (Lynen, 1969) but may also be present in the cytosol (Audley and Archer, 1988).

The primary isoprenic unit is then built up to stage C_{20}, totally or partly in the trans form (geranyl geranyl pyrophosphate). The chain is then lengthened by the terminal addition of IPP units (up to 10,000 units) by means of a transferase which is also at the surface of the rubber particles. However, the question of the origin of the first isoprenic particle remains to be resolved (Archer and Audley, 1987).

The metabolic problem posed by rubber synthesis

Three questions related to rubber synthesis need to be clarified: the production of acetate or acetyl-Coenzyme A, the energy supply and the regeneration of the reduction potential required for the synthesis of the polymer.

Production of acetate or acetyl-Coenzyme A

This depends essentially on the functioning of glycolysis. The factors liable to modifications in the glucidic catabolism have an effect on isoprenic anabolism and hence on rubber production.

The pyruvate-acetyl CoA transformation is still not well understood and may be a limiting factor in polyisoprene regeneration. Indeed, incorporation of ^{14}C pyruvate in rubber is always lower than that of ^{14}C acetate or of ^{14}C acetyl CoA in in vitro studies (Kekwick, 1988).

Archer and Audley (1967) consider that there may be two pools of acetyl-CoA in the laticiferous tissue, one mitochondrial pool which would be oxidised preferentially in these organelles and a cytosol pool which would be used mainly in isoprenic anabolism. A cytosolic pyruvate decarboxylase has also been revealed and indicates that acetate is produced from pyruvate (Tupy and Primot, 1976).

Energy supply

On the basis of theoretical calculations, d'Auzac (1965) showed that glycolytic functioning alone would be sufficient to cover the energy requirement involved in the synthesis of rubber. However, other processes also require energy and in particular ATPase transmembrane transfers (d'Auzac, 1988) and the other cell syntheses which occur at the same time as isoprenic anabolism; these notably include synthesis of proteins and fatty acids connected with the regeneration of enzyme molecules and subcellular structures. It is probable that metabolic pathways such as oxidative phosphorylation producing ATP, may also participate in the production of biochemical energy in the laticifers (Tupy and Primot, 1976). However, it is difficult to evaluate the activity in vitro since the mitochondria, which are the sites of these reactions, remain in the laticiferous tissues when the trees are tapped (Dickenson, 1965).

Reduction potential

This is essentially linked with the availability of NADPH, a specific cofactor in the functioning of HMG-CoA reductase (Wititsuwannakul et al., 1986) which catalyses a limiting stage in isoprenic anabolism (Lynen, 1969). The very low NADP content in cytosol (less than 1 μm) (Lorquin et al., 1987) should be emphasized. In addition to the lutoid phosphatases which can hydrolyse NADP(H) into NAD(H) (Jacob and Sontag, 1974), there is a cytosol hydrolase which is specific to the cofactor: 2'nucleotidase (Jacob and Sontag, 1973). A NAD kinase, likely to synthesize NADP, has been found in latex (Lorquin et al., 1987). This is membrane bound in relation to particles of the heavy fraction of latex similar to that of HMG-CoA reductase (Hepper and Audley, 1969). The very low availability in the cytosol of the cofactor already mentioned leads to the hypothesis of a compartmentalisation of the reactions directly connected with the production of HMG-CoA. Nevertheless, the cytosolic dehydrogenases which regenerate NADPH from its oxidised form are present in the cytosol: a fructose-6-phosphate dehydrogenase (Arreguin and Rock, 1967), a non-phosphorylating glyceraldehyde 3-phosphate dehydrogenase which is part of glycolysis (Jacob and d'Auzac, 1971) and above all a malic enzyme (with a very low Km value) (Jacob and Prevot, 1981) having very high affinity for NADP and with very low Km value.

Metabolic processes other than the isoprenic metabolism

Although the isoprenic metabolism dominates among the overall metabolic pathways in laticiferous tissue, the other pathways nevertheless add to the complexity of the metabolism in laticifers. A review of the knowledge acquired in this domain has been published recently (Jacob et al., 1988b) covering the synthesis of quebrachitol, quantitatively the most important molecule after rubber and probably having a role in the osmoticum in laticiferous tissue (Low, 1978), synthesis of proteins whose functioning is crucial for the regeneration of latex between two tappings, synthesis of free isoprene compounds which can compete with isoprenic anabolism, and synthesis of lipids which play a major role in the membrane structure of cell organelles.

FACTORS REGULATING METABOLISM IN LATEX

Studies on latex metabolism have revealed the significant influence of certain parameters on regulation and hence on the production of latex.

Availability of sucrose in laticiferous tissue

The sugar content of latex is the result of carbohydrate loading to laticiferous tissue and its use at cell level. In a healthy tree, the availability of sucrose in latex is a prime factor in the metabolic activity of the laticifers and of production (Tupy, 1988a). Indeed sucrose catabolism supplies the acetate molecules which initiate the isoprene chain and provide the biochemical energy necessary for the functioning of the laticifers (Jacob et al., 1988b). Positive, highly significant correlations have been established between sugar concentration and latex production (Jacob et al., 1986).

Sugar supply

Sucrose loading of the laticifers is thus an extremely important phenomenon. Recent work has shown the existence of an intermembrane transport process which requires energy. It operates at laticifer plasmalemma level and involves a contrasport H^+-sucrose energized by an electrochemical proton gradient set up by an ATPase proton pump (Lacrotte et al., 1985; Lacrotte et al., 1988a,b).

Levels and compartmentalisation in latex of various ions, organic acids and thiols.

Contents of ions such as Mg^{2+}, PO^- or other products such as citrate, thiols (R-SH) and their distribution in the latex cytosol or lutoid compartments, influence the activity of certain key enzymes in the metabolism as activators or inhibitors. This is the case, for example, of invertase (Conduru Neto et al., 1984) which controls glycolytic activity (Tupy, 1973), pyruvate kinase and phosphoenolpyruvate carboxylase on which depends the orientation of glucidic catabolism (Jacob et al., 1983).

The pyrophosphate metabolism also deserves mention. Pyrophosphate is released into cytosol during isoprenic synthesis each time an IPP link is joined to the polymer chain to lengthen it (Fig. 2) (Lynen, 1969) and is thus produced in large quantities. Harmful accumulation is prevented by the presence of an efficient cytosol alkaline pyrophosphatase (Jacob et al., 1988a).

The cytosolic concentration of certain ions (Mg^{2+}, Pi, Ca^{2+}), organic acids (citrate) and basic amino acids is regulated by the process of active lutoid loading (Ribaillier et al., 1971) and thus detoxifies the cytosol of certain ions which could be powerful enzymatic inhibitors like citrate. These transport phenomena are linked with the functioning of the lutoid membrane ATPase proton pump (d'Auzac, 1975) which, by inducing an electrochemical proton gradient, enables the cotransport of molecules with H^+ symport or antiport (Fig. 3) (Marin et al., 1981). A lutoid membrane pyrophosphatase, which is also a proton pump (Prevot et al., 1988), probably plays a similar role to that of the ATPase mentioned above.

The functioning of these lutoids requires their membrane structures to be completely intact. However, senescence processes lead to production of O^{2-} which can attack this structure by oxidizing phospholipids. This toxic molecule is produced under certain conditions by an NADH quinone reductase (Chrestin, 1985; d'Auzac et al., 1986) located on the lutoid membranes. There are physiological protection mechanisms which involve associated enzymatic systems (Chrestin, 1985) such as super oxide dismutase and catalase and molecules with high reducing potential such as ascorbic acid or thiols, which neutralize toxic forms of oxygen in the cytosol. If there is insufficient catalase, H_2O_2 formed by SOD may be converted into aggressive or flocculant substances (OH or quinone molecules). These can also be trapped by what are referred to as scavenger molecules (Chrestin, 1985; Chrestin et al., 1986).

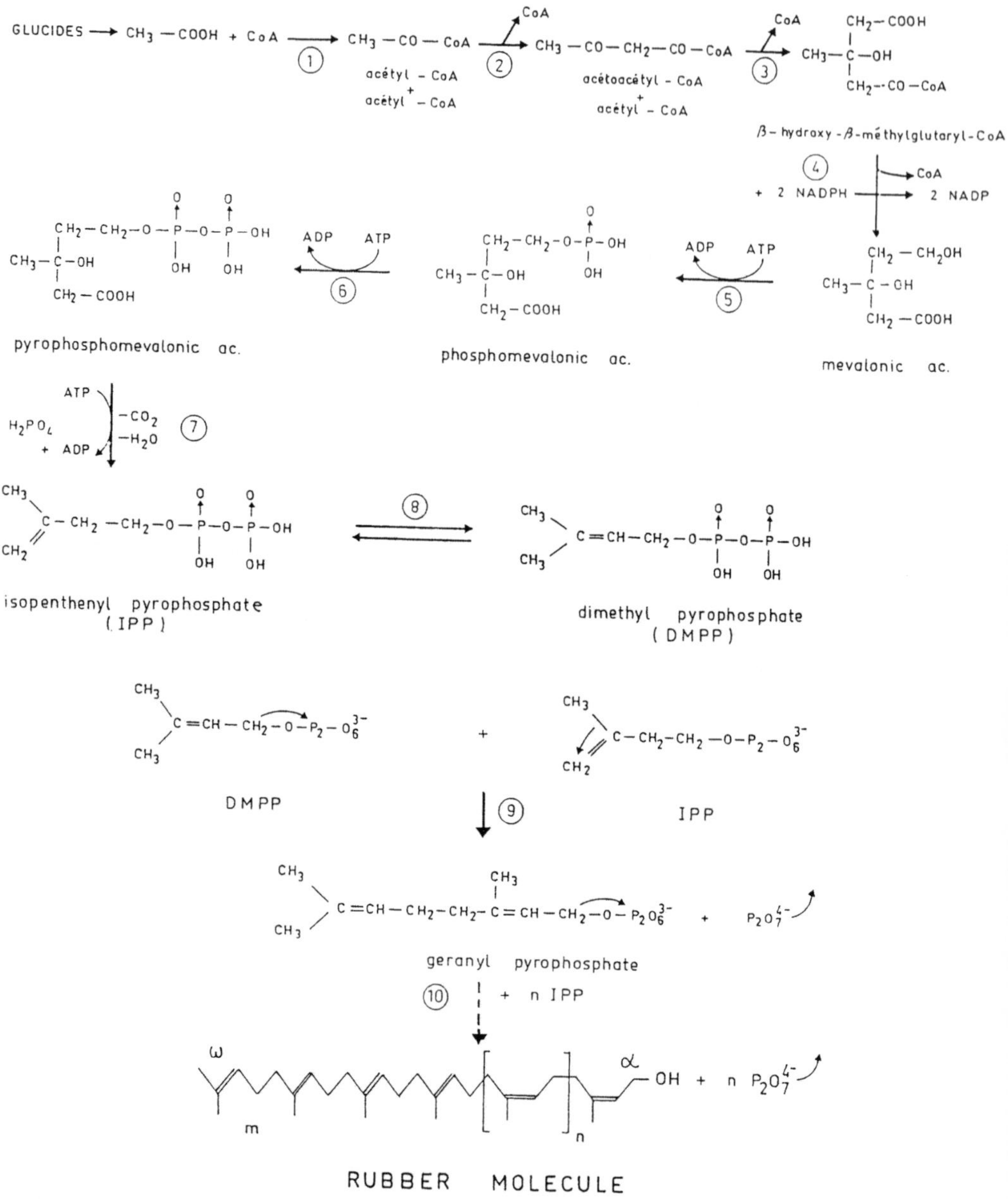

Fig. 2.   Scheme of cis-polyisoprenic metabolism
(for details, see end of text).

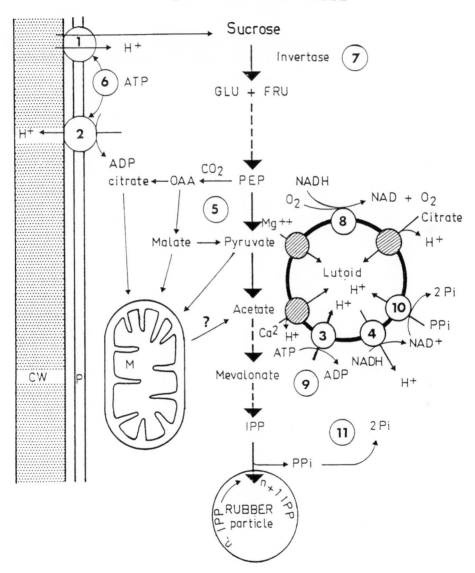

Fig. 3. Absorption of sucrose into laticiferous cells
(for details, see end of text)

pH
—

The pH is an essential factor in the regulation of the laticiferous metabolism through the control that it exerts in particular, on glycolysis and its orientation. Indeed, two key enzymes in sugar catabolism are extremely sensitive to physiological variations in pH. These are invertase, whose activity determines the rate of glycolysis as a whole (Tupy, 1973), and phosphoenolpyruvate carboxylase (PEPcase) which governs the diversion of glycolysis to the organic acid pathways (Oxaloacetate, citrate, malate) (Jacob et al., 1983). These organic acids are not connected with isoprenic synthesis (Fig. 1), but may be connected with the energy-producing oxidation reactions. The pyruvate-acetate sequence, a key point in isoprenic anabolism, is also dependent on the pH (Tupy and Primot, 1976). Distinction must be made between cytosol pH and the pH of the lutoid compartment. The former, which corresponds in general to the pH measured in whole latex, is approximately neutral and varies between 6.5 and 7.4. Lutoid pH is much more acidic and lies between 5.2 and 5.8 (Brzozowska-Hanower et al., 1979).

Numerous factors regulate the pH of the cytosol, where the isoprenic metabolism occurs. Some of these are not well known, such as ion and proton transfers at laticifer plasmalemma level. These are probably connected with the functioning of specific ATPases whose importance in the carbohydrate supply to the laticifers is suspected (Lacrotte et al., 1985; Lacrotte et al., 1988a,b). Others are more easily accessible to experimentation. The existence of a biochemical 'pH-stat' in the cytosol (Chrestin et al., 1985) (Fig. 1) has been shown; this involves the functioning of PEPcase. The functioning of this enzyme, which diverts glycolysis to the synthesis of C_4 and C_6 organic acids (Oxaloacetate, malate, citrate), tends to produce molecules which, by acidifying the medium, inhibit it very strongly in return. A malic enzyme might also be involved in this regulation (Jacob and Prevot, 1981). However, one of the major mechanisms in cellular homeostasis in this biological medium is the intercomparatmental relationship between cytosol and lutoids involving a bio-osmotic 'pH-stat' (Chrestin, 1985; Chrestin et al., 1985) (Fig. 3). Demonstration of functional proton pumps: ATPases (d'Auzac, 1975; Marin et al., 1981; Chrestin, 1985) and pyrophosphatase (Prevot et al., 1988) easily account for the movement of H^+ ions to the lutoid serum. In addition, by generating a membrane protomotive force, it also enables the transport from cytosol to lutoids of substances having an acidifying potential such as citrate. Acidification of the lutoids is thus accompanied by alkalinization of the cytosol.

The importance of pH on latex production has been demonstrated; highly significant correlations have been found in intra-clonal experiments between cytosol or latex pH and production, together with highly significant negative correlations between lutoid pH and production (Brzozowska-Hanower et al., 1979). The difference between cytosol and lutoid pH values (ΔpH) has also been positively and significantly correlated with production (Marin and Chrestin, 1984).

INFLUENCE OF STIMULATION WITH ETHYLENE ON THE METABOLISM OF LATICIFEROUS CELLS

Stimulation of latex production by the application of an ethylene generator (chloroethylphosphonic acid) to the bark is very widely used today in rubber estates. The increase in the quantity of latex obtained on tapping when this technique is used, is the result on the one hand of easier, longer flow and on the other of activated in situ generation of the cell contents. Studies on the mechanisms induced by ethylene have made it possible to shed light on certain phenomena which might explain the metabolic modifications which are responsible for stimulation of production in Hevea (Fig. 4). It must indeed be stressed that although latex regeneration appears to be dependent on the cell metabolism, it has also been shown that the flow itself is partly related to the availability of biochemical energy in the laticifers (Jacob et al., 1988).

Response to stimulation is extremely rapid at membrane level. Acceleration of the kinetics of penetration of tritiated water and ^{14}C sucrose is observed within a few hours after treatment with ethrel (Tupy, 1984). Nevertheless, the increase in the sucrose content in latex, reflecting a sink effect, with carbohydrate assimilates moving to the stimulated zone, appears distinctly only after 48 hours and may be preceded by discrete decrease in concentration, indicating initiation of metabolic activation (Prevot et al., 1986). There is a significant increase in pH in the cytosol 14 to 24 hours after treatment and the peak is reached after 3 to 5 days (Coupe, 1977; Prevot et al., 1986). There are several reasons for this alkalinisation. In plasmalemma, ATPase proton pump activity (Lacrotte et al., 1985; Lacrotte et al., 1988a) connected with the loading of sugars, may be activated resulting in excretion of more cytosol H^+. It has also been shown unambiguously that the functioning of the proton pumps of lutoid membranes is accelerated by de novo synthesis of their active sites and adenosine phosphates (particularly ATP) (Chrestin, 1985; Gidrol et al., 1988) after about twelve hours. This activation results in the transfer of cytosol protons to lutoid serum and an increase of the pH gradient between the two

128

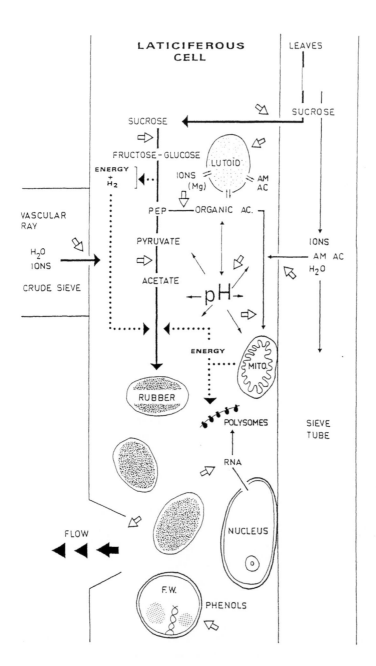

Fig. 4. Influence of ethylene on yield (for details, see end of text)

compartments. Alkalinisation of the cytosol results in its metabolic activation.

The rRNA content displays the same evolution pattern as sucrose, increasing very noticeably 48 hours after treatment (Tupy, 1988a), as does the ribosome polymerisation index (Coupe, 1977); these features support the hypothesis of the acceleration of certain specific protein syntheses such as lutoid ATPases (Gidrol et al., 1988). The total protein content does not vary (Tupy, 1988b). Certain parameters are modified by direct effect of stimulation as in the case of pH, proton pumps activities and the synthesis of certain molecules (adenosine phosphates, rRNA). Changes in certain other factors however are the indirect consequence of stimulation treatment (Lacrotte et al., 1988a). Thus, the total solids content (TSC) falls very slightly after stimulation even before the next tapping; this is a sign of translocation of water to the laticifers. This fall in TSC, indicating an acceleration of translocation, is considerably accentuated during the 2 or 3 subsequent tappings and accounts, in part, for the process of increased production. The important role played by the availability of laticifer biochemical energy in active water-solute transport to the laticifers is another point to be stressed, even though it remains obscure in certain respects (Jacob et al., 1988b).

The other example involves Pi, which reflects the metabolic activity of the laticiferous cells (Jacob et al., 1986). Concentration is not strongly modified by stimulation alone. However, subsequent tappings result in a considerable increase in Pi. There is a simple explanation for this phenomenon: increased latex production requires increased in situ regeneration and hence an accelerated Pi metabolism (Prevot et al., 1988).

Thiols (R-SH) should also be taken into consideration, particularly as these molecules form part of the anti-senescence system of latex cells (Chrestin, 1985). It would appear that the enzymes likely to generate toxic forms of oxygen, and in particular NADH oxidases (NADH quinone reductase), are activated by stimulation, like the proton pump ATPases (Chrestin, 1985). This is followed by over-consumption of thiols, whose content falls steeply (Lacrotte et al., 1988b). It is easy to understand here that over-stimulation may cause a very large decrease in the thiols content. This can result in inadequate antioxidant protection of subcellular membranes and hence in the degeneration of laticifers which may cause bark dryness (Chrestin, 1985; Chrestin et al., 1986). In a normal situation, considerable rate of synthesis of RSH occurs after a fall in the thiols content, observed two or three tappings after the application of ethrel. This varies according to clone, and may result in large increases in the content of these molecules

130

compared with the initial level (Eschbach et al., 1984). This phenomenon remains to be fully elucidated.

The richness of latex in cyclitol molecules (mainly quebrachitol), which form one of the major components of the osmoticum of the medium, tends to decrease with stimulation (Low, 1978).

Stimulation also influences active migration of ions; the K^+ content tends to increase in latex cytosol (Yip and Chin, 1977). This is expected as this cation is associated with inter-cellular translocation (lauchli, 1979); the same applies to Cu^{++} at lutoid level (Ribailler, 1972). In contrast, a fall in overall Ca^{++} content and cytosol and lutoid Mg^{++} levels has been observed but has not yet been explained (Yip and Chin, 1977).

Synthesis of phenolic compounds is also disturbed by stimulation, which may cause the appearance, increase, decrease or disappearance of some of them. In general terms, an increase in total phenols and a decrease in ortho-diphenoloxidases are still observed after many post-stimulation tappings (Hanower et al., 1979; Chrestin, 1978).

Nevertheless, all the metabolic modifications caused by ethylene treatment are transitory. They peak from the second or third tapping after stimulation and subsequently decrease fairly rapidly (Prevot et al., 1986) except in cases where over-exploited trees are unable to recover their physiological balance at laticifer level. The sucrose content is drastically reduced since there are no reserves (there is no sink effect in this case). Fatigue of the laticiferous system is reflected in the physiological parameters of latex. Senescence mechanisms accelerate and the laticifers soon degenerate, leading to the dry bark syndrome (Van de Sype, 1988; Chrestin, 1985).

REFERENCES

Archer, B.L. and Audley, B.G. 1967. Biosynthesis of rubber. Adv. Enzymol., 29: 221-257.
Archer, B.L. and Audley, B.G. 1987. New aspects of rubber biosynthesis. Bot. J. Linnean Soc., 94: 181-196.
Archer, B.L., Audley, B.G. and Bealing, F.J. 1982. Biosynthesis of rubber in Hevea brasiliensis. Plastics and Rubber International. 7: 109-111.
Arreguin, B. and Rock, M.C. 1967. Ciclo de la pentosas en el latex de Hevea brasiliensis Muell. Arg. Inst. Oim. Univ. Nac. Aut. Mexico., 19: 58-73.
Audley, B.G. and Archer, B.L. 1988. Biosynthesis of rubber. In: A.D. Roberts (Ed.), Natural Rubber Science and Technology. Oxford United Press, pp. 16-62.
Bandurski, R.S. and Teas, H.J. 1957. Rubber biosynthesis in latex of Hevea brasiliensis. Plant Physiol., 32: 643-648.
Bealing, F.J. 1975. Quantitative aspects of latex metabolism : possible involvement of precursors other than sucrose in the biosynthesis of

Hevea rubber. In: Proc. Int. Rubb. Conf., Kuala Lumpur 1975. Rubber Research Institute of Malaysia, Kuala Lumpur, Vol. 2, pp. 543-563.

Bouchardat, A. 1875. Structure du caoutchouc naturel. Bull. Soc., Chim. Paris, 24: 109.

Brzozowska-Hanower, J., Cretin, H., Hanower, P. and Michel, P. 1979. Variations de pH entre compartiments vacuolaire et cytoplasmique au sein du latex d'Hevea brasiliensis. Influence saisonniere et action du traitement a l'Ethrel generateur d'ethylene. Repercussion sur la production et l'apparition d'encoche seche. Physiol. Veg., 17: 851-857.

Chesterton, C.J. and Kekwick, R.G.O. 1968. Formation of 3-isopentenyl pyrophosphate by Hevea brasiliensis. Biochem. J., 98: 26 p.

Chrestin, H. 1985. La Vacuole dans l'homeostasie et la senescence des cellules laticiferes d'Hevea. O.R.S.T.O.M. Collection Etudes et Theses, Paris, pp.575.

Chrestin, H., Jacob, J.L. and d'Auzac, J. 1986. Biochemical basis for cessation of latex flow and occurence of physiological bark dryness. In: J.C. Rajarao and L.L. Amin (Eds.) Int. Rubb. Conf. Proc. Kuala Lumpur 1985. Rubber Research Institute of Malaysia, Kuala Lumpur, Vol. 3, pp.20-42.

Chrestin, H., Gidrol, X., d'Auzac, J., Jacob, J.L. and Marin, B. 1985. Cooperation of a "Davies type" biochemical pH-stat and the tonoplastic bioosmotic pH-stat in the regulation of the cytosolic pH of Hevea latex. In: B. Marin (Ed.). Biochemistry and Functions of Vaculoar Adenosine Triphosphatase in Fungi and Plants. Springer Verlag, Berlin, Heidelberg, New York, Tokyo, pp.245-249.

Compagnon, P. 1986. Le caoutchouc naturel. Maisonneuve G.P. et Larose, Paris, pp.595.

Conduru Neto, J.M. Jacob, J.L., Prevot, J.C. and Vidal, A. 1984. Some properties of key enzyme in the metabolism of latex : invertase. In: C.R. Coll. Expl. Physiol. Amelioration Hevea, Montpellier 1984. IRCA-CIRAD, Montpellier, pp. 121-134.

Coupte, M. 1977. Etudes physiologiques sur le renouvellement du latex d'Hevea brasiliensis. Action de l'ethylene. Importance des polyribosomes. These d'Etat de Sci. Nat., Universitie des Sciences et Techniques du Languedoc. Montpellier II, Montpellier.

Coupe, M., Lambert, C. and d'Auzac, J. 1976. Etude comparative des poly-ribosomes foonctionnes du latex d'Hevea brasiliensis sous l'action de l'Ethrel et d'autres produits augmentant l'ecoulement du latex. Physiol. Veg., 14: 391-406.

Cretin, H. 1978. Contribution a l'etude des facteurs limitant la production du latex d'Hevea brasiliensis. Rapport d'Eleve plus 2 annexes, O.R.S.T.O.M., Paris.

D'Auzac, J. 1965. Sur quelques relations entre la composition, l'activite biochimique du latex et la productivite de l'Hevea brasiliensis. These d'Etat de Sciences Naturelles, Universite de Paris.

D'Auzac, J. 1975. Characterization of a membrane ATPase with an acide phosphatase in the latex from Hevea brasiliensis. Phytochem., 14: 671-675.

D'Auzac, J. 1988. Transmembrane transport mechanisms. Application to the laticiferous system. C.R. Coll. Expl. Physiol. Amelioration Hevea, Montpellier 1988. In: J.L. Jacob and J.C. Prevot (Eds.). IRCA-CIRAD, Montpellier, pp. 73-89.

D'Auzac, J., Sanier, C. and Chrestin, H. 1986. Study of NADH quinone reductase producing toxic oxygen from Hevea latex. In: J.C. Rajarao and L.L. Amin (Eds.). Int. Rubb. Conf. Proc. Kuala Lumpur 1985. Rubber Research Institute of Malaysia, Kuala Lumpur, Vol. 3, pp. 122-144.

132

Dickenson, P.H. 1965. The ultrastructure of vessel of Hevea brasiliensis. In: L. Mullins (Ed.), Proc. Nat. Rubb. Res. Ass. Jub. Conf., Cambridge 1964. MacLaren and Sons Ltd., London, pp.52-66.

Eschabach, J.M., Roussel, D., Van de Sype, H., Jacob, J.L. and d'Auzac, J. 1984. Relationships between yield and clonal physiological characteristics of latex from Hevea brasiliensis. Physiol. Veg., 22: 295-304.

Gidrol, X., Chrestin, H., Mounoury, G. and d'Auzac, J. 1988. Early activation by ethylene of the tonoplast H$^+$ pumping ATPase in the latex from Hevea brasiliensis. Plant Physiol., 86: 899-903.

Gomez, J.B. and Moir, G.F.J. 1979. The ultracytology of latex vessels in Hevea brasiliensis. Monography No. 4, Malaysia Rubb. Res. Dev. Board, Kuala Lumpur.

Hanower, P., Brzozowska-Hanower, J., Cretin, H. and Chezeau, R. 1979. Composes phenoliques du latex d'Hevea brasiliensis. Aglycones. Phytochemistry, 18: 686-687.

Hebant, C. 1981. Ontogenie des laticiferes du systeme primaire de l'Hevea brasiliensis : une etude structurale et cytochimique. Can. J. Bot., 59: 974-985.

Hepper, C.M. and Audley, B.G. 1969. The biosynthesis of rubber from hydroxymethylglutaryl CoA in Hevea brasiliensis latex. Biochem. J., 114: 379-386.

Jacob, J.L. 1970. Particularites de la glycolyse et de sa regulation au sein du latex d'Hevea brasiliensis. Physiol. Veg., 8: 395-411.

Jacob, J.L. and d'Auzac, J. 1971. La glyceraldehyde-3-phosphate deshydrogenase du latex d'Hevea brasiliensis, comparison avec son homologue phosphorylante. Eur. J. Biochem., 31: 255-265.

Jacob, J.L. and Prevot, J.C. 1981. Mise en evidence d'une enzyme malique dans le latex d'Hevea brasiliensis. C.R. Acad. Sci. serie 3, Paris, 293: 309-312.

Jacob, J.L. and Sontag, N. 1973. Une nouvelle enzyme : la 2'nucleotidase du latex d'Hevea brasiliensis. Dur. J. Biochem., 40: 207-214.

Jacob, J.L. and Sontag, N. 1974. Purification et etude de la phosphatase acide lutoidique du latex d'Hevea brasiliensis. Biochimie., 56: 1315-1322.

Jacob, J.L., Prevot, J.C. and d'Auzac, J. 1983. Physiological regulation of PEPcase from a non-chlorophyllian system : the Hevea latex. Physiol. Veg., 21: 1013-1019.

Jacob, J.L., Prevot, J.C. and Kekwick, R.G.O. 1988a. General metabolism of Hevea brasiliensis latex. In: J. d'Auzac, J.L. Jacob and H. Chrestin (Eds.). Physiology of rubber tree latex. C.R.C. Press, Boca raton, Florida.

Jacob, J.L., Prevot, J.C., Eschbach, J.M., Lacrotte, R., Serres, E. and Clement-Vidal, A. 1988b. Metabolism of laticiferous cell and yield of Hevea brasiliensis. In: J.L. Jacob and J.C. Prevot (Eds.). C.R. Coll. Explo. Physiol. Amel. Hevea, Montpellier. IRCA-CIRAD, Montpellier, pp. 217-270.

Jacob, J.L., Eschbach, J.M., Prevot, J.C., Roussel, D., Lacrotte, R., Chrestin, H. and d'Auzac, J. 1986. Physiological basis for latex diagnosis of the functioning of the laticiferous system in rubber trees. In: J.C. Rajarao and L.K. Amin (Eds.). Proc. Int. Rubb. Conf. Kuala Lumpur 1985. Rubb. Res. Inst. Malaysia, Kuala Lumpur, Vol. III. pp. 43-65.

Kekwick, R.G.O. 1988. The formation of polyisoprenoids in Hevea latex. In: J.d'Auzac , J.L. Jacob and H. Chrestin (Eds.). Physiology of rubber tree latex. C.R.C. Press, Florida.

Lacrotte, R., Van de Sype, H. and Chrestin, H. 1985. Influence de l'ethylene sur l'utilisation du saccharose exogene par les laticiferes d'Hevea brasiliensis: proposition d'un mechanisme d'action. Physiol. Veg., 23: 187-198.

Lacrotte, R., Serres, E., d'Auzac, J., Jacob, J.L. and Prevot, J.C. 1988a. Sucrose loading in laticiferous cell. In: J.L. Jacob and J.C. Prevot (Eds.). C.R.Coll. Expl. Physiol. Aml. Hevea, Montpellier 1988. IRCA-CIRAD, Montpellier, pp. 119-135.

Lacrotte, R., Cornel, D., Monestier, C., Chrestin, H., d'Auzac, J. and Rona, J.P. 1988b. Electrophysiological studies of sugar transport mechanisms through plasmalemma of latex vessels of Hevea brasiliensis. In: J.L. Jacob and J.C. Prevot (Eds.). Addendum of C.R. Coll. Expl. Physiol. Amel. Hevea, Montpellier 1988. IRCA-CIRAD, Montpellier.

Lauchli, A. 1979. Potassium through plant cell membranes and metabolic role of potassium in plant. In: Potassium Research Review and Trends. Proc. 11th Congress of International Potash Institute 1978 "Der bund" AG, Bern, pp. 111-163.

Lorquin, J., Francey, J. and Chrestin, H. 1987. Mise en evidence d'une NAD kinase dans le compartiment membranaire du latex d'Hevea brasiliensis, C.R. Acad. Sci., serie 3, Paris, 305: 187-191.

Low, F.C. 1978. Distribution and concentration of major soluble carbohydrates in Hevea latex. The effect of Ethephon stimulation and the possible role of these carbohydrates in latex flow. J. Rubb. Res. Inst. Malaysia, 26: 21-32.

Lynen, F. 1963. La biosynthese du caoutchouc dans les plantes. Rev. Gen. Caout. Plast., 40: 83-90.

Lynen, F. 1969. Biochemical problems of rubber synthesis. J. Rubb. Res. Inst. Malaya, 21: 389-406.

Marin, B. and Chrestin, H. 1984. Citrate compartimentation in the latex from Hevea brasiliensis. Relationship with rubber production. In: C.R. Coll. Expl. Physiol. Amel. Hevea Montpellier 1984. IRCA-CIRAD, Montpellier, pp. 169-182.

Marin, B., Marin-Lanza, M. and Komor, E. 1981. The proton motive potential difference across the vacuolysosomal membrane-bound adenosine triphosphatase. Biochem. J., 198: 365-372.

Prevot, J.C., Jacob, J.L., d'Auzac, J., Clement-Vidal, A., l'Huillier, L. and Chrestin, H. 1988. Pyrophosphate metabolism in Hevea brasiliensis latex. In: J.L. Jacob and J.C. Prevot (Eds.). C.R. Coll. Expl. Physiol Amel Hevea, Montpellier, 1988. IRCA-CIRAD, Montpellier, pp. 195-215.

Prevot, J.C., Jacob, J.L., Lacrotte, R., Vidal, A., Serres, E., Eschbach, J.M. and Gigault, J. 1986. Physiological parameters of latex from Hevea brasiliensis. Their use in the study of laticiferous system. Typology of functioning production mechanisms. Effect of stimulation. In: Y. Pan and C. Zhao (Eds.). Proc. Rubb. Physiol. Expl. Meeting. 1986 Hainan. Hainan Chian, South Academic of Tropical Crops. pp. 136-157.

Pujarnizcle, S. 1968. Caractere lysosomal des lutoides du latex d'Hevea brasiliensis. PHysiol. Veg., 6: 27-46.

Ribaillier, D. 1972. Quelques aspects du role des lutoides dans la physiologie de l'ecoulement du latex d'Hevea brasilinesis. These d'Etat de Sci. Nat. Universite d'Abidjan.

Ribaillier, D., Jacob, J.L. and d'Auzac, J. 1971. Sur certains caracteres vacuolaires des lutoides du latex d'Hevea brasiliensis Mull. Arg. Physiol. Veg., 2: 423-437.

Sethuraj, M.R. and Raghavendra, A.S. 1987. In: M.R. Sethuraj and A.S. Raghavendra (Eds.). Tree Crops Physiology. Elsevier, Amsterdam, Oxford, Tokyo, pp. 213-224.

Sipat, A.B. 1982. Hydroxymethylglutaryl coenzyme A reductase NADPH dependant (Ec 1.1.1.34) in the latex of Hevea brasiliensis. Phytochem., 21: 2613-2618.

Skilleter, D.N., Willianson, I.P. and Kekwick, R.G.O. 1966. Phosphomevalonate kinase from Hevea brasiliensis. Biochem. J., 98: p.279.

Southorn, W.A. 1969. Physiology of Hevea latex flow. J. Rubb. Res. Inst. Malaya, 23: 492-512.

Tupy, J. 1973. The activity of latex invertase and latex production of Hevea brasiliensis. Physiol. Veg., 11: 633-641.

Tupy, J. 1984. Translocation, utilization and availability of sucrose for latex production in Hevea. In: C.R. Coll. Expl. Physiol. Amel. Hevea Montpellier 1988. IRCA-CIRAD, Montpellier, pp. 135-154.

Tupy, J. 1988a. Sucrose supply and utilization for latex production. In: J.d'Auzac, J.L. Jacoba nd H. Chrestin (Eds.). Physiology of Rubber Tree Latex. C.R.C. Press, Florida.

Tupy, J. 1988b. Ribosomal and polyadenylated RNA content of rubber tree latex, association with sucrose level and latex pH. Plant Sci., 55: 137-144.

Tupy, J. and Primot, L. 1976. Control of carbohydrate metabolism by ethylene in latex vessels in Hevea brasiliensis in relation to rubber production. Biol. Plant., 18: 374-384.

Tupy, J. and Primot, L. 1982. Sucrose synthetase in the latex of Hevea brasiliensis. J. Exp. Bot., 33: 988-995.

Van de Sype, H. 1988. The dry cut syndrome of Hevea brasiliensis. Evolution, Agronomical and Physiological aspects. In: C.R. Coll. Expl. Physiol. Amel. Hevea, Montpellier 1988, IRCA-CIRAD, Montpellier, pp. 249-272.

Williamson, I.P. and Kekwick, R.G.O. 1965. The formation of 5-phospho-mevalonate by mevalonate kinase in Hevea brasiliensis. Biochem. J., 96: 862-871.

Wititsuwannakul, R., Sukonrat, W. and Chotephiphatworakul, W. 1986. 3-Hydroxy-3-methylglutaryl Coenzyme A reductase from latex of Hevea brasiliensis. In: Y. Pan and C. Zhao (Eds.). Proc. IRRDB Rubb. Physiol. Expl. Meet. Hainan 1986. SCATC, Hainan, China, pp. 47-58.

Yip, E. and Chin, H.C. 1977. Latex flow studies X. Distribution of metallic ions between phases of Hevea latex and the effects of yield stimulation on this distribution. J. Rubb. Res. Inst. Malaysia, 25: 31-49.

CAPTIONS TO FIGURES

Fig. 1. Glucidic catabolism, phosphoenolpyruvate crossing pathway scheme and summarized rubber synthesis in Hevea brasiliensis latex. Enzyme reference number from enzyme nomenclature (1979) : 1. fructofuranosidase (invertase), EC 3.2.1.26; 2. UDPG glucose fructose glucosyl transferase (sucrose synthetase), EC 2.4.1.13; 3. hexokinase, EC 2.7.1.1 and fructokinase, EC 2.7.1.4; 4. UDPG pyrophosphorylase, EC 2.7.1.9; 5. phosphoglucomutase, EC 5.4.2.2; 6. phosphohexose isomerase, EC 5.3.1.19; 7. ATP phosphofructokinase, EC 2.7.1.11; 8. pyrophosphate: fructose-phosphate-6-phosphotransferase, EC 2.7.1.90; 9. fructose diphosphate aldolase, EC 4.1.2.13; 10. phosphotriose isomerase, EC 5.3.1.1; 11. glyceraldehyde phosphate dehydrogenase (non phosphorylating), EC 1.2.1.9; 12. glyceraldehyde phosphate dehydrogenase (phosphorylating), EC 1.2.1.12; 13. 3-phosphoglycerate kinase, EC 2.7.2.3; 14. 1,2 phosphoglycerate mutase, EC 5.4.2.1; 15. phosphopyruvate hydratase (enolase), EC 4.2.1.11; 16. pyruvate kinase, EC 2.7.1.40; 17. phosphoenol pyruvate carboxylase, EC 4.1.1.31; 18. malate dehydrogenase, EC 1.1.1.37; 19. malic enzyme, EC 1.1.1.40; 20. lactate dehydrogenase, EC 1.1.1.27; 21. pathway not yet completely elucidated; 22. non oxidative pentose phosphate pathway; 23. glucose-6-phosphate dehydrogenase, EC 1.1.1.49; 24. gluconate-6-phosphate dehydrogenase, EC 1.1.1.44; reactions indicated by dotted lines (i.e. 20, 23, 24) are nonfunctional in situ; TCA : tricarboxylic acids cycle; NDP, NTP : nucleotides di-, tri-phosphates; DPGA : diphosphoglycerate; PGA : phosphoglycerate; PEP : phosphoenolpyruvate; AOA : oxaloacetate.

Fig. 2. Scheme of cis-polyisoprenic metabolism. Enzyme reference number from enzyme nomenclature (1979) : 1. acetyl coenzyme A synthetase, EC 6.2.1.1; 2. thiolase, EC 2.3.1.9; 3. hydroxymethylglutaryl coenzyme A synthetase, EC 4.1.3.5; 4. hydroxymethylglutaryl coenzyme A reductase, EC 1.1.1.34; 5. mevalonate kinase, EC 2.7.1.36; 6. phosphomevalonate kinase, EC 2.7.4.2; 7. pyrophosphomevalonate decarboxylase, EC 4.1.1.33; 8. isopentenylpyrophosphate isomerase, EC 5.3.3.2; 9. dimethylallyl transferase producing rubber synthesis initiator : geranylgeranylpyrophosphate, EC 2.5.1.1; 10. rubber transferase, EC 2.5.1.20; structure of rubber molecule : diagram from Tanaka (1983).

Fig. 3. Absorption of sucrose into laticiferous cells, isoprenic anabolism and dual role of lutoidic tonoplast in the control of cytosolic homeostasis and 'detoxicating traps' compartmentalising inhibitory products of cytosolic metabolism. Adapted from Jacob et al. (1985).

CW : cell wall; P : plasmalemma; M : mitochondria; GLU : glucose; FRU : fructose; IPP : isopentenyl pyrophosphate; PEP : phophoenol pyruvate.

1. proton-sucrose symport; 2. plasmalemmic ATPase (?), EC 3.6.1.3; 3. tonoplastic ATPase, EC 3.6.1.3; 4. NADH cytochrome c oxidoreductase, EC 1.6.19.3; 5. biochemical pH-stat; 6. active influx of sucrose into latex cell which may be activated by ethylene; 7. invertase, EC 3.2.1.26; step pH-regulated and limiting factor of sucrose catabolism; 8. production of toxic oxygen by NADH dehydrogenase (quinone reductase, EC 1.6.99.2), leading to a degradation of lutoids and a malfunctioning of cytosol metabolism;

9. specific activity of tonoplastic ATPase increases 12 h after ethe-phon treatment causing a cytosol alkalinisation and subsequently an increase of the glycolytic activity; 10. tonoplastic alkaline pyro-phosphatase functioning as protons pump, EC 3.6.1.1; 11. cytosolic soluble alkaline pyrophosphatase avoiding accumulation of PPi produced during polymer chain elongation, EC 3.6.1.1. Enzyme reference number from enzyme nomenclature (1979).

Fig. 4. Influence of ethylene treatment for increasing latex yield, on the metabolic transport and flow mechanisms of laticiferous vessel. This influence is visualized by empty arrows.

CHAPTER 7

YIELD COMPONENTS IN HEVEA BRASILIENSIS

M.R. SETHURAJ
Rubber Research Institute of India, Kottayam-686009, Kerala, India.

The economic product from rubber tree (Hevea brasiliensis) is latex, contained in an anastomosing latex vessel system situated in the bark. Latex is obtained by wounding the bark by a process termed tapping. Latex is a specialised cytoplasm of the laticiferous tissue and contains 30-40 per cent rubber, which chemically is cis-polyisoprene (see Chapter 6, Eds.). Usually a half spiral cut is made on the trunk with a tapping knife to extract latex. The latex flows out when a tapping cut is made, mainly because of the very high turgour pressure in the latex vessels. The initial flow of latex is due to elastic contraction of walls after a sudden release of turgour as a result of tapping. After a while, the flow is regulated by capillary forces until it ceases as the latex coagulates and plugs the vessels (Boatman, 1966; Milford et al., 1969) (see Chapter 14, Eds.). A budded tree is opened for tapping when it attains a girth of 50 cm at a height of 125 cm from the bud union. In most of the rubber growing countries, the trees are tapped alternate daily though lower frequencies have been adopted in many plantations (see Chapter 12, Eds.). On the contrary, many small farmers tap the trees even daily. The total period of exploitation of a tree is about 25 years.

In this chapter, an attempt is made to analyse the components of yield at whole-tree, cellular and sub-cellular levels. For a general understanding of the physiology of production, a reference to earlier reviews by Blackman (1965), Sethuraj (1968, 1985, 1987), Southorn (1969), Boatman (1970), Buttery and Boatman (1976), Moraes (1977), Gomez (1983) and d'Auzac et al. (1989) is recommended.

A comprehensive analysis of the components of yield should separately consider the factors influencing (i) the yield per tree each time it is tapped, (ii) annual yield from unit area of land, and (iii) the cumulative yield through the economic life span of the plantation.

YIELD FROM A TREE EACH TIME IT IS TAPPED

The yield of rubber from a tree on tapping is determined by the volume of latex and the percentage of rubber it contains. The relationship of yield with its main components is presented by the following formula derived by Sethuraj (1981):

$$y = \frac{F.1 \; C_r}{p} \qquad \qquad \ldots \ldots (1)$$

Where y = Yield of rubber tree^{-1} tap^{-1} (obtained from a tree each time it is tapped)

  F = the average initial flow rate per cm of tapping cut during the first 5 min after tapping

  l = the length of the cut (cm)

  p = the plugging index, which is a measure of the extent of latex vessel plugging (Milford et al., 1969)

  C_r= the rubber content (% by weight)

  Of the four major components in the formula, the length of tapping cut for a given system of tapping is determined by the girth and thus the growth vigour, which is predominantly a clonal character and is influenced by the total biomass production and partitioning between growth and latex production. The other three main components also are influenced by the inherent characteristics, exploitation systems and environment. The anatomical, physiological and biochemical sub-components influencing the main components are listed below. This list is only indicative and is not considered complete.

Sub-components of major components

Initial flow rate (F)
* Number of latex vessel rings
* Diameter and other anatomical characteristics of latex vessels
* Turgor pressure at the time of tapping

Length of the cut (l)
* Average annual biomass increment (a function of photosynthesis and translocation)
* Partitioning coefficient between growth and latex production

Rubber content (C_r)
*   Rate of biosynthesis of rubber
*   Intensity of exploitation

Plugging index (p)
*   Stability of rubber particle
*   Stability of lutoid particle
*   Flocculation potential of lutoid serum
*   Antagonising effect of C-serum on lutoid serum activity
*   'Dilution reaction' after tapping
*   'Drainage area' and related biophysical aspects
*   Whole-tree water relations
*   Mineral composition of latex

Initial flow rate (F)

The influence of F on yield has been established although its relative importance as a factor determining yield is less than that of plugging index (Sethuraj et al., 1974; Yeang and Paranjothy, 1982). Seasonal variation in initial flow rate which was related to variation in yield in different clones was reported by Saraswathy Amma and Sethuraj (1975).

Sub-components of F

Bark anatomy

The number of latex vessel rings, diameter of latex vessels and many other structural characteristics of the laticiferous system are genetically determined and clonal variational in this regard has been widely reported. Narayanan et al. (1973) could correlate yield with most of the anatomical characteristics of bark. Henon and Odier (1983) also have studied relations between yield and structural features such as bark thickness, phloem thickness, total number of latex vessel rings etc. They found significant correlations between yield and the number of latex vessel rings. Gomez (1983) has extensively reviewed the influence of anatomical characters on yield. Gomez (1982) and Henon and Nicolas (1989) have reported distinct clonal variations in latex vessel diameter. The effect of the number of latex vessel rows and density on yield can be expected to be mediated through their direct effect on the initial rate of flow. The diameter of latex vessels, in addition to its direct effect on initial flow rate, may also influence the duration of flow as was indicated by Frey-Wyssling (1932) and Riches and Gooding (1952). Premakumari et al. (1985) have indicated that orientation of latex vessel also might influence the initial flow rate and suggested further studies on this aspect. Specific studies on the direct effect of these anatomical characters on the components of latex flow are however meagre. Sethuraj et al. (1984) reported that initial flow rate is positively correlated with the number of latex vessel rows, in a population of cross-pollinated families. The rate of growth can influence some of these structural characters. The effect of these anatomical sub-components should preferably be studied in relation to the initial rate of flow rather than in relation to yield, because the other factors influencing yield can mask its effect.

Turgor pressure (P_{LV})

Turgor pressure is a dynamic factor easily influenced by the components of water relations of the plant (Buttery and Boatman, 1964; Gomes, 1983; Pakianathan et al., 1989). Nevertheless, clonal variations in turgor pressure have been reported by Buttery and Boatman (1967), Gururaja Rao et al. (1988), Devakumar et al. (1988) and Chandrashekar et al. (1990). Ninane (1970) realised the importance of saturation deficit of air in regulating water relations of the plant and elucidated the relationship between saturation deficit of air and variations in trunk diameter, as influenced by the time of the day. Using the data of Buttery and Boatman (1966), he could associate variations in turgor pressure with those of vapour pressure deficit of the air.

The effect of time of tapping on yield is basically determined by the latex vessel pressure existing at that time. Devakumar et al. (1988) and Gururaja Rao et al. (1990) obtained positive correlations between pre-tapping latex vessel turgor and initial flow rate (Fig. 1). Chandrashekar et al. (1990), comparing two clones, RRIM 600 and GT 1, observed that both pre-tapping and post-tapping latex vessel turgor pressure (P_{LV}) were higher in RRIM 600 throughout the year, more so during the dry months. In both clones, the lowest values were obtained in May and the highest in June. Pre-dawn leaf water potential was also higher in RRIM 600 which also maintained a significantly higher latex solute potential(Latex π) during summer months compared to that of GT 1. The mechanism of osmotic adjustment has been observed in RRII 105 (Devakumar et al., 1988). They recorded a high P_{LV} despite a low latex solute potential. They could also relate the high moisture status in RRII 105 with higher stomatal resistance and higher xylem sap flow. The comparatively higher yield of RRII 105 during drought could be partially ascribed to the low transpiration coefficient as well. Relationship among transpiration, turgor pressure and yield as well

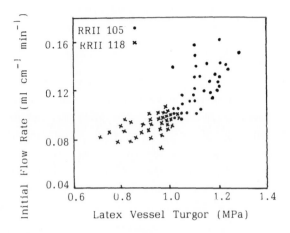

Fig. 1.   Relationship between latex vessel turgor and initial flow rate of latex in RRII 105 and RRII 118 (observations taken from different seasons)

as the correlation between turgor pressure and saturation deficit have been reviewed by Pakianathan et al. (1989). It is thus evident that the turgor pressure and other components of water relations do play a major role in determining the yield obtained.

Length of the tapping cut (l)

The length of the cut is determined by the girth of the tree for a given system of tapping (S, 1/2S, 1/3S or 1/4S). The factors influencing the girth of the plant and the annual girth increment will be discussed in detail in the section dealing with the factors influencing the cumulative yield through the economic life span of the tree.

Plugging index (p)

Plugging index (Milford et al., 1969) indicates the intensity of flow restriction mechanism operating in the latex vessel after tapping. Lutoid particle stability may be the predominant factor influencing plugging index. Clonal variations in plugging index have been well established (Milford et al., 1969; Paardekooper and Somosorn, 1969; Saraswathy Amma and Sethuraj, 1975; Sethuraj et al., 1974). Though a strong clonal character, plugging index is easily influenced by environmental factors, especially soil moisture content (Milford et al., 1969; Saraswathy Amma and Sethuraj, 1975; Sethuraj and George, 1976; Gomez, 1983).

142

Sub-components of p

Lutoid stability

The presence of lutoids in latex was discovered by Homans et al.
(1948). Ruinen (1950) coined the term lutoids for this particle. Lutoids
form 10-25% by volume of fresh latex. The lysosomal nature of lutoids was
clearly demonstrated by Pujarniscle (1968). The lutoid membrane is
osmotically sensitive (Pakianathan et al., 1966; Pujarniscle, 1968; 1969).
Studies on the chemical composition of lutoid membrane revealed that
phospholipid content form 37.5% of the weight of proteins (DuPont et al.,
1976) and that the main phospholipid component is phosphatidic acid (82%
of the total phospholipids). The importance of lutoids in the coagulation
of latex was first recognised by Paton (1953). A clear demonstration of
the role of lutoids was later made by Southorn (1968) and Southorn and
Yip (1968b). Ribailler (1968) perfected a test (bursting index) to measure
lutoid stability based on the ratio of free acid phosphatase in C-serum
and the total acid phosphatase in latex after ample destruction of lutoids
with a detergent and obtained indications of clonal variation in lutoid
stability. Sherief and Sethuraj (1978) could relate clonal variation in lutoid
stability to phospholipid content of the membrane. The latex vessel plugging
during latex flow was found to be negatively correlated with lutoid
stability. Inverse relationship between yield and bursting index also has
been established (Jacob et al., 1985). The seasonal variability in plugging
index also could be related to changes in the phospholipid content of the
lutoid membrane (Premakumari et al., 1980). These authors could explain
clonal variation in tolerance to soil moisture stress in terms of variation
of lutoid phospholipid content.

Flocculating potential of B-serum

The content of lutoid (B-serum), which consists of an acid serum
enriched with divalent cations (Mg^{++} and Ca^{++}) and cationic proteins is
capable of provoking the formation of micro-flocs of rubber particles (Southorn
and Yip, 1968a). The acidic pH, divalent cations and positively charged
proteins in the B-serum may neutralise the negative charge of rubber
particles. Also, some of the acid hydrolases trapped in lutoid may also
attack the protective coating of rubber particles. Employing a microscopic
method, Usha Nair et al. (1978) measured the coagulating potential of
B-serum from different clones, using a dilute suspension of rubber particles.
While a distinct clonal variation was apparent, they could not relate this
with plugging index recorded for these clones. The authors surmised that

while the flocculating potential of B-serum per se might be important in neutralising the charge of rubber particles, the plugging index of a clone would be determined by the total B-serum released in vivo during latex flow and that would depend on the stability of lutoid membrane. In addition to this, the antagonising properties of C-serum (cytosol of latex) against the reaction of B-serum may also play a role. Several authors (Karunakaran et al., 1960; Moir and Tata, 1960 and Tata and Moir, 1964) using starch gel electrophoresis proved that C-serum proteins are anionic while those in lutoid serum are cationic. The interaction of these two sera, after disruption of lutoids, will naturally destabilise the colloidal stability of latex. Southorn and Edwin (1968) and Southorn and Yip (1968b) endeavoured to elucidate the mechanism of action of B-serum. The role of divalent cations was found to be of minor importance. Their studies indicated a major role for the cationic proteins, though not exclusive in the destabilising activity of B-serum. In addition to this, they also demonstrated that C-serum opposes the destabilising action of B-serum in vitro when varying proportions of these two sera are allowed to interact with a dilute suspension of rubber particles. These authors concluded that breakage of lutoids within latex vessels during latex flow, inducing the formation of microflocs inside the latex vessel might play an important role in the cessation of latex flow. While the electrostatic mechanism described above can account for the destabilisation of latex and formation of microflocs, the complete process of coagulation might involve enzymic process as well. The breakdown of lutoids liberates hydrolytic enzymes capable of acting on the phospholipoprotein film which protects the stability of rubber particles. Among the lutoid enzymes characterised by Pujarniscle (1968; 1969) only a protease, Cathepsin was found to be involved in this process. However, lyzosyme, an important hydrolytic lutoid enzyme was unable to coagulate a suspension of rubber particles (Woo, 1973). Involvement of lutoid enzymes in latex coagulation and the process of plugging remains to be confirmed. Further studies on clonal variations in lutoid serum properties and their relationship with yield would be rewarding.

Stability of rubber particle

Electronegative charge on rubber particles is primarily responsible for the colloidal stability of latex (Southorn and Yip, 1968). Comparing the data of Milford et al. (1969) on plugging index with that of Subramaniam (1975) on acetone extract of rubber, Gomez (1983) suggested that an inverse correlation may exist between the acetone extract of rubber and plugging index. Continuing the investigation, Ho et al. (1975) could relate levels of acetone extract with levels of neutral lipids, especially of triglycerides associated with rubber particles. Subramaniam (1975) further suggested that tryclycerides of rubber particles might influence colloidal stability and thus plugging index. Later, Sherief and Sethuraj (1978) could obtain negative correlation between neutral lipid content of rubber particles and plugging index. Premakumari et al. (1980) obtained evidence for lower levels of neutral lipids in rubber particles of latex from drought susceptible clones during periods of drought. Usha Nair et al. (1978) could detect distinct clonal variation in stability of rubber particles. They have also made an interesting observation that tapping rest leads to a fall in the stability. As stability of rubber particle is an integral part of latex stability, it is only rational to assume that this character can directly influence the process of plugging.

Drainage area

The term 'drainage area' is used to delineate the portion of bark which predominantly contributes latex after each day's tapping. Considerable work has been done on 'drainage area' by early workers such as Vischer (1920; 1922) and Bobilioff (1921). Detailed studies by Fre-Wyssling (1932) revealed a relation between the extent of drainage area and yield. More recent studies also have established correlations between total potential displacement area (corresponds to 'drainage area') and yield (Gomez, 1983). These results are indicative of a negative relationship between drainage area and plugging index. Comparing tapping cuts of different length, Southern and Gomez (1970) found that shorter the cut, higher the plugging index and smaller the drainage area. Sethuraj et al. (1974), by artificially manipulating the drainage area, could establish correlations between drainage area and plugging index. There could also be clonal variations in the easiness with which the drainage area develops when a tree is brought under tapping. This factor which might be determined by anatomical, bio-chemical and biophysical factors of the laticiferous system might influence latex flow pattern and plugging index.

Mineral elements

The effect of mineral elements in latex and their ratio on plugging index or initial flow rate has not been studied in detail. High magnesium content and Mg/P ratio are reported to be associated with frequent premature coagulation at the tapping panel (Beaufils, 1957; d'Auzac, 1960). It has been observed that clones with low P/Mg ratios in latex tend to have a higher plugging index (Yip and Gomez, 1980). Ribaillier (1968) and

Ribaillier and d'Auzac (1970) have associated high magnesium level in latex with short flow time, i.e., with high plugging index. Ribaillier (1971) however has noted exceptions; clone LCB 1320 with high magnesium content in latex shows a low plugging index. High potash content in latex is believed to increase production and Pushpadas et al. (1975) related incidence of brown bast with high potassium fertilization. Watson (1980) found that both volume yield of latex and rate of flow were enhanced by the application of K. Increase in yield as a result of stimulation is also associated with an increase in K content of latex (Tupy, 1973). The increase in yield after stimulation is basically a function of reduced plugging index and the above result indicates a relationship between latex K and plugging index.

Water relations

Soil moisture content, which influences water relations of the plant including the dilution reaction during tapping is known to influence yield. Plugging index tends to increase during soil moisture stress, but the magnitude of this effect varies in different clones (Saraswathy Amma and Sethuraj, 1975). The increase in plugging index by moisture stress can be considerably reduced by irrigating the plants (Sethuraj and George, 1976; Mohankrishna et al., 1991). The data of Haridas (1980) also indicate the influence of irrigation on plugging index and yield.

Rubber content in latex (C_r)

In untapped trees the rubber content reaches a high level of 60% or above and presumably by feed back inhibition, further biosynthesis is arrested. When latex is extracted by tapping, the latex vessels from where the latex is lost start to re-synthesise latex. The rubber content in latex stabilises at a particular level with regular tapping. This level is determined by the system of tapping and thereby, the extent of extraction of rubber and the inherent biosynthetic capacity of the laticiferous system. The rubber content of latex, each time the tree is tapped is determined basically by the period which elapsed after the previous tapping as well as the inherent rubber regenerating capacity of the laticiferous cells. The rubber content tends to be lower in clones with low plugging because more latex is lost every time the tree is tapped. While theoretically, rubber content is a factor determining yield, the differences encountered in rubber content is much smaller than the differences in the volume of latex obtained on tapping. The normal variation in rubber content is only to the extent of 5% and may not exceed more than 10%, whereas differences in the volume of latex collected in each tapping can vary from 50 to 100% between trees of the same clone and between clones. In other words, the direct effect of rubber content on yield can be masked by the large differences in the volume of latex obtained.

As a matter of fact, negative correlation between yield and rubber content has been established (Paardekooper and Sookmook, 1969). This only indicates that the high yielding character of a clone is a consequence of the low plugging and that because of the higher extraction of latex the rubber content can be maintained only at a lower level under a given re-generative capacity. Seasonal variation in yield also is mainly a function of volume of latex and not that of rubber content. In this context, it is relevant to state, that rubber content of latex tends to become higher during the period of the lowest yield in summer. The direct effect of climate on biosynthesis has not been studied in detail. However, it has been observed in North-East India (RRII, unpublished data) that the rubber content falls steadily when the minimum temperature drops below 15°C in winter. The observation of a fall in plugging index in winter in China is indicative of a fall in rubber content as well (Xu Wenxian and Pan Yanging, 1990). This drop in rubber content is presumably a direct effect of temperature on the biosynthetic rate. Clonal variation in biosynthetic ability, while expected, requires further experimental support. Clonal variations have been reported in HMG CoA reductase activity and this has been related to yield variations (Wititsuwannakul and Sukonrat, 1984). Usha Nair et al. (1990) reported significant difference in HMG CoA reductase activity in the bark between high yielding and low yielding clones. There are experimental limitations in accurately measuring the biosynthetic rate in vitro. Studies conducted with in vitro systems using latex do not simulate the in vivo conditions, because, in spite of the presence of the required enzyme systems, the cellular composition of the collected latex differs drastically from the cellular composition in vivo. Perfection of a method to study the biosynthetic capacity using bark samples might be useful. Information on the inherent biosynthetic capacity of laticiferous system of a clone would be highly helpful to determine the optimum exploit-ation levels.

Other metabolic factors

The level of sucrose, inorganic phosphorus (Pi), magnesium (Mg^{++}), thiols (R-SH), pH of latex and the redox potential of latex have been implicated in the biosynthesis of rubber. Many authors have demonstrated an important role for sucrose in latex production (d'Auzac and Pujarniscle,

1961; Eschback et al., 1986; Low, 1978; Tupy, 1969; Tupy and Primot, 1976). A high sucrose content in latex may indicate good loading of this precursor to the laticifers cells, which may be accompanied by active metabolism. But a high sucrose content in latex may also indicate low metabolic utilisation of sucrose and hence low productivity (Prevot et al., 1984). Under conditions in which sucrose is a limiting factor, positive correlations between sucrose content and rubber biosynthesis can be expected. However, when different clones of varying productivity are compared, this relationship becomes more complicated; the sucrose content at any given time would be determined by the combined effect of the rate of loading of sucrose in the laticiferous system, its utilisation in metabolism and the level of latex production.

The cytosol pH is considered to be an important factor regulating the metabolic activity of the laticiferous system and highly significant positive correlations have been obtained between pH and latex production under certain conditions (Brozozowska-Hanower et al., 1979; Cretin et al., 1980; Eschback et al., 1984). Increase in yield after Ethephon stimulation is also associated with alkalinisation of cytosol pH (Gidrol and Chrestin, 1984).

The inorganic phosphorus content indicates the energy metabolism of latex. It influences glucidic catabolism, systhesis of nucleotides, involved in energy transfer and isoprene synthesis (Jacob et al., 1989). Significant relations between availability of labile energy phosphates and biosynthetic rates measured in vitro, have been obtained by d'Auzac (1964). Direct correlations between the Pi content of latex and rubber production by certain clones have also been reported by Eschback et al. (1986) and Subronto (1978). While Subronto reported a significant inverse correlation between magnesium content and production, Eschback et al. (1989) demonstrated a positive correlation between magnesium and production. This would indicate that in the latter case, the role of magnesium as an activator of enzymes in the laticiferous system outweighed its destabilising role.

The role of thiols as an activator of invertase (Jacob et al., 1982) and of pyruvate kinase (Jacob et al., 1981) has been established indicating that thiol can affect the metabolic rate and hence the regeneration of latex.

Inverse relations between redox potential of latex and rubber production have also been demonstrated (Prevot et al., 1984).

FACTORS INFLUENCING ANNUAL YIELD FROM UNIT AREA

Yield of rubber per unit area (ha) per year (y_h) is determined by the average yield tree^{-1} tap^{-1} ($\bar{y}$), the number of trees (N) and number of tappings per year (n_t):

$$y_h = \bar{y} \, N \, n_t \qquad\qquad \ldots\ldots (2)$$

$\bar{y}$ is influenced by seasonal variation and system of tapping.

Effect of seasonal variations

Yield is influenced by the effect of different environmental parameters on the components of yield. A distinct clonal variation is observed in this regard. In regions with a distinct period of drought, only such clones which possess some degree of drought tolerance can maintain high annual yields. Tjir 1 is a typical example of drought susceptible clone; seasonal variation in yield is more pronounced in this clone. Saraswathyamma and Sethuraj (1975) have studied clonal variation in tolerance to soil moisture stress. Environmental constraints such as drought, high or low temperature and low relative humidity influence annual yield and more so in regions where such constraints persist for long periods. In India, the period of peak yield is from September to January. Seasonal variation in yield has been ascribed to variation in both initial flow rate and plugging index (Milford et al., 1969; Paardekooper and Somosorn, 1969; Sethuraj, 1977; Saraswathyamma and Sethuraj, 1975). Ninane (1970) unambiguously demonstrated that the effect of drought and high temperature is mediated mainly through changes in plugging index than through changes in initial flow rate. Ribaillier (1968) recorded comparatively higher bursting index during periods of low yield. Premakumari et al. (1980) could relate decreasing production during dry periods with decrease in lipid content affecting the stability of lutoid membrane, resulting in high plugging index. These authors ascribed this to an increase in the rate of phospholipase activity during the period of drought.

The positive relation between C_r and y, as depicted in formula (1), is normally not realised because of its interactions with p, volume of latex lost and rate of regeneration.

As a matter of fact, the lowest values in C_r are encountered during the monsoon season and the maximum values during the dry season, which are periods of high and low yield respectively. Negative relations between yield recorded over a period of one year and C_r has been established (Brzozowska-Hanower et al., 1979; Heuser and Holder, 1931; Wiltshire, 1934).

In summary, annual yield depends upon the average values of F, p and C_r through different seasons of a year. The environmental situations leading to lower F and higher p will reduce yield. The widest fluctuations are seen in plugging index, which can be considered the most sensitive yield component affected by environmental factors. The variation in C_r is related to its interaction with plugging and volume of latex lost. When yield is reduced due to high plugging, C_r increases. This aspect is discussed in detail elsewhere in this chapter. The extent and duration of environmental constraints affecting p and F thus predominantly determine annual yield. The relationship between seasonal variation in yield and variations in yield components and sub-components is presented in Table 1.

TABLE 1

Yield, yield components and components of water relations in Gl 1 and Tjir 1 during dry and wet seasons at Malankara

Parameter	Dry Season		Wet Season	
	Gl 1	Tjir 1	Gl 1	Tjir 1
Yield (g $tree^{-1}$ tap^{-1})	35.92	17.56	96.50	99.20
Initial flow rate (ml cm^{-1} min^{-1})	0.089	0.062	0.091	0.093
Rubber content (%, w/v)	43.60	41.90	39.30	36.90
Plugging index	5.49	8.08	1.89	1.81
Pre-tapping P_{lv} (MPa)	0.95	0.89	1.02	1.08
Minimum P_{lv} (MPa)	0.194	0.196	0.246	0.274
Latex solute potential (-MPa)	1.02	1.01	0.92	0.91
Pre-dawn ψ leaf (-MPa)	0.23	0.25	0.14	0.15
Mean ψ leaf (MPa)	1.90	1.94	1.32	1.27
After-noon ψ leaf (-MPa)	2.32	2.64	1.66	1.58
Minimum r_s (scm^{-1})	1.42	1.66	1.08	1.50
Mean r_s (scm^{-1})	9.36	9.42	7.92	8.70
Transpiration (mm day^{-1})	0.832	1.331	4.516	4.549
Transpiration coefficient	0.151	0.224	1.039	1.130
Xylem sap speed (cm/12 h)	89.70	97.76	108.76	109.58

N : Yield and number of trees ha^{-1}

Increased density of planting results in lower tree girth, biomass and crown, higher crotch height and lighter branching (ng et al., 1979; Setheesan et al., 1982; Webster and Baulkwill, 1989; Napitupulu, 1977; Leong and Yoon, 1982). Virgin bark and renewed bark also become thinner with higher stand per hectare. The reduction in thickness is more pronounced in renewed bark (Ng et al., 1979). Because of these effects, yield per tree tends to be lower with increased number of trees per unit area. In addition to this, percentage tappability in a field, during the initial years of tapping, also decreases with increasing density, thus affecting yield per unit area.

Ng et al. (1979) studying the economic aspects of high density planting, concluded that under Malaysian conditions the cumulative yield increased as the planting density was increased from 211 to 741 trees ha^{-1}, but declined with further increase in density. However, as the density increased from 211 to 741 trees ha^{-1} the production cost (as determined by yield per tapper) was found more than the revenue, and as a result, the highest cumulative profit was realised with a planting density of 399 trees ha^{-1}. The highest yield per tapper was also obtained with a density of 399 trees ha^{-1}.

Influence of number of tappings on yield (n_t)

The annual average yield per tap will also depend on the frequency of tapping. At least for the first few years of tapping, the annual yield tends to be higher with increased intensity of tapping. Intensity of tapping is determined by the length of the tapping cut (l) and the frequency of tapping. It has been well established that a lower frequency of tapping, i.e., permitting a longer interval between tappings, results in higher rubber production every time the tree is tapped (Ng et al., 1969; Sethuraj, 1977). This phenomenon indicates the need for providing sufficient time for the regeneration of latex. This is confirmed by a higher C_r and total solids content observed with longer intervals between tappings (Jacob et al., 1989). Nevertheless, yield tap^{-1} tree^{-1} reduces when the interval exceeds seven days. This could be due to an increase in the total solids content as well as BI which slows down latex flow. Usha Nair et al (1980) have reported a fall in the stability of rubber particles in latex collected after prolonged tapping rest. A higher sucrose content in latex in low frequency systems also indicates a lower anabolic activity. Higher intensities of tapping result in higher incidence of dry trees (brown bast) in a field, thus reducing the number of productive trees in unit area (Bealing and Chua, 1972). The relationship among frequency of tapping, yield per tap and certain latex physiological parameters has been discussed in detail by Jacob et al. (1989) and Eschback et al. (1989). The effect of increasing intensity of tapping and thereby extraction of more latex on C_r is illustrated in Table 2.

It is thus evident that annual yield from unit area is influenced by factors related to agronomic practices planting density, exploitation system and seasonal fluctuations. The extent of realisation of the yield potential of a clone is regulated by these extraneous factors.

FACTORS INFLUENCING CUMULATIVE YIELD THROUGH THE ECONOMIC LIFE SPAN OF THE TREE

The first two half spiral tapping panels in the virgin bark (Panel (BO-1 and BO-2, see Chapter . Eds.) are consumed in 10 to 12 years. Renewed bark is tapped for another 10 to 12 years followed by upward

Table 2. Effect of system of tapping on rubber content of latex
8 weeks after imposition of tapping treatments*

System of tapping	Clone RRIM 602		Clone RRIM 603		Clone RRIM 612	
	Total latex extracted	C_r (%)	Total latex extracted	Cr (%)	Total latex extracted	C_r (%)
½S d/2	903	30.0	3283	34.5	3203	35.6
¼S d/1	2721	31.7	3999	30.0	2116	32.0
S d/1	5528	17.4	4079	19.0	8345	22.0
½S 2d/1	6867	23.1	8105	27.0	9187	26.9

*Adapted from Sethuraj (1977).

and 'slaughter' tappings. In general, there is a steady increase in yield from the first year of tapping and it gets stabilised in BO-2 panel. Certain high yielding clones, however, record a decline in yield in renewed bark.

The annual increment in yield has to be ascribed basically to changes in the four major components of yield, i.e., F, l, C_r or p with successive years of tapping. Girth increment after tapping is always found to be less compared to that observed before the tree is opened for tapping. There is distinct clonal variation in this regard. Clones with comparatively higher girth increment in tapped trees may maintain better growth of bark as well. This will entail these clones to have a higher F. Clones with lower p are known to maintain a reduced girth increment.

Annual yield increment observed, however, is proportionately much more than what could be accounted for by an increase in l. It thus follows that there should also be either an increase in F or C_r, or a reduction in p to account for the annual yield increment index. Sethuraj et al. (1974) proposed a formula to relate yield increment with girth increment. These authors observed clonal variation in this character. Their data indicated that for a given increase in l (or girth), the percentage increase in yield varied considerably between clones.

The substantially higher yield increment in clones RRII 102 and RRII 208 resulted in a very poor girth increment after the trees were opened for

Table 3. Comparison between yield increment and girth increment.

Clone	Girth, cm			Yield, g tree^{-1} tap^{-1}		
	1st year	5th year	Increase (%)	1st year	5th year	Increase (%)
RRII 102	61.1	73.4	20.1	26.0	94.3	262.6
RRII 105	58.3	76.1	30.5	36.8	70.0	90.1
RRII 115	66.5	86.0	29.3	26.3	51.7	96.5
RRII 208	77.2	87.7	13.6	43.0	100.1	132.5
Tjir 1	50.7	69.4	36.8	22.3	34.1	52.9

tapping. Various authors have found that the amount of rubber produced in a year is less than the 'unrealised' shoot dry weight increment. Thus, the production of a tree depends upon the total annual biomass increment, the partitioning ratio and the rate at which the unrealised dry weight is converted into rubber (Simmonds, 1982). A comprehensive analysis of the relations among the potential biomass increment, the partitioning ratio and the loss in biomass which can be and cannot be accounted for by rubber yield will be very useful (Sethuraj, 1985).

The annual biomass produced by a tree subjected to regular tapping (W_a), is substantially low compared to that by an untapped tree (W_m). Marked clonal variation also is reported in this regard. This reduction in biomass is only partially accounted for by yield of rubber. The proportion (1-k) of the biomass potential that is not realised in a tapped tree and not accounted for by rubber yield is an important factor determining annual biomass increment after a tree is opened for tapping. By reducing this proportion, W_a can be increased:

$$W_a = W_m (1-k) \qquad \qquad \qquad \cdots \cdots (3)$$

Simmonds (1982) assumed that the loss in biomass by a tree subjected to regular tapping can be accounted for by rubber and other products removed in latex. Reported data, however, indicate that the loss in biomass by a tapped tree, compared to that in an untapped tree, may vary almost seven times between clones for comparable rubber yield (Templeton, 1969).

Calculations based on energy value of rubber and other substances lost in latex can hardly account for the huge loss in biomass owing to tapping. It is, therefore, preferable to treat the part of the biomass loss that cannot be accounted for by rubber yield separately from that which can be accounted by for it.

The total biomass utilisation by a tree subjected to regular tapping (W_a), can be partitioned to shoot plus root biomass (W_g) and annual yield of rubber (y_a):

$$W_a = W_g + 2.5 \, y_a \qquad \ldots \ldots (4)$$

The factor 2.5 accounts for the higher calorific value of rubber.

The relationship among the annual biomass increment of an untapped tree (W_m), the biomass increment of a tapped tree (accounting for also the biomass of rubber) (W_a) and its biomass increment (W_g) can be expressed by the following formulae:

$$W_a = W_g + 2.5 \, y_a$$

$$W_a = W_m \, (1-k)$$

$$W_g = [W_m(1-k)] - 2.5 \, y_a \qquad \ldots \ldots (5)$$

Thus the annual biomass increment of shoot and root is influenced mainly by the extent of biomass lost due to tapping and rubber yield obtained.

In an untapped tree, biosynthesis of rubber is almost nil in the laticifers of the trunk and there is no translocation of latex. Tapping induces an abnormal physiology by removal of latex. Laticiferous system in the 'drainage area' is then triggered for the resynthesis of lost latex. Latex flow and the consequent regeneration of latex as a result of tapping should be considered an abnormal physiological phenomenon. The ratio of rubber yield tree^{-1} year^{-1} (y_a) to the total biomass tree^{-1} year^{-1} (W_a) is defined as harvest index (c). In calculating the harvest index, only the rubber extracted is considered and the high calorific value of rubber (2.5 times that of carbohydrate) is accounted for:

$$c = \frac{2.5 \, y_a}{W_a} \qquad \ldots \ldots (6)$$

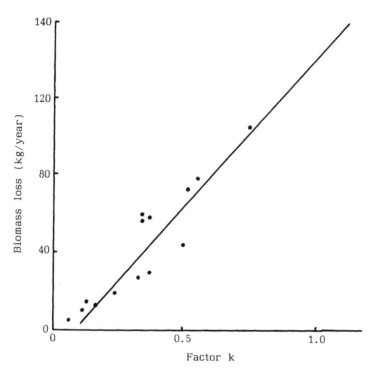

Fig. 2. Relationship between biomass loss and factor k.

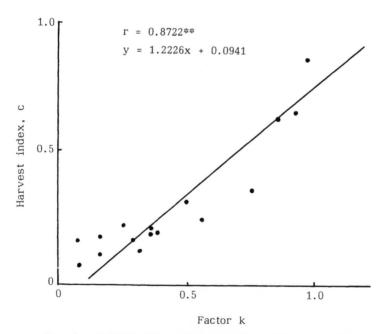

Fig. 3. Relationship between harvest index c and factor k.

and therefore

$$y_a = \frac{W_a \, c}{2.5} \qquad \qquad \cdots\cdots (7)$$

As W_a is depended upon k and W_m,

$$y_a = \frac{W_m \, (1-k) \, c}{2.5} \qquad \qquad \cdots\cdots (8)$$

Sethuraj et al. (1985) compared different exploitation systems and established that the biomass loss on tapping is positively correlated with the factor 'k' (Fig. 3). On the other hand, when different clones were compared, the k factor varied significantly for comparable yield levels.

The physiological reasons for the biomass loss of a tapped tree, which is not accounted for by rubber yield, have not been elucidated. Enhanced respiratory activity and the consequent loss of biomass could be one of the causes. Clonal variation in this respect also can be expected. It has been established that tapping and stimulation result in an increase in respiratory activity of the bark (Sethuraj et al., 1974). Besides, direct effects of wounding and extraction of latex may also lead to higher metabolic activities. The positive correlation between the 'k' factor and harvest index obtained is indicative of the correctness of this assumption. Clonal variation in the rate of increase in respiration in response to tapping, though not studied, is a theoretical possibility. Significant variation in leaf respiratory rate among clones have already been reported (Ceulemans et al., 1984). While the impediment to translocation by tapping can also be considered as one of the possible factors the wide clonal variation observed in biomass loss, under the same system of tapping, discount this as a major factor. Basic studies to elucidate the physiological reasons for biomass loss in tapped trees as well as the basis of clonal variation in this regard would be rewarding. A low 'k' value should be considered a desirable character to be aimed at by breeders.

Yield increment can only be partly influenced by girth increment. Annual increase in yield is clearly related to increased partitioning (harvest index) (Sethuraj et al., 1974a). This can be ascribed only to changes in either F or p, with successive years of tapping. Data comparing F and p of different years of tapping is scarce and it would be interesting to study the changes in the yield components during successive years of tapping. A re-evaluation of published data on yield and yield components of clones of different ages would indicate that there can be a gradual reduction in plugging index with successive years of tapping. Our present knowledge of the physiology of latex flow is totally inadequate to satisfactorily explain this phenomenon. It would be rewarding to examine if, with years of tapping any gradual increase in drainage area occurs. If this happens, a hypothesis can be put forward that the enlarging drainage area, per se, may be directly responsible for the gradual reduction in plugging index with successive years of tapping.

While the harvest index in most of the agricultural crops is a function of genetically determined factors, in rubber, harvest index can also be altered at will, by changing the exploitation system or by using yield stimulants. When the exploitation system employed in a particular year leads to higher harvest index and thus a yield level beyond the inherent latex regenerating capacity of the clone, the physiological damage caused can influence the yield in subsequent years. Tapping panel dryness (brown bast) is generally considered to be a reaction of the tree to over-exploitation. It has been repeatedly documented that incidence of tapping panel dryness always increases with increased intensity of exploitation (Eschbach et al., 1989). When trees are rendered unproductive by the onset of brown bast, the number of productive trees in a plantation will decline affecting the cumulative production from unit area.

Yield from renewed bark

Another important aspect which determines the cumulative yield through the economic life span, is the yield obtained from renewed bark as compared to that from virgin bark. Clones exhibit substantial variation in this regard. Certain clones exhibit an increasing trend in yield in renewed bark (eg. GT 1, PB 217) while certain other clones (eg. RRII 105, PB 235) show a declining trend. Though a thorough analysis is lacking, the general trend in different clones seems to indicate that precocious high yielding clones present a declining trend in renewed bark. An analysis of clonal variations at yield component level is difficult in the absence of adequate data. However, the number of latex vessel rows in renewed bark as compared to that in virgin bark may be one of the influencing factors.

In summary, the cumulative yield will depend upon girth and yield increments with successive years of tapping and may be determined by the factors included in the following formulae:

$$W_g = [W_m(1-k)] - 2.5 \, y_a$$

and

$$y_a = \frac{W_m (1-k) \; c}{2.5}$$

The harvest index 'c' can be regulated by adopting the most physiologically sound exploitation level. This will, in turn, control the factor 'k' and thus the biomass loss. Experiments to determine the optimum value of c for different clones would be most rewarding. In order to facilitate computations of these factors, it is necessary to have one experimental treatment with untapped trees to know the value of W_m.

REFERENCES

Auzac (d'), J. 1960. Sur la signification du rapport Mg/P des latex frais vis-a-vis de leur stabilite, in Opusc. Technol. Inst. Rech. Caoutch. Vietnam. 40/60, 1.

Auzac (d'), J. and Pujarniscle, S. 1961. Apercu su l'etude des glucides de l'Hevea et sur leur variation, Rev. Gen. Caoutch., 38(7-8), 1131.

Auzac (d'), J. 1964. Disponibilite en phosphore energetique, biosynthese du caoutchouc et productivite de l'Hevea brasiliensis, C.R. Acad. Sci. Paris, 258, 5091.

d'Auzac, J. and Pujarniscle, S. 1961. Apercu sur l'etude des glucides de l'Hevea et sur leur variation, Rev. Gen. Caoutch., 38: 1131.

d'Auzac, J., Jacob, J.L. and Chrestin, H. 1989. Physiology of Rubber Tree Latex. The Laticiferous Cell and Latex - A model of cytoplasm, CRC Press Inc., Boca Raton, Florida.

Bealing, F.J. and Chua, S.E. 1972. Output, composition and metabolic activity of Hevea latex in relation to tapping intensity and the onset of brown bast. J. Rubber Res. Inst. Malaya, 23: 204.

Beaufils, E.R. 1957. Research on rational exploitation of Hevea using a physiological diagnosis based on mineral analysis of the various parts of the plant. Fertilite, 3: 27-37.

Blackman, G.E. 1965. Factors affecting the production of latex. In: L. Mullins, (Ed.), Proceedings of the Natural Rubber Producers Research ASsociation Jubilee Conference, Cambridge, 1964. Maclaren, London, pp.43-51.

Boatman, S.G. 1966. Preliminary physiological studies on the promotion of latex flow by plant growth regulators. J. Rubb. Res. Inst. Malaya, 19, 243.

Boatman, S.G. 1970. Physiological aspects of the exploitation of rubber trees. In: Luckwill, L.C. and Cutting, C.V. (Eds.). Physiology of tree crops. London and New York, Academic Press, pp. 323-333.

Bobilioff, W. 1921. Onderzoekingen over den oorsprong van latex big Hevea brasiliensis. Archf. Rubbercult. Ned-Indie., 5: 95.

Brzozowska-Hanower, J., Cretin, H., Hanower, P. and Michel, P. 1979. Variations du pH entre compartiments vacuoline et cytoplasmique au sein du latex d'Hevea brasiliensis; Influence saisonniere et action due traitment par l'Ethrel generateur d'ethylene, Repercussion sur la production et l'apparition d'encoches seches, Physiol. Veg., 17: 851.

158

Buttery, B.R. and Boatman, S.G. 1964. Turgor pressures in phloem: measurements on Hevea latex. Sicence, 145-285.

Buttery, B.R. and Boatman, S.G. 1966. Manometric measurement of turgor pressures in laticiferous phloem tissues. J. Exp. Bot., 17: 283-296.

Buttery, B.R. and Boatman, S.G. 1967. Effects of tapping, wounding and growth regulators on turgor pressure in Hevea brasiliensis Muell. Arg. J. Exp. Bot., 18: 644.

Buttery, B.R. and Boatman, S.C. 1976. Water deficits and flow of latex. In: T.T. Kozlowski (Ed.). London and New York: Academic Press, 4: 223 p.

Chandrashekar, T.R., Jana, M.K., Joseph Thomas, Vijayakumar, K.R. and Sethuraj, M.R. 1990. Seasonal changes in physiological characteristics and yield in newly opened trees of Hevea brasiliensis in North Konkan. Indian J. Nat. Rubb. Res., 3: 88-97.

Ceulemans, R., Gabriels, R., Impens, I., Yoon, P.K., Leong, W. and Ng, A.P 1984. Comparative study of photosynthesis in several Hevea brasiliensis clones and Hevea species under tropical field conditions. Trop. Agric., Trin., 61: 273.

Cretin, H., Jacob, J.L., Prevot, J.C. and d'Auzac, J. 1980. Le pH du latex: son influence sur la production et les elements due sa regulation. Rev. Gen. Caoutch. Plast. 603: 111.

Devakumar, A.S., Gururaja Rao, G., Rajagopal, R., Sanjeeva Rao, P., George, M.J., Vijayakumar, K.R. and Sethuraj, M.R. 1988. Studies on soil-plant-atmosphere system in Hevea. II. Seasonal effects on water relations and yield. Indian J. Nat. Rubb. Res., 1: 45-60.

Dupont, J., Moreau, F., Lance, C. and Jacob, J.L. 1976. Phospholipid composition of the membrane of lutoids from Hevea brasiliensis latex. Phytochemistry, 15: 1215.

Eschbach, J.M., Roussel, D., Van de Sype, H. and Jacob, J.L. 1984. Relationships between yield and clonal physiological characteristics of latex from Hevea brasiliensis. Physiol. Veg., 22: 295.

Eschbach, J.M., Tupy, J. and Lacrotte, R. 1986. Photosynthate allocation and productivity of latex vessels in Hevea brasiliensis. Biol. Plant., 28: 321.

Eschbach, J.M., Lacrotte, R. and Serres, E. 1989. Conditions which favour the onset of brown bast. In: d'Auzac, J., Jacob, J.L. and Chrestin, H. (Eds.). Physiology of rubber tree latex. CRC Press Inc., Boca Raton, Florida. pp. 443-458.

Frey-Wyssling, A. 1929. Microscopish endersoch Laar Let nerkomen van Larsen in latex van Hevea, Arch. Rubbercult., 13: 394.

Frey-Wyssling, A. 1932. Investigations on the dilution reaction and the movement of the latex of Hevea brasiliensis during tapping. Arch. Rubbercult., 16: 285.

Gidrol, X. and Chrestin, H. 1984. Lutoidic ATPase functioning in relation to latex pH regulation and stimulation mechanisms in C.R.Coll. Exp. Physiol. Amel. Hevea, IRCA-CIRAD, Montpellier, France, 1984, 81 p.

Gomez, J.B. 1982. Anatomy of Hevea and its influence on latex production. MRRDB Monograph No. 7, Kuala Lumpur.

Gomez, J.B. 1983. Physiology of latex (Rubber) production. MRRDB Monograph No. 8, Kuala Lumpur.

Gururaja Rao, G., Devakumar, A.S., Rajagopal, R., Annamma, Y., Vijayakumar, K.R. and Sethuraj, M.R. 1988. Clonal variation in leaf epicuticular waxes and reflectance: Possible role in drought tolerance in Hevea. Indian J. Nat. Rubb. Res., 1: 84-87.

Gururaja Rao, G., Sanjeeva Rao, P., Rajagopal, R., Devakumar, A.S., Vijayakumar, K.R. and Sethuraj, M.R. 1990. Int. J. Biometeorol., 34: 175-180.

Haridas, G. 1980. Soil moisture use and growth of young Hevea brasiliensis as determined from lysimeter studies. J. Rubber Res. Inst. Malays., 28: 49 p.

Henon, J.H. and Nicolas, D. 1989. Relation between anatomical organisation of the latex yield: Search for early selection criteria. In: d'Auzac, J., Jacob, J.L. and Chrestin, H. (Eds.). Physiology of rubber tree latex. CRC Press Inc., Boca Raton, Florida, pp. 31-50.

Henon, J.H. and Odier, F. 1983. Premier bilan sur l'etude des criteres anatomiques pour la selection precoce, unpublished report, Int. Rech. Caoutch. (IRCA), 1983.

Heuser, C. and Holder, H.J. 1931. Tappings results with the new double cuts tappings system, Hevea buddings in 1930, Arch. Rubbercult., 15: 246.

Ho, C.C., Subramaniam, A. and Young, U.M. 1975. Lipids associated with the particles in Hevea latex. Proc. Int. Rubb. Conf. 1975. Kuala Lumpur, 11: 441 pp.

Homans, L.N.S., Van Dalfsen, J.W. and Van Gils, G.E. 1948. Complexity of fresh Hevea latex. Nature, London, 161: 177.

Jacob, J.L., Prevot, J.C. and Primot, L. 1981. La Pyruvate kinase du latex d'Hevea brasiliensis. Rev. Gen. Caoutch. Plast. 612, 89.

Jacob, J.L., Prevot, J.C. and Auzac (d'), J. 1982. Physiological activators of invertase from Hevea brasiliensis latex. Phytochemistry, 21: 851.

Jacob, J.L., Eschbach, J.M., Prevot, J.C., Roussel, D., Lacrotte, R., Chrestin, H. and d'Auzac, J. 1985. Physiological basis for latex diagnosis of the functioning of the laticiferous system in rubber trees. Proc. Int. Rubber Conf. 1985, Rubber Research Institute of Malaysia, Kuala Lumpur, p. 43.

Jacob, J.L., Prevot, J.C., Roussel, D., Lacrotte, R., Serres, E., d'Auzac, J., Eschbach, J.M. and Omont, H. 1989. Yield limiting factors, latex physiological parameters, latex diagnosis and clonal typology. In: d'Auzac, J., Jacob, J.L. and Chrestin, H. (Eds.). Physiology of Rubber Tree Latex. CRC Press Inc., Boca Raton, Florida. pp.345-382.

Karunakaran, A., Moir, G.F.T. and Tata, S.J. 1960. The proteins of Hevea latex: ion exchange chromatography and starch gel electrophoresis. In: Proc. Nat. Rubber Res. Conf. 1960, Rubber Research Institute of Malaya, Kuala Lumpur, 1960, 798 p.

Leong, W. and Yoon, P.K. 1982. Modification of crown development of Hevea brasiliensis by cultural practices. II. Tree density. Journal of the Rubber Research Institute of Malaysia, 30, 128-130.

Low, F.C. 1978. Distribution and concentration of major soluble carbohydrates in Hevea latex. The effects of Ethephon stimulation and the possible role of these carbohydrate in the latex flow. J. Rubber Res. Inst. Malays. 26: 21.

Milford, G.F.J., Paardekooper, E.C. and Ho Chai Yee. 1969. Latex vessel plugging, its importance to yield and clonal behaviour. J. Rubber Res. Inst. Malaya, 21: 274.

Mohankrishna, T., Bhaskar, C.V.S., Sanjeeva Rao, P., Chandrashekar, T.R., Sethuraj, M.R. and Vijayakumar, K.R. 1991. Effect of irrigation on physiological performance of immature plants of Hevea brasiliensis. Indian J. Nat. Rubb. Res. 4 (in press).

Moir, G.F.J. and Tata, S.J. 1960. The proteins of Hevea brasiliensis latex. III. The soluble proteins of bottom fraction. J. Rubb. Res. Inst. Malaya, 16: 155.

Moraes, V.H.F. 1977. Rubber. In: Alvim, P. de T. and Kozlowski, T.T. (Eds.). Ecophysiology of Tropical Crops. Academic Press, New York, pp. 315-331.

Narayanan, R., Gomez, J.B. and Chen, K.T. 1973. Some structural factors affecting the productivity of Hevea brasiliensis. II. Correlation studies

160

between structural factors and yield. J. Rubber Res. Inst. Malaya, 23: 285.

Napitupulu, L.A. 1977. Planting density experiment on rubber clone AVROS 2037. Bulletin Balai Penelitian Perkebunan Medan 8: 99-104.

Ng, E.K., Abraham, P.D., P'ng, T.C. and Lee, C.K. 1969. Exploitation of modern Hevea clones. J. Rubb. Res. Inst. Malaya, 21: 292.

Ng, N.P., Abdullah Sepien, C.B., Ooi and W. Leong. 1979. Report on various aspects of yield, growth and economics of a density trial. Proceedings of the Rubber Research Institute of Malaysia Planters Conference, 1979. Kuala Lumpur. pp. 303-331.

Ninane, F. 1970. Les Aspects Ecophysiologiques de la Productivite chez Hevea brasiliensis an Cambodge, These, Universite Catholique de Louvain, 1970.

Paardekooper, E.C. and Somosorn, S. 1969. Clonal variation in latex flow pattern. J. Rubb. Res. Inst. Malaya, 21: 264.

Paardekooper, E.C. and Sookmark, S. 1969. Diurnal variation in latex yield and dry rubber content, and relation to saturation deficit of air. J. Rubb. Res. Inst. Malaya, 21: 341.

Pakianathan, S.W., Boatman, S.G. and Taysum, D.H. 1966. Particle aggregation following dilution of Hevea latex: A possible mechanism for the closure of latex vessels after tapping. J. Rubb. Res. Inst. Malaya, 19: 259.

Pakianathan, S.W., Haridas, G. and d'Auzac, J. 1989. Water relations and latex flow. In" Jacob, J.L. and Chrestin, H. (Eds.). Physiology of Rubber Tree Latex. CRC Press Inc., Boca Raton, Florida, pp. 233-256.

Paton, F.J. 1953. The importance of yellow fraction in spontaneous coagulation. Arch. Rubbercult. Extra No. 2, 93.

Premakumari, D., Sherief, P.M. and Sethuraj, M.R. 1980. Lutoid stability and rubber particle stability as factors influencing yield depression during drought in rubber. J. Plant. Crops., 8: 43.

Premakumari, D., Joseph, G.M. and Panikkar, A.O.N. 1985. Structure of the bark and clonal variability in Hevea brasiliensis Muell. ARg. (Willd. ex A. Juss.). Annals of Botany, 56: 117-123.

Prevot, J.C., Jacob, J.L. and Vidal, A. 1984a. Le Potentiel d' Oxydoreduction. Etude de sa Signification Physiologique dans le Latex de l'Hevea brasiliensis. Son Utilisation comme critere Physiologique, Rapprt 2/84, Inst. Rech. Caoutch. CIRAD, Montpellier, 1984.

Prevot, J.C., Jacob, J.L. and Vidal, A. 1984b. The redox potential of latex criteriam of the physiological state of laticiferous system in C.R. Coll. Exp. Physiol. Amel. Hevea, IRCA-CIRAD, Montpellier, France, 227 pp.

Pujarniscle, S. 1968. Caractere lysosomal des lutoides due latex d'Hevea brasiliensis Muell. Arg. Physiol. Veg., 6: 27.

Pujarniscle, S. 1969. Etude de quelques facteurs intervenant sur la permeabilitie et la stabilite de la membrane des lutoides due latex d' Hevea brasiliensis Muell. Arg. Physiol. Veg., 7: 391.

Pushpadas, M.V., Kochappan Nair, K., Krishnakumari, M. and Karthikakutty Amma, M. 1975. Brown bast and nutrition: a case study. Rubber Board bull. 12: 83-88.

Ribaillier, D. 1968. Action in vitro de certains ions mineraux et composes organiques sur la stabilite des lutoides du latex d' Hevea. Rev. Gen. Caoutch. Plast., 45: 1395.

Ribaillier, D. and d'Auzac, J. 1970. NOuvelles perspectives de stimulation hormonale de la production chez l' Hevea brasiliensis. Revue. Gen. Caoutch. Plastiq., 47: 433.

Ribaillier, D. 1971. Etude de la variation saisonniere de quelques proprietes du latex d' Hevea brasiliensis. REvue. Gen. Caoutch. Plastiq., 48: 1091.

Riches, J.P. and Goodding, E.G.B. 1952. Studies in the physiology of latex. I. Latex flow on tapping - theoretical considerations. New Phytol., 51: 1.

Ruinen, J. 1950. Microscopy of the lutoids in Hevea latex. Ann. Bogoriensis, 1: 27.

Saraswathy Amma, C.K. and Sethuraj, M.R. 1975. Clonal variation in latex flow characteristics and yield in rubber tree (Hevea brasiliensis). J. Plant. Crops, 3: 14-15.

Satheesan, K.V., Gururaja Rao, G. and Sethuraj, M.R. 1982. Clonal and seasonal variations in osmotic concentration of latex and lutoid serums of Hevea brasiliensis Muell. Arg. Proc. PLACROSYM V 1982, Kasargod, ISPC, 1984. pp. 240-246.

Sethuraj, M.R. 1968. Studies on the physiological aspects of rubber production. 1. Theoretical consideration and preliminary observations. Rubb. Board bull., 9: 47.

Sethuraj, M.R., Joseph G. Marattukalam, George, P.J. and Markose, V.C. 1974a. Certain physiological parameters to determine the high yielding characteristics in rubber. Paper presented at the IRRDB Symp., Cochin, 1974.

Sethuraj, M.R., Usha Nair, N., George, M.J. and Sulochanamma, S. 1974b. Studies on the physiological aspects of yield stimulation in Hevea brasiliensis Muell. Arg. Int. Rubb. Res. Dev. Bd. Symp. Cochin, 1974.

Sethuraj, M.R., Sulochanamma, S. and George, P.J. 1974. Influence of the initial flow rate, latex vessel rows and plugging index on the yield of hand pollinated clones of Hevea brasiliensis Muell. Arg. with Tjir 1 as the female parent. Ind. J. Agric. Sc., 44: 354-356.

Sethuraj, M.R., George, M.J. and Sulochanamma, S. 1975. Physiological studies on yield stimulation of Hevea brasiliensis. Proc. Int. Rubb. Conf. 1975. Kuala Lumpur, RRIM, 1976. pp. 280-289.

Sethuraj, M.R. 1977. Studies on the physiological factors influencing yield in Hevea brasiliensis Muell. Arg. Ph.D. thesis, Banaras Hindu University, Banaras, India.

Sethuraj, M.R. and George, M.J. 1976a. Drainage area of the bark and soil moisture content as factors influencing latex flow in Hevea brasiliensis. Indian J. Plant Physiol., 19: 1.

Sethuraj, M.R. and George, M.J. 1976b. Multiple band application of ethepon to increase the drainage area and yield. Paper presented at the IRRDB Symp., Cisaura, Indonesia, 1976.

Sethuraj, M.R. 1981. Yield components in Hevea brasiliensis. Theoretical considerations. Plant Cell Environ., 4: 81.

Sethuraj, M.R., Gururaja Rao, G. and Raghavendra, A.S. 1984. The pattern of latex flow from rubber tree (Hevea brasiliensis) in relation to water stress. Paper presented at the symposium on Cellular and Molecular Biology of Plant Stress, Keystone, Colorado, 1984.

Sethuraj, M.R. 1985. Physiology of growth and yield in Hevea brasiliensis. In: Proc. Int. Rubb. Conf., Vol. 3, Rubber Research Institute of Malaysia, Kuala Lumpur, 1986, pp. 3-19.

Sethuraj, M.R. 1987. Rubber, In: Sethuraj, M.R. and Ravhavendra, A.S. (Eds.). Tree Crop Physiology. Elsevier, Amsterdam, pp. 193-224.

Sherief, P.M. and Sethuraj, M.R. 1978. The role of lipids and proteins in the mechanism of latex vessel plugging in Hevea brasiliensis. Physiol. Plant. 42: 351-353.

Simmonds, N.W. 1982. Some ideas on botanical research on rubber. Trop. Agric. Trin., 59: 1.

Southorn, W.A. 1968. Latex flow studies. I. Electron microscopy of Hevea brasiliensis in the region of the tapping cut. J. Rubb. Res. Inst. Malaya, 20: 176.

162

Southorn, W.A. and Edwin, E.E. 1968. Latex flow studies. II. Influence of lutoids on the stability and flow of Hevea latex. J. Rubb. Res. Inst. Malaya, 20: 187.

Southorn, W.A. and Yip, E. 1968a. Latex flow studies. V. Rheology of fresh Hevea latex flow in capillaries. J. Rubb. Res. Inst. Malaya, 20: 236.

Southorn, W.A. and Yip, E. 1968b. Latex flow studies. III. Electrostatic consideration in the colloidal stability of fresh Hevea latex. J. Rubb. Res. Inst. Malaya, 20: 201.

Southorn, W.A. 1969. Physiology of Hevea (latex flow). J. Rubb. Res. Inst. Malaya, 21: 494.

Southorn, W.A. and Gomez, J.B. 1970. Latex flow studies. VII. Influence of length of tapping cut on latex flow pattern. J. Rubb. Res. Inst. Malaya, 23: 15.

Subramaniam, A. 1975. Molecular weight and other properties of natural rubber: A study of clonal variations. Proc. Int. Rubb. Conf., 1975. Kuala Lumpur, IV, 3.

Subronto, 1978. Correlation studies of latex flow characters and latex mineral content, in Int. Rubb. Res. Dev. Board Symp. Rubber Research Institute Malaysia, Kuala Lumpur, 1978.

Tata, S.J. and Moir, G.F.J. 1964. The proteins of Hevea brasiliensis latex. V. Starch gel electrophoresis. J. Rubb. Res. Inst. Malaya, 21: 477.

Templeton, J.K. 1969. Where lies the yield summit for Hevea? Plrs' Bull. Rubb. Res. Inst. Malaya No. 104, 220 pp.

Tupy, J. 1969. Stimulatory effects of 2,4-dichlorophenoxyacetic acid and 1-naphtylacetic acid on sucrose level, invertase activity and sucrose utilization in the latex of Hevea brasiliensis. Planta, 88: 144.

Tupy, J. 1973. Possible mode of action of potassium on latex production. Symposium of the International Rubber Research and Development Board, Puncak, Indonesia, 2-4 July, 1973.

Tupy, J. and Primot, L. 1976. Control of carbohydrate metabolism by ethylene in latex vessels of Hevea brasiliensis Muell. Arg. in relation to rubber production. Biol. Plant. 18: 373.

Usha Nair, N., Sherief, P.M., Sulochanamma, S. and Sethuraj, M.R. 1978. Clonal variations in the colloidal properties of latex and its relation to latex flow in Hevea brasiliensis. Proc. PLACROSYM I, Kasargod, ISPC, 1978, 00 307-315.

Usha Nair, N., Saleenamma Mathew, Sulochanamma, S. and Sethuraj, M.R. 1980. Clonal variation in lutoid stability, B-serum activity and C-serum activity in Hevea brasiliensis. Paper presented at the Int. Rubb. Conf. India, 1980.

Usha Nair, N., Molly Thomas, Sreelatha, S., Sheela, P.S., Vijayakumar, K.R. and George, P.J. 1990. Clonal variations in the activity of 3-Hydroxy-3-Methyl Glutaryl-CoA reductase in bark of Hevea brasilinesis. Indian J. Nat. Rubb. Res., 3: 40-42.

Vischer, W. 1920. De anatomische bouw van het latex Vatenstelsel bij Hevea in verband met de latex productie. Archf. Rubbercult. Ned-Indie, 4: p. 473.

Vischer, W. 1922. Over een proef om de latex beweging in de latex vaten van Hevea brasilinesis bij het tappen experimenteel aan te toenen. Archf. Rubbercult. Ned-Indie., 6: 444 pp.

Watson, G.A. 1989. Nutrition. In: Webster, C.C. and Baulkwill, W.J. (Eds.). Rubber. Longman Scientific and Technical, England. pp. 291-348.

Webster, C.C. and Baulkwill, W.J. 1989. Rubber. Longman Scientific and Technical, England, 604 pp.

Wittshire, J.L. 1934. Variations in the composition of latex from clone and seedling rubber. Bull. Rubb. Res. Inst. Malaya, 5: 1.

Wittsuwannakuk, R. and Sukonrat, W. 1984. Diurnal variation of 3-Hydroxy-3-methyl glutaryl coenzyme A reductase activity in latex of *Hevea brasiliensis* and its relation to rubber content. Compte-Rendu da Colloque Exploitation Physiologie at Amelioration de l'*Hevea*, Montpellier, France. 115-120 pp.

Woo, C.H. 1973. Rubber coagulation of *Hevea brasiliensis* latex. J. Rubber Res. Inst. Malaya, 23: 323.

Xu Wenxian and Pan Yanging. 1990. Progress in cold tolerance physiology of *Hevea brasiliensis* in China. Proc. of IRRDB Symposium on Physiology and Exploitation of *Hevea brasiliensis*, Kunming, China, 6-7 Oct. 1990.

Yeang, H.Y. and Paranjothy, K. 1982. Initial physiological changes in *Hevea* latex and latex flow characteristics associated with intensive tapping. J. Rubb. Res. Inst. Malaya, 30: 31.

Yip, E. and Gomez, J.B. 1980. Factors influencing the colloidal stability of fresh Hevea lattices as determined by the Aerosol OT test. J. Rubb. Res. Inst. Malaysia, 32: 1-19.

CHAPTER 8

PROPAGATION AND PLANTING

JOSEPH G MARATTUKALAM and C.K. SARASWATHYAMMA
Rubber Research Institute of India, Kottayam-686009, Kerala, India.

In the beginning of the rubber plantation industry in South-East Asia more than a century ago, the main source of propagation was unselected seeds. Later, using selected seeds from best yielding trees, appreciable yield increase could be achieved (Dijkman, 1951).

The vegetative propagation of selected trees by budding on the seedling materials was developed by Van Helten (Dijkman, 1951). This method attained commercial acceptability in the early 1920s. Among the many vegetative methods of propagation possible, budgrafting alone has gained commercial adoption.

PROPAGATION MATERIALS

Seeds

Rubber trees generally produce seeds once in a year. In some countries such as Malaysia, a second round of seed production is also noticed (Paranjothy, 1980). Shape, size, weight, pattern of mottlings etc., vary very widely in seeds of different clones, and these characteristics are used as clues for the identification of cultivars.

Seeds collected from clonal trees are called clonal seeds while those produced by seedling trees are known as ordinary seeds. Ordinary seeds, if collected indiscriminately from seedling areas, are termed 'unselected' and those collected from selected seedling trees are known as 'selected'. Similarly, clonal seeds are also of two types, namely monoclonal and poly-clonal. Monoclonal seeds are collected from monoclonal areas whereas, polyclonal seeds are produced in specially raised polyclonal seed gardens. Among the different categories of seeds, polyclonal seeds are considered superior due to their hybrid nature.

Clones show very wide variation in the number of seeds produced. Natural fruit setting is very low (1-2%). Seed production of about 150 kg ha^{-1} is considered reasonable though different estimates are available

(Simmonds, 1989).

Heavier seeds produce more vigorous seedlings and hence selecting such seeds can give better results (Saraswathy Amma and Nair, 1976). Seeds are germinated in raised germination beds. Beds of convenient length, 90 cm width with walking space in between and their surfaces raised 10 to 15 cm above the soil surface are made for this purpose. A free-draining friable material like river sand is the common medium for germination. This is spread above the bed to a thickness of about 5 cm. While germinating very valuable seeds, containers filled with soil and a 5 cm thick top layer of germinating medium are used (Subramaniam, 1980). Seeds are spread horizontally in a single layer touching one another, and pressed lightly into the rooting medium until the micropyle is buried under it (Dijkman, 1951). Then the beds are covered with a layer of gunny bags, coir matting or similar material to prevent loss of too much moisture from the rooting medium. Beds are irrigated lightly every day, to keep up a high level of moisture.

Seeds begin to germinate about seven days after sowing, and those which do not germinate within 14 to 21 days should not be used as they are likely to give rise to weak seedlings (Edgar, 1958; Potty, 1980). Eighty per cent germination is considered good. Rubber shows hypogeal germination (Kuruvilla, 1965). On germination, the radicle pushes open the cap which closes the micropyle and emerges as a small white stump with a flattened end. As it elongates, the end becomes conical and a number of minute points develop around its margin, which in turn develop in to lateral roots. The conical tip grows into the tap root. Laterals develop more quickly than the tap root. Though the cotyledons remain inside the seed and absorb food materials stored in endosperm, their stalks are pulled outwards and the young shoot is seen in between them with the plumule just inside the seed. When the tap root has developed, the epicotyl emerges as a loop, and as it grows, the plumule is pulled out of the seed. The shoot subsequently straightens up and becomes vertical (Edgar, 1958).

Beds should be inspected every day and the germinated seeds should be removed as soon as the radicle emerges (Johnson, 1959). If the root is allowed to elongate chances of damage during transfer from the beds are greater.

Germinated seeds are planted in seedling nurseries, polybags or in the field to raise seedlings. Seedling nurseries are established in well drained soils with enough water for irrigation. Easy access, proper protection and facilities for effective supervision are also necessary for the proper upkeep of the nursery. If the land is not flat, bench terracing is necessary. First, the soil is loosened by digging to a depth of 60 to 75 cm. All stones, stumps and roots are removed and the soil is made into a fine tilth (Potty, 1980). If the nutrition level of the soil is low, it has to be improved by mixing fertilizers with the soil. Beds either sunken or raised are prepared to convenient length and to a width of 90 to 120 cm. Wherever necessary, drainage and pathways should be provided. Germinated seeds are then planted in this nursery. For planting, small pits wide enough to accommodate the seeds in a horizontal position and about 5 cm deep are made. Seeds are gently placed in these pits with the radicle turned downwards and then covered with soil. Different spacings are adopted for the planting of seeds in the nursery depending on the type of planting material intended to be produced. To produce pencil stumps and green budded stumps spacing adopted is 23x23 cm (Edgar, 1958). For brown budded and seedling stumps spacing of 30x30 cm or 60x15 cm is preferable. A spacing of 60x60 cm or 90x30 cm is required to produce mini stumps. For maxi stumps and three part stumps (stumped three part plants) convenient distances are 90x90 cm or 90x60 cm. Plants intended for soil core plants and small polybag plants are raised at a spacing of 60x60 cm. Large polybag plants are prepared by keeping them in double rows, the distance between rows of bags being 30 cm and that between two bags of the same row being 20 cm. Two such double rows are kept 75 cm apart. To raise brown budwood plants, the spacing can be 120x60 cm or 90x60 cm. For green bud-shoot nursery a spacing of 120x120 cm is adequate. Thinning is carried out to get rid of very weak plants and boost the growth of the remaining plants. It can be done in two stages, at four and six months from planting. By this practice the remaining seedlings will show a high degree of uniformity in growth (Rubber Research Institute of Malaya, 1959a). The seedlings are nursed carefully by manuring, mulching, weeding, irrigation, control of pests and diseases etc. at the appropriate time.

If properly attended to, the seedlings grow vigorously. First, the shoot pushes upwards vertically at a slow pace for the first two or three days followed by rapid growth. After attaining some height, vertical growth temporarily ceases and leaves develop. When the leaves have fully grown the terminal bud resumes its growth and the second cycle of growth commences resulting in the formation of another storey of leaves. This process is repeated. A complete cycle takes about 36 days, 18 days for the elongation of shoot and the remaining period for the development of leaves. Plants under proper care, attain a girth of 7 to 10 cm after one years growth. Then they are pulled out either before or after budding to prepare seedling stumps and budded stumps respectively.

Seedling stumps

A seedling made to a convenient size by pruning the stem and the roots is called a seedling stump. Only healthy and vigorous seedlings attaining a girth of at least 7.5 cm above the collar and developing brown coloured bark upto a height of 45 cm or more from the collar, in one year are selected for preparing seedling stumps. First, the seedlings are cut back at the point where the brown colour ends, preferably above a whorl of buds by a slanting cut. While cutting back, green or partially brown stem should never be retained on the stump. Such regions are capable of transpiration causing loss of water which may result in the death of plants after planting in the field, especially under unfavourable conditions (Edgar, 1958). Cutting back is best done seven to ten days before planting. During this period, a few buds below the cut end become activated and begin to swell. Then the plants are pulled out carefully, causing minimum damage to the root system and bark of the stem. In the case of larger plants, and when the soil is dry or stony, digging is carried out to loosen the lateral roots and then are pulled out more easily. After pulling out, the tap root is pruned to the maximum possible length, but not more than 60 cm for the sake of convenience and not less than 45 cm (Ramakrishnan, 1987) for the sake of better establishment. Lateral roots are also pruned to a length of about 10 to 15 cm. Plants with deformed taproots, or those showing symptoms of root diseases etc., should be discarded. If multiple taproots are present, all except the largest one should be cut off. The cut end of the stem is sealed by dipping in melted wax to a length of about 2 cm.

Budded stumps

Budded stumps are prepared by bud grafting the stock plants and then pruning their roots and stem to the required specifications. Among the different types of vegetative propagation methods, budding alone is now adopted for the commercial production of propagating materials. This technique consists of taking a lateral bud of a plant along with a small piece of bark and attaching it to another plant. The plant which donates the bud is called the scion and the patch of bark taken from it is the bud patch. The plant to which the bud is attached is known as the stock plant. After the bud has become firmly attached to the stock, the portion of the stock above the bud is removed, and the bud develops and takes its place, resulting in the formation of a two-part plant comprising of a root system belonging to the stock plant and a shoot system belonging to the scion donor. The parts developing from the bud will possess all the characteristics of the plant from which the bud was taken. The method of budding adopted for _Hevea_ is a modified form of Forkert method of patch budding (Teoh, 1972).

Based on the colour and age of the buds used, budding is classified mainly into two, green budding and brown budding. For green budding tender green coloured buds are used whereas, brown budding is done with more mature brown coloured buds. Depending on the position at which the budding is done on the stock, three types of budding are identified. They are base budding (budding just above collar), crown budding (budding crown portion) and over budding (budding in between base and crown).

Budding is done with a specially designed knife known as budding knife (Joseph et al. 1980). It is a kind of folding knife having two blades about 7 cm long. One of the blades has a groove near the free end on the back side. While budding, the index finger can be kept in this groove, thus giving more support to the blade as well ensuring precise manoeuvring. This blade, which is always kept very sharp, is used to make the incisions as well as trimming the edges of the budpatch. The other blade which is less sharp is used for stripping the bark as well as for other uses that may arise. Keeping the first blade very sharp is essential for carrying out the budding operation smoothly and quickly. An ordinary pen knife with a folding blade about 7 cm long can also be used for budding.

After placing the budpatch on the stock plant it is kept in position; until it becomes firmly attached, by bandaging with a suitable material. Transparent polythene film of about 62.5 microns (250 gauge) thickness is the most commonly used material for this. Waxed cloth, jute twine, coconut leaf, coir, wire ring etc also could be used as alternate binding materials (Rubber Research Institute of Malaya, 1947). When polythene film and waxed cloth are used they are cut into strips of suitable size before using.

Another important requirement for budding is the wiping material. Different materials such as rags, cloth cuttings, cotton waste and cotton wool could be used for this. During the whole process of budding, utmost cleanliness should be maintained. Budding knife, hands of the budder, bandaging material, budwood, stock plant, etc should be kept clean by wiping with a suitable material whenever required. It is advisable to carry all the materials required for budding in a small box, to ensure cleanliness and to avoid damage and loss.

Green budded stumps

In green budding, young stock plants, two to eight months old, are used (Hurov, 1960). Healthy and vigorous plants having a girth of at least 2.5 cm above the collar with brown bark upto a height of 15 cm or more is suitable for this type of budding. Under normal conditions of growth, three to five months are required for stock plants to attain this stage. With special cultural operations, such as intensive manuring, they could be brought to buddable size within two months (Leong et al. 1985). Stocks are used for budding when the bark peels off very easily. This generally happens when the top whorl of leaves are fully expanded but not hardened (Joseph et al. 1980). In case of doubt, peeling quality can be ascertained by test peeling a small patch of bark about 15 cm above the collar. All stock plants may not be ready for budding simultaneously and hence only a part of them can be budded at a time, necessitating several rounds of budding. The first step involved in budding is the thorough cleaning of the base of the stock plant upto a height of 15 cm. Then a vertical incision 5 cm long and starting from a point about 2.5 cm above the collar is made. This is followed by another similar incision, parallel to the first one and 1 cm apart. Lower ends of these incisions are connected by a horizontal one. The plant is left in this condition for a few minutes until the oozing of latex from the wounds is over. During this time, a few more plants are marked. When the dripping ceases, latex is removed by wiping. The piece of bark marked out on the seedling is now gently lifted upwards. The opening formed is the budding panel where the budpatch is to be placed. The stripped flap of bark is cut and removed leaving a short tongue about 1.5 cm long at the top. The stock is now ready to receive the budpatch.

Buds used for green budding are collected from young green shoots with one whorl of leaves. Depending on the vigour of growth, six to eight weeks are required for the shoots to attain this stage. Green bud shoots are produced on specially raised source bush plants in a green bud shoot nursery. For raising a source bush plant, the scion of a well established budded plant is cut back at a height of about 75 cm. From among the several shoots that spring up from below the cut, three to five of the most vigorous ones are retained and the others removed (Rubber Research Institute of Malaya, 1964). When they develop brown colour to a length of a few centimeters, they are again cut back at the top of the brown portion. Vigorous shoots growing from these branches are allowed to develop. By repeating this process a bush having many branches is formed (Tinley, 1962). When these branches produce one whorl of leaves they are harvested by cutting at the base with a sharp knife and fresh shoots are again allowed to develop. Since only six to eight weeks are required for the development of one set of shoots, several rounds of harvesting are possible in the same year. Green bud shoots are harvested when the bark peels easily. Generally, peeling is poor in the case of very tender bud shoot. More mature shoots with pinkish green to dark green leaves show better peeling of the bark. After harvesting, the top portion along with the leaves are cut off. The non-leafy part bears three to five scale buds which are used for green budding. Buds in the axils of normal leaves are not generally used for budding because they are more difficult to peel and give less budding success. However, in special cases, such buds, in the axils of the lower leaves of the whorl which are widely spaced can be used. Before stripping such buds, the leaf stalk subtending the bud should be cut flush with the stem. Doing this about seven days before budding gives better budding success (Togun and Ajibike, 1964). They should also be peeled from bottom to top. Exposing these buds by clipping the leaves two to three weeks before peeling may cause the falling of the stump of the leaf stalk. The cutting of the petiole flush to the stem could be avoided by this. However, this practice is likely to delay the sprouting of the bud after cutting back the stock plant. Sprouting of buds is comparatively slow in the case of scale buds than leaf buds. However, the shoots emerging from scale buds exhibit more vigorous growth during the initial stages.

Two methods can be adopted for stripping the budpatch from the green bud shoot. In the first method, the budpatch is marked out by four incisions each 5 cm long and parallel to each other are made on the two sides of the bud at a distance of 0.5 cm from the bud. Two transverse incisions are then made connecting the adjacent ends of the vertical cuts. When the flow of latex ceases, it is wiped and the budpatch is removed carefully. After taking out, the budpatch is examined to ascertain the presence of core (eye) of the bud on the inner side. The core appears as a small protuberance below the bud which has the appearance of a small pin-head like projection externally. Budpatches not having the core are discarded. The four edges of the budpatch are now trimmed to remove damaged tissues if any, and its length and breadth made slightly less than 5 cm and 1 cm respectively. In the other method, the budpatch is first taken from the shoot along with a thin slice of wood. This is called the budslip. Two sides of the budslip are then trimmed to a width of little less than 1 cm. The budpatch and slice of wood are now separated from each other gently by pulling them apart. While doing so, the budpatch should not be bent as it may damage the tissues and reduce budding success. If necessary the slice of wood can be bent. Both ends of the budpatch are now trimmed to make it slightly shorter than 5 cm. While handling the budpatch, care should be taken not to touch its inner side. It must be held by the edges only. Foreign bodies such as sand, dust particle, water drop, sweat etc. should not be allowed to fall on this side. Neither should this side be exposed to sunlight for long, as otherwise the delicate cambium may get damaged resulting in failure of bud union. If any dust particle falls on the cambium it can be removed very carefully with the pointed tip of the budding knife without causing any damage to the cambium.

The budpatch is then placed in the budding panel very gently taking care to ensure that the relative positions of leaf scar and bud remain the same as they were on the bud stick. The upper end is inserted under the tongue of bark of the stock plant after lifting it a little. Then the remaining part of the budpatch is placed in the budding panel and held in position by gently pressing against the stock stem. Once placed in the budding panel, the budpatch should on no account be allowed to slip away from position as it may cause rubbing of the cambial layers of the stock and budpatch resulting in their damage. Now the budpatch is fixed in its position by bandaging with a strip of transparent polythene. Transparent strip is preferred because it allows sunlight to fall on the budpatch which in turn favours better budtake. However, very bright sunlight may cause damage to the buds and hence under such conditions it is preferable to carry out budding on the shaded side of the stock plant (Tinley, 1962). Dimensions of the tape can be 25x2 cm for a stock plant of average size. For large stock plants, length of the strip may have to be increased. Bandaging should commence at the bottom and move upwards in close spirals, taking care to prevent slipping of the budpatch. Bandaging should be fairly tight. Tie pressure of 4 kg or more per square centimeter is necessary for good callusing (O.F.W. 1931). After covering the budpatch completely, the end of the strip is kept intact at the top by a tight knot.

The plant is left in this position undisturbed for 20 days. During this time, cambium of the stock and budpatch become united and the budpatch permanently establishes itself as an inseparable part of the stock. After this period the bandage is removed and the budpatch is observed carefully. If it retains green colour the budding is a success. If it has failed, the colour will be dark brown. In the case of successful budded plants, the tongue is cut and removed carefully. Budding on the opposite side can be attempted in case of the failed stocks. Successful plants are retained in the nursery for ten more days to ascertain whether any of the successful buds may fail subsequently. Those which are found to retain green colour at the time of this observation can be considered as final success.

Budded plants should be preferably transplanted soon after bud take. If delayed very much, it may prolong the dormancy of buds after transplanting. Successfully budded plants are cut back at a height of about 7.5 cm above the upper end of the budpatch, giving a 45° slant to the cut. The slant should be from the side of bud towards the opposite side. Cutting back is done about seven days before pulling out. During this period the bud begins to swell, which will result in better success after transplanting. Plants are then pulled out and the roots are pruned as in the case of seedling stumps. While gripping the stem of the plant for pulling out, care should be taken not to subject the bud to too much pressure. Pulling out the plant after cutting back is sometimes found very difficult because the stock stem is too short to provide a good grip. To avoid this, plants can also be pulled out before cutting back followed by cutting back of stem. Roots are then pruned as in the case of seedling stumps. However, for planting in bags the tap root should be pruned to 15 cm less than that of the soil core. A budded plant prepared in this way is called a green budded stump. To prevent loss of moisture, the cut end of the stem is sealed with melted paraffin wax. If cutting back is delayed very much, dormancy of bud after transplanting may be prolonged.

The budpatch of each plant is protected by covering it with a piece of banana sheath. As in the case of seedling stumps, plants with good tap root alone should be used as planting material. If the stump is to be planted immediately, no other treatment is necessary. Pulling out budded plants, where green colour of the budpatch is retained, three weeks after budding is also practiced. In the case of these plants, polythene bandage is not removed before pulling out. It is retained on the plant to protect the bud on transit, and is removed just before planting (Chandra Samaranayake, 1984).

Brown budded stump

Brown budded stumps, produced by brown budding, were traditionally adopted in Hevea until the technique of green budding was developed. This budding method was perfected in Indonesia by the horticulturist Van Helten in 1916 in collaboration with Bodde and Tass, two planters. The first hand book on the subject was published in 1918 by Bodde (Dijkman, 1951).

Older stock plants aged ten months or more having a basal girth of around 7.5 cm, are used as stock plants for this type of budding (Rubber

Research Institute of Malaya, 1939). However, more than the age, size of the plant is the criterion adopted for the selection of stock. As in the case of green budding, brown budding also is carried out on vigorously growing healthy plants. Peeling quality of the bark should be very good, which usually happens when the top whorl of leaves is growing vigorously and the leaf whorl well formed, but before further extension growth commences (Edgar, 1958). However, test peeling is the sure method to ascertain the peeling quality.

Brown buds are obtained from mature shoots which are about one year old and brown in colour. Slightly younger and older shoots also could be used. These shoots, called brown budwood are collected from plants named brown budwood plants maintained in a brown budwood nursery. Each plant is usually allowed to develop one or two shoots depending on the vigour and the spacing adopted for their planting. Each healthy shoot, after one years growth, will yield two to three metres of brown budwood. It is not economical to harvest budshoot which has not developed atleast 1.2 m of brown wood (Mann, 1929). Buds found in the axils of fallen leaves alone are used for budding. If leaves have failed to fall off from any of the brown coloured portion of the budwood they are clipped off leaving the leaf stalk on the plant. This is done four to seven days before harvesting the budwood. Within a few days these stalks will fall off rendering these portions of the budwood also usable. However, budwood made usable in this manner should be used within one week after the clipping of leaves. If not, buds may sprout and the budwood will become unusable. Girdling the budwood by removing a ring of bark near the base, nine to fourteen days before harvesting, increases budding success (White and Imle, 1944).

Budwood is harvested by sawing the shoot with a pruning saw at about 15 cm above the base. The basal 15 cm is left to facilitate the development of fresh bud shoots for the next season. Harvesting should be done only when the peeling quality of the bark is good, which is often so when the topmost flush of leaves have fully expanded but not hardened (Joseph et al. 1980). In case of doubt, test peeling can be carried out to assess the peeling quality. Budwood should be cut in the morning, if for the same days use and in the evening if for transit, so that they could be transported at night avoiding the heat of the day. Harvesting during the hot hours of the day, when the water content in the tissues is very low, should be avoided as it will adversely affect its quality. After harvesting, the brown portion is cut into pieces of convenient length usually one metre. The partially brown or green top portion is discarded as buds taken from this portion generally give comparatively less budding success (Sharp, 1932). The pieces of budwood are then properly labelled and used for budding. The process of brown budding is same as that for green budding. However, there are two main differences. Width of budding panel is 1.5 cm and the flap of the bark stripped from the stock is not cut and removed but placed back on the budpatch after its insertion and secured by bandaging. In this case, opaque materials also could be used for bandaging. Since the stock plants are comparatively larger, bigger strips of about 45x2.5 cm have to be used. In the case of plants which are exposed to strong sunlight this portion can be protected by shading with a rubber leaf which is tied over this part.

Twenty days after budding, the bandage and flap are removed. If opening is delayed, a thin white layer of callus may completely cover the budpatch which will also have to be removed very carefully. The outer brown coloured tissue of the budpatch is gently scraped at a point above or below the bud. If green colour appears, budding can be considered a success. Ten more days are allowed, after which one more round of inspection is made. If green colour is still retained, final confirmation can be made. Failed plants can be budded on the opposite side.

Successfully budded plants are pulled out and budded stumps are prepared more or less in the same way as that of green budded stumps. However, advance cutting back of the stem is done about 10 days before extraction. For making the pulling out easy, cutting back can be done after pulling out also. Cut end of the stem is sealed with melted paraffin wax. Budpatch of the stump is protected using a piece of banana sheath.

A comparison of green budding and brown budding in terms of advantages and disadvantages would be useful. Green budding is simpler and faster and hence more budding could be done within a given time than brown budding. Since several crops of green budshoots can be taken from the same nursery in a year, availability of green buds is two to three times greater than that of brown buds, which could be harvested only once in a year. Size of the polythene strip used for green budding is much smaller than those used for brown budding which results in savings in material cost. Wastage due to cutting back of the stock stem is less in the case of green budding as the plants are cut back when they are very young. Green budding technique is more suited for the production of advanced planting materials such as bag plants. A reduction in the immaturity period is sometimes obtained by adopting the green budding technique. Green bud shoots cannot be retained in the source bush nursery in a viable stage for long. After the first whorl of leaves have developed, the shoots will remain in that condition only for a few weeks. They have to be harvested at this stage, otherwise the second whorl of leaves will begin to develop and the buds become unusable. In the case of brown buds, however, their viability is maintained for several months even after it has become usable. Green budshoots after harvesting can be stored for a short duration only because they are tender, and chances for loss of moisture is more, whereas, brown budwood which is more mature could be preserved for longer time. Similarly, green budded stumps also can be stored for a shorter duration than brown budded stumps. Chances for diseases is more in green budwood nurseries as the shoots are always in a tender stage compared to brown budwood nursery. Due to its smaller size green budded plants contain limited quantity of food materials and as such the scion is smaller and weaker during the initial stages of development. Their ability to withstand adverse conditions is also therefore limited (Rubber Research Institute of Malaya, 1963). As a result, loss of plants due to die back and other reasons, after planting is more (Rubber Research Institute of Malaya, 1965). Hence special attention has to be paid for their protection during the early parts of growth (Rubber Research Institute of Malaya, 1963). Brown budded stumps on the other hand produces more healthy scion shoot which grow more vigorously and hence are more hardy in resisting adverse conditions. Green budding is more suited for crown budding as it reduces loss of crown by wind snap and die back (Rubber Research Institute of Malaya, 1968b). Green budding generally gives more success than brown budding. The differences are more pronounced during summer (Joseph and Premakumari, 1981). As a result, budding can be carried out on young stock plants during summer and bagplants raised soon, so that they can be planted out in the field during the ensuing planting season. If brown budding technique is adopted, budding operation is possible only during the first rainy season and further waiting of one more year is necessary for field planting of the bagplants. This aspect has special significance in countries like India where the seedfall is during July-September and the planting season June-July.

There is no definite budding season. Budding can be carried out at any time of the year. However, periods of showery weather generally gives best budding success (Edgar, 1958). Too dry or too wet weather is unsuitable for budding (De Silva, 1957) as budding success will be less during these periods. During summer budtake is adversely affected mainly by the poor growth of plants and low moisture content of the plant tissues, two factors which are essential for the budtake. During heavy rain maintenance of proper cleanliness is difficult which in turn causes budding failures.

Depending on the nature of soil and its water-holding capacity, effect of drought may vary. It is more pronounced in light sandy soils. It is advisable not to do budding during dry periods even if high budding success is obtained initially, because subsequent damage to buds will be more when exposed to hot weather. Since young stock plants and budwood do not show wintering there is no harm in carrying out budding during the general wintering period of the estate provided other conditions are favourable. However, if either of them show any symptom of wintering budding should not be attempted as the peeling quality of bark is poor at this stage. Mornings and evenings are the best time for budding. As far as possible field budding should be avoided during the hot hours of the day between 10 am and 3 pm. However, in nurseries where the plants are protected, budding could be carried out at any time (Chandra Samaranayake, 1984). Studies conducted in India have shown that the important climatic factors having influence on budding success are rainfall, relative humidity and maximum temperature (Joseph and Premakumari, 1981).

"Young Budded" plants

This type of propagating material is produced by budding very young stock plants, less than two months old, with green buds. Seedlings which are seven to eight weeks old are usually used for this type of budding (Ooi et al. 1976). Buds are taken from young green bud shoots of comparable girth. Stock plants are raised in small polythene bags with 33x15 cm layflat dimensions containing about 2.5 kg of soil. Special attention is given to ensure quick growth of the plants. Foliar applications of nutrients and fungicide mixtures are carried out twice weekly for a period of two months (Leong and Yoon, 1985). From second to sixth week, weekly application of 15:5:6:4 NPKMg compound fertilizer with soluble P in slurry form also is undertaken. Green shoots required for young budding are produced in the same way as that of green budding. However, since the stock is smaller, budshoots as small as 6 mm diameter could also be used. Budding can be done at anytime irrespective of growth condition of the top whorl of leaves of stock. Opening of the budding panel is also similar to that for green budding. However, its width should not be more than half the circumference of the stock. Preparation of budpatch and process of budding also are the same as those adopted for green budding. But width of the polythene strip need be only half of that used for green budding. Four weeks after budding, successfully budded plants are cut back. Unlike in the case of green budding, a longer snag of 20 to 25 cm should be retained. A long snag with more food reserves increases the vigour of scion and reduces dieback of the emerging scion. One disadvantage of the long snag is that more stock shoots are likely to develop which have to be pruned repeatedly. This can be avoided to some extent if cutback is done just below one whorl of leaves provided the snag is sufficiently long. Nicking all the snag buds is a sure method that can be adopted to avoid the development of stock shoots. This also results in earlier and more uniform sprouting of buds (Yoon et al. 1987). Application of 'Atrinal' also enhances sprouting of buds. Plants are retained in the bags until the scion shoot produces two to six whorls of leaves and are then transplanted in the field.

Crown budded plants

The crown of several high yielding clones are found to be very much susceptible to diseases, wind damage etc. (Rubber Research Institute of Malaysia, 1980, 1983a). Due to this drawback, the yield potential of these clones is often not exploited fully. If the defective crown were to be replaced by a resistant one, this problem could be solved. Using crown budded plants, which are produced by crown budding, this could be attained. Crown budding technique was first adopted in South America in the 1930s to safeguard the trees from the dreaded South American Leaf Blight (Yoon, 1973). By adopting this technique, ultimately a three-part plant, comprising of the root system of the stock plant, trunk of a high yielding clone and crown of a resistant one is formed.

Crown budding is carried out when the scion has grown to a height of 2.4 to 3.0 m. Depending on the growth vigour, one to two years are generally required for this. Budding is carried out at a point between 2.1 to 2.4 m above the bud union, on the interwhorl region between the top two whorls of leaves. Budding is done only when the top flush of leaves is fully mature. If a plant has grown over 2.4 m and yet is not buddable because the top whorl of leaves has not hardened enough, budding could be attempted below the second whorl of leaves which has already matured fully. However, lowering the height of budding will ultimately cause more loss of plant tissues during cutting back, which subsequently will retard growth of the plant initially. At the site of budding stem should be green or dark green in colour, to ensure maximum budding success. This will also ensure rapid emergence of the crown shoot after cutting back, reduce dormancy of crown bud and enhance crown trunk union which in turn will reduces loss of crown shoot due to dieback. Breakage at the bud union caused by wind is also minimised by this. Method of budding adopted for crown budding is green budding (Rubber Research Institute of Malaysia, 1986), because this is more suited for the green parts of the trunk. Too tender or too mature stem tissues adversely affect budding success. If brown budding is adopted it has to be done much below where, the trunk is more mature and bigger. This will result in loss of a considerable portion of the trunk while cutting back. Due to this loss, as well as the maturity of the bud, dormancy of buds after cutting back is enhanced. However, after emergence, the crown shoot grows very vigorously producing large leaves and heavy flushes, even before the bud union is sufficiently hardened. This results in more incidence of die back and snapping off of the crown shoot at the union. Therefore, in case taller plants have to be budded, it is preferable to raise the height of budding the green regions rather than budding at the brown portion to keep up the prescribed budding height. Since all the plants in an area may not be in the buddable stage at the same time, several rounds of crown budding may be necessary. The first round of budding should be carried out when atleast 50% of the plants are buddable. If plants in a particular area are small in general, due to poor growth, they need not be taken into consideration for calculating percentage buddability of the entire field. As more and more plants become buddable, additional rounds of budding are undertaken until all plants are budded. Climatic requirements, time of budding, etc. of crown budding are similar to those of green budding. Budding is done standing on a self supporting ladder, 'crown budding stool' or box. Cutting back of successful buddings is carried out at a height of about 5 cm above the budpatch. After cutting back, leaves and branches that may be present on the snag are pruned off. Failed plants can be budded again on the opposite side of the stem about 5 cm above or below the first budding point. Budpatches showing even partial drying also should be counted as failures and rebudded, because such cases often fail completely. Cutting back the crown budded plant during the dry spell is not advisable and if necessary, cutting back can be delayed by a few weeks to obtain favourable weather conditions. After cutting back, the cut end may be sealed with a proper wound dressing material to prevent loss of too much moisture. Trunk shoots that may emerge are pruned close to the trunk with a sharp knife or lopper, fitted with long handle. Pruning should be done meticulously every fortnight as it is very essential for the development and proper growth of the crown shoot. Emerging crown shoot should be properly maintained. If more than one crown shoots sprout, only the most healthy one should be retained and others have to be pruned off. Any branches that may come below the third whorl of leaves are also to be removed. In case any 'cluster' branching occurs, only the most healthy one need be retained. Take necessary steps to protect the shoot from diseases and wind damage as and when required. If the crown bud dries after cutting back, two or three trunk shoots are allowed to develop and the most vigorous among them is again crown budded when it produces two whorls of hardened leaves. Budding is done on the side of the branch facing the trunk. If all attempts to crownbud this new shoot also fail, it could be allowed to develop as the crown.

In the case of plants where the crown shoot has emerged, two or three trunk shoots are also allowed to develop till the crown shoot has grown to a height of more than 2.5 cm. They are meant as reserve shoot, to be used in case the crown shoot is lost later on. They are retained at about 15 to 40 cm below the crown shoot and should be spaced apart facing different directions. Only the remaining trunk shoots are pruned further. If trunk shoots are not present at this height, others arising from below could be retained. But those arising from any point below 1.7 m from the basal bud union should not be retained. They should not be allowed to grow taller than the crown shoot. Whenever they overtake the crown shoot their height should be reduced by pruning below the top whorl of leaves. If the crown shoot is lost due to any reason, crown budding can be attempted on any one of these shoots, preferably the most vigorous one. After the crown shoot has grown for about nine months and has firmly established the crown-trunk union, trunk shoots can be completely removed. If all attempts to crown bud the plants fail, one trunk shoot could be allowed to develop as the crown.

Some of the clones used for crown budding are RRIM 612, GT 1, PR 261, AVROS 1279, AVROS 2037 (Rubber Research Institute of Malaysia, 1986), RRII 33, F 4542 (Hevea benthamiana), FX 516 (Hevea benthamiana x Hevea brasiliensis) (Radhakrishna Pillai et al. 1980) and LCB 870 (Chandrasekera, 1980).

The following important precautions have to be taken in connection with crown budding. Do not carry out any routine pruning of branches two weeks before crown budding. Plants should not be bent for budding, cutting back, pruning or any other operations. Fertilizer application should be avoided during the two months before budding. Similarly, manuring should not be done until the first whorl of leaves of the crown shoot has hardened. Herbicides with long persistance in soil should not be used before crown budding. After crown budding certain herbicides such as MSMA, 2,4-D, sodium chlorate, paraquat, diuron etc. could be used in appropriate concentrations and combinations without socrching the shoot.

Important advantages of crown budding are reduction of wind damage and incidence of diseases of crown. This results in savings in the cost of plant protection operations. Production of crop also is increased. Wider choice of planting material is possible. Yield depression occurring during refoliation following wintering is also reduced.

Over budding

This is usually done for converting the upper parts of a budwood plant of one clone to another clone. By adopting this technique, budwood nursery of one clone could be made that of another clone without replanting it. In the case of green budwood nursery, green budding is adopted while brown budding is followed in the case of brown budwood nursery. Budding is carried out at the base of the brown budwood/green budshoot before harvesting them. Cut end of the stump left on the budwood plant is treated with a wound dressing compound. All branches produced by the basal plant are pruned off to facilitate the development of the 'over bud'. After the overshoot had developed a minimum of one whorl of leaves, a ring is marked at the base of this with paint. This paint marking is renewed regularly to indicate the point of over budding. Future collection of budwood is always made from above this region. Any shoot that may come below this ring is removed as soon as they are noticed. Since the 'over shoot' is growing on a well established plant it grows very vigorously. By adopting this method uninterrupted supply of budwood could be ensured from the same nursery in spite of converting one clone to another. Further, the heavy expenditure involved in replanting the nursery also is avoided.

ADVANCED PLANTING MATERIALS

Planting materials such as seedling stumps and budded stumps are usually planted in the field before their buds develop. As a result, entire development and growth of the buds take place after transplanting. Because of this the plants have to be maintained in the field during their entire immature phase. If a part of this immature phase could be covered in the nursery itself, it will be very advantageous, as the cost of maintenance of plants in the nursery is much less than that in the field. For this, budded plants are allowed to grow to some extent in the nursery and are then transplanted to the field in this developed condition. Since the plants are transplanted in an advanced stage of growth they are known as advanced planting materials. As the plants have already covered a part of their growth in the nursery, only the remaining part need be covered in the field, which results in a reduction in the field immaturity period. Uniform growth, less casualty, early establishment, cost reduction, less weed growth

etc. are some other advantages that could be obtained by the use of appropriate advanced planting materials.

Bag plants

This is the most commonly used type of advanced planting material.
They are mostly produced by raising plants in bags made of polyethylene
film, commonly referred to as polybags. The bags can be either black or
transparent if they are to be kept buried in soil. If kept exposed to sun,
black polythene is preferred as it facilitates better growth of roots. Depending on the size of the plants intended to be produced, bags of different
dimensions are used. Small bags of lay flat size 56x25 cm, capable of holding
about 9 kg (Rubber Research Institute of Malaya, 1973) soil are used to
raise plants for four to five months so as to produce two or three whorls
of leaves. Larger plants with six or seven whorls of leaves are produced
by retaining them in bags for eight to nine months. For this larger bags
of dimensions 65x35 cm (Potty, 1980) or 64x38 cm holding about 23 kg of
soil are used (Rubber Research Institute of Malaya, 1973). Low density
polyethylene (LDPE) sheet of 400 gauge and 500 guage thickness are usually
used for making small bags and large bags respectively. High molecular
high density polyethylene (HMHDPE) sheets also could be used for this
purpose (Narayanan, 1988). A few perforations are made in the lower half
of the bags before filling to facilitate easy drainage of excess water.

Soil used for filling the bags should possess good retention capacity
for moisture and nutrients, promote root development and bind the roots
firmly to prevent damage during transplanting. A soil with fairly heavy
clay loam texture, good structure and friability is ideal for this purpose.
Fertile top soil collected after removing the surface vegetation and leaf
litter, is usually used for filling the bags. Before filling, the soil should
be cleaned of roots, stones, stubbles etc. Large clods of soil should be
broken to small pieces, and if too wet, partially dried. If fertility level
of the soil is low it can be raised by adding compost or old cow dung
(Mainstone, 1962). Soil prepared in this way is ready for filling into the
bags. While filling, the bag should be gently tapped to ensure that no
cavities are formed in the bag and the soil is packed somewhat firmly.
Filling is done upto about 2 cm below the brim of the bag. Powdered rock
phosphate at the rate of 25 g for small bag and 75 g for large bag is
then incorporated into the top layer. Polybags can be kept in the nursery
either in trenches or within wooden frames. The former method is better
for the protection of bags as well as better growth of plants. Trenches,
having a width equal to the diameter of the bag are usually dug in pairs.

In the case of small bags, depth of the trench should be about 20 cm (Potty, 1980), distance between trenches 15 cm and gap between bags of the same trench 10 cm. Corresponding figures for large bags are 30 cm, 30 cm and 20 cm (Joseph, 1981). Foot path of 75 cm width have to be left between two pairs of trenches. After placing the bags in the trench, excavated soil is filled in the gap between them. Balance soil is mounded around the bags so as to protect their sides from sun. Plants can be raised in bags either by planting budded stumps or by planting seeds for raising seedlings and subsequently budding them. The former method is definitely superior due to several reasons, and hence it is usually followed. Plants in the bags are carefully nursed by proper manuring, watering, weeding, shading, disease and pest control etc. Manuring schedule recommended by the RRII for budded stumps planted in bags is monthly application of 10:10:4:1.5 NPKMg mixture, incorporating a soluble phosphate. Quantity applied per plant is 10 g during first month which is gradually increased to 30 g in four months time (Potty, 1980). Quantity recommended by the RRIM during the first nine months is 14, 22, 28, 28, 28, 28, 42, 42 and 43 g respectively. It is also advisable to use a fertilizer with nitrogen in the ammonium form rather than a nitrate as it causes severe scorching of the young rubber plants even if a slight excess is added. Fertilizer application should be avoided when the leaves are very tender. While applying the fertilizer, care should be taken to prevent it from coming in contact with the young plant as it will cause scorching. Watering should be carried out soon after manuring (Rubber Research Institute of Malaya, 1973). Irrigation has to be done regularly especially during dry periods. Too much watering is not advisable as it may cause water logging. In small nurseries it can be done manually while in large nurseries using a sprinkler or drip irrigation system is more economical. Weeding has to be done periodically in polybag nursery. When the plants are tender, only manual weeding should be adopted. After the stem develops brown bark, chemical weeding can be employed. During summer months, providing partial shade to the plants by erecting overhead shade is advisable. Bag plants are infested by several fungal diseases and pests. Appropriate prophylactic and curative measures have to be taken against these maladies.

Stumped buddings

This is another type of advanced planting material. For their production, seedlings are first budded as in the case of budded stumps. The budded plants are not pulled out, but cut back above the bud and the scion is allowed to develop in the nursery itself. After growing for some time they are pulled out and transplanted to the field after appropriate pruning of roots and shoots. Depending on the size and form of the material at the time of transplanting they are classified into several categories. Those having the scion shoot developed to 60 cm height are called mini stumps while the term used for bigger ones with 240 cm long shoot is maxi stumps. When they are raised in polybags and transplanted along with the soil core they are referred to as core stumps. They can also be crown budded in the nursery, the crown shoot allowed to develop and transplanted after pollarding the crown shoot. Such materials are known as 'three part stumps' or stumped-three-part-plants.

To produce mini stumps, seedlings are raised in seedling nursery at a spacing of 90x30 cm or 60x60 cm. When they reach green buddable size, they are green budded and the stock stem is cut back to develop the scion. When the scion produces brown colour upto 60 cm from the bud union, they are cut back at the top of the brown region just below a whorl of buds. The cut end is treated with wound dressing compound and the stem is white-washed with hydrated lime. Plants are left in this stage for about 10 days to activate the buds. Then they are pulled out and planted in field after pruning the roots as in the case of budded stumps. For easy pulling out, tap root can be severed with a crowbar which is inserted through an opening made on one side of the plant by removing soil. This is called tailing and is carried out just before extraction.

For preparation of maxi stumps, seedlings are raised at a wider spacing of 90x90 cm, 90x60 cm or 68x68 cm (Strivens, 1962). Budding and cutting back are similar to that of mini stumps. Plants are grown in the nursery until the bud shoots develop brown colour to a height of 2.40 m. This may require about 18 months (Rubber Research Institute of Malaya, 1959a). Tailing is done five weeks before pulling out at a depth of 45 to 60 cm. In addition to making the pulling out easier it also reduces transplanting shock and enhances the development of new roots after transplanting. After tailing, trenches/pits taken for this purpose are filled up with the soil removed from them. Pollarding of stem is done at a height of about 2.40 m from the bud union (Rubber Research Institute of Malaysia, 1976) just below a cluster of buds. Wound dressing and white-washing are done as in the case of mini stumps. Pulling out and pruning of lateral roots, done 10 days later, are similar to that for mini stumps. Treating the roots with growth harmones is helpful in the development of roots and better establishment of plants after field planting.

Preparation of 'three-part-stumps' is similar to that of maxi stumps in the early phases, i.e., until the scion grows to the height required for crown budding. At this stage crown budding is carried out and the trunk shoot is cut back above the crown bud to allow the development of crown shoot. When this shoot develops brown colour upto the second whorl of leaves or more, stumping is done below the second whorl of buds. Tailing, pulling out, root pruning, white washing, wound treatment, transporting, planting etc. are similar to that for maxi stumps (Rubber Research Institute of Malaysia, 1976).

Core stumps

It is a kind of stumped budding which is planted out in the field with an intact soil core and active root mass. Seedlings are raised in poly-bags of sizes ranging from 45x20 cm to 50x25 cm lay flat dimensions and 0.14 mm thickness (Leong and Yoon, 1988). When the seedlings are big enough they are green budded and cut back. Alternatively green budded stumps produced in ground nurseries also could be planted in bags. They are planted with the cut end of the tap root at least 8 cm above the bottom of the bag. The scion is allowed to develop two or three whorls of mature leaves. By this time the plants produce sufficient active root mass to bind the soil core together. However, they should be transplanted only after ensuring that this has taken place. For transplanting, base of the polybag is first cut and removed to enable the unrestricted development of roots after transplanting. Roots that may remain coiled at the bottom of the soil core are cut off. They are then planted in the nursery at any spacing between 90x90 cm or 120x120 cm. While planting, the top 8 cm of the bag should not be buried. Leaving this portion above ground is essential for the development of a good root system. The plants are given regular foliar spray with a mixture of foliar feed and fungicide. Nutrition is also provided by applying NPKMg mixture with soluble P, in slurry form. Alternatively slow release fertilizers such as 'Nurseryace' also could be used. Regular pruning of branches also is carried out. In this manner plants are maintained until they reach the maxi stump stage (Leong and Yoon, 1987). Then their stems are pollarded and given other treatments similar to that of maxi stumps. The plant is now dug out with a back-hoe or oil-palm harvester after tailing the tap root. They can be transplanted to the field at any stage, from bud-break to two whorls of hardened leaves. Transporting can be done safely without much damage to the soil core as the soil is bound intact by the well developed root system. Transporting and field planting is done in a way similar to that adopted for large bag plants. Stumps with broken soil core and damaged or improperly developed root system should be discarded.

Soil core plants

In this method of propagation, buddings growing in nurseries are directly transplanted to the field with most of its root system in a block of soil. Budded materials ranging in size from budded stumps to five whorl plants excepting one whorl plants can be successfully handled in this manner (Sergeant, 1967). However, transplanting with three whorls of mature leaves is more desirable because larger plants suffer more set back in growth after transplanting. Nurseries for producing soil core plants are established in places where the soil has sufficient clay to produce a firm core which will not disintegrate when extracted, handled or transported. To prepare soil core plants stock plants are first raised at a spacing of 60x60 cm (Rubber Research Institute of Malaysia, 1976), 38x30 cm (Shepherd, 1967) or (35x35 cm) 53 cm (35x35 cm) (Sergeant, 1967). When five months old, they are either green budded, or brown budded when seven to nine months old. Plants are extracted with special tools such as a metal cylinder with two arms, double headed hammer, piston and crowbar. The cylinder is placed carefully around the plant and driven down into the soil with the double headed hammer until its arms reach the ground level. With the help of the crowbar the cylinder is levered upwards, simultaneously pulling the plant also gently upwards. The cylinder with the plant and soil core is placed on the head of the piston after pruning of the tap root that may protrude beyond the bottom of the cylinder. By pushing the cylinder downwards the core of soil is separated without causing any damage to it. The soil core is then wrapped with polythene sheet or paper and tied with string. They are then kept in light shade for one or two days and watered before planting in the field. If the plants are extracted before the scion has produced leaves, they are kept under shade for longer periods until a few shorls of leaves are produced. Extraction of soil core plants from nursery can be mechanised using the 'Plant Extractor', a special devise mounted on tractor and operated hydraulically (Stephens, 1966). Transporting and field planting of soil core plants are carried out in the same way as that for bag plants.

Crown budded bag plants

These materials are prepared by crown budding seedlings or budded plants raised in small polybags of size varying from 38x15 cm to 50x20 cm. Seedlings are crown budded initially at a height between 91 to 122 cm. In the case of failures second round of budding is carried out at a height of around 76 cm. In the case of budded plants, height of budding ranges between 183 cm and 213 cm above the stock scion union. Second round of budding is done at about 152 cm height. Budding method adopted for this is green budding. The bandage is removed four weeks after budding. Plants budded during the different rounds of budding are cut back simultaneously for uniform growth. Pruning of snag shoots/trunk shoots, protection of crown shoot etc. are carried out in the usual way. When the second whorl of leaves of the crown shoot is fully hardned, plants are tailed. They are then kept under mist sprinklers for a week for hardening, after which they are transplanted to the field as two-part or three-part plants. Plants show a high rate of establishment, fast recovery, good growth, and suffer very little from shock. These materials can be considered as advanced container plants (Yoon et al. 1988).

Other methods of propagation

In addition to propagation using seeds and buds, several other methods of propagation have also been developed in this crop. Important among these are rooting of cuttings, approach grafting, cleft grafting, root grafting, layering and micropropagation. However, due to various drawbacks, these techniques are at present not commercially adopted.

Rooted cuttings

Cuttings of rubber are rooted in mist propagation units. A typical unit consists of a raised bed made of the rooting medium such as river sand, mist producing atomisers kept above the bed, and proper cover on all sides of the bed to provide shade as well as to prevent drifting of mist (Tinley, 1960). Healthy terminal cuttings, about 30 cm long with fully expanded leaves and dormant terminal buds are used for rooting. Cuttings are planted in the bed after treating the cut ends with fungicide - stimulants (Leong et al. 1976; Tinley, 1961). Mist is applied continuously during day time and at night nutrients are provided through the atomisers. After five to nine weeks cuttings strike roots. When roots grow to a length of 1.25 cm, the cuttings are carefully transferred to polythene bags filled with John Innes type potting mixture. They are then hardened by gradually reducing the duration of mist as well as increasing exposure to light. Hardening usually requires three to six weeks. By this time the root system develops well, and frequently the terminal buds also may begin to grow. They are then transplanted to the field, like normal bagplants. As a planting material rooted cuttings have no scope because they do not produce tap roots which are essential for the proper anchorage of the rubber trees.

Approach grafted plants

This technique involves the grafting of the scion without severing it from the source plant, to the stock plant raised in bags. Scion shoots with one whorl of leaves and stock plants having five or six whorls of leaves and a basal girth comparable to that of the scion are used for this type of grafting. The first step involved in the process is placing the stock plants along with the container near scion in such a way that the portions of the scion and stock intended to be grafted remain parallel to each other. A strip of bark along with a thin slice of wood is then removed from the facing sides of the stock and scion over a length of about 18 cm. Exposed portions are now pressed together and kept in this position with a bandage as in the case of budding. Stock-scion union will be over in seven to eight weeks. Now the scion is severed from the source plant by cutting at a point about 5 cm below the graft union. The stock is then cut back at about 5 cm above the approach union. Grafted plants are nursed well for one or two months more for recovering from the shock caused by the separation of the scion from the source plant. Then they are transplanted to the field (Ooi et al. 1976).

Cleft grafted plants

In this method the shoot apex is grafted to the decapitated stem of the stock plant (Webster, 1989). Two to three week old stocks raised in small polybags (Teoh, 1972), and apical portion of young shoots having a length of 2.5 to 5.1 cm and a girth exactly the same as that of the stock are made use of in this type of grafting. The stock is decapitated at a height of about 4 cm above the collar by two slanting cuts from opposite directions so as to make the cut end wedge shaped. Then the stock is split into two along the pointed edge to a depth of about 1.5 cm, in such a way that each slanting cut is on each half. Basal end of the scion shoot is also shaped into a wedge of about 1.3 cm with two opposing sloping cuts. Base of the scion is then inserted into the split made on the stock and kept in position by bandaging with clear polythene tape. The plant is then kept in a mist propagation unit and mist applied continuously for two weeks or kept under dense shade, with troughs of water around to provide humid atmosphere, for four weeks. Then it is transferred to the hardening frame under light shade and hardened over a period of four weeks, followed by field planting.

Root grafted plants

Grafting of roots to green scion shoots is termed root grafting. Roots of both seedlings and rooted cuttings could be used for this. Roots of seedlings are grafted after removing the scion from the source plant (Ooi et al. 1976) where as clonal roots are grafted without severing the shoots (Leong et al. 1976). Healthy green shoots about 35 cm long with one whorl of fully expanded leaves and dormant terminal buds are used as scion. One week old seedlings with well developed radicle and plumule are ideal stocks. A grafting panel is opened at the base of the scion by stripping a flap of bark of dimensions 2.5x0.5 cm. Stock is prepared by cutting the epicotyl at a distance of about 2 cm from the collar with a very slanting cut so as to produce a large cut surface. Grafting is done by placing the epicotyl in the grafting panel in such a way that its cut surface touches the exposed wood of the scion in the panel. The root is kept in this position by firmly bandaging with polythene strips. The grafted plant is immediately planted in the mist chamber and mist applied continuously for three weeks for the completion of the graft union. Grafted plants are then transferred to polybags for further growth and hardening for a period of three to four weeks.

Layers

Propagation of rubber is carried out by 'air layering'. Under this system, roots are developed on a stem while it is still attached to the parent plant (Hartman and Kester, 1968). Young branches are generally used for layering (Rubber Research Institute of Malaya, 1960). First step is the girdling of the stem at the point to be rooted by removing the bark around it to a width of about 5 cm. The exposed wood is scraped to remove the cambium and phloem completely. To enhance rooting, growth harmones can be applied at this region. This portion is then covered with a ball of rooting medium, which is usually moist soil, containing plenty of organic matter. The ball is then covered with polythene sheet (Yoon and Ooi, 1976) to keep it intact as well as to prevent loss of moisture. Ringing prevents the downward translocation of organic products from the region above the ring to other parts of the plants, resulting in their accumulation above the ring. Rooting medium covering the stem prevents light from falling on it which in turn decreases deposit of materials on the cell wall and thereby increases the number of parenchyma cells in this part (Mecdonald, 1986) resulting in callus formation above the ring (Rubber Research Institute of Malaya, 1959b). Rooting medium also helps in maintaining the optimum levels of moisture, temperature and aeration required for rooting. Due to the combined action of all these factors roots develop from the upper edge of the ring within a few weeks and grow into the rooting medium. When the roots are well developed, the branch is severed from the source plant by cutting below the soil ball. Then the polythene cover is removed carefully without causing any damage to the soil ball and the plants are ready for planting.

Micropropagated plants

Rubber can be propagated by micropropagation (tissue culture) techniques also. This is the technique of growing plants from small (micro) pieces of plant tissues (Hartman and Kester, 1968; Macdonald, 1986). Different parts of the rubber plant can be used as explants for this purpose. Rubber Research Institute of Malaysia produced plantlets from embryonic tissues and successfully established them in soil (Paranjothy and Ghandimathi, 1976; Rubber Research Institute of Malaysia, 1983b). Plants with 2-4 cm long shoot and roots were developed from stem cuttings 3-4 cm long (micro-cuttings) bearing one or more axillary buds (Enjalric and Carron, 1982). Plantlets with tap roots and trifoliate leaves have been produced from anther wall cultures (Rubber Research Institute of Malaysia, 1984). Baoting Research Institute of Tropical Crops, China, have developed plantlets from pollen in 1977 and have successfully transplanted them in the field (Zhenghua, 1984). Rubber Research Institute of India has recently perfected a technique for large scale in vitro propagation of Hevea from shoot apices of clone GT 1. In this method 3-5 mm long shoot apices after sterilisation are placed in the culture medium AH-I taken in culture tubes. Concentration of Bacto Agar (BA) was 8.0 gl^{-1} and pH was 5.7. Autoclaving was done for 15 min at 1.01 kg cm^{-2} and 121°C. Kinetin and indole acetic acid (IAA) were added to the medium in appropriate concentrations. Cultures were maintained at 23°C (±2) for three months under a light regime of 16 h (1.5 klx), using cool white fluorescent bulbs. Explant enlargement was observed 3-5 days after inoculation. Leaf and shoot elongation occurred during subsequent weeks. Rhizogenesis was usually observed 6-8 weeks after inoculation. Tap root emerged first and laterals developed 4-5 weeks later. Plants were hardened off over a period of 3-4 weeks by gradually reducing the humidity and increasing the temperature. Potting mixture was composed of equal sand and soil (v/v). Hardened plants were transplanted in the field where the survival percentage was 93 (Asokan et al. 1988). About 500 numbers of plants produced in this way have already been successfully established in the field.

PACKING AND TRANSPORTING

Seeds

Rubber seeds loss viability if exposed to the sun for more than three days and hence are collected every day or at least every alternate day. Seeds after selection, are put for germination immediately after collection. Viability of fresh seeds can be retained for about seven days by keeping them under shade (Joseph et al. 1980). Viability of seeds could be maintained by soaking them in water for five days so as to increase their water content to about 32% (Nurita Toruan, 1983). If seeds are packed in wet charcoal having 40% moisture, in well aerated containers, 70% viability could be retained upto 30 days (Eikema, 1941).

Retention of viability upto four months is possible by storing the seeds at 4°C in sealed plythene bags, which helps in inhibition of respiration (Wycherley, 1971). Seeds immersed in Captan 75 at 0.2% concentration for 10 min and kept in perforated transparent plastic bags of 0.5 mm thickness are reported to have retained good viability even after 285 days storage (Silvio Moure Cicero et al. 1986). Packing with wet charcoal powder and wet saw-dust also prevents loss of moisture from seeds. Charcoal also helps in absorbing carbon dioxide and other gases produced by the seeds due to metabolic activities (Nurita-Toruan, 1983). Storing seeds at low temperature inhibits respiration (Subramaniam, 1980) a process that consumes the food materials stored in the seed, and thereby conserves energy. Removing the microplyar cap before sowing speeds up germination (Dijkman, 1951). Harvesting seeds when the pericarp of the fruit is greenish yellow in colour gives a significantly higher percentage of germination and produces more vigorous seedlings (Premakumari and Bhaskaran Nair, 1980). For packing seeds, different types of containers such as wooden boxes, cardboard boxes, double gunny bags, gunny bags lined with polythene and polythene bags are used. While packing, seeds are arranged in layers with at least 2 cm thick layers of the packing materials in between (Joseph et al. 1980) and about 2.5 cm thick layer on all sides of the container (Subramaniam, 1980). Enough perforations should be provided for aeration (Joseph et al. 1980). While transporting by air, seeds should be kept in warm pressurised compartments to avoid low temperature and pressure (Subramaniam, 1980).

While collecting and transporting germinated seeds, special care should be taken to avoid damage to the tender radicle. For short distance transport, carrying sprouted seeds in buckets or trays half filled with water is very helpful. Transfer of seeds should be done preferably before

10 AM and after 4 PM and should be provided with shade to prevent drying up (Edgar, 1958). For transportation over long distance, special packing techniques have to be adopted. However, packing and despatch of germinated seeds should be attempted only if the period of transit is less than three days as otherwise the emerging tender roots may get damaged. Germinated seeds are packed in boxes or cartons lined inside with grease-proof paper. Aged saw dust dampened to 85% moisture by weight is the commonly used packing material. Charcoal powder and coconut fibre also can be used for this purpose (Joseph et al. 1980). Seeds are arranged in layers with 2.5 cm thick layers of the packing material in between and also providing a layer of this material with the same thickness around all sides of the container (Subramaniam, 1980). Seeds should be placed in the container with their radicle pointed downwards and this position must be maintained when handling the cartons to avoid, distortion of the developing roots. Leaving a little space at the top of the box will provide room for the expansion of the contents of the box caused by the elongation of the roots. On un-packing, the saw dust must be gently removed either by brushing or by spraying water.

Seedling stumps

If the material is to be planted soon after pulling out no treatment other than sealing the cut end of the stem is necessary (Edgar, 1958). If the stumps have to be stored overnight they should be left standing in fresh water, without sealing the cut end of tap root. For transporting over short distances they are tied into bundles of fifty and each bundle then covered with a layer of grass or leaves. While transporting, the bundles should not be exposed to hot sun. If the stumps are to be stored for longer periods or transported over very long distances they should be packed in boxes along with wet saw dust (Rubber Research Institute of Malaya, 1968a).

Green bud shoot and green budded stumps

Since green bud shoots are tender, they should be used as quickly as possible after collection and handled very carefully. They are preferably harvested in the morning when the water content of the tissues is very high. For use on the same day or transporting over short distances green bud shoots are kept rolled in wet jute sac or cotton cloth without touching each other and always kept in shade. Alternatively they can be carried in a tray with water and protected from strong sunlight. If properly packed with aged wet saw dust in boxes after sealing both ends of bud stick with wax, the viability could be maintained upto six days.

Green budded stumps can be stored overnight kept standing in water as in the case of seedling stumps. If they are to be kept for two days or transported over short distances they are made into bundles after sealing the cut end of the tap root also. Bundles are covered with banana sheath, grass or leaves for protecting the stumps from damage caused by transporting as well as limiting dehydration. For transporting over long distances or preserving for longer periods they are packed in boxes with wet saw dust. In this method, the viability can be maintained upto six days.

Brown budwood and brown budded stumps

For the same day's use or transporting over short distances brown budwood need only be kept wrapped in wet sacking. If to be stored over night, it should be kept standing in a tray with 2.5 cm of water with the lower end immersed in it. For longer storage, cut ends of the budwood should be first sealed with melted paraffin wax. Then each piece is covered with banana sheath, grass leaves, wet coconut fibre or wet sacking. Budwood pieces covered in this way are tied into bundles and transported. Viability of buds can be retained for about three days in this way. For storing the budwood for ten to 14 days and transporting over long distances each piece is first enclosed in a perforated polythene sheath. They are then packed in boxes in layers alternating with wet saw dust or wet well teased coconut fibre so that each piece is covered on all sides by the packing medium. This arrangement prevents the loss of moisture from the tissues very effectively and also avoids mutual rubbing which can damage the buds.

Overnight storing of brown budded stumps is done by standing the stumps in water. For storing upto three days and transporting over short distances, cut end of the tap root also is sealed and the stumps are tied into bundles, each bundle being protected by banana sheath, or grass leaves. By packing the stumps in boxes with wet saw dust viability could be retained upto 30 days. Rolling the root portion alone in polythene sheet lined with paper using wet saw dust as packing medium prevents the desiccation of these materials for 20-30 days. However, callus appears on roots after 10 to 15 days and hence they have to be planted 10 to 12 days after packing (Rubber Research Institute of Malaysia, 1982). By treating the stumps with Captan, covering with polythene film and then packing with jute hessine they can be stored upto four weeks with good results (Premakumari and Nair, 1974).

Bag plants

Transplanting of bag plants to the field is carried out during rainy periods. Plants with the top whorl of leaves in fully matured condition alone are transplanted. Bags are dug out from the trenches after removing the soil mounted around them on both sides. If any root is seen protruding from the bag it is pruned off. Bags are transferred to the field immediately, without causing any damage to the soil core and roots. Short distance transporting is usually done by workers.For long distance transfer, trucks and trailers could be made use of. While loading on vehicles, bags should be arranged compactly to avoid shaking. Other container plants like core stumps, soil core plants and crown budded bagplants are also transported in a similar fashion.

Stumped buddings

Packing and transporting of mini stumps are similar to that for seedling stumps. Maxi stumps are not usually made into bundles for transporting because of their large size. They can be packed head-to-tail onto lorry or trailer using grass or leaves in sufficient quantities as packing medium to prevent bruising and drying.

When planting materials are to be sent abroad, before packing they should be properly disinfected by fumigation, sulphur treatment etc. Phytosanitary certificate issued by the concerned authority, should be sent along with the material. The materials should be properly packed and sent by air, sea or land. If transporting over very long distance is involved, airways may be better to avoid delay.

PLANTING IN FIELD

Seeds

Only germinated seeds are used for planting in the field. Field planting of germinated seeds is carried out in pits or planting holes. These are prepared at a size of 60x40x60 cm, 60x30x30 cm, 45x45x45 cm or 45x30x30 cm (Edgar, 1958). If the soil is very hard, the dimensions of pits have to be increased. In very good types of soil, for economy, pits can be dug wider at the top and tapering towards the bottom. Distance between pits vary depending on the system of planting. Important planting systems adopted are square, rectangular, triangular, quincunx, avenue and bedge. Distance between pits also is altered depending on whether the seedlings will be allowed to develop into trees or will be converted to budded trees by budgrafting. If seedlings are proposed to be grown into trees the planting density is fixed in such a way to accommodate 445 to 520 pits per hectare. Convenient planting distances that can be adopted for this are 670x340 cm (445 plants per hectare), 640x340 cm (487 plants per hectare), 460x460 cm (479 plants per hectare), 550x370 cm (499 plants per hectare), 490x400 cm (516 plants per hectare). In case the seedlings are to be budgrafted in the field, the distances are adjusted in such a way so that the stand per hectare is between 420 to 445. To obtain this stand, 490x490 cm (420 plants per hectare), 550x430 cm and 640x370 cm (427 plants per hectare), 670x340 cm (445 plants per hectare) can be adopted (Rubber Board, 1989). Planting pits are prepared by excavating them first and then filling with top soil cleared of stones, roots, etc. and mixed with manure, wherever necessary. Planting of seeds is done in the same way as in the seedling nursery. Usually, three seeds are planted in a pit, in a line or triangle at a distance of 15 to 23 cm. The weakest among the three is removed after a few weeks and the second one is removed after a few months or after budding, leaving the most vigorous one in the planting point.

Seedling stumps

Planting of seedling stumps, as in the case of other planting materials, should be done during favourable weather, i.e. at the beginning of the rainy season when adequate rainfall could be expected at least for a few weeks after planting. However, it is not advisable to undertake planting when very heavy rains are being received, as the soil in the planting pits becomes slushy and could adversely affect the establishment of plants resulting in high casualty. The most favourable time for planting can be assessed by carefully studying the rainfall records of previous years (Potty, 1980). In addition to weather conditions, quality of the planting materials and the care with which the planting operation is done also influence the success of planting.

The first step involved in the planting of a seedling stump is the making of a shallow pit, in the centre the planting pit. This should have a depth of 5 cm and a width, sufficient enough to accommodate the lateral roots just below the collar region. Using a crowbar (alavango) or a pointed stick, a hole (planting cavity), is made in the centre of this shallow pit to accommodate the tap root. Tap root of the stump is carefully inserted into the planting cavity and pushed down gently until its lower end firmly touches the bottom of the planting cavity. The lateral roots below the collar region, are accommodated in the shallow pit. Any gap left between the tap root and the planting hole is filled firmly with soil. This can be conveniently done by thrusting the crowbar into the pit at a point about 15 cm from the stump in a slanting manner until the lower end of the crowbar just touches the end of the tap root. Then the crowbar is drawn towards the stump, holding it firmly in position until the crowbar becomes vertical. This process will fill the cavity at that side. Repeating this process on all the four sides of the stump, will ensure the filling up of the gap by soil. The lateral roots are then placed at the bottom of the shallow pit properly without any twisting and bending. The pit is now refilled with the soil, removed earlier and firmly pressed. Surface of the pit is levelled with a gentle slope so that water does not collect near the stump. When planting is over, the collar of the plant should be at the original soil level of the pit or a little below.

Budded stumps

Budded stumps are planted either in polythene bags or in the field. Method of planting adopted is similar to that of seedling stumps, but adequate care should be taken not to damage the bud while compacting the soil around the tap root with the crowbar. When planting is over the bud patch should be just above the soil level. The stumps should be planted with the budpatches facing the same direction. Direction is decided taking into account the local conditions. Where incidence of Phytophthora infection is likely, buds are faced towards east so that during rainy season the young roots dry up more quickly due to exposure to the morning sun. In regions where southwest effect of the sun is pronounced buds are faced towards the northeast so that base of the scion is shaded by the snag during day time. Brown budded stumps are planted in the same way as green budded stumps.

Bag plants

If planting is done in a planting pit already filled up, a planting hole slightly bigger than the size of the bag is dug first. Bag is then inserted into this hole, after slashing the bottom. After positioning the plant properly in the hole sides of the bag are also slashed and the bag is carefully pulled up without breaking the soil core. Gap around the soil core is filled with soil and then compacted by pressing. Since these plants possess already developed root systems they establish in the field easily and grow vigorously without any growth retardation.

Other container plants such as core stumps, soil core plants and brown budded bag plants are also planted like bag plants. Studies conducted by the Rubber Research Institute of India have indicated that by adopting this technique trees could be brought into tapping one year earlier, compared to brown budded stumps (Joseph and Nair, 1984).

Stumped budding

Their field planting is done in the same way as that of budded stumps, burying the tap root upto the collar only. However, for planting maxi stumps, shallow pits of 30 cm depth also can be used. In such cases a small cavity is made at the bottom of the planting hole with a crowbar to accommodate the end of the tap root (Rubber Research Institute of Malaysia, 1976). After planting, the soil surface is thickly mulched with dry leaves upto a distance of 75 cm from the plant.

REFERENCES

Asokan, M.P., Sobhana, P., Sushama Kumari, S. and Sethuraj, M.R. 1988. Tissue culture propagation of rubber [Hevea brasiliensis (Willd. ex Adr. de Juss.) Muell. Arg.] clone GT (Gondang Tapen) 1. Indian J. of Nat. Rubb. Res., 1(2): 10-12.
Chandra Samaranayake, C. 1984. Budgrafting. In: Liyanage, A. de S. and Peries, O.S. (Eds.). A Practical Guide to Rubber Planting and Processing. Rubber Research Institute of Sri Lanka, Agalawatta, pp. 19-25.
Chandrasekera, L.B. 1980. Crown budding with clones LCB 870. Rubber Research Institute of Sri Lanka Bulletin, 15: 24-27.
De Silva, F.L. 1957. First aid in rubber planting. The Colombo Apothecaries Co. Ltd., Colombo, pp. 257.
Dijkman, M.J. 1951. Hevea Thirty Years of Research in the Far East. University of Miami Press, Florida, pp. 329.
Edgar, A.T. 1958. Manual of Rubber Planting. The Incorporated Society of Planters, Kuala Lumpur, pp. 705.
Eikema, J.S. 1941. Germination and viability tests with Hevea seeds. Berg-cultures, 15: 1049-1060.
Enjalric, F. and Carron, M.P. 1982. Microbouturage in vitro des jeunes plants d' Hevea brasiliensis. Comptes rendus des S'eances´de l' Acad' emic des Science III. Science de la vie 295: 259-264.
Hartman, H.T. and Kester, D.E. 1968. Plant propagation principles and practices. Prentice-Hall, Inc., New York. pp. 702.
Hurov, H.R. 1960. Green bud strip budding of two to eight months old rubber seedlings. Proceedings of the Natural Rubber Research Conference, Kuala Lumpur, 1960, pp. 419-428.
Joseph G Marattukalam, Saraswathy Amma, C.K. and Premakumari, D. 1980. Methods of propagation and materials for planting. In: Radhakrishna Pillai, P.N. (Ed.). Handbook of Natural Rubber Production in India. Rubber Research Institute of India, Kottayam, pp. 63-83.
Joseph G Marattukalam and Premakumari, D. 1981. Seasonal variation in budding success under Indian conditions in Hevea brasiliensis Muell. Arg. Proc. Fourth Annual Symposium on Plantation Crops, Mysore, 1981. pp. 81-86.
Joseph G Marattukalam. 1981. Reduce the immaturity period of the rubber tree, Hevea brasilienis. Rubber Reporter, 6: 113-114.

Joseph G Marattukalam and Bhaskaran Nair, V.K. 1984. Comparative growth performance of polybagged plants and brown budded stumps. Proc. Fifth Annual Symposium on Plantation Crops, 1982. Kasaragod. pp. 158-162.

Kuruvilla, O.J. 1965. A Manual of Botany. Joice & Laila, Changanacherry. pp. 240.

Leong, S.K., Ooi, C.B. and Yoon, P.K. 1976. Further development in the production of cuttings and clonal root stocks in Hevea. Proc. National Plant Propagation Symposium, 1976, Kuala Lumpur. pp. 154-165.

Leong, S.K., Yoon, P.K. and P'Ng, T.C. 1985. Use of young budding for improved Hevea cultivation. Proc. International Rubber Conference, 1985. Vol. 3. pp. 555-577.

Leong, S.K. and Yoon, P.K. 1987. Development of 'core' stumped buddings to reduce immaturity period in Hevea. Proc. Rubber Growers' Conference, 1987. Desaru, Johore. pp. 133-153.

Leong, S.K. and Yoon, P.K. 1988. Production of 'core' stumps for rubber cultivation. Planters' Bulletin, 195: 57-63.

Macdonald, B. 1986. Practical Woody Plant Propagation for Nursery Growers. B.T. Batsford Ltd., London. pp. 669.

Mainstone, B.J. 1962. Dunlop polythene bag planting technique. Planters' Bulletin, 63: 154-161.

Mann, C.E.T. 1929. Care of multiplication nurseries. Rubber Research Institute of Malaya Quarterly Journal, 1: 226-229.

Narayanan, P.K. 1988. Koodathaikal thayarakkumpal sradhikkuka (take care while preparing bag plants). Rubber, 271: 6-7.

Nurita-Toruan, M. 1983. Some factors affecting the viability of seeds. Balai Penelitican Perkebunan Bogor, 51: 149-155.

Ooi, C.B., Leong, S.K. and Yoon, P.K. 1976. Vegetative propagation techniques in Hevea. Proc. National Plant Propagation Symposium, 1976, Kuala Lumpur. pp. 116-131.

O.F.W. 1931. Nieuwe Oculatiemethoden voor Hevea. (New budding methods for Hevea). De Bergcultures, 5: 1372-1375.

Paranjothy, K. and Ghandimathi, H. 1976. Tissue and organ culture of Hevea. Proc. International Rubber Conference, 1975, Kuala Lumpur, Vol. 2. pp. 59-84.

Paranjothy, K. 1980. Physiological aspects of wintering, flower induction and fruit-set in Hevea (Lecture Notes, Hevea Breeding Course). Rubber Research Intitute of Malaysia, Kuala Lumpur.

Potty, S.N. 1980. Nursery establishment and field planting. In: Radhakrishna Pillai, P.N. (Ed.). Handbook of Natural Rubber Production in India. Rubber Research Institute of India, Kottayam, pp. 113-131.

Premakumari, D. and Bhaskaran Nair, V.K. 1980. Possibilities of collecting rubber seeds at the yellow pod stage. Paper presented at the International Rubber Conference, Kottayam, India, 1980.

Premakumari, D. and Nair, V.K.B. 1974. Storage of budded stumps and their effect on bud viability. Proc. International Rubber Research and Development Board Scientific Symposium, Cochin, pp. 67-72.

Radhakrishna Pillay, P.N., George, M.K. and Rajalakshmy, V.K. 1980. Leaf and shoot diseases. In: Radhakrishna Pillay, P.N. (Ed.). Handbook of Natural Rubber Production in India. Rubber Research Institute of India, Kottayam. pp. 249-273.

Ramakrishnan, A. 1987. Rubber Nurserykal (Rubber nurseries). Rubber Vithu Muthal Vipani Vare (Rubber from Seed to Market). Rubber Board, Kottayam, pp. 31-35.

Rubber Research Institute of Malaya. 1939. Budgrafting practical instructions. Planting Manual, Rubber Research Institute of Malaya, 8: 1-2.

Rubber Research Institute of Malaya. 1947. Budgrafting without waxed budding tape (Circular). Rubber Research Institute of Malaya, 25: 3.

Rubber Research Institute of Malaya, 1959a. Transplanting stumped buddings. Planters' Bulletin, 44: 116-119.

198

Rubber Research Institute of Malaya. 1959b. Developments in the propagation of Hevea. Planters' Bulletin, 45: 143-146.

Rubber Research Institute of Malaya, 1960. Annual Report, 1959: 41-43.

Rubber Research Institute of Malaya. 1963. Advantages of green budding. Planters' Bulletin, 67: 77-79.

Rubber Research Institute of Malaya. 1964. Instructions for green budding rubber trees. Planters' Bulletin, 72: 54-60.

Rubber Research Institute of Malaya. 1965. Green budding: result of a survey of estate experience. Planters' Bulletin, 78: 99-103.

Rubber Research Institute of Malaya. 1968a. Despatch of planting materials. Planters' Bulletin, 97: 114-117.

Rubber Research Institute of Malaya. 1968b. Annual Report 1968: 9.

Rubber Research Institute of Malaya. 1973. Establishment and maintenance of nursery for raising planting materials in polybags to an advanced stage. Planters' Bulletin, 129: 173-178.

Rubber Research Institute of Malaysia. 1976. Nursery practices and planting technique. Planters' Bulletin, 143: 25-49.

Rubber Research Institute of Malaysia. 1980. RRIM Planting Recommendations 1980-82. Planters' Bulletin, 162: 4-22.

Rubber Research Institute of Malaysia. 1982. A technique for improved field planting of Hevea budded stumps for small holdings. Planters' Bulletin, 172: 79-84.

Rubber Research Institute of Malaysia. 1983a. RRIM Planting Recommendations 1983-85. Planters' Bulletin, 175: 37-55.

Rubber Research Institute of Malaysia. 1983b. Annual Report 1983: 31.

Rubber Research Institute of Malaysia. 1986. RRIM Planting Recommendations 1986. Planters' Bulletin, 186: 4-22.

Rubber Board. 1989. Rubber and its Cultivation. The Rubber Board, Kottayam. 91 pp.

Saraswathy Amma, C.K. and Nair, V.K.B. 1976. Relationship of seed weight and seedling vigour in Hevea. Rubber Board Bulletin, 13: 28-29.

Sergeant, C.J. 1967. The soil core method of transplanting young rubber as practiced on Kirby Estate. Planters' Bulletin, 92: 256-263.

Sharp, C.C.T. 1932. Some variations of budding technique on big stock. Journal of the Rubber REsearch Institute of Malaya, 4: 39-45.

Shepherd, R. 1967. Commercial experience with a plant extraction technique. Planters' Bulletin, 92: 264-269.

Silvio Moure Cicero, Julio Marcos Filho and Francissco Ferra de Toleo. 1986. Effects of the fungicide treatment and of three storage conditions on the quality of the Hevea seeds. Anais da Escola superior de agricultura "Luiz de Queiroz", 43: 763-787.

Simmonds, N.W. 1989. Rubber breeding. In: C.C. Webster and W.J. Baulkwill (Eds.). Rubber. John Wiley & Sons, Inc., New York. pp. 102.

Stephens, P.R.O. 1966. The plant extractor. Planters' Bulletin, 85: 90-96.

Strivens, L.V. 1962. Planting stumped buddings. Planters' Bulletin, 62: 148-151.

Subramaniam, S. 1980. Collection, handling and planting of propagation materials of para rubber Hevea brasiliensis. (Lecture NOtes, Hevea Breeding Course). Rubber Research Institute of Malaysia, Kuala Lumpur.

Teoh, K.S. 1972. A novel method of rubber propagation. Proc. RRIM Planters' Conference, 1972, Kuala Lumpur, pp. 59-72.

Tinely, G.H. 1960. Vegetative propagation of clones of Hevea brasiliensis by cuttings. Proc. Natural Rubber Research Conference, 1960, Kuala Lumpur, pp. 409-416.

Tinely, G.H. 1961. Effect of ferric dimethyl dithiocarbamate on the rooting of cuttings of H. brasiliensis. Nature, 191: 1217-1218.

Tiney, G.H. 1962. Propagation of Hevea by budding young seedlings. Planters' Bulletin, 62: 136-147.

Togun, S. and Ajibike, B. 1964. Preliminary evaluation of green budding of rubber in Nigeria. Proc. Agricultural Society, Nigeria, 3: 17-20.

Webster, C.C. 1989. Propagation, planting and pruning. In: Webster, C.C. and Baulkwill, W.J. (Eds.). Rubber. Longman Scientific and Technical, UK. pp. 195-244.

White, A. and Imle, E.P. 1944. Hevea budgrafting improved. Agriculture Americas, 4: 62.

Wycherley, P.R. 1971. Hevea Seed, Part II. Planter, 47: 405-410.

Yoon, P.K. 1973. Technique of Crown budding. Rubber Research Institute of Malaya, Kuala Lumpur, pp. 27.

Yoon, P.K. Leong, S.K. and Hafsah Jaafar. 1987. Some improvements to young budding technique. Proc. Rubber Growers' Conference, 1987, Desaru, Johore. 90-123.

Yoon, P.K., Leong, S.K., Leong, H.T., Phun, H.K. and Chiah, H.S. 1988. Preparation of crown budding in nursery. Planters' Bulletin, 55: 64-70.

Yoon, P.K. and Ooi, C.B. 1976. Deep planting of propagated materials of Hevea - its effects and potentials. Proc. National Plant Propagation Symposium, 1976. pp. 273-293.

Zhenghua, C. 1984. Rubber (Hevea). In: William, R., David, A.E., Philip, V. and Yasuyuki Yamada, A. (Eds.). Handbook of Plant Cell Culture. Volume 2, Crops Species. Macmillan Publishing Co., New York, pp. 546-557.

CHAPTER 9

CLIMATIC REQUIREMENTS

P. SANJEEVA RAO and K.R. VIJAYAKUMAR
Rubber Research Institute of India, Kottayam-686009, Kerala, India.

Hevea brasiliensis is indigenous to the rain forests of the Amazon basin, situated generally within 5° latitudes of the equator and at altitudes below 200 m. The climate of this region is wet equatorial type (Strahler, 1969), characterized by a mean monthly temperature of 25 to 28°C, ample rainfall with no marked dry period, and mild breezes round the year (Bradshaw, 1977). The species, evolved in this environment, has developed an ecological preference for warm, breezy, humid weather and fertile soil (Polhamus, 1962; Opeke, 1982). Under commercial cultivation, Hevea performs best in climates closely resembling that in its centre of origin. But with the increase in global demand for natural rubber, plantations have been extended to less suitable regions beyond the traditional latitudes, in India, China, Burma and Brazil (Dijkman, 1951; Omont, 1982; de Barros et al., 1983; Pushparajah, 1983; Zongdao and Xueqin, 1983; Sethuraj, 1985; Sethuraj et al., 1989).

The optimum climatic requirements of Hevea are:

A rainfall of 2000 mm or more, evenly distributed without any marked dry season and with 125-150 rainy days per annum,
A maximum temperature of about 29-34°C, minimum of about 20°C or more with a monthly mean of 25-28°C,
High atmospheric humidity of the order of 80 per cent with moderate wind, and
Bright sunshine amounting to about 2000 hours, at the rate of six hours per day in all months.

Only a few rubber growing regions in the world quality to fall within such a climatic profile (Yew, 1982; Domroes, 1984; Chan et al., 1984). The substantial differences in productivity observed among different regions can be ascribed to the variations in the climatic composition (Table 1 and Fig. 1).

Table 1. Climatic characteristics of certain major rubber growing regions in the world.

Region	Location	Air Temperature (°C)		Rainfall		R.H. (%)	Mean wind speed m s^{-1}	Sunshine duration (h)
		Annual Mean	Range	Total (mm)	Days			
Agartala (NE India)	25°53'N,91°15'E,21M	24.9	10.1	1932	119	77	1.9	2049
Jinghong (China)	21°52'N,101°04'E,553M	21.7	9.9	1209	--	83	1.6	2153
Danxian (China)	19°30'N,109°35'E,169M	24.1	11.4	1826	164	83	2.4	2020
Qiong Hai (China)	19°14'N,110°28'E,24M	24.1	12.1	2006	168	86	2.8	2116
Chum Phon (Thailand)	10°30'N,99°11'E,10M	27.4	5.1	1888	188	73	13.3	--
Kottayam (South India)	09°32'N,76°36'E,73M	26.6	3.1	3171	139	76	1.0	2546
Phuket (Thailand)	07°53'n,98°24'E,3M	28.0	1.7	2150	170	72	8.5	2163
Dartonfield (Sri Lanka)	06°32'N,80°09'E,66M	27.3	1.7	4129	219	79	0.6	1773
Kepala Batas (Malaysia)	06°12'N,100°25'E,4M	27.7	1.6	1977	163	79	1.6	2586
Anguededou (Ivory Coast)	06°N, -	27.4	2.6	1867	129	83	--	1649
Kuala Lumpur (Malaysia)	03°07'N,101°42'E,39M	27.1	1.1	2499	195	77	--	2230
Senai (Malaysia)	03°08'S,104°18'E,10M	27.2	1.1	2361	191	88	2.4	2049
Manaus (Brazil)	03°08'S,60°01'E,48M	27.4	1.9	2101	171	81	1.6	2097

202

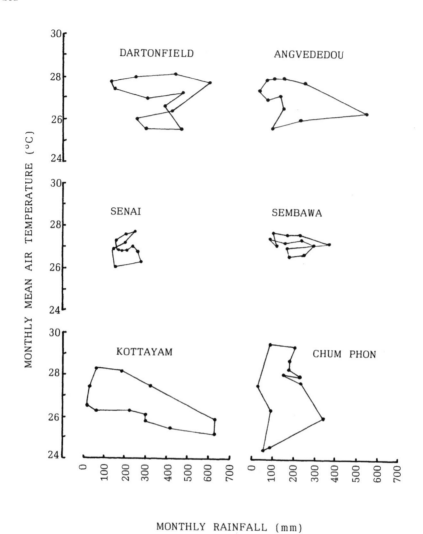

Fig. 1. Climatograms showing the distribution of mean monthly rainfall and temperature at certain rubber growing regions.

In the traditional rubber growing areas, the total rainfall ranges between 2000 and 4000 mm, spread over 140-220 days. It is evenly distributed throughout the year, with not more than one to four dry months. The mean annual temperature is 27±1°C, and the difference between the highest and the lowest monthly temperature is within 3°C. The mean annual maximum and minimum temperatures are 30-33°C and 22-23°C, respectively. The diurnal temperature range is in between 7-10°C. Correlation of mean monthly temperature with rainfall distribution (Fig. 1) indicates that the seasonal changes are minimum at Senai (South Malaysia) and Sembawa (South Sumatara, Indonesia). In these regions conventional planting techniques give satisfactory establishment of rubber.

In non-traditional rubber growing regions (above 10°N latitude), even where annual rainfall is sufficient, there can be periods of severe moisture deficits for four to six months. The mean annual temperature is less than 25°C (around 20°N latitude) and the difference between the highest and the lowest monthly temperatures is 10-12°C. The mean annual maximum and minimum temperatures are 30-37°C and 11-24°C, respectively. Thus, away from the equator, most of the regions in North-East India, China, Bangladesh and Vietnam experience severe cold and dry conditions. In these areas the diurnal temperature differences are high and are in the range of 15-31°C during winter months. Further, in some areas in eastern and western India as well as northern parts of Thailand there is a marked dry season of 6-7 months, with severe moisture deficit. The temperatures may range between 14 and 38°C, with a rainfall of 1500-2500 mm per annum. All these conditions will also demand appropriate location specific planting practices as well as harvesting techniques.

EFFECT OF CLIMATIC COMPONENTS ON GROWTH AND YIELD

Rainfall and water balance

Rubber plantations are traditionally raised under rainfed conditions except in nurseries where essential irrigation is given. Ideally, the monthly rainfall should be sufficient to meet the water requirement of the plantation. In the tropical monsoonal climate, the potential evapotranspiration rate is around 4 mm day^{-1} (Montieth, 1977). Therefore, a rainfall of 125 mm per month with equal distribution is considered essential to maintain optimum gaseous exchange. The amount and distribution of rainfall in the rubber growing regions show considerable variation (Table 2).

Rain interception by the Hevea canopy causes direct evaporation from the leaves and this amount of water is lost. Rain water reaching the ground is the net rainfall (stem-flow and throughfall). This is not fully absorbed by the soil. Depending on the terrain and permeability of the soil, a certain amount of water reaching the ground is lost as surface run-off. This occurs when the net rainfall is in excess of the infiltration capacity of the soil. Part of the water entering the soil is lost through deep percolation and seepage. Measurements by Toeh (1971) showed that in an eight year old plantation, about 83% of the rain reaches the ground as throughfall and 2% as stem-flow. Between 0.5 and 1.6% of the gross rainfall is lost as

Table 2. Rainfall distribution at different rubber growing regions (millimeters)

Region	Jan	Feb	Mar	Apr	May	Jun	Jul	Aug	Sep	Oct	Nov	Dec
Manaus (Brazil)	276	277	301	287	193	98	61	41	62	112	165	228
Senai (Malaysia)	151	142	192	236	200	149	166	159	204	229	257	275
Kepala batas (Malaysia)	14	41	105	204	236	150	200	199	279	280	203	67
Sembawa (Indonesia)	180	163	366	227	167	100	122	85	164	241	301	245
Phuket (Thailand)	10	20	47	139	279	261	338	250	352	260	130	64
Surat thani (Thailand)	75	30	27	91	194	159	137	155	150	258	455	180
Chumphon (Thailand)	62	95	28	89	209	181	177	228	155	235	341	88
Tombokro (Ivory Coast)	23	54	99	119	187	180	101	95	214	130	35	16
Abidjan (Ivory Coast)	39	71	101	145	232	561	239	60	94	154	137	73
Anguededou (Ivory Coast)	35	69	100	144	247	545	225	62	91	151	128	69
Dartonfield (Sri Lanka)	137	125	246	432	601	416	251	295	466	391	472	297
Kottayam (India)	18	32	62	180	320	629	631	417	304	305	221	54
Punalur (India)	14	42	74	228	294	460	434	295	242	355	217	37
Mudigere (India)	3	14	42	81	111	501	619	496	236	162	66	35
Dapchari (India)	0	0	0	0	9	597	936	544	426	94	9	0
Agartala (India)	4	19	62	185	307	348	232	303	270	150	41	11
Xaun Loc (Vietnam)	10	15	22	96	241	294	355	362	376	286	95	36
Cox's Bazar (Bangladesh)	92	98	125	135	140	107	107	97	95	91	82	93
Qiong Hai (China)	38	40	69	126	185	240	190	289	331	286	135	76
Danxian (China)	22	24	40	93	211	217	229	306	352	208	87	37

surface run-off. A canopy of 15 year old LCB 1320 trees intercepts about 18-24% of the rainfall (Haridas and Subramaniam, 1985). The percentage of throughfall is bound to decrease with increasing density of planting.

Positive correlations between rainfall deficit and cumulative production loss during a season have been established (Ninane, 1970; Cretin, 1978). At low soil moisture levels, the rate and duration of latex flow as well as yield are reduced (Buttery and Boatman, 1976; Sethuraj and Raghavendra, 1984; Devakumar et al., 1988; Gururaja Rao et al., 1988, 1990; Vijayakumar et al., 1988). Clonal variation in tolerance to moisture stress also has been observed (Saraswathyamma and Sethuraj, 1975).

Soil moisture stress has significant effect on the yield components such as initial flow rate, plugging index and dry rubber content. Besides the direct effect on turgor pressure, water deficit also triggers a series of biochemical changes in latex (Premakumari et al., 1980). The turgor pressure of the laticiferous system is influenced by the evapotranspiration demand, soil moisture availability and root resistance. Under prolonged dry conditions, however, fall in turgor pressure is countered to a great extent by osmotic regulation by the plant (Chandrashekar et al., 1990). High rainfall for a prolonged period, however, can also have a negative effect on yield. Shorter sunshine duration, associated with high rainfall, results in low photosynthetic efficiency. Higher moisture status in soil may also lead to dilution of latex.

Soil moisture never becomes a limiting factor for growth during wet season. In heavy soils, however, excess rainfall may lead to problems associated with water logging. Heavy rainfall also causes nutrient loss by erosion and leaching. In a study conducted in Sri Lanka (Yogarathnam, 1985) about 62 t ha^{-1} of top soil was found to be lost during the initial three years, when the land was clean weeded and kept bare. However, it was possible to reduce this loss to 1.3 t ha^{-1} by mulching or by growing leguminous cover crop.

Rate of evapotranspiration (ET) is determined by the availability of soil moisture in the root zone, plant water status, stomatal conductance and atmospheric demand (Devakumar et al., 1988). Atmospheric demand is regulated by radiation, temperature, vapour pressure deficit and wind speed. The relation between soil moisture content and transpiration rate in the available moisture range is complex. The root characteristics such as density and distribution also affect transpiration rates, primarily as they influence water uptake by the plant.

In Malaysia, the daily ET rate of the clone RRIM 600 grown under glass house was found to vary from 2.1 to 6.9 mm d^{-1} (Haridas, 1980) and under field conditions this was found to be 4.4 mm d^{-1} when averaged over 21 months. While measuring stream-flow in a small watershed, Haridas (1985) found close agreement between annual ET and pan evaporation, and the daily ET fell between 2 to 8 mm d^{-1}. In Ivory Coast, while measuring the energy balance components of a homogenous plot of 25 ha with 18 year old rubber trees, Monteny et al. (1984, 1985) observed that rubber plantations give out 4 to 6 mm of water vapour daily into the atmosphere when the soil moisture availability is adequate and only 2 to 4 mm when it is inadequate.

Measurement of transpiration rates in mature trees of different Hevea clones (RRII 105, RRII 118,Tjir 1 and Gl 1) under South Indian conditions by Devakumar et al. (1988) and Gururaja Rao et al. (1990) indicated that the relative transpiration rate varies from 0.11 during dry period to 1.13 during the peak production period. Assuming the crop coefficient to be 1.0, Vijayakumar et al. (1988) estimated the water requirement of Hevea at different ages under South Indian conditions. The estimated mean water requirement of rubber tree was found to vary from 10 l per plant per day for the plant in the first year to 100 l per plant per day for the mature plant.

The concept of water balance model is useful in assessing the suitability of areas for planting rubber and for analysing the causes of yield fluctuations in already established regions. Application of the Thornthwaite's climatic water balance model (Thornthwaite and Mather, 1955) to the established rubber growing regions indicates that some of the regions are free from water deficit (Table 3). However, annual water deficits of the order of 200 to 350 mm are found in marginal areas brought under plantations, indicating the adaptability of the plant to dry conditions (Moraes, 1977). In such regions, however, there could be water surplus during rainy season.

At present, cultivation of rubber is being extended even to regions with well defined and prolonged dry season (five to seven months) in many countries. With a marked dry season of six months in Thailand, around 15% growth inhibition was recorded (Saengruksowong et al., 1983). In Ivory Coast, growth of clone GT 1 under dry conditions was found to be reduced significantly (Omont, 1982). The extent of growth retardation was, however, smaller with increase in age of the tree. It is probable that under severe moisture deficit conditions, moisture absorption by the tap root occurs at 2-3 m depth. Under such conditions water stored in the trunk of the plant also might be utilised (Monteny et al., 1985).

In certain dry regions in India, where seven to eight rainless months

Table 3. Components of climatic water balance from various rubber growing regions of the world (millimeters).

Region	Rainfall	Evapotranspiration		Water deficiency	Water excess
		Potential	Actual		
Dartonfield	4129	1066	1066	0	3063
Siantar*	2705	1665	1665	0	1040
Singapore*	2282	1716	1715	1	567
Kuala Lumpur*	2499	1709	1705	4	794
Yangamki*	1840	1333	1329	4	511
Malaca*	2190	1665	1655	10	535
Medan*	1931	1579	1560	19	371
Kotatingi	2828	1669	1639	30	1189
Sembawa	2361	1437	1407	30	924
Chumphon	1889	1334	1229	105	660
Djakarta*	1797	1540	1308	232	489
Anguededou	1866	1658	1405	253	461
Kottayam	3170	1411	1147	264	2023
Cox's Bazar	2923	1262	974	288	1949
Phuket	2150	1807	1480	327	670
Altor Star	1770	1575	1229	346	541

*Adapted from Moraes (1977).

Table 4. Girth (cm) of Hevea clones in the fifth year after planting under different agroclimatic conditions in India.

Clone	Chethackal (9°N,50M)	Poonoor (11°N, 75M)	Mudigere (13°N,950M)	Dapchari (20°N, 58M)
RRII 105	30.9	27.8	20.0	18.7
RRIM 612	32.2	27.5	21.9	25.3
RRIM 501	24.3	26.4	20.1	26.1
Gl 1	26.1	21.3	16.7	24.4
PR 107	23.1	20.2	17.3	28.7
GT 1	27.8	28.3	21.8	26.4
RRIM 600	31.6	25.5	23.1	28.5
Tjir 1	33.1	20.9	21.1	26.0
RRII 118	32.2	33.0	20.1	26.4
Mean	29.0±3.6	25.7±4.0	20.2±2.0	25.6±2.9

Source: Rubber Research Institute of India, Annual Reports 1986-87 & 1987-88.

(Latitude and altitude of the locations are given in parenthesis.)

208

are common (Dapchari, Konkan region of Maharashtra) the overall growth inhibition recorded by rubber plants, maintained with life saving irrigation during summer, was in the range of 14-20% by the fifth year of planting (Table 4). Clones PR 107 and RRIM 600 were found to perform better under these conditions. In addition to loss of leaf area, severe reduction in photosynthetic rate achieved was only 50% of that recorded during the wet season. However, Chandrashekar et al. (1990) reported reasonably good yield in the first year of tapping from clones RRIM 600 and GT 1 under rainfed conditions in this region. Tapping rest for 3-4 months has been suggested considering the low latex vessel turgor pressure observed during summer months.

Pushparajah and Haridas (1977) showed that moisture deficit of about 180-220 mm in two consecutive years severely affected growth of young rubber in a well drained clay soil. However, the effect was not so severe even in an alluvial soil where the water table was high. Haridas (1984) showed that irrigation can increase yield of GT 1, RRIM 612 and RRIM 703.

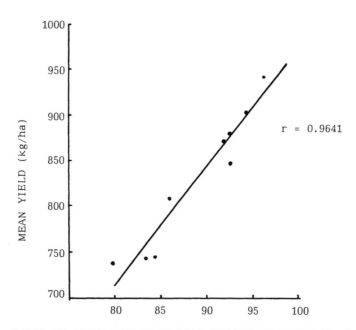

Fig. 2.   Estimated soil moisture availability from meteo-
          rological data and its relationship with yield
          under South Indian conditions.
          (Source: Sanjeeva Rao et al., 1990).

In Ivory Coast, irrigation at the rate of 0.9 times pan evaporation could reduce the immaturity period by 18 months (Omont, 1982). By the fifth year of tapping, even with irrigation, a yield of only 1650 kg ha^{-1} could be obtained compared to that of 1900 kg ha^{-1} expected under favourable rainfed conditions.

In South India, regional yield fluctuations were found to be related to the moisture availability in soil (Sanjeeva Rao et al., 1990) and is shown in Fig. 2. By increasing the availability of moisture to the plant by way of irrigation, or by adopting moisture conservation techniques like silt pits, biological bund and mulch, problems of initial establishment, retarded growth and low yield can be avoided to some extent (Haridas et al., 1987; Vijayakumar et al., 1988). However, more data have to be generated to evaluate the effectiveness and economics of irrigation during pre and post tapping phases. Another approach is to select drought tolerant clones for predominantly dry regions.

RAINFALL AND TAPPING

Rainfall exceeding 9-11 mm per day is not congenial to high yield owing to difficulties in harvesting and other operations (Liyanage et al., 1984; Haridas and Subramaniam, 1985). A larger than optimum number of rainy days in a year decreases the total yield expected for that year. It is difficult to operate a plantation economically with more than 150 rainy days due to loss of tapping days, unless tapping is done with the help of rain guard and panel protectants. Rainfall higher than 34 mm in 24 h may make tapping operation difficult. Such conditions may also promote soil erosion.

The diurnal pattern of rain also has a marked influence on crop harvesting. Rainfall in the early hours of the day or just before the normal time of tapping, makes the bark wet and untappable. Such a condition may necessitate late tapping and will be reflected in a pronounced decrease in the total volume of latex, but may result in increased dry rubber content (d.r.c.) due to increased rate of evaporation caused by high radiation and vapour pressure deficit (v.p.d.). Rain during tapping interferes in the operation and causes spillage and washout. Heavy rainfall before latex collection leads to washout even with the skirt type of rain guard. When rain occurs after latex collection, an increase is generally noticed in the volume of latex obtained on the next day. However, d.r.c. will be below normal indicating dilution of latex due to increased moisture availability and higher plant moisture status.

In Malaysia, of the total rainfall, 21% occurs between midnight and 0600 h, 16% in the normal tapping period of 0600-1200 h, 35% during 1200-1800 h and 28% during the remainder of the day (Wycherley, 1963; Watson, 1989; Daud et al., 1989).

The response of trees to chemical yield stimulation also depends on climate and soil moisture. A highly significant positive correlation has been demonstrated between response to stimulation and cumulative rainfall during the months preceeding stimulation (Abraham and Tayler, 1967). During the dry season, stimulation may not only be ineffective but may also be harmful to the tree.

TEMPERATURE

Temperature is one of the key environmental factors influencing plant growth. Hevea, being a species adapted to moderate temperatures, naturally gets affected by extreme temperatures. Mean monthly temperatures of 25 to 28°C has been found to be the optimum. The mean monthly maximum and minimum temperatures prevailing in the established rubber growing regions of the world are presented in Table 5. The temperature records from Indonesia and Brazil indicate a range from 29.9 to 33.1°C as mean maximum temperature and 22.7 to 24.1°C as mean minimum temperature. In the tropics, the mean annual air temperatures are in the range of 27 to 28°C with a diurnal variation of 4 to 10°C. In the drier parts of Malaysia, the maximum temperature is more than 34°C for two months but the minimum is always above 20°C. In Thailand, the mean maximum temperature is more than 34°C for about four months and the minimum less than 20°C for about five months. In South India, temperature range is 20 to 34°C, whereas in the Konkan region (North West India) the maximum is 34°C and above for three months and the minimum 20°C or below for five months. Occasionally, temperature rises above 40°C in this region.

High temperature conditions result in higher rates of evapotranspiration, leading to severe soil moisture stress in the absence of rainfall. High temperatures above 37°C, coupled with soil moisture stress, result in injury to leaf and killing of leaf margins (Chandrashekar et al., 1990). During planting, thermal injury results in increased casualty. Clonal variation in susceptibility to thermal injury has been reported (Rajagopal et al., 1988). However, drying of leaf margins due to the combined effect of drought and high temperature could be prevented by adequate irrigation in the North Konkan region of India (Mohankrishna et al., 1991). Contact shading of leaves by spraying a suspension of China clay was also found to be effective in mitigating thermal injury (Rubber Research Institute of India, 1991).

Table 5.   Temperature at different regions of rubber cultivation.

Region	Jan	Feb	Mar	Apr	May	Jun	Jul	Aug	Sep	Oct	Nov	Dec
				Mean Maximum ($^{\circ}$C)								
Manaus	30.0	29.9	30.0	29.9	30.7	31.1	31.6	32.7	33.1	32.7	32.0	31.7
Sembawa	30.0	30.8	31.1	31.7	31.6	31.8	31.5	32.0	31.7	31.9	31.4	30.5
Senai	30.7	32.0	32.3	32.6	32.3	31.9	31.2	31.4	31.3	31.6	31.2	30.3
Kepala batas	32.8	34.4	34.5	33.6	32.4	31.9	31.5	31.6	31.2	31.5	31.4	31.5
Anguededou	31.9	32.9	32.5	33.0	32.4	29.8	29.0	27.9	28.2	30.3	31.5	31.2
Dartonfield	32.7	33.6	33.8	33.0	31.8	30.5	30.3	30.4	30.5	31.2	31.6	32.2
Phuket	33.4	34.9	35.3	35.6	34.3	33.5	33.3	33.8	33.1	33.3	32.9	33.1
Chum phon	31.8	33.8	35.6	36.7	35.8	34.5	33.9	33.4	33.7	33.4	32.4	31.7
Kottayam	31.9	32.7	33.4	32.8	31.4	29.2	28.4	28.6	29.3	29.9	30.4	31.1
Dapchari	30.5	32.3	35.3	36.1	35.9	32.3	29.3	28.5	31.8	33.5	33.0	30.8
Agartala	25.7	27.8	32.0	33.5	32.2	31.6	31.4	31.5	31.1	31.4	29.6	26.2
Qiong Hai	30.6	32.6	34.8	35.2	38.9	37.1	39.0	37.1	36.0	33.5	31.3	30.4
Danxian	34.3	36.5	37.8	38.6	38.1	38.9	37.7	36.1	36.0	34.3	34.5	32.3
				Mean Minimum ($^{\circ}$C)								
Manaus	23.3	23.2	23.3	23.3	23.6	23.4	23.2	23.5	23.9	24.1	24.0	23.5
Sembawa	23.1	23.1	23.2	23.4	23.6	23.5	22.7	22.9	22.7	22.8	23.0	22.8
Senai	21.4	21.7	22.1	22.7	22.9	22.6	22.3	22.2	22.2	22.4	22.4	22.2
Kepala batas	21.6	22.2	23.0	23.8	24.3	24.0	23.4	23.5	23.4	23.5	23.2	22.6
Anguededou	22.9	23.5	23.4	23.3	23.5	23.0	22.9	23.3	23.1	22.8	23.1	22.9
Dartonfield	21.6	21.7	22.4	23.4	23.7	23.6	23.3	23.1	22.9	22.5	22.1	22.1
Phuket	21.1	21.6	21.9	22.4	22.6	22.7	22.4	22.4	22.3	22.2	22.1	21.5
Chum phon	17.0	19.0	19.3	22.3	22.9	22.9	22.4	22.7	22.5	21.9	19.5	17.4
Kottayam	21.1	22.1	23.2	23.6	23.5	22.6	22.1	22.2	22.4	22.3	22.2	21.5
Dapchari	14.0	15.8	19.7	22.9	26.0	26.0	24.7	24.6	23.8	21.6	17.4	14.9
Agartala	10.4	12.6	19.0	21.8	22.6	24.7	24.6	24.6	23.8	21.3	16.1	11.6
Qiong Hai	5.9	7.2	9.2	10.6	18.8	20.8	21.0	21.2	16.6	14.8	9.7	5.3
Danxian	2.9	5.8	8.6	9.2	17.0	19.6	20.4	20.6	16.1	11.7	7.2	3.4

In Malaysia, even under marginal management, rubber plantations can be brought into production in about six years, while in Bangladesh (23°N) the immaturity period could be seven years or more (Pushparajah, 1983). In the traditional rubber growing regions in India, the gestation period is normally about six to seven years and in the non-traditional regions it could be eight years or more (Sethuraj et al., 1989). Notwithstanding this general trend, growth of certain clones in the non-traditional region is comparable to that in the traditional region (Table 4).

The rubber growing regions in China are exposed to extreme temperature variations, maximum temperatures of 36 to 39°C and minimum temperatures of 2.9 to 14.8°C are experienced. The mean minimum temperature drops below 20°C for about nine to ten months and the mean maximum temperature rises above 34°C for about seven to eleven months.

In North East India, the minimum temperature recorded is less than 20°C for five months and the maximum temperature throughout the year is less than 34°C. During periods of low temperature, growth retardation has been observed both in North East India and in China. At Agartala (North East India) during winter months clones RRII 118, RRII 300 and RRIM 600 performed better than the other clones tested (Sethuraj et al., 1989). In China, growth of rubber is retarded drastically during winter, the growing period being limited to between June and October (Zongdao and Xueqin, 1983). The higher the temperature the higher the growth rate and the threshold temperature for growth is believed to be in the range of 20°C (Jiang, 1988). Clonal variation in susceptibility to occasional low temperature had been reported (Polhamus, 1962).

It has been demonstrated in China that a ten day mean temperature of over 22°C at 6 a.m. during July-September was donducive to latex regeneration but unfavourable for latex flow. A temperature of 18-21°C at 6 a.m., on the other hand, favoured latex flow (Tropical Crops Research Institute, 1986). According to Shangpu (1986), optimum temperature condition for latex production is 27-28°C, while an ambient temperature of 18-22°C is most ideal for latex flow. High temperature was found to retard latex flow and reduce yield (Lee and Tan, 1979). A significant positive correlation, between the temperature at 8 a.m. and plugging index, was reported (Tropical Crops Research Institute, 1986).

At altitudes higher than 200 m, for every 100 m increase in altitude, a six month delay in reaching tappable trunk size has been reported (Dijkman, 1951; Moraes, 1977). This corresponds to approximately a 0.6°C decrease for every 100 m increase in altitude. About 26% growth inhibition has been recorded by different clones under tropical high elevation conditions (Table 4). RRII 105 was found to perform worse than RRIM 600 and GT 1. In China, GT 1, Haiken 1, PR 107 and RRIM 600 were found to be better performers in the cold-ridden high altitude sub-tropical monsoon regions.

In the tropical low elevation regions such as Kottayam a monthly mean temperature of 26-28°C with adequate soil moisture and sunshine are associated with high production. During the peak production period of November, the daily temperature varies from 22-31°C with a mean of 26.3°C. During this period adequate soil moisture is available from 200-250 mm rainfall received during 11 days. The average duration of sunshine during this period is 6.5 h d^{-1}. In July, on the other hand, heavy rainfall of the order of 600 mm is received during 24 days. In this month the mean minimum and maximum temperatures are 22°C and 29°C respectively. Average daily sunshine duration is only 2.3 h. These conditions are not congenial for good productivity and the latex yield is lower during this period. Temperatures of 23-34°C during March with 9.4 h of sunshine and low rain-fall of around 60 mm also result in low yields. This may be due to the combined effect of high temperature and soil moisture stress in addition to the effects of annual defoliation.

RELATIVE HUMIDITY

Transpiration rate is influenced by temperature and relative humidity (R.H.) of the surrounding atmosphere. The moisture exchange capacity of the atmosphere is indicated by the R.H. Conditions of high and low R.H. occur during rainy and dry seasons, respectively. In the rainy season irradiation is low and duration of leaf wetness is high. Relative humidity is an indirect measurement of v.p.d. at ambient temperature. High to moderate R.H. prevails in most of the rubber growing regions (Table 6).

Monteny et al. (1985) observed that with low soil moisture availability and a weak surface to air water vapour gradient, ET rate is reduced progressively by the activity of stomata. Variations in v.p.d., wind and radiation results in differences in the water requirements of plantations in different locations.

It was demonstrated that diurnal variation in latex yield is inversely related to saturation deficit of the air (Paardekooper and Sookmark, 1969; Ninane, 1970). Conducting tapping at different hours of the day, it was found that latex yield was maximum and constant between 8 p.m. and 7 a.m., and decreased gradually to a minimum of 70% of the maximum at around 1 p.m. The decrease in yield during the course of the day is related to increased loss of water due to transpiration and the resultant drop in

Table 6.   Mean   relative   humidity   (per   cent)   at   different   rubber   growing
regions.

Region	Jan	Feb	Mar	Apr	May	Jun	Jul	Aug	Sep	Oct	Nov	Dec
Manaus	88	88	88	88	86	83	80	77	78	79	82	85
Senai	81	79	80	81	82	82	82	82	82	82	83	84
Kuala Lumpur	83	81	82	85	84	84	83	83	84	85	87	85
Kepala batas	72	70	73	78	82	83	83	83	83	83	82	77
Sembawa	90	90	90	90	90	88	87	85	87	87	89	90
Dartonfield	77	75	75	80	81	82	82	80	80	80	81	78
Anguededou	80	78	83	80	82	85	86	90	89	84	82	79
Surat thani	86	85	81	82	84	79	86	87	86	87	88	86
Phuket	67	67	66	70	74	75	76	73	76	75	73	70
Chumphon	72	71	67	70	73	74	75	75	76	76	75	73
Kottayam	64	66	68	73	77	83	85	83	80	79	77	70
Agartala	71	68	66	72	79	83	85	85	86	82	75	73
Danxian	84	84	82	79	80	81	81	85	87	85	85	84
Qiong Hai	86	88	87	86	83	84	83	86	87	86	85	86

pressure potential in the latex vessels (Buttery and Boatman, 1976; Devakumar
et al., 1988). The recovery of yield in the late afternoon is correlated
to reduction in v.p.d.   A sharp decline in yield was also found whenever
v.p.d. reached 8 mm and restoration of yield in the afternoon lagged behind
the decrease in v.p.d. (Ninane, 1970).

SUNSHINE
        Though photosynthetic light conversion efficiencies of agricultural
crops with closed canopy are in the range of only 7.4-10.2%, positive correlation exists between radiation and biomass production. Effect of sunshine
hours on crop growth and productivity is often mediated through its effects
on photosynthesis and crop water requirements. Under limited soil moisture
availability, sunshine duration will have a negative effect on photosynthesis
and growth. Any condition contributing to good supply of water to tissues
or limiting loss of water by ET are favourable for prolonged flow of latex.
Seasonal variation in the availability of water and sunlight cause a change
in d.r.c.

Most of the agrometeorological observatories in the rubber growing regions have only sunshine recorders and the data are presented in Table 7.

Table 7.   Mean sunshine duration at different rubber growing regions (hours per day).

Region	Jan	Feb	Mar	Apr	May	Jun	Jul	Aug	Sep	Oct	Nov	Dec
Danxian	4.3	4.5	4.9	6.0	7.1	6.7	7.3	6.3	5.7	5.2	4.4	4.1
Qiong Hai	4.2	3.5	4.7	6.3	7.7	7.2	8.2	6.9	6.2	5.5	4.5	3.9
Senai	5.7	6.4	6.0	5.7	5.8	5.7	5.4	5.4	4.5	4.7	4.2	4.5
Kuala Lumpur	6.2	7.4	6.5	6.3	6.3	6.6	6.5	6.3	5.6	5.3	4.9	5.4
Kepala batas	8.8	8.4	8.4	8.3	7.1	6.6	6.6	6.3	5.7	6.0	5.9	7.0
Sembawa	4.1	5.0	5.1	6.1	6.5	6.7	6.6	7.0	5.6	5.2	5.1	4.3
Dartonfield	5.5	7.0	6.6	5.1	4.2	3.2	3.8	4.5	3.7	5.0	4.8	5.0
Anguededou	6.0	5.5	5.5	6.9	5.1	3.1	2.6	2.0	2.4	4.0	5.6	5.6
Kottayam	9.4	9.4	9.3	8.6	7.0	4.5	3.9	4.9	6.2	6.2	6.5	8.0
Agartala	7.8	8.3	8.5	7.8	6.3	5.4	5.8	5.6	4.8	7.5	8.5	8.3

Detailed radiation data are available from only a few stations. The duration and intensity of sunshine should have a significant influence on latex sucrose levels (Tupy, 1989). An increase in sunshine duration towards the end of the rainy season is often associated with an increase in latex production. Also, lower latex production during rainy season can be attributed to reduced sunshine hours.

High radiation and its long duration from December to April cause scorching of bark in young rubber plants. To protect the bark, contact shading of stem with reflectants is adopted in South India. Contact shading of leaves of young plants of rubber was found beneficial in the Konkan region (North West India). Monthly variation in yield performance may be influenced by differences in hours of sunshine, but the exact requirement for optimum yield is yet to be quantified.

WIND

Wind is another important climatic factor having a tremendous influence on the performance of rubber plantations. High frequency of gale can cause considerable damage to plantations by promoting branch snap, trunk snap, uprooting etc. Morphological and anatomical deformations are reported to be usually associated with high wind velocities. In addition to the mechanical effects, advective cold and dry winds affect physiological processes. One of the notable features of trees in windy places is the deformation of their canopies to produce an asymmetric structure in which the branches appear to be swept to the leeward side (Grace, 1977). In valleys where the wind direction is often upslope in the day and downslope in the night due to cold air drainage, flagging is in the direction of night winds.

Windiness, whether indicated by mean wind velocity or the incidence of gale has a tendency to increase near the coast and at higher elevations. In general, the mean wind speeds in the rubber growing areas are in the range of 1-3 m s^{-1} and at times 4 m s^{-1} (Oldeman and Frere, 1982). Wind damages, mechanical and physiological, often lead to low productivity (Yee et al., 1969).

In general, young plantings with heavy canopy may show stem bending and require corrective pruning and roping. Susceptibility to wind damage is the greatest at the time of maximum girthing and canopy development. Trees with narrow crotches are more prone to wind damage (Dijkman, 1951). Tracks of strong wind should be avoided for cultivation of rubber.

In China, an annual mean wind velocity below 1 m s^{-1} has a favourable effect on the growth of rubber trees. At a velocity of 1.0-1.9 m s^{-1} no retardation in growth was observed and at 2.0-2.9 m s^{-1} both growth and latex flow are affected. At a velocity of above 3 m s^{-1}, however, growth and latex flow are severely inhibited (Zongdao and Xueqin, 1983). Strong wind of 8-14 m s^{-1} causes crinkling or laceration of young leaves. A cold wave with strong wind will aggrevate the damage. When wind velocity is beyond 17 m s^{-1}, wind susceptible clones are subjected to branch break and trunk snap. At wind speeds of over 24.5 m s^{-1}, most of the rubber trees are uprooted. Clonal variation in susceptibility to wind damage is generally observed.

Shelter belts are widely used in China to protect rubber trees in highly wind prone areas. In such areas, shelter belts of 20-25 m width are made, consisting of main, secondary and undergrowth trees forming dense mixed forest belts. Main trees shall be fast growing and wind resistant species such as Eucalyptus with Accacia confusa, Homalium hainanesis and Michelia macclurei as secondary trees and Camellia oleifera as under growth trees (Zongdao and Xueqin, 1983). High density planting with wind resistant clones will provide mutual shelter and will tend to limit crown size leading to reduced chances of wind damage.

ACKNOWLEDGEMENTS

The authors thankfully acknowledge the encouragement received from Dr. M.R. Sethuraj and Dr. K. Jayarathnam, Rubber Research Institute of India, Kottayam. Thanks are also due to the Directors of other Rubber Research Organizations for the generous supply of some of the climatological data presented here.

REFERENCES

Abraham, P.D. and Tayler, R.S. 1967.   Stimulation of latex flow in Hevea brasiliensis. Exp. Agric., 3:1.

Bradshaw, M.J. 1977.   Earth, the living Planet. Hodder and Stoughton, London, 302 pp.

Buttery, B.R. and Boatman, G.G. 1976.   Water deficits and flow of latex, In: T.T. Kozlowski (ed.) Water deficits and plant growth. Vol. IV, pp. 233-289.

Chan, H.Y., Yew, F.K. and Pushparajah, E. 1984.   Approaches towards land evaluation systems for Hevea brasiliensis cultivation in Peninsular Malaysia. Proceedings Int. Rubb. Conf. Colombo, pp. 545-562.

Chandrashekar, T.R., Jana, M.K., Joseph Thomas, Vijayakumar, K.R. and Sethuraj, M.R. 1990.   Seasonal changes in physiological characterstics and yield in newly opened trees of Hevea brasiliensis in North Konkan. Indian J. Nat. Rubb. Res., 3: 88-97.

Cretin, H. 1978.   Influence de quelques parametres ecoclimatiques et de la stimulation a l'Ethrel sur la production et certaines characteri-stiques physcoclimiques du latex d' Hevea brasiliensis en basse cote d'Ivoire, DEA, Abidjan University.

Daud, M.N., Nayagam, J. and Veramuthoo, P. 1989.   Effects of selected Environmental and Technological Factors on Rubber Production - A case study of RRIM Economic Laboratory. J. Nat. Rubb. Res., 4: 66-74.

de Barros, J.C.M., de Castro, A.M.G., Miranda, J., Schenkel, C.S. 1983. Development and present status of rubber cultivation in Brazil. Proceed-ings of the RRIM Planters' Conference, Kuala Lumpur, pp. 18-30.

Devakumar, A.S., Gururaraja Rao, G., Rajagopal, R., Sanjeeva Rao, P., George, M.J., Vijayakumar, K.R. and Sethuraj, M.R. 1988.   Studies on soil-plant-atmosphere system in Hevea: II. Seasonal effects on water relations aand yield. Indian J. Nat. Rubb. Res., 1: 45-60.

Dijkman, M.J. 1951.   Hevea - Thirty years of Research in the Far East. University of Miami Press, Florida, 329 p.

Domoroes, M. 1984.   The 'Rubber Climate' of Sri Lanka: Observations on an Agroclimatic land classification for rubber cultivation in Sri Lanka. Proceedings Int. Rubb. Conf., Colombo, pp. 381-387.

Grace, J. 1977.   Plant response to wind. Academic Press, London, 204 pp.

Gururaja Rao, G., Rajagopal, R., Devakumar, A.S., Sanjeeva Rao, P., George, M.J., Vijayakumar, K.R. and Sethuraj, M.R. 1988. Study on yield and yield components of Hevea clones during water stress in 1987. Proceedings of the International Congress of Plant Physiology, New Delhi, India, pp. 499-503.

Gururaja Rao, G., Sanjeeva Rao, P., Rajagopal, R., Devakumar, A.S., Vijayakumar, K.R. and Sethduraj, M.R. 1990.   Influence of soil, plant and meteorological factors on water relations and yield in Hevea bra-siliensis. Int. J. Biometeorol., 34: 175-180.

Haridas, G. 1980.   Soil moisture Use and Growth of young Hevea brasiliensis as determined from Lysimeter study. J. Rubb. Res. Inst. Malaysia, 28: 49-60.

Haridas, G. 1984. The influence of irrigation on latex flow properties and yield of different Hevea cultivars. Proceedings of the Int. Conf. on Soils and Nutrition of Perennial Crops, Kuala Lumpur.

Haridas, G. 1985. Stream-flow measurements in a small watershed to estimate evapotranspiration from a stand of rubber. Proceedings of the Rubb. Conf. 1985, Kuala Lumpur, pp. 670-681.

Haridas, G. and Subramaniam, T. 1985. A critical study of the hydrological cycle in a mature stand of rubber (Hevea brasiliensis Muell. Arg.), J. Rubb. Res. Inst. Malaysia, 33: 70-82.

Haridas, G., Eusof, Z., Mohd. N.S., Abu, T.B. and Raymond, Y.Y. 1987. Soil moisture conservation in rubber. Proceedings RRIM Rubber Growers' Conference, 1987. pp. 154-166.

Jiang, A. 1988. Climate and natural production of rubber (Hevea brasiliensis) in Xishuangbanna, southern part of Yunnan Province, China. Int. J. Biometeorol., 32: 280-282.

Lee, C.K. and Tan, H. 1979. Daily variations in yield and dry rubber content in four Hevea clones. J. Rubb. Res. Inst. Malaysia, 27: 117-126.

Liyanage, A. de S., Gibb Ann and Weerasinghe, A.R. 1984. A crop-weather calander for rubber. Proceedings Int. Rubb. Conf., Colombo, 1984, pp. 354-366.

Mohankrishna, T., Bhasker, C.V.S., Sanjeeva Rao, P., Chandrashekar, T.R., Sethuraj, M.R. and Vijayakumar, K.R. 1991. Effect of irrigation on physiological performance of immature Hevea brasiliensis. Indian J. Nat. Rubb. Res. (In press).

Monteny, B.A., Barbier, J.M. and Omont, C. 1984. Micrometeorological study of an Hevea forest plantation. In: R. Lal, P.A. Sanchez and R.W. Cunnings, Jr. (Eds.) Land Clearing and Development in the Tropics. A.A. Balkema Publishers Inc., pp. 203-214.

Monteny, B.A., Barbier, J.M. and Bermos, C.M. 1985. Determination of the energy exchanges of a forest-type culture: Hevea brasiliensis. In: B.A. Hutchison and B.B. Hicks (Eds.) Reidel Publishing Company, Dordrecht, pp. 211-233.

Montieth, J.L. 1977. Climate. In: P. de T. Alvim and T.T. Kozlowski (Eds.) Ecophysiology of Tropical Crops. Academic Press, New York.

Moraes, V.H.F. 1977. Rubber. In: P. de T. Alvim and T.T. Kozlowski (Eds.) Ecophysiology of Tropical Crops. Academic Press, New York, pp. 315-331.

Ninane, F. 1970. Les Aspects Ecophysologiques de la productivite chez Hevea brasiliensis Muell. Arg. au cambodge, Teses Doct. Sci. Agron., Louvain.

Oldeman, L.R. and Frere, M. 1982. A study of the Agroclimatology of the Humid Tropics of south east Asia. FAO/UNESCO/WMO Interagency project on Agroclimatology, Technical Report, FAO, Rome. 230 p.

Omont, H. 1982. Plantation d'heveas en zone climatique marginale. Revue Generale des Caoutchoues et Plastiques, 625: 75-79.

Opeke, L. 1982. Tropical tree crops. John Willey and Sons, New York, pp. 215-241.

Paardekooper, E.C. and Sookmark, S. 1969. Diurnal variations in latex yield and dry rubber content and relation to saturation deficit of air. J. Rubb. Res. Inst. Malaya, 21: 341-347.

Polhamus, L.G. 1962. Rubber - botany, production and utilisation. Interscience Publishers Inc., New York, 449 p.

Premakumari, D., Sherif, P.M. and Sethuraj, M.R. 1980. Variations in lutoid stability and rubber particle stability as factors influencing yield depression during drought in Hevea brasiliensis. J. Plant. Crops, 8: 43-47.

Pushparajah, E. 1983. Problems and potentials for establishing Hevea under difficult environmental conditions. Planter, Kuala Lumpur, 59: 242-251.

Pushparajah, E. and Haridas, G. 1977. Developments in reduction of immaturity period of Hevea in Peninsular Malaysia. J. Rubb. Res. Inst. Sri Lanka, 54: 93-105.

Rajagopal, R., Devakumar, A.S., Vijayakumar, K.R., Annamma, Y., Gururaja
    Rao, G. and Sethuraj, M.R. 1988. Variation in leaf tissue membrane
    thermostability among clones of Hevea brasiliensis. Indian J. Nat.
    Rubb. Res., 1: 79-81.
Rubber Research Institute of India, 1991. Annual Report 1989-90, Kottayam,
    India.
Saengruksowong, C., Dansagoonpan, S. and Thammarat, C. 1983. Rubber
    planting in the North Eastern and Northern Regions of Thailand. Proceed-
    ings Symposium IRRDB, Beijing, China.
Sanjeeva Rao, P., Jayarathnam, K. and Sethuraj, M.R. 1990. Water balance
    studies of the rubber growing regions of South India. J. Applied
    Hydrology, 3: 23-30.
Saraswathyamma, C.K. and Sethuraj, M.R. 1975. Clonal variation in latex
    flow characteristics and yield in the rubber. J. Plant. Crops, 3:14-15.
Sethuraj, M.R. 1985. Physiology of growth and yield in Hevea brasiliensis.
    Proceedings Int. Rubb. Conf., Kuala Lumpur, Malaysia, pp. 3-19.
Sethuraj, M.R. and Raghavendra, A.S. 1984. The pattern of latex flow from
    rubber tree Hevea brasiliensis in relation to water stress. J. Cell
    Biochem., Supl. 8B: 236.
Sethuraj, M.R., Potty, S.N., Vijayakumar, K.R., Krishnakumar, A.K.,
    Sanjeeva Rao, P., Thapaliyal, A.P., Mohankrishna, T., Gururaja Rao,
    G., Chaudhury, D., George, M.J., Soman, T.A. and Meenattoor, J.R.
    1989. Growth performance of Hevea in the non-traditional regions of
    India. Proceedings RRIM Planters' Conference, Malaysia, pp. 212-227.
Shangpu, L. 1986. Judicious tapping and stimulation based on dynamic analysis
    of latex production. Proceedings IRRDB Rubber Physiology and Exploit-
    ation meeting, SCATC, Hainan, China, pp. 230-239.
Strahler, A.N. 1969. Physical Geography, 3rd edn. Wiley, New York.
Thornthwaite, C.W. and Mather, J.R. 1955. The water balance. Publ. in
    Clim., Drexel Inst. Tech. Lab. of Clim., 8: 1-104.
Toeh, T.S. 1971. Where does all the rainfall go? Plant. Bull., Rubber Res.
    Inst., Malaysia, 115: 215.
Tropical Crops Research Institute. 1986. Latex producing pattern under earlier
    reopening in Yunnan Province. Proceedings IRRDB Rubber Physiology
    and Exploitation meeting, SCATC, Hainan, China. pp. 209-222.
Tupy, J. 1989. Sucrose supply and utilisation for latex production. In:
    d'Auzac, J., Jacob, J.L. and Chrestin, H. (Eds.) CRC Press Inc.,
    Florida, pp. 179-218.
Vijayakumar, K.R., Gururaja Rao, G., Sanjeeva Rao, P., Devakumar, A.S.,
    Rajagopal, R., George, M.J. and Sethuraj, M.R. 1988. Physiology
    of drought tolerance of Hevea. C.R. du Colloque Exploitation Physiologie
    et Amelioration de l'Hevea, pp. 269-281.
Watson, G.A. 1989. Climate and Soil. In: Webster, C.C. and Baulkwill, W.L.
    (Eds.) Longman, Singapore, pp. 125-164.
Wycherley, P.R. 1963. Variation in the performance of Hevea in Malaya.
    J. Trop. Geogr., 17: 143.
Yee, H.C., Peng, N.A. and Subramaniam, S. 1969. Choice of Clones. Planters'
    Bulletin, 104: 226-247.
Yew, F.K. 1982. Contribution towards the development of a land evaluation
    system for Hevea brasiliensis Muell. Arg. cultivation in Peninsular
    Malaysia. Doctoral thesis, State University of Ghent, Belgium.
Yogarathnam, N. 1985. Management of soil resources to meet the challenge
    of the future. RRIC Bulletin, 30: 29-32.
Zongdao, H. and Xueqin, Z. 1983. Rubber cultivation in China. Proceedings
    RRIM Planters' Conference, pp. 31-43.

CHAPTER 10

RUBBER CULTIVATION UNDER CLIMATIC STRESSES IN CHINA

HUANG ZONGDAO and PAN YANQING

South China Academy of Tropical Crops, Baodao Xincun, Danxian, Hainan, People's Republic of China.

INTRODUCTION

It was only since the founding of the People's Republic of China in 1949 that China endeavoured to expand its rubber industry. By 1987, the area under rubber had increased to 580,000 ha, which is 200 times larger than that in the pre-1949 period. Production registered an increase of 1,200 times during this period; the production in 1987 was 240,000 tonnes. This made China one of the major natural rubber producing countries in the world, ranking fifth among the NR producers (Huang Zongdao and Pan Yanqing, 1979; He Kang and Huang Zongdao, 1987).

CHINA'S RUBBER TRACT AND ITS CLIMATIC CHARACTERISTICS

China's rubber growing area stretches over five provinces of south China: Hainan Province, Guangdong Province, Fujian Province, Yunnan Province and Guangxi Autonomous Region extending upto the southern-most city, Sanya, and to Yunxiao County of Fujian Province and Ruili County of Yunnan Province in the north, situated between 97°31'-121°E and 18°09'-24°N. This rubber tract can be divided into four regions: Hainan region, West Guangdong and Southeast Guangxi region, East Guangdong and Fujian region and Yunnan region (Fig. 1). China's rubber growing tract, located between 18-24°N, is entirely beyond the traditionally-recognized northern limit. Rubber plantations at higher latitudes beyond the traditional northern limit are often subjected to the perils of typhoons and cold waves. Scientists and growers in China thus have to devise unique cultural techniques, including selections and breeding of stress-tolerant clones (Huang Zongdao et al. 1980).

The major rubber tract lies in the tropics and south subtropics, having a monsoonal climate. The annual mean temperature ranges from 20°C to 25.5°C (Fig. 2) and annual rainfall varies from 1,000-2,500 mm (Fig. 3). Pronounced dry and wet seasons prevail with rainy season from May to November and dry season from December to April. Hainan Island, West Guangdong and East

221

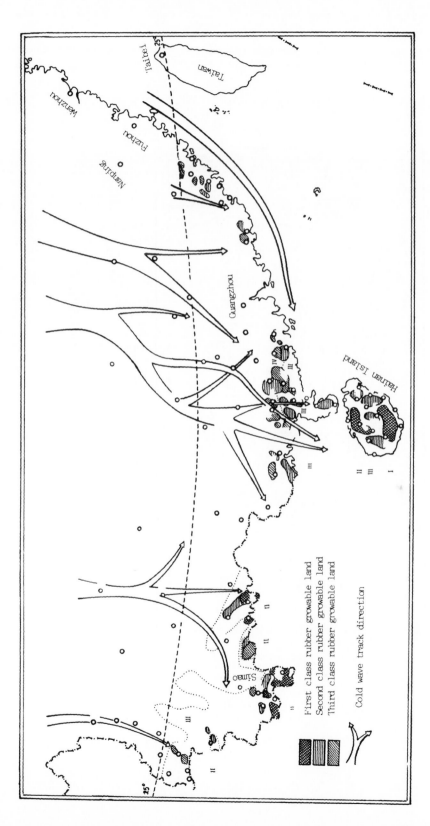

Fig. 1. Diagram of rubber growable land distribution, cold wave invasion and climatic zones in China.

I. Most optimum climatic area.   II. Optimum climatic area.
III. Below optimum climatic area. IV. Localized optimum climatic area.

222

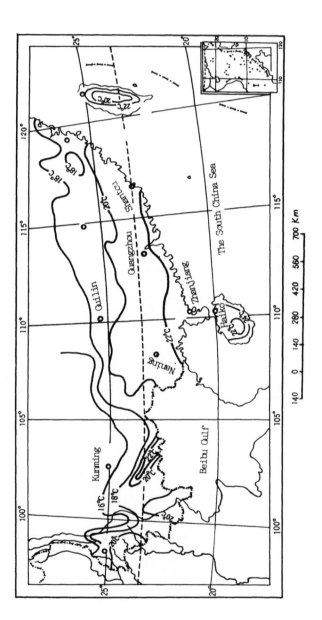

Fig. 2. Annual mean temperature over 1950–1970 in China's rubber areas.

223

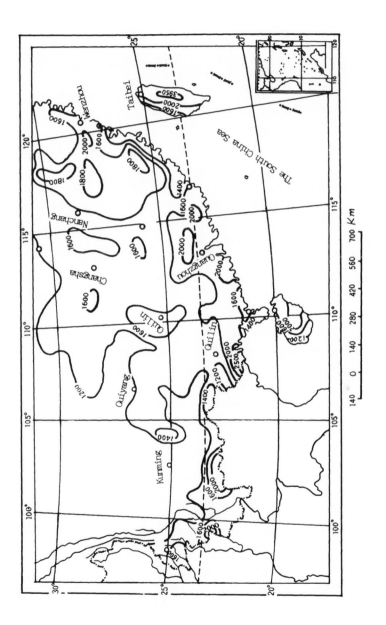

Fig. 3. Annual precipitation in China's rubber areas over 1951-1970.

Guangdong-Fujian rubber regions are also exposed to typhoon while Yunnan, West Guangdong and East Guangdong-Fujian rubber regions are liable to cold wave invasion in the winter. It can thus be seen that China's rubber industry is constrained by typhoons and cold waves that handicap the growth and latex regeneration of rubber trees.

Tropical storms and typhoons

In East Asia, typhoons and tropical storms are brewed over the ocean surface 5-20°N, to the east of the Philippines. On their way moving toward the west, they are subjected to the influence of the position of subtropical air pressure ridge over the Pacific Ocean, and take three tracks, i.e. westward track, northwest track and parabolic (recurving) track. Among them the westward track plays the most terrible havoc with rubber plantations in China. Typhoons or tropical storms along this tract, having crossed the northern part of the Philippines and the South China Sea, make their landfall on the South China coastal areas, more frequently in June, September and October (Lian Shihua, 1984). It is recorded that 154 typhoons and tropical storms have made their landfall on the regions south to 25°N during the last three decades, with a yearly mean of 4.5 cases. However, variations existed between years with the highest number of nine in one year and the minimum of one in another year. The violence of the landed typhoons or tropical storms is drastically reduced by landforms and ground covers. Typhoons are not only devastating per se to crop but also are carriers of heavy rain. Typhoons reaching Hainan Island sweep ashore on the east coast and that explains why more rainfall is received in the east coastal area than in the west coastal area (Yin Xiuchun, 1987).

Cold waves

In winter, cold waves of Siberian origin move southward with great momentum and enter South China up to the north of Hainan Island. The cold waves that pass the Great Bend of the Huanghe River, between 105-115°E and move along the border between Hunan and Guangxi Provinces before finding their way southward through the gap of Nanling Mountains, divide into branches for invading the West Guangdong, Southwest Guangxi, and Hekou and Xishuangbanna of Yunnan (Fig. 4). A sudden fall in temperature by 10-15°C is common in these areas on arrival of the cold waves. Meteorological records over the last three decades showed the frequency of the cold waves invading South China's rubber plantations at an interval of 2-3 years, e.g., 1954-55, 1960-61, 1962-63, 1966-67, 1967-68, 1973-74, 1975-76, 1976-77, 1983-84.

225

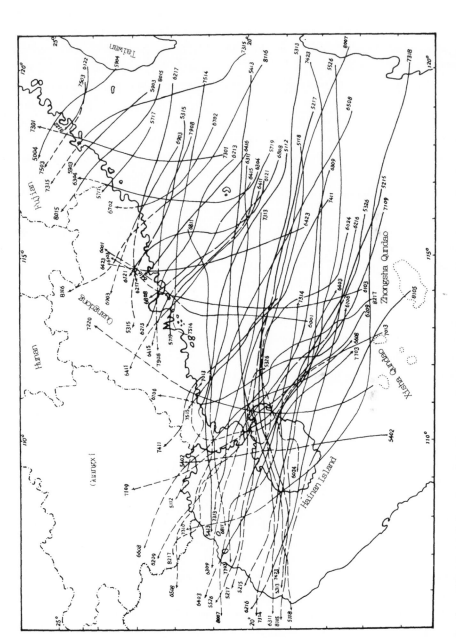

Fig. 4. Diagram of tracks of typhoons visiting coastal rubber plantations in China between 1949-1982.

Other climatic conditions

Annual total solar radiation ranges between 110-152 Kcal/cm^2, with the highest received in summer. The yearly sunshine hours amount to 1700-2250. The seasonal variation of this factor shares the same pattern with the former. The annual accumulated temperature of daily means 10°C accounts for 6000-9000°C with a duration of 300-365 days, while the annual accumulated temperature of the daily means $\geqslant$15°C totals 5000-9000°C with a duration of 230-330 days, and the annual accumulated temperature of the daily means $\geqslant$18°C sums to 3700-8800°C with a duration of 170-350 days. The extreme minimum temperature averages over 2°C, with a diminishing tendency from the south to the north. The extreme minimum temperature recorded in Sanya City, the southernmost part of Hainan Island, is only 10°C.

Soils and fertility status

The soils of rubber plantations in China, in general, consists of latosol and lateritic red soil. Latosol is derived from basalt, granite, sandy shale, shallow coastal deposit, etc. and is found mainly in hilly, rolling, terrace and bench areas in Hainan Island, Leizhou Peninsula and South Yunnan Province. The fertility is variable with parent rocks and vegetation, usually higher in basalt-derived soil and lower in shallow coastal deposit derived soils.

Lateritic red soil is developed from granite, gneiss, sandstone and schist and the fertility differs greatly with the vegetation. The difference between latosol and lateritic red soil is in the lower eluviation and weathering state of the latter (Liang Jixing, 1988).

WIND DAMAGE AND WIND-PROOF PRACTICES

Wind damage is caused to rubber trees by the action of tropical cyclones or typhoons. The severity of wind damage is determined by the wind force, wind-endurance of the trees and cultural practices.

Wind force is measured based on Beaufort Scale, e.g. Beaufort Scale 8 (17.2-20.7 m s^{-1}), Beaufort Scale 9 (20.8-24.4 m s^{-1}), Beaufort Scale 17 (56.1-61.2 m s^{-1}), etc.

The criteria for grading of wind damage severity were set up by the Ministry of Agriculture, PRC, based on years of survey and observations (Table 1).

In determining the seversity of wind damage three formulae are used:

TABLE 1

Grading criteria for wind damage of rubber trees.

Grade	Unbranched young trees	Branched trees
0	No inury	No inury
1	Leaves lacerated, 1/3 trunk snapped	Foliage lacerated, tender shoots broken, <1/3 leaders broken or <1/3 crown foliage lost.
2	1/3-2/3 trunk snapped	1/3-2/3 leaders broken or 1/3-2/3 crown foliage lost
3	>2/3 trunk sanpped with scion left	>2/3 leaders broken or >2/3 crown foliage lost
4	Scion split, unable to reshoot	All leaders broken or one leader split or trunk snapped above 2 m from ground
5		Trunk snapped less than 2 m from found
6		Scion broken entirely
Leaning		Trunk slanted <30°
Half-prostrated		trunk slanted 30-45°
Prostrated		Trunk slanted >45°

Note: Number of trees suffering from snapping and prostrating = Total number of trees in grades 4, 5 and 6.

$$\text{Wind damage percentage} = \frac{\text{sum of all damaged trees}}{\text{total trees surveyed}} \times 100$$

$$\text{Mean grade of wind damage} = \frac{\leq(\text{each grade value x number of trees of the grade})}{\text{total trees surveyed}}$$

(Leaning trees are considered as grade 1, half-prostrated as grade 3 and prostrated as grade 5).

$$\text{Percentage of broken, snapped and prostrated trees} = \frac{\text{No. grade 4 trees + No. grade 5 trees + No. grade 6 trees + No. prostrated trees}}{\text{total trees surveyed}} \times 100$$

As rubber trees possess very good regenerative characters, those trees suffered from wind damage not higher than grade 3 could be restored, to its original shape usually three years after wind injury with only insignificant loss of yield, while those which succumb to damage at grade 4 and above could hardly resume its original size and shape.

Wind damage surveys for many years in China disclose the correlation between wind damage severity and wind forces (Table 2).

TABLE 2

Correlation between wind force and wind damage severity of rubber trees.

Beaufort scale	8	9	10	11	12	13	14	15	16	17
Wind speed (m s^{-1})	17.2-20.7	20.8-24.4	24.5-28.4	28.5-32.6	32.7-36.9	37.0-41.4	41.5-46.1	46.2-50.9	51.0-56.0	56.1-61.2
Wind damage percentage (%)	2-5	5-10	10-16	16-24	24-33	33-45	45-55	55-56	66-80	80

Clones in relation to wind damage

Wind endurance varies remarkably with cultivars or clones, and may be divided into three groups based on their performance in China:

Better wind-bearer	Haiken 1, PR 107, Youxian 5-2, Hongxing 1;
Moderate wind-bearer	Nanyang 12-2, Nanqiang 1-97, GT 1, Shenjing 1, Qingwanpo 17-12;
Poor wind-bearer	RRIM 600, PB 86, Nanhua 1, Hekou 3-11, Guangxi 6-68, Tjir 1, RRIM 513, RRIM 501.

As a general rule, seedlings are endowed with better wind fastness than buddings owing to their stronger taproot that penetrate deep into the subsoil and anchor the tree against uprooting.

The difference in wind endurance between clones is dependent upon the tree crown shape, wood properties and branching habit. An open conical crown (eg., clones Haiken 1 and PR 107) is rendered with lower wind damage because its horizontal branches are evenly arranged on the central leader.

A hemispherical or fan crown (eg., clones RRIM 600 and PB 86) will have greater wind damage as a consequence of its dense canopy consisting of bunch branches. Smaller branching angle (eg., clone GT 1) however is liable to split at the joint with the trunk; a serious wind damage pattern.

Wind-proof practices

In wind-prone areas, the first choice to reduce wind damage is to use wind-endurant clones. Nevertheless, various wind-proof cultural practices are still necessary. These include:

Establishment of windbreak network: The principle of alleviation of wind damage by wind shelter-belts lies in their ability to reduce the kinetic energy of air streams sweeping through the belts and to modify their structure. It was demonstrated that windbreaks could decrease wind speed at a certain scope twice the tree height from the windward side and 10-15 times the tree height from the leeward side of the shelter-belts. The wind-protection efficiency of shelter-belt is associated with its structure, height, width, orientation, size of sheltered block and whether in network or not. Structurally, windbreaks are three types: (1) dense type - consists of upper, middle and lower storeys with different tree species in association; (2) loose type - consists of upper and lower storeyed tree species with certain permeability; (3) permeable type - shelter-belts with upper storey but without middle and lower storey tree species. In mild wind-swept areas loose type shelter-belts are preferred, while in highly wind-prone areas dense type is adopted. Under China's climatic conditions, the main trees recommended for planting in the shelter-belts are: Eucalyptus excerta F. Muell., Casuarina equisetifolia L., Eucalyptus 12 ABL (Eucalyptus tereticornis x E. camaldulensis), Michelia macclurei Dandy, Schima superba, and secondary trees in the belts may include Acacia confusa Merr. and Camellia oleifera Abel. The size of sheltered block is advisable to be within 1-2 ha depending upon the wind-proneness of the location. Each block should be rectangular with its long side facing the direction. In hilly areas windbreaks are established according to the relief of the location, eg., either set-up of shelter-belts along the ridge of the hills or preservation of a lump of forest left uncut when clearing. In some cases, where the slope is more, one to two contour shelter-belts are established along the slope where the slope is too wide for one block, and longitudinal belts are set up at intermediate places. Individual shelter-belts, 10-20 m in width, are connected to each other to form a windbreak network that could offer better wind protection.

Dense planting: This practice is workable in wind-prone areas, because a denser population gives better self-protection and higher survival rate that leads to higher yield per unit area or alleviate yield loss due to wind damage. In this method the planting density is on an average 600 trees ha^{-1}.

Pruning: It has been established that wind damage could be diminished by pruning, which reforms the crown shape, enabling the trees to have better wind endurance. However, as pruning operation is somewhat difficult and labour-intensive, it is impractical in most cases. In the event that pruning is imperatively necessary in highly wind-exposed areas, the amount of pruning shall not exceed one third of the crown, as otherwise, pronounced drop in rubber yield and girthing can be expected.

COLD DAMAGE AND MEASURES TO ALLEVIATE IT

Cold damage types: Cold damage on rubber trees is the result of sudden drop in temperature or excessive accumulated low temperature, unbearable to rubber trees. Thus the type of cold damage is related to the type of cold current, cold hardiness of the clones and the location. Essentially two types of cold currents are found in China.

Advective type: Under the reign of cold front or stationary front, the combination of long spell of bleak weather, insufficient sunshine and cold wind results in chilling injury to rubber trees. As a rule, a temperature of $<10°C$ lasting for 20 days would be accompanied by the occurrence of cold damage at Grade 4-6 for over 30% of the trees.

Radiative type: When a clear but breezy weather occurs after the arrival of a cold wave, a sharp fall in night temperature (to 5°C or lower) would appear owing to static heat radiation, while the day temperature remains quite high, bringing about a diurnal range of over 15°C or 20°C. Rubber trees thus get exposed to extreme cold and hot conditions on the same day and this results in acute radiation cold injury. However, in situations where the night temperature is not so low and the daily range of temperature is not so wide, cold damage can still take place with a longer duration of cold radiation, which nevertheless is usually recorded in partly shady locations such as northward or sunless hillslopes, constantly fogging valleys and fully crown-closed rubber plantations (Jiang Ailiang and Qian Ping, 1985).

Symptoms of cold damage

Crown: Shrivelling of young leaves, dieback of old leaves and die back of shoots are the main symptoms of cold damage. The severity is correlated with the intensity of cold waves and the endurance of clones. In some cases, leaves become wilted and fall while in other cases leaves wither but do not abscise.

Trunk: Main symptoms which develop on tree trunk are black spots on green parts and drying of outher bark in brown parts, and necrosis of inner and outer bark, bark bleeding, stem base rot, etc.

Roots: Symptoms noted on roots are the bark cracking of tap root, partial or complete drying of lateral roots, die back of conductive roots. These symptoms appear only in soils within about 20 cm below the ground surface.

Cold damage symptoms on rubber trees do not manifest themselves on all parts except crown until 1-2 months after cold waves cease. The symptoms develop in such a gradual way that a cold damage survey is only reliable when it is done two months after the occurrence of cold waves.

Grading of cold damage

The severity of cold damage on rubber trees is classified according to a fixed criteria for different ages. That for trees under tapping is shown in Table 3.

TABLE 3

Classification criteria of cold damage on trees under tapping.

Grade	Crown	Trunk	Base
0	No inury	No inury	No inury
1	1/3 injured	1/6 trunk bark injured	1/6 girth mortified
2	2/3 injured	2/6 trunk bark injured	2/6 girth mortified
3	2/3 injured	3/6 trunk bark injured	3/6 girth mortified
4	Completely shrivelled	4/6 trunk bark injured or circumferential bark dried 1 m from ground	4/6 girth mortified
5		5/6 trunk bark injured or circumferential bark dried above 1 m from ground	5/6 girth mortified
6		Circumferential bark dried 1 m from ground	5/6 girth moritified (excluding the stock)

Source: Ministry of Agriculture, Draft Technical Regulations for Rubber Cultivation (1979).

Factors related to cold damage

Meteorological factors which influence cold damage are: temperature, sunshine, wind velocity and humidity.

Temperature: Low temperature is the key factor responsible for cold damage for rubber trees. Variations in temperature patterns bring forth different detrimental levels of low temperature. For advective cold damage, the daily mean temperature over the advective period is the criterion while for the radiative type, the extreme minimum temperature in the radiative spell is the factor. At critical temperatures, lower temperature is inevitably associated with severe cold injury. The relation among them is tabulated in Table 4.

TABLE 4

Relationship between low temperatures and cold injury of rubber trees.

Advective temperature falling		Radiative temperature falling	
Daily mean temperature of coldest days (°C)	cold injury severity (grade)	Extreme minimum temperature (°C)	Cold injury severity (grade)
>8	1-2	>3	a few leaves withered
6-7.5	2-3	2- -1	1-2
6-4	>3	-1- -3	3
<4	4 to destructive	-3- -4.5	4 to destructive

Sunshine: Sunshine also is known to influence cold damage, especially with radiative type injury on tree base. It has been demonstrated that an effective daily sunshine duration of 1.5 h would result in grave injury to stem base as opposed to a situation when the duration is 2-4 h. However, no injury is observed when the duration is over 4 h.

Wind velocity and humidity: In the process of advective temperature drop that often accompanies strong wind, high velocity of wind, coupled with rapid temperature drop would cause severe cold damage to the tree. Likewise, at certain low temperatures, higher humidity is bound to give rise to greater cold damage to the crop. Freezing tests indicated that at -1° to -5°C for three consecutive days and nights, the clones tested, 93-114, PB 86 and Nanhua 1 showed no symptom of cold damage when the humidity was 48%. However, when humidity was over 80%, grade 5 inury was recorded.

Terrain and relief

The terrain and relief have a redistributing effect on such meteorological elements as temperature, moisture, sunshine hours and wind, resulting in the reformation of different climatic conditions and thus incurring different severities of cold damage to rubber tree.

Meso-terrain environment: It refers to a meso-environment for rubber created by knolls, hills, terrace, river basins and valleys with diverse structure and composition. Any closed space, poor outlet for cold air, higher and longer cold air stagnation or cold current channelling relief will result in severe cold damage.

Micro-terrain environment: This type of environment is hinged on the slope direction, slope position, slope form, terrace height, etc. Rubber trees on southward slopes are normally affected by mild cold injury while those on northward slopes have to encounter grave cold damage. In advective type, rubber trees on higher slopes are likely to bear severe cold injury while those on lower slopes get comparatively mild cold injury, and in the radiative types it is vice versa. On straight or convex slopes where it is favourable for cold air to move out, rubber trees show only mild cold damage while those on concave slopes are subjected to serious cold damage due to stagnation of cold air.

TABLE 5

Cold endurance of some clones in China

Clone	Tolerant to advective low temperature	Tolerant to radiative low temperature	Yield potential
Wuxing I_3	***	*****	*
93-114	****	***	**
Nanhua 1	****	****	**
GT 1	***	****	****
Haiken 1	***	***	***
RRIM 600	*	*	*****
PB 86	**	*	****
PR 107	*	**	****
RRIM 623	**	***	****

Cold endurance: ***** very strong. **** strong. *** average. ** poor. * very poor.

Yield potential: ***** very high. **** high. *** average. ** fair. * poor.

Clone factors

Hevea clones vary greatly in cold endurance, and even the same clone has different responses to advective low temperature and radiative low temperature. For instance, clone 93-114 shows better tolerance to advective low temperature while Wuxing I_3 displays greater endurance to radiative low temperature. Examples of clones exhibiting better cold hardiness are Tianren 31-45, 93-114, GT 1, Wuxing I_3, IAN 873, etc. Cold tolerance performance of some popular clones in China is given in Table 5.

Approaches to mitigate cold damage

Presently only two approaches are found effective to mitigate cold damage to rubber trees: one by carefully choosing comparatively suitable environments for rubber and the other in the discriminatory use of the existing clones. Certain cultural practices were also tried but they failed to give satisfactory results and were impracticable on a commercial scale. The most effective method, however, is to obtain new clones endowed with much better cold hardiness and high yield potential.

Choice of macro and meso environment to escape cold damage

The first step in the process of picking out cold-insensitive environments is to get a clear picture of cold current sources, its intensity and the impact of meso-terrains on meteorological factors. On this basis, demarcation is made of macro and meso environments for rubber such as highly cold-susceptible area, moderately cold-susceptible area and mildly cold-susceptible area. Each meso type can be further divided by taking into consideration the slope direction and position, slope gradient, slope forms and specific relief. The finalized micro-environments serve as the basic unit for growing proper clones in optimum locations.

Selection of clones for optimum performance

Selection of clones for cold-susceptible areas shall be made on a location specific basis to suit each meso-environment. The principle is to give preference to clones with high yield and poor to moderate cold tolerance for mildly cold-susceptible areas. In moderately cold-susceptible areas, clones with good yield potential and moderate cold endurance are preferentially selected over clones with strong cold resistance and moderate yield potential. In highly cold-susceptible areas clones with strong cold hardiness and moderate yield potential are preferred. An appropriate mix up of clones considering the climatic situation, on the basis of the above principles, is generally attempted.

Treating cold-damaged trees

Stem and shoots: It is unnecessary to give any treatment to the trunk and shoot with grade 1-2 cold injury. For those with grade 3 injury, the spoiled shoots should be removed with a saw at the interconnection between the damaged and the healthy parts when the first leaf whorl is fully mature. The saw cut should then be dressed with coal tar. For grade 4-5 injury, rebudding or crown grafting is recommended with clones possessing better cold endurance in the case of immature plants upto 3 years old or alternatively, replanting may be carried out.

Bark: No treatment is necessary when cold induced dryness is confined to outer bark that would fall off naturally and be replaced by new growth. However, as the dryness goes further deep into the wood, all mortified bark should be excavated and the dilapidated surface be scraped out, and dressed with 'colophony' compound to induce a gradual healing of the wound. On cold-induced bark-burst, latex coagulum should be removed and mortified tissues cut off with wound dressing provided.

GROWTH AND YIELD PERFORMANCE OF RUBBER TREES IN CHINA AND MEASURES FOR HIGHER YIELD

Growth

Rubber trees in China are deprived of conditions for normal growth all the year round by the prevailing low temperature and drought during winter that give rise to a distinctive and long defoliation period of about three months (Hu Yaohua and Xie Haisheng, 1985). The growable months within a year stretch from March to November, and the main period of growth is from June to October with an annual increment in girth of 6-8 cm. Consequently, rubber trees cannot reach their tappable size until seven to nine years after planting.

To reduce the unproductive phase in the field, some cultural practices are being followed. In field transplanting, planting materials used include budded stumps and advanced planting materials such as stumped buddings or polythene-bagged plants. Field planting in early spring is a special practice in China for the purpose of obtaining maximum growth in the planting year and for reducing loss of young transplants from cold in the first winter in cold-susceptible areas. Some of the package of practices followed for good management are: (1) split fertilizer application during the hot and growing period and applying liquid manure in the drought spell; (2) special care for slow growing trees and plantations; (3) spraying with glyphosate to eradicate weeds and to control all undergrowth for minimizing their competition with young rubber plants; (4) mulching the tree base all the year round and establishment of cover crops in interrows; (5) controlled pruning of young plants to form a wind-resistant crown.

Yield

Under the climatic conditions in China, only 180-270 tappable days are obtained annually, which is much less compared to that in equatorial countries or regions. Correspondingly, the yield per unit area in China is somewhat lower. However, trial plots and commercial plantation experience showed that it is possible to obtain a dry yield of 1500-3000 kg ha^{-1} y^{-1} in China once some integrated practices for higher and stable yield are adopted to incorporate exploitation with maintenance and recuperation. These integrated practices have proved effective under China's local conditions and they conform well with the latex regeneration patterns within the trees. The following are some highlights of these practices conducive to high and stable yields.

Good upkeep practices

A unique practice in China's rubber plantations is to apply organic manures or green dressing into an already dug ditch 50 cm in depth in the interrows, followed by covering with a thick layer of soil or polyethylene film. This is intended to enhance soil physical properties and available nutrient supplying capacity which, in turn, will boost the growth of root systems and its nutrient uptake efficiency. At the same time fertilizers are also supplied to the trees on the basis of nutrient diagnosis to enable them maintain a relatively high nutrient level and satisfying the requirements of latex regeneration (Lu Xingzheng and He Xiangdong, 1982). This lays a good foundation for sustained higher yield.

Exploitation based on analysis of physiological status of rubber trees

The dynamic analysis of physiological status of rubber trees is to study the external dynamic expression of latex regenerating physiology in different periods within a year. To cite an instance, during the period between refoliation and full maturity of the new flushes each year, almost all nutrient reserves are mobilized for the development of the first leaf whorl with very little left for use in rubber formation (Hu Yaohua and Xie Haisheng, 1985). Tapping, if done at this stage, could not only give a lower yield but, could also hamper latex regeneration of the tree over the whole year. In consideration of this drawback, it is most desirable not to reopen the trees until 15 days after the first leaf storey attains full maturity. Even one month prior to the full maturity of the second leaf whorl, intensive tapping should be avoided. During July to mid-October when the well-developed crown begins to work under the stimulus of sufficient solar radiation, rainfall and heat, rubber trees are potentially at their peak latex regeneration period if nutrient status is satisfactory. Exploitation measures with any possible technique including use of stimulants, that could tap fully the yield potentials, are applied to intensify latex flow. Thereafter, air temperature tends to fall and the photosynthetic capacity of the trees slows down whilst the trees start to accumulate starch. This situation is most conductive to promotion of latex flow and thus it is likely to prompt late dripping and hence brown bast (see also Chapter 9. Eds.). The exploitation system should be so adjusted, taking these aspects into account. In the second half of the year the variation in dry rubber content should be monitored to ensure that d.r.c. does not fall below 30% during the peak yielding period and 28% in October-November.

Judicious exploitation

Tapping operations should keep in tune with the dynamic physiological status of rubber trees. Reopening shall be done only 15 days after the first leaf whorl attains full maturity. Shallow tapping (ca. 2 mm from the cambium) shall be practised in the period upto the maturing of the second leaf whorl. Proper application of yield stimulants (2-4% ethephon), reduced tappings (ca. 60 t/y) and stoppage of tapping when atmospheric temperature is 15°C, are also regular practices. Tapping intensities are judiciously adjusted in the case of wind and cold stunted trees depending upon the severity of incidence. Those trees with some prognostication of brown bast should be given rest from tapping.

SUMMARY

Extracted from scientific experiments and commercial experience under the constraints of typhoon and cold, China's unique package of cultural techniques for rubber plantation have proved very effective and successful on a commercial scale over the last three decades. However, it does not mean that the problems from the constraints are entirely relieved. On this account, in-depth studies on all related aspects are needed, especially breeding programmes. To free the industry from cold hazard, much better cold-tolerant clones with high yield are required. Fortunately, 6-7 clones from the 1981 IRRDB/Brazilian collection are found to be better than clone 93-114 in phytotron tests in reference to cold hardiness. This signifies that

238

some superior cold resistant germplasm have been obtained. Triploids derived appear to be useful in breeding programmes aimed at evolving cold tolerant clones (Zheng Xueqin et al. 1983). In wind-resistance breeding programme, the main objective is to get dwarf varieties, since shorter tree trunk was noted to associate with less wind damage. Some dwarf germ-plasm have been collected for use in the programme. Presently, the rapidly developing molecular biology will make possible the breeding of good stress-tolerance clones with high yield potentials by way of genetic engineering. It thus follows that China's rubber industry has a very bright future.

REFERENCES

He Kang and Huang Zongdao. 1987). Rubber cultivation in the North Part of Tropical Area. Guangdong Science Press. pp. 498.
Huang Zongdao and Pan Yanqing. 1979. Rubber Cultivation Science. Agriculture Press. pp. 303.
Hu Yaohua and Xie Haisheng. 1985. A study on the growth of new plant body and the law of growth and decline of stored matter in the plant body during the sprouting of the first leaf storey of Hevea. Chinese Journal of Tropical Crops, Vol. 6: 29-38.
Jiang Ailiang and Qian Ping. 1985. A study of the relationship between natural cooling process in winter on different facing slopes in Xishang-banna and the chilling injury of rubber trees. Chinese Journal of Tropical Crops. Vol. 6: 1-12.
Liang Jixing. 1988. Major soil types of Hainan Island. Chinese Journal of Tropical Crops. Vol. 9: 53-72.
Lian Shihua. 1984. A preliminary study on causes of typhoon damage on rubber trees. Chinese Journal of Tropical Crops. Vol. 5: 59-72.
Lu Xingzheng and He Xiangdong. 1982. Fertilizer application based on nutrient diagnosis of rubber trees. Chinee Journal of Tropical Crops. Vol. 3: 27-39.
Yin Xiuchun. 1987. Major ditrimental wind directions of typhoons landed on Hainan Island. Chinese Journal of Tropical Crops. Vol. 8: 87-96.
Zheng Xueqin, Zeng Xiansong, Chen Xiangmin and Yang Guangling. 1983. A new method for inducing triploid of Hevea. Chinese Journal of Tropical Crops. Vol. 4: 1-4.

CHAPTER 11

NUTRITION OF HEVEA

A.K. KRISHNAKUMAR and S.N. POTTY
Rubber Research Institute of India, Kottayam-686009, Kerala, India.

Judicious nutrient management has long been recognised as the surest means of sustaining high levels of productivity and rubber is no exception to this. Nutrient demands of Hevea were generally believed to be modest mainly because of the fact that earlier plantations were mostly in newly cleared forest soils, rich in plant nutrients. Moreover, rubber plantations present almost a closed ecosystem, in a near steady state, during their life span. Nutrient management has gained greater importance in recent years because of two reasons: firstly, Hevea plantations are no longer raised in virgin forests and secondly, most of the plantations are either in the second or third cycle of replantation. Even though nutrients removed through the crop are negligible, large amounts of mineral elements are locked up in the process of biomass accumulation and are lost through timber during replanting. Gradual depletion of mineral resources through cycles of re-plantation warrants appropriate nutrient management. Of late, more and more marginal and depleted soils are being brought under rubber cultivation to meet the increasing demand and under such situations proper soil and nutrient management is essential to sustain productivity at economic levels.

Response of a perennial crop like rubber to nutrition is influenced by the nutrient supplying capacity of the soil on the one hand and factors like clonal variation, stage of growth, intensity of exploitation and ground cover management on the other. A wealth of information has emerged on this aspect from extensive studies conducted in almost all rubber growing countries, the major contribution being from Malaysia. General fertilizer schedules for local conditions have been perfected. In this chapter, an attempt is made to review studies on nutrition conducted in the major rubber producing countries.

SOILS UNDER HEVEA

Hevea, a native of Amazon tropical forests, can grow on a wide range of soils, but deep well drained soils of pH below 6.5 and free from underlying sheet rocks are well suited for its performance as a commercially viable plantation crop (Pushpadas and Karthikakutty Amma, 1980). However, as opined by Webster (1989) most of the plantations in Africa and Asia are located in areas chosen solely on grounds of availability and convenience. Nevertheless, such areas satisfied most of the minimum requirements of soil conditions except in situations where the planters were forced to accept marginal sites.

Soils under Hevea, in general, have developed either under warm humid equatorial monsoon climate with a little or no dry spell or under tropical wet-dry monsoon climate with variable duration of dry season, extending from three to five months. The Indonesian archipelago, Malaysia and southern part of Sri Lanka fall under the former climatic zone while India, northern part of Sri Lanka, Burma, Thailand, Vietnam and the Philippines archipelago and southern part of Indonesia fall under the latter (Pushpadas and Karthikakutty Amma, 1980). In Brazil, the soils are mostly red yellow podsols or latosols, formed under warm humid equatorial climate. The alternating wet and dry monsoon climate prevailing in most of the rubber growing countries including India favour high degree of laterization.

The soils under rubber in Malaysia have originated mainly from igneous, metamorphic, argillaceous and arenaceous sedimentary rocks. Of the seventeen series, seven have shales as parent rocks. The other parent rocks are granite, granodiorite, basalt, andesite, rhyolites/volcanic tuff, dacite, quartzite/shale and sandstone (Chan et al. 1977). Considerable areas also represent soils originated from marine sediments and volcanic ashes as in the case of Indonesia. The major rock types in the traditional rubber growing regions in India are charnockite, pyroxene, gneiss, khondalite etc., of the precambian metamorphic complex and the soils are laterites, lateritic and red soils in catenary sequence with laterites (Krishnakumar, 1989). The parent rocks in North East India are however sandstone and shales.

Influence of soil properties on crop performance

The rubber tree can withstand soil physical conditions ranging from stiff clayey with impeded drainage to well drained sandy loam. However, its performance is affected by adverse soil conditions. Physical, chemical, physico-chemical and physiographic features of soil also influence the growth and productivity of rubber to a considerable extent.

Physiographic features: Physiographic features such as degree of slope, aspect, soil depth, rockiness etc., have been reported to have profound influence on growth and yield of rubber (Chan et al. 1972). Soil depth is considered to be an important parameter influencing growth and yield of this perennial crop. Shallow soils restrict development of tap root affecting anchorage of trees. Deep soils with large quantities of clay, which serves as a reservoir of moisture, help to tide over drought situations. A minimum depth of 100 cm has been considered essential for successful rubber cultivation (Pushpadas and Karthikakutty Amma, 1980). Intervening hard pans, if present, should be well below 1.5 m. So also, it is not desirable to have underlying sheet rocks in the profile within a depth of 2 m. Depth of more than 125 cm increases growth, yield and leaf nutrient content (Chan et al. 1974). These authors reported that slope also affects growth of Hevea. A slope upto 26% is reported to favour growth and productivity. Nevertheless, rubber is grown satisfactorily in much steeper slopes.

Aspect is found to influence growth and performance of Hevea significantly. Aliang Jiang (1981) reported that in China, rubber trees on the leeward slopes suffered less damage due to cold. Studies on microclimatic observations conducted in China by the above author revealed that trees on south and west slopes suffer less cold damage. Aspect coupled with soil properties also influence growth of rubber trees. Either water stress or excess stagnant water in the root zone, will weaken cold hardiness of rubber trees. The authors of this chapter have observed that the initial establishment and growth of immature Hevea are better on the southern slopes than on other aspects in Tura (North East India) situated at an altitude of 600 m MSL. This observation is quite contradictory to the experience in the traditional rubber growing regions in South India experiencing warm humid climatic conditions.

Physical properties: A wide range of textures from clay to sandy loam has been reported in the soils under rubber. Soils of loamy texture are best suited for cultivation of rubber (Pushpadas and Karthikakutty Amma, 1980). However, within loams higher clay content has been reported to promote growth as well as yield. In Malaysia, lack of adequate clay has been found to affect the growth of rubber in some of the soil series due to poor nutrient retention and this has warranted rescheduling of the fertilizer application (Sivanadyan, 1972). Feeder root development has been reported to be affected by soil texture (Soong, 1971). He observed a positive correlation of root development with sand content and negative correlation with clay content in the soils of Malaysia. Though clay content in the rubber growing soils of India has been reported to be relatively high, it is moderated by the presence of high amount of sesquioxides, thus reducing the adverse effect of high clay content. Other physical parameters like bulk density and porosity also affect the growth of Hevea indirectly through their influence on soil erosion and consequent root development.

The productive potential of a soil is influenced by its moisture retention characteristics and rubber being a rainfed crop, this factor assumes greater importance. A wide range of available water content has been reported from soils under Hevea in Malaysia and India (Soong and Lau, 1977; Krishnakumar et al. 1990). The amount of moisture retained at various tensions was also found to vary depending on the nature and content of clay. At field capacity (-0.033 MPa) 19.5 to 37.8% soil moisture was found to be retained in the surface soils of the west coast of India (Krishnakumar, 1989). In these soils, 75% of moisture was desorbed at -0.5 MPa and hence this tension can be considered critical as far as soil moisture availability in rubber growing soils is concerned. The Kaoline-ironoxide aggregates present in the high clay tropical soils tend to hold more moisture at lower tensions behaving like sands, and at the same time these aggregates behave like clays at higher tensions, enabling retention of sizeable quantity of water. In the latter case the aggregate size becomes irrelevant because it is the micropores within aggregates that get drained.

Ground covers help in improving soil physical properties. Aggregation of soil particles under both natural and legume covers was found to be better when compared to soils under clean cultivation. Between the legume and natural covers, the former influences soil physical properties more favourably than the latter (Duley, 1952; Bremner, 1956; Cornfield, 1955 and Watson, 1957). A better soil structure in terms of mean weight diameter, rate of infiltration and percentage of water stable aggregates, has been reported in India, when rubber is grown in association with legume cover than with natural cover (Krishnakumar, 1989).

In denuded forests as well as in areas subjected to continuous shifting cultivation the soil physical properties have been found to be improved considerably once a rubber plantation is established. Twenty years after planting of rubber the moisture retained at field capacity was significantly higher vis-a-vis the moisture retained in an adjacent field under continuous shifting cultivation (Krishnakumar et al. 1990a).

Physico-chemical properties: The minerology of clays present in the soils largely influences their physical and physico-chemical behaviour. Cations such as calcium, magnesium and potassium play an important role in the nutrition of rubber and dynamics of these mineral elements is governed by the type of clay minerals. Rubber growing soils of Malaysia are rich in Kaolinite (1:1 layer silicate clays). Of the 23 soil series, 14 series have been reported to have kaolinite content of more than 50% and other seven series have 20-50% (Chan et al. 1977). Illite also has been reported to be present in all the above soil series. In the clay fraction of soils of basaltic origin from eastern and south eastern Nigeria, kaolinite was identified to be the dominant clay mineral (Eshett and Omutei, 1989). Eshett and Omutei (1989) reported the presence of hematite, lepidocrocite, mica, smectite and goethite in the soils of south eastern Nigeria. The clay fraction in the soils under Hevea in India is also dominated by kaolinite. The presence of the oxides of iron and aluminium in the clay fraction leads to poor release of nutrients through mineral weathering to meet the demand of rubber for rapid growth and high yield (Eshett and Omutei, 1989). This situation warrants judicious fertilizer application and proper establishment of leguminous ground covers. Appreciable amount of illite, mostly of degraded nature, has been identified in the soils along the west coast of India. Soils with appreciable amount of this mineral would render potassium unavailable by fixation. Inconsistent response to potassium observed by Ananth et al. (1966) in immature phase and by Potty et al. (1976) in mature areas could be attributed to the diverse minerology of soil clay fractions especially of illite group. The range of aluminium in some of the soils is so high that it could lead to toxicity problems. The optimal pH for rubber is reported to be in the range of 4 to 6.5. The possibility of aluminium toxicity in the above range is very remote. Even though Hevea is reported to tolerate a pH range of 3.8 to 7.0, the extremes could affect its growth and productivity. In soils with pH less than 4, it is not the low pH per se but the toxicity or deficiency of mineral elements that often limits crop production (Marscher Horst, 1986).

Rubber is grown in soils with a wide range of CEC. While CEC of 2.05 to 15.96 m eq $100g^{-1}$ is reported in Malaysia, it ranges from 3.55 to 18.02 meq 100 g^{-1} in soils under Hevea in India. In the soils of basaltic origin in Nigeria the CEC is relatively low (Eshett and Omueti, 1989). Soils in the rubber growing countries, in general, have been found to be low in exchangeable bases especially in the upper horizons and this could be attributed to leaching appreciable quantities of metallic cations down the profile due to high rainfall. Next to calcium, aluminium is found to be the dominant cation of the exchangeable elements. In soils with a pH value of less than 5, preponderance of exchangeable aluminium is generally observed. A drop in aluminium to calcium ratio in the soils adversely affects the growth in most of the crops. Presence of organic matter also is known to reduce the toxic effect of aluminium by chelation (Mutatkar and Pritchelt, 1967). In the case of rubber soils, maintenance of a relatively higher organic matter status through organic matter recycling, coupled with calcium enrichment indirectly by the addition of rock phosphate, could lead to a favourable condition to alleviate the toxic effect of aluminium through chelation and optimum calcium to aluminium ratio. Moreover, the rubber tree itself most probably has a certain degree of aluminium tolerance (Krishnakumar, 1989).

Fertility status: In general, soils under Hevea have been found to be rich in organic matter compared to other agricultural soils in the same region. The organic matter distribution down the profile shows an enrichment in the surface horizons and a decline in the sub-surface horizons. This high accumulation of organic matter in the top soil is due to maintenance of a luxurient leguminous ground cover which adds about six tonnes of organic matter per hectare during the pre-tapping phase of rubber. The organic matter enrichment, however, depends upon the nature of the ground cover. The most commonly grown, Pueraria phaseoloides, has been reported to add about 3.0 tonnes of organic matter during the first four years whereas Mucuna bracteata, a species introduced from the north eastern parts of India, adds 5.6 tonnes during the same period (Kothandaraman et al. 1990). Addition of litter through annual leaf fall also helps build up of organic matter. On an average, about six tonnes of organic matter is added every year through annual leaf fall. The slow pace of oxidation inside the closed canopy of rubber plantations helps to maintain the high organic matter status. The cultural operations with nearly zero tillage also favour stabilisation of organic matter at a relatively high level. Since most soils under Hevea have an abundance of aluminium, the organic matter is complexed with this element, thereby leading to a reduction in decomposition. Inter-action of oxides/allophane with the organic matter in the tropical soils again render the organic matter relatively resistant to mineralisation. Extreme phosphorus deficiency consequent to the presence of higher amount of soluble aluminium inhibits microbial growth and these factors also result in low mineralisation of organic matter (Munevar and Wollum, 1977). A range of organic carbon content from 1.11 to 3.76% has been reported in the soil under Hevea in India. A comparable range has also been reported from Malaysia. The C/N ratio was around 10 in the rubber growing soils of India suggesting the stable nature of organic matter in these soils. However, in the rubber growing soils of north east India, an extreme deficiency of organic matter has been encountered consequent to the traditional practice of shifting cultivation (Krishnakumar and Potty, 1989). Planting of rubber has been reported to enrich organic matter status in depleted soils. Studying the influence of rubber plantation on the eco-system of north east India, Krishnakumar et al. (1990b) reported an increase of 0.6% in the organic matter content of soils in a 10 year old rubber plantation compared to a 'jhume' cultivated field in the same location.

An available nitrogen content of 11.6 to 38.8 ppm has been reported in the soils under Hevea in India (Krishnakumar, 1989). The organic carbon content has a positive correlation with available nitrogen. Rao et al. (1990) reported a low rate of nitrification in soils under rubber plantations of north east India where low temperatures prevail for nearly four months.

In general, rubber growing soils are rich in iron and aluminium which render phosphorus sparingly available. A wide-spread deficiency of available phosphorus and potassium in soils under Hevea in the traditional rubber growing regions in India has been reported by George (1961b). Extreme deficiency of available soil phosphorus has also been reported in soils of non-traditional rubber growing regions in north east India.

The available calcium content, generally in the lower range, varied widely in the profiles studied in the west coast of India. Rubber trees appear to have a degree of adaptability to low calcium environment. The influence of calcium is however vital mainly because of its role in alleviating aluminium toxicity thereby resulting in increased availability of phosphorus and potassium. The ratio of calcium to total cation should be around 0.15 for the roots to grow uninhibited. Al/Ca molar activity ratios also influence root development and growth in acid soils. A ratio of 0.02 is considered to be the upper limit beyond which growth will be affected (Marschner Horst, 1987).

The available magnesium content exhibits varied distribution in soils under Hevea in various countries. In India, even within the traditional rubber growing region, distinct delineation has been attempted with respect to available magnesium content (Pushpadas and Ahmed, 1980) and the variation has been attributed to the parent material.

Pedogenesis and classification

Based on the cation exchange capacity, Al_2O_3, SiO_2 / R_2O_3 ratio, nature of the underlying rocks and also the clay mineral assemblage, Krishnakumar (1989) suggested a probable pedogenesis of the soils under Hevea in the west coast of India. He suggested a weathering sequence where a rapid leaching of silica takes place as a consequence of alkaline hydrolysis of the parent rocks of mixed chemical composition, resulting in a higher iron and aluminium content. Since iron could get transformed into oxides more rapidly than aluminium, there is more iron in the sesquioxides and a good quantity of it also appears as free iron oxides. Aluminium released due to hydrolytic decomposition reacts with silica to form mostly kaoline and other layer silicates. A part of unreacted aluminium crystallises as gibbsite, creating a kaoline-gibbsite system that is stable in the prevailing acid conditions. Presence of aluminium in the exchange sites indicates intense tropical weathering. An oxide mixed minerology of the soil is therefore a logical outcome. A lower SiO_2/R_2O_3 ratio encountered suggests a highly weathered condition of the soil. R_2O_3 is dominated by iron and this has resulted in relatively higher SiO_2/R_2O_3 ratios.

International system of soil classification had been attempted to characterise rubber growing soils. The soils in Malaysia have been classified upto series level under the orders Entisols, Inceptisols, Ultisols and Oxisols (Chan, 1977). A soil suitability classification scheme also has been suggested by him wherein five classes (I to V) have been described on the basis of limitation to rubber cultivation. Separate classification systems have been adopted by different countries. Da Costa (1968) developed a system in Brazil for the local red-yellow podzols and latosols. Similarly the Commission for Technical Cooperation in Africa (CCTA) have published a soil map of Africa following French and Belgian systems (d'Hoore, 1964). The classification systems developed by FAO and UNESCO 1974) also are being followed in some countries. A land capability classification has been suggested for rubber by Sys (1975) as per FAO guidelines. Broadly classifying the soils under Hevea in India, Krishnakumar (1989) ascribed three large groups in the major rubber growing belt viz; Paleudalfs (Kanyakumari and Calicut region and Goa); Paleudlts (Central Kerala) and Paleustalfs (Karnataka).

Differential performance of clones in different series has been observed in Malaysia and specific recommendations for different series with respect to clones have been evolved. For instance, Chan and Pushparajah (1972) showed that RRIM 600 gave the highest yield in Holirood and Munchung series but its performance was inferior in Rengam series. Thus an understanding of the taxonomic unit would help in offering proper recommendation with respect to the clone to be planted. In view of this an environmax approach has been evolved in Malaysia, wherein clonal recommendations are made considering the constraints existing in a particular environment. A detailed account of the soil capability and clonal effects, soil-clone interaction and the environmax approach for rubber has been given by Watson (1989).

Response to nutrients

Though the need for manuring rubber has long been recognised this aspect has not received the required attention compared to other agricultural and commercial plantation crops. Nutritional studies on rubber received low priority as the crop was normally treated as a forest species.

Moreover, the initial plantations were on rich newly cleared forest soils. Nevertheless, manurial trials on rubber, started in the early 1900s in Malaysia and Indonesia confirmed good response of the tree to the application of fertilizers (Penders, 1940). The effect of various nutrient elements on the growth of Hevea was also established from the studies conducted in Malaysia (Bollejones, 1954). The mineral composition of Hevea was reported to be influenced by soil fertility status (Dijkman, 1951).

The trees immobilise substantial quantities of nutrients in the trunks, branches and roots of which about half get immobilised during the pre-tapping phase. During the immature phase the nutrient requirement is estimated to be much greater than the fertilizer input. For instance, the total nitrogen immobilised during the first six years is estimated to be 728 kg ha^{-1} which works out to be on an average 120 kg ha^{-1} yr^{-1}. However, the recommended level of nitrogen application is far below this. The deficit has to be met from nutrient reserves of the soil. The resultant nutrient depletion leaves the soil less fertile and this situation can be overcome only by ensuring establishment and maintenance of a luxurient leguminous ground cover which will gradually release substantial quantities of nutrients once the canopy of rubber closes. Full compensation of the nutrients immobilised by the growing trees may lead to the development of heavy crown rendering them vulnerable to the risk of wind damage. However, adoption of cultural practices such as branch induction could be beneficial to a certain extent (Watson, 1989).

Nutrition during immature phase

Most of the earlier studies were confined to the pre-tapping phase. Studies in Indonesia indicated that the effect of fertilizers was the highest in the pre-tapping stage and the trees attained tapping girth two to three years earlier with proper fertilizer application (Dijkman, 1951). Owen et al. (1957), reviewing the results of 17 trials conducted in Malaysia, found only negligible response to applied nitrogen during early immaturity phase and the response was evident from the sixth year only. On the other hand, significant effect of phosphorus was obtained at fifth, sixth and seventh year after commencement of manuring programme. No significant response to potassium was obtained in most of the trials. The higher level of soil fertility in the newly created plantations could be the main reason for the lack of response in the early stage.

In the laterite soils of South West India, rubber plants respond positively in terms of girth increment to application of phosphatic fertilizers at lower levels. A positive benefit from nitrogen manuring also has been reported. However, no significant girth increase has been shown by the application of potassium alone. The pattern of response indicated a positive interaction between nitrogen and potassium (George, 1963). Results of multi-locational trials in India at pre-tapping stage revealed that the response to applied fertilizers during the first four years of immaturity is dependent on the initial soil fertility status (Ananth et al. 1966). However, lack of response to nutrients, particularly to nitrogen and phosphorus was reported from the fifth year onwards. This was attributed to the large quantities of nutrients released by the leguminous cover. Response of rubber to fertilizer application depends on the type of ground covers also (Potty et al. 1978).

The extent of response to fertilizers depends on the type of soil as well as the clone (Bolton, 1960b; Krishnakumar and Potty, 1989). On sandy latosols in Malaysia, marked response to soluble phosphatic fertilizers, and lower response to nitrogen were reported. The effect of potassium was not significant (Bolton, 1960b). Response of Hevea to potassium has been reported to be influenced by potassium status of the soil (Pushparajah and Guha, 1969). In the highly depleted soils of North East India, Krishnakumar and Potty (1989) observed a marked increase in the girth of plants at higher levels of nitrogen, phosphorus and potassium.

Systematic application of fertilizers throughout the pre-tapping phase leads to build up of plant nutrient reserves. Application of phosphorus at the rate of 30 kg P_2O_5 ha^{-1} increased the soil phosphorus status from 0.47 mg 100 g^{-1} to 2.43 mg 100 g^{-1} and by further raising the level to 60 kg P_2O_5 ha^{-1} soil phosphorus increased to 6.8 mg 100 g^{-1} over a period of nine years. The nutrient concentration in leaf also registered a corresponding increase. Application of rock phosphate at the rate of 30 and 60 kg P_2O_5 ha^{-1} raised the leaf values to 0.19 and 0.22% respectively from 0.15% in the control plot. Similarly, addition of muriate of potash helped in raising the available potassium level in soil to 5.94 mg 100 g^{-1} at 20 kg K_2O ha^{-1} and to 10.93 mg 100 g^{-1} at 40 kg K_2O ha^{-1} (Krishnakumar and Potty, 1990). Application of rock phosphate increased the available calcium in soil as well as the leaf calcium level (RRII, 1988).

During the early immature phase, the nutrient demand of Hevea has been found to vary with the type of planting material. For instance, application of higher doses of fertilizers was needed when polybag plants were used (Krishnakumar and Potty, 1989).

Mature phase

Loss of nutrients such as nitrogen, phosphorus, potassium and magnesium as a result of latex extraction is negligible. Major nutrients, nitrogen, phosphorus, potassium and magnesium have positive effect on rubber yield. This could be a direct effect or mediated through their effect on growth of bark, bark renewal etc (Samsidar et al. 1975; Pushparajah, 1969). Experiments in India (George, 1962) that application of nitrogen, phosphorus and potassium could substantially increase yield. Owen et al. (1957) however, found that nitrogen had no significant influence on yield during the first four years. There was, however, evidence of response to application of phosphorus, which influenced yield significantly during the first four years. In the red loam soils of South India, Punnoose et al. (1978) reported lack of any response to major nutrients applied from the fifth year of planting upto commencement of tapping except for a marginal increase obtained by increasing potassium from 50 to 100 kg ha^{-1}. Towards the later stage of tapping however, residual effect of potassium was noticed, which could be attributed to the minerology of soil permitting fixation of potassium. Presence of appreciable amount of illite in the clay minerals lock up potassium through fixation which gets released with progress of time and results in delayed response.

Philpot and Westgarth (1953) found beneficial effect of phosphorus and potassium on the stability of latex. The positive effect due to combination of phosphorus and potassium was ascribed to a more balanced Mg/P ratio in the latex. Ram Beaux and Danjard (1963) suggested application of potassium for reducing the Mg/P ratio. Addition of potassium in Terrarosa soils of Vietnam has increased the yield by decreasing the Mg/P ratio in clone Gl 1, which has high latex magnesium (Beaufils, 1954). According to Owens (1957) if proper initial management is given, especially with respect to optimum nutrition during the pretapping phase, plantations would continue to perform well even if a period of neglect sets in. Therefore, supply of nitrogen, phosphorus and potassium during pretapping phase is highly essential. The effect of phosphorus applied during immature phase continues to the early years of tapping, as reported by Haines and Crother (1940).

Nutrient requirement with stimulation

Application of yield stimulants such as 2,4-D and 2,4,5-T (De Jonge, 1955) and Ethrel (2-chloroethyl phosphonic acid) (Abraham et al. 1968; Abraham, 1970) results in higher yield and subsequently more drainage of nutrients through latex extracted. Pushparajah (1966) and Lustinec et al. (1967) reported higher levels of extraction of major nutrients as a consequence of stimulation. Stimulation also resulted in lower potassium levels in leaves (Puddy and Warrior, 1961). Pushparajah et al. (1972) found an increase in concentration of potassium removed per unit weight of rubber in stimulated trees. Application of stimulants was reported to affect concentration of calcium and pH of the serum as well. An increase in yield by 1150 kg on stimulation resulted in increased drainage of nitrogen, phosphorus, potassium and magnesium by 14, 5, 14 and 2 kg ha^{-1} respectively in clone PB 86 (Pushparajah et al. 1972). In the experiment with 2,4,5-T, it was noticed that application of potassium had beneficial effect, but nitrogen gave a negative effect over a period of six months following stimulation. The study also revealed that, where the response to fertilizer under normal exploitation system was negligible, increase in fertilizer dose with Ethrel stimulation gave higher yield in clone GT 1. While Ethrel application in unfertilized plot gave an increase of 41% in yield, it gave 50% increase with application of nitrogen. The corresponding figures for clone LCB 1320 were 49 and 127% respectively. Potassium application also improved the response of GT 1 and LCB 1320. Similar results were obtained for magnesium also. Taking into consideration the nutrients immobilised in the tree, those removed through latex, leaching losses and depletion of soil reserves, it would be established that additional quantity of nutrients are required where stimulation is practiced (Pushparajah et al. 1972). A budget for nitrogen and potassium in the case of clone RRIM 600 is given in Table 1.

TABLE 1

Budget for nitrogen and potassium required by RRIM 600 in panel C.
Adapted from Pushparajah et al. (1972).

| Soil series | Added as fertilizer | | Deficit | | | |
| | | | Unstimulated | | Ethrel stimulated | |
	N	K	N	K	N	K
Rengam (Typic Paleudult)	46	45	-15	-5	-75	-62
Holyrood (Oxic Dystropept)	36	41	-25	-9	-85	-66
Malacca (Plinthic Haplorthox)	51	39	-10	-11	-70	-68
Munchong (Tropeptic Haplorthox)	41	36	-20	-14	-80	-71

Nutrient interactions

A highly significant correlation between magnesium and manganese concentration in Hevea leaves was reported by Beaufils (1955) and Bollejones (1957). A higher level of magnesium induces manganese deficiency symptoms. A wide-spread magnesium deficiency has been reported from Malaysia. However, the case is not so under Indian conditions in spite of the highly acidic nature of soils. Increased supply of manganese also could reduce magnesium concentration as reported by Bollejones (1957). An enhanced manganese content not only influences magnesium status in Hevea laminae but also reduces soil pH and concentration of potassium ions. The close relationship between manganese and phosphorus in the mineral nutrition of rubber also has been reported by him. Bollejones (1954) had shown that increased magnesium supply resulted in an increase in the concentration of phosphorus in the leaf. Also, in the early stages of seedling development the growth seemed to be governed more by magnesium than by phosphorus.

Moisture relations

A low soil moisture level has been reported to reduce uptake of potassium and phosphorus and leads to absorption of more calcium and magnesium (Talha et al. 1979). In areas with long dry spell larger quantities of exchangeable potassium are required to be added to compensate for the deficiency of moisture (Paauw, 1978). Though natural rubber producing countries are mostly in the humid tropics, there are areas where long dry spells are encountered particularly in the non-traditional rubber growing regions in India. Gander and Tanner (1976) stated that when plants were under moisture stress, uptake of nutrients usually decreased and this was particularly the case with phosphorus, potassium, calcium and magnesium. Water relations govern yield of rubber. Ionic balance, especially of the cations, is known to play a significant role in the growth of rubber. Studies conducted on the influence of soil moisture on cation uptake revealed that application of higher levels of potassium helps in maintaining higher exchangeable potassium and magnesium content in the latex. During stress period, higher potassium application was found to maintain higher water potential, longer flow and thereby higher yield (Krishnakumar, 1989).

Fertilizer recommendations

Importance of regular manuring of rubber is evident from field experiments conducted in most of the rubber growing countries. In Malaysia soil classification upto series level has been attempted long back and therefore specific fertilizer recommendations have been evolved to suit different soil series. In other countries, information available from location specific fertilizer trials have formed the basis for evolving general fertilizer schedules. In India, Nair (1956) suggested a blanket recommendation based on the soil fertility status and the observations from the fertilizer trials on rubber conducted elsewhere. Simultaneously, multi-locational field trials were started to provide necessary information to revise the recommendations for rubber at different stages of growth. These experiments also revealed that response of rubber is directly related to soil available nutrients and leaf nutrient status (Ananth et al. 1966; Potty et al. 1976). A discriminatory approach was therefore proposed as the most efficient and economic method for optimum fertilizer usage.

A general outline of the fertilizer recommendation for rubber followed by Rubber Research Institute of India in the seedling nursery, budwood nursery, polybag nursery, immature phase and mature phase is given in Table 2. Through the fertilizer trials conducted at different agroclimatic regions and for different clones, this general fertilizer recommendation will be updated from time to time to suit the agroclimatic and soil conditions and the clonal characters.

For the north eastern India separate fertilizer recommendation was evolved by Krishnakumar and Potty (1989) considering the physico-chemical characteristics of the soil. Pushparajah and Low (1977) have given the detailed fertilizer schedule followed in Malaysia for immature rubber. Pushparajah (1983) compared the fertilizer recommendation for immature rubber between countries. In Malaysia based on the parent material, nature of the soil, clonal characters, management etc. modified fertilizer schedule was evolved separately for estate sector and small holders (Watson, 1989).

Method and time of fertilizer application

The application of fertilizer is to be undertaken taking into consideration the stage of growth of plant, ground cover management, soil, climatic factors and type of fertilizer. During the initial phase, growth is active and hence split applications are desirable. Similarly on soils of light texture also, higher frequency is warranted. The type of fertilizer also decides the method of application. In Malaysia split application is recommended in series with light textured soils and also in heavy textured soils with high rainfall (Pushparajah et al. 1974).

For mature rubber fertilizers are to be applied in rectangular or square patches in between four trees and are to be gently forked in. In slopy areas, where cover crop has not completely died out or wherever

TABLE 2

General fertilizer recommendations for rubber

Stage	Fertilizer (quantity and time)
1. Seeedling nursery	Organic manure 2 tonnes ha^{-1} basally. 700 kg P_2O_5 ha^{-1} and rock phosphate as basal dressing. Rock phosphate needs to be applied once in three years if the bed is in continuous use. 10:10:4:1.5 NPKMg mixture 2.5 tonnes ha^{-1} as basal dose 6-8 weeks after planting. 550 kg ha^{-1} urea top dressing 6-8 weeks after basal dressing.
2. Budwood nursery	30 g of P_2O_5 as rock phosphate basal dose 250 g $plant^{-1}$ 10:10:4:1.5 NPKMg mixture in two doses of 125 g each and from 2nd year onwards 125 g $plant^{-1}$ 2-3 months after cutting back.
3. Polybag nursery	10:10:4:1.5 NPKMg mixture 10-30 g $plant^{-1}$ depending on the age.

4. Immature rubber*		N	P_2O_5	K_2O	MgO
			(kg ha^{-1})		
	1st year	10	10(5)	4	1.5
	2nd year	40	40(20)	50	40.0
	3rd year	50	50	20	7.5
	4th year	40	40	16	6.0
5. Mature rubber		30	30	30	--

Figure in parentheses are the water solube form of phosphorus.

* From 5th year onwards the recommendation for mature rubber is followed in areas where good leguminous ground cover is maintained. If not 60:40:20 N:P_2O_5:K_2O ha^{-1} is recommended.

** In Mg rich areas no MgO is recommended.

bench terracing is practised, broadcasting in inter-row space can be resorted to (Pushpadas and Ahmed, 1980). If the fertilizer mixture contains urea, it is essential that the mixture be forked in to prevent losses.

Fertilizers can also be applied as foliar sprays. Wherever a quick result is needed as for correcting a deficiency this method can be practised. Spraying urea, zinc sulphate, ammonium phosphate etc. under certain conditions yield good results.

Fertilizer has to be applied when the soil moisture is optimum. Periods of heavy rainfall and dry spells have to be avoided. In India the recommended time of manuring is during the premonsoon period, ie. April/May and postmonsoon period, ie. September.

Discriminatory fertilizer usage

The concept of discriminatory fertilizer recommendation envisages supply of adequate quantity of nutrients to the plants taking into consideration the nutrient reserves and the available nutrient content in the soil, plant nutrient status, site characteristics and other specific parameters. This practice has been widely accepted and extensively used in Malaysia (Chang and Teoh, 1982) and in India.

The differential response of Hevea to fertilizers has already been established based on long term field trials. The difference in response of rubber to fertilizer application among different soil types have also been observed (Silva, 1976). Hence, correlation of results of field experiments with data from soil and leaf analyses only would help to overcome this difficulty. Relationship between soil and leaf nutrient levels also has been confirmed by Owen (1953) and Lau et al. (1977). Critical soil nutrient content for Hevea in some soil series also has been reported by Guha (1969). Various improvisations in the diagnostic techniques of soil analysis paved the way for more authentic soil nutrient assessment methods (Singh and Talibudin, 1969; Singh, 1970). Soil analysis is influenced by many site specific factors that have to be accounted for before offering any fertilizer recommendation.

The assessment of nutrient requirement through leaf analysis was reported by Shorrocks (1961; 1962a,b; 1965a,b; Chapman, 1941; Beaufils, 1955). The analytical values, however, largely depend on sampling (Chang and Teoh, 1982). The critical leaf nutrient content for Hevea in shade leaves, as reported by Watson (1989), is reproduced in Table 3.

TABLE 3

Range of leaf nutrient content at optimum age in shade leaves (%)

Nutrient	Clonal group*	Low	Medium	High	Very high
N	1	<3.21	3.21-3.50	3.51-3.70	>3.70
	2	<3.31	3.31-3.70	3.71-3.90	>3.90
	3	<2.91	2.91-3.20	3.21-3.40	>3.40
K	I	<1.26	1.26-1.50	1.51-1.65	>1.65
	II	<1.36	1.36-1.65	1.66-1.85	>1.85
P		<0.20	0.20-0.25	0.26-0.27	>0.27
Mg		<0.21	0.21-0.25	0.26-0.29	>0.29
Mn (ppm)		<45	45-150	>150	

* For N: Group 1 clones are all clones except those in group 2 and 3.
Group 2 clones are RRIM 600 and GT 1.
Group 3 clones are all wind-susceptible clones.
eg. RRIM 501, RRIM 513, RRIM 605, RRIM 623, etc.

For K: Group I are all clones except those in group II.
Group II are RRIM 600, PB 86, PB 5/51.

Low: Well below optimum, tending to visual deficiency.

Medium: Suboptimal.

High and very high: Levels above which responses are unlikely.

Voluminous work has been conducted in Sri Lanka in refining the sampling of leaf and analytical methods (Silva, 1976). In India, leaf sampling season starts from August and extends upto October, for the routine analysis for offering fertilizer recommendations. The critical levels fixed for soil and leaf for discriminatory fertilizer recommendations are given in Table 4.

Watson (1989) has summarised the usefulness of leaf nutrient content as an indicator of fertilizer requirement. While dealing with commercial experience in the field of leaf analysis for diagonising nutritional requirement of Hevea, Chang and Teoh (1982) reported much variation in nutrient concentrations particularly for nitrogen and phosphorus, with age of the plant.

TABLE 4

Soil fertility standard and critical leaf nutrient levels.
(Adapted from Pushpadas and Ahmed, 1980).

Nutrient	Standard		
	Low	Medium	High
Soil			
Organic carbon (used as a measure of available nitrogen) (%)	<0.75	0.75-1.50	> 1.50
Available phosphorus (mg 100 g^{-1})	<1.00	1.00-2.50	> 2.50
Available potassium (mg 100 g^{-1})	<5.00	5.00-12.50	>12.50
Available magnesium (mg 100 g^{-1})	<1.00	1.00-2.50	> 2.50
Leaf			
Nitrogen %	<3.00	3.00-3.50	> 3.50
Phosphorus %	<0.20	0.20-0.25	> 0.25
Potassium %	<1.00	1.00-1.50	> 1.50
Magnesium %	<0.20	0.20-0.25	> 0.25

Beaufils (1954, 1957) studied the possibility of using nutrient ratio in leaf and latex as a guide to evolve fertilizer recommendation. The work was mainly based on the studies in the rubber growing soils of Vietnam and Cambodia. This system was further studied under different situations (Fallows, 1961) and it was found that the Beaufils system was not readily applicable in Malaysia.

Role of mineral nutrients and deficiency symptoms

Hevea grows well on a wide variety of tropical soils and responds to any deficiency or excess of nutrients. The studies conducted by Bollejones (1954) and Shorrocks (1965) confirmed the essentiality of nutrients in the growth of rubber. The characteristic leaf deficiency symptoms shown by both rubber and cover plants are illustrated in the book 'Mineral Deficiencies in Hevea and Associated Cover Plants' (Shorrocks, 1964). Mineral deficiencies, apart from their influence on the growth of rubber plants and yield, have been reported to affect the ultrastructure and stability of latex (Gomez, 1978).

In India deficiency symptoms are generally observed for N, K and Mg. Micronutrient deficiencies are seldom encountered. Zinc deficiency symptoms are noticed in the seedling nursery plants and also in the early immature phase in the main field, quite often as a result of heavy application of phosphatic fertilizers. The symptoms of zinc deficiency are usually transient.

Nitrogen: The characteristic yellowing is first observed in the lower whorls and as the intensity of deficiency increases it affects the younger leaves. Symptom appears in sun leaves in mature plantations and results in overall retardation of growth.

Phosphorus: In seedlings the typical symptoms are yellowish-brown discolouration of the upper surface and purpling of the undersides of upper and middle leaves. Acute deficiency results in the laminae bending upwards and tips becoming scorched. Deficiency in the mature phase does not produce any visual symptoms, and is detected only by leaf analysis. It results in retardation of growth and affects stability of latex.

Potassium: Development of marginal and tip chlorosis followed by necrosis is the characteristic symptom of K deficiency. On young unbranched trees the symptoms will appear first on the older whorls of the plant. In mature plantations K deficiency is revealed by the appearance of a butter yellow colour over the canopy. Leaf size also will be reduced considerably.

Magnesium: The early stage of magnesium deficiency is a pale green interveinal mottling which changes into bright yellow and spreads towards the margin. In the case of severe deficiency, yellowing is often followed by interveinal and marginal scorch of leaves (brown necrotic patches). In the young unbranched trees the symptoms are usually seen in the lower storeys. In mature trees symptoms appear in fully exposed sun leaves.

Zinc: The leaves of the upper stories of the tree become reduced in breadth, relative to length, with wavy undulating margins. Chlorosis of the leaves also is noticed with the mid-rib and main veins remaining dark green in colour. In young unbranched trees, the symptoms are seen in the top storeys resulting in death of apical meristem and development of auxillary meristems.

Calcium: Deficiency of calcium is first manifested as a scorching of the tip or margin of leaf. Since the element is immobile in young trees, deficiency symptoms are first noticed in the younger leaves and under severe deficiency, the point may die back. The deficiency symptom is generally seen on shade leaves in mature trees. Calcium also plays an immense role in the stability and flow of latex.

Manganese: In sand culture, manganese deficiency causes the laminae of middle and upper whorls to develop a diffuse pale yellow green interveinal region. Veins remain markedly white but are surrounded by green strips of tissue and thus could be distinguished from the uniform pale green interveinal region (Bollejones, 1954). The deficiency symptoms in unbranched trees first appear in the lower leaves and could extend to all leaves in acute cases. The deficiency appears in the shaded leaves in the initial phase in mature trees but could spread to exposed branches (Watson, 1989).

Boron: Deficiency of boron in young plants appears with young leaves surrounding the growing point of the plants being killed and turning black at the tips. In the initial phase boron deficient plants showed a darker green foliage. Shedding of the youngest laminae before the expanded one accompanied by a pale yellow diffuse chlorosis of the topmost expanded laminae also were manifested when boron deficiency was induced in plants. In young plants, as a result of the deficiency, the terminal leaf whorls are produced without discrete internodes giving a bottle brush effect. Death of apical meristems have also been observed simultaneously leading to development of auxillary meristems.

REFERENCES

Abraham, P.D. 1970. Field trials with Ethrel. Planters Bulletin, Rubb. Res. Inst. Malaya, 3: 366.
Abraham, P.D., Wycherley, P.R. and Pakianathan, S.W. 1968. Stimulation of latex flow in Hevea brasiliensis by 4-amino, 3,5,6-trichloropicolinic acid and 2-chloroethenephosphonic acid. J. Rubb. Res. Malaya, 20: 291.
Aliang Jiang. 1981. Temperature inversion and vegetation inversion in xi-shuangbanna. Mountain Research and Development 1. No. 3-4, 275-280.
Aliang Jiang. 1983. A geo-ecological study of rubber tree cultivation at high altitude in China.
Ananth, K.C., GEorge, C.M., Mathew, M. and Unni, R.G. 1966. Report of the results of fertilizer experiments with young rubber in South India. Rubber Board Bull., 9: 30-42.
Beaufils, E.R. 1954. Contribution to the study of mineral elements in field latex. Proceedings of the third Rubber Technology Conference, London, 1953. pp. 87-98.
Beaufils, E.R. 1955. Mineral diagnosis of some Hevea brasiliensis. Archs. Rubber Cult., 32: 1.
Beaufils, E.R. 1957. Research for rational exploitation of the Hevea using a physiological diagnosis based on mineral analysis of various facts of the plant. Fertilite, 3: 27-38.
Bollejones, E.W. 1954. Nutrition of Hevea brasiliensis. II. Effect of nutritional deficiencies of growth, chlorophyll, rubber and mineral contents of Tjir and ji 1 seedlings. J. Rubb. Res. Inst. Malaya, 14: 209-230.
Bollejones, E.W. 1957. A magnesium-manganese inter-relationship in the mineral nutrition of Hevea brasiliensis. J. Rubb. Res. Inst. Malaya, 15: 22-28.
Bolton, J. 1960). The effective of fertilizers on pH and the exchangeable cation of some Malayan soils. Proc. Nat. Rubb. Res. Conf. Kuala Lumpur, 1960. p. 70.
Bolton, J. 1960). The response of Hevea to fertilizers on a sandy latosol. J. Rubb. Res. Inst. Malaya, 16: 178-190.
Bremner, J.M. 1956. Some soil organic matter problems. Soils and Fert. 19: 115.

Chan, H.Y. 1977. Soil classification. In: E. Pushparajah and L.L. Amin (Eds.). Soils under Hevea in Peninsular Malaysia and their management. Rubber Research Institute of Malaysia, Kuala Lumpur. pp. 57-54.

Chand, H.Y. and Pushparajah, E. 1972. Productivity of mature Hevea in West Malaysian soils. Proc. Rubb. Res. Inst. Malaya Planters Conf. Kuala Lumpur, 1972. 97 pp.

Chan, H.Y. and Pushparajah, E. 1972. Productivity potentials of common clones on common rubber growing soils. Proc. Rubb. Res. Inst. Malaya Planters Conf. Kuala Lumpur, 1972. 99 pp.

Chand, H.Y., Pushparajah, E. and Sivanadhyan, K. 1972. A preliminary assessment of the influence of soil morphology and physiography on the performance of Hevea. Proc. 2nd ASEAN Conference, Djakarta, 1972.

Chan, H.Y., Wong, C.B., Sivanadhyan, K. and Pushparajah, E. 1974. Influence of soil morphology and physiography of leaf nutrient content and performance of Hevea. In: Proc. RRIM Planters Conference, Kuala LUmpur. pp. 115-126.

Chand, H.Y., Pushparajah, E., Mohd. Nordin bind Wan dand., Wong, C.B. and Zainol Eurof. (1977). Parent materials and soil formation. In: E. Pushparajah and L.L. Amin (Eds.). Soil under Hevea in Peninsular Malaysia and their management. Rubber Research Institute of Malaya, Kuala Lumpur. pp. 1-19.

Chang Ahkow and Teoh Cheng Hai. 1982. Commercial experience in the use of leaf analysis for diagnosing nutritional requirement of Hevea. In: Proc. RRIM Planters Conference, Kuala Lumpur. pp. 220-231.

Chapman, G.W. 1941. Leaf analysis and plant nutrition. Soil Sci., 52: 63.

Constable, D.H. 1953. Manuring replanted rubber. 1938-1952 Quarterly circular of the Rubber Research Institute of Ceylon, 29: 16.

Cornfield, A.H. 1955. The measurement of soil structure and factors affecting it: A review. J. Sci. Fd. Agric., 6: 356.

daCosta, L. 1968. The main tropical soils of Brazil. In: Approaches to soil classification, FAO World Soil Resources Report. No. 32. pp. 95-106.

Dettasen, I. 1950. Deficiency symptoms in Hevea brasiliensis. Arch. Rubber Cult., 27: 107.

De Jong, P. 1955. Stimulation of yield in Hevea brasiliensis. III. Further observation on the effects of yield stimulants. J. Rubb. Res. Inst. Malaya, 14: 385.

D'Hoore, J.L. 1964. Soil Map of Africa - exploratory monograph, CCTA Publication 93, Lagos.

Dijkman, M.J. 1951. Hevea - Thirty years of Research in the Far East. University of Miami Press, Florida.

Duley, F.L. 1952. Relationship between surface covers and water penetration, run off and soil losses. Proc. Sixth Int. Grassl. Congr., 2: 942.

Eshett, R.T. and Omutei, J.A.I. 1989. Potential for rubber (Hevea brasiliensis) cultivation on the basaltic soils of south eastern Nigeria; climatic, pedochemical and minerological considerations. Indian J. Nat. Rubb. Res. 2: 1-8.

Fallows, J.C. 1961. The major elements in the foliage of Hevea brasiliensis and their inter-relation. In: Proc. Natural Rubber Research Conference, 1960, Kuala Lumpur. pp. 142-153.

FAO-UNESCO 1974. In: Dudal et al. (Eds.). Soil Map of the World. Vol. I. UNESCO, Paris, Legend.

Garder, P.W. and Tanner, C.B. 1976. Leaf growth, tuber growth and water potential in potatoes. Crop Sci., 16: 534-538.

Gallez, J., Juo, A.S.R., Herbillon, A.J. and Moormann, F.R. 1975. Clay minerology of selected soils in southern Nigeria. Soil Science Society of America Proceedings 39, pp. 577-585.

George, C.M. 1961a. A short account of rubber soils in India. Rubber Bd. Bull., 5: 92-95.

George, C.M. 1961b. A note on manuring of rubber seedling nursery. Rubber Bd. Bull. 5: 21.

George, C.M. 1962. Mature rubber manuring: Effect of fertilizers on yield. Rubber Bd. Bull., 5: 202.

George, C.M. 1963. A preliminary report of the permanent manurial experiments on rubber. Rubber Bd. Bull., 7: 37-40.

Gomez, J.B. 1978. Effects of mineral deficiencies on the ultrastructure of Hevea leaves. I. Palisade tissues. IRRDB Symposium, 1978, Kuala Lumpur, pp. 1-12.

Guha, M.M. 1969. Recent advances in fertilizer usage for rubber in Malaya. J. Rubb. Res. Inst. Malaya, 21: 207-216.

Haines, W.B. and Crowther, E.M. 1940. Emp. J. Exp. AGric., 8: 169.

Huang Zongdao and Zheng Xueqin. 1983. Rubber cultivationin China. Proc. RRIM Planters Conference, 1983. pp. 31-42.

Jeevaratnam, A.J. 1969. Relative importance of fertilizer application during pre and post tapping phases of Hevea. J. Rubb. Res. Inst. Malaya, 21: 175-180.

Jungerius, P.D. and Levelt, T.W.M. 1964. Clay minerology of soils over sedimentary rocks in eastern Nigeria. Soil Science, 97: 89-95.

Juo, A.S.R. and Moormann, F.R. 1980. Characteristics of two soil toposequences in south eastern Nigeria and their relation to potential agricultural land use. Nigerian Journal of Soil Science, 1: 47-61.

Krishnakumar, A.K. 1989. Soils under Hevea in India: A physical chemical and mineralogical study with a reference to soil moisture cation influence on yield of Hevea brasiliensis. Ph.D. Thesis, Indian Institute of Technology, Kharagpur.

Krishnakumar, A.K. 1990. Role of rubber plantation in ecological and socioeconomic development of north eastern region. Paper presented at Symposium on Bioresources of Tripura in March 1990, University of Tripura, Agartala.

Krishnakumar, A.K. and Potty, S.N. 1990. Response of Hevea planted as polybag plants to fertilizers during immature phase in soils of Tripura. Indian J. Nat. Rubb. Res., 2: 143-146.

Krishnakumar, A.K. and Potty, S.N. 1989. A new fertilizer recommendation for NE REgion. Rubber Bd. Bull., 24: 5-8.

Krishnakumar, A.K., Thomas Eappen, Rao, N., Potty, S.N. and Sethuraj, M.R. 1990. Ecological impact of rubber (Hevea brasiliensis) in north east India: I. Influence on soil physical properties with special reference to moisture retention. Indian J. Nat. Rubb. REs., 3: 53-63.

Krishnakumar, A.K., Datta, B. and Potty, S.N. 1990. Moisture retention characteristics of soils under Hevea in India. Indian J. Nat. Rubb. Res., 3: 9-21.

Kothandaraman, R., Jacob Mathew, Krishnakumar, A.K., Kochuthesiamma Joseph, Jayarathnam, K. and Sethuraj, M.R. 1989. Comparative efficiency of Mucuna bracteata D.C. and Pueraria phaseoloides Benth. on soil nutrient enrichment microbial population and growth of Hevea. Indian J. Nat. Rubb. REs., 2: 147-150.

Lau, C.H., Pushparajah, E. and Yap, W.C. 1977. Evaluation of various soil-P indices of Hevea. Proc. Conference on Chemistry and Fertility of Tropical Soils, 1973, Kuala Lumpur. pp. 103-111.

Lustinec, J., Langlois, S. Resing, N.L. and Chat Kim Chun. 1967. La stimulation de l'Hevea par les acides chlorophenoxy - a ectoques et son influence sur l' aire drainee. Revue gen. Plastq., 44: 635.

Marschner, Horst. 1986. Mineral Nutrition of higher plants. Academic Press (London) Ltd.,

Munevar, F. and Wollum, A.G. II (1977). Effects of the addition of phosphorus and inorganic nitrogen on carbon and nitrogen mineralisation in andepts and Colombia. Soil Sci. Soc. Am. J., 41: 540-545.

261

Owen, G. 1953. Determination of available nutrients in Malayan soils. J. Rubb. Res. Inst. Malaya, 14: 109-120.

Nair, C.K.N. 1956. Fertilisers for rubber. Rubb. Board Bull., 4: 7-16.

Owen, G., Westgarth, D.R., Iyer, G.C. 1957. Manuring of Hevea - effects of fertilizers on growth and yield of mature rubber trees. J. Rubb. Res. Inst. Malaya, 15: 29-52.

Paauw, F. van der. 1978. Relation between potash requirement of crops and meteorological condition. Pl. Soil, 9: 254.

Penders, J.M.A. 1940. Groundverbetering en bemesting bij rubber. In Dictaat van den cursus over de rubber-culture. Buitenzorg Den Dienst van den Landbonw. p. 173-225.

Philpot, M.W. and Westgarth, D.R. 1953. Stability and mineral composition of Hevea latex. J. Rubb. Res. Inst. Malaya, 14: 133.

Potty, S.N., Abdulkalam, M., Punnoose, K.I. and George, C.M. 1976. Response of Hevea to fertilizer application in relation to soil fertility characters. Rubb. Board Bull., 13: 48-54.

Potty, S.N., Mathew, M., Punnoose, K.I. and Palaniswamy, R. 1978. Results of fertilizer experiments on young rubber trees grown with legume and nature ground covers. Proc. 1st Annual Symposium on Plantation Crops (PLACROSYM-I), Kottayam. pp. 141-147.

Paddy, C.A. and Warriar, S.M. 1961. Yield stimulation of Hevea brasiliensis by 2,4 dichlorophenoxyacetic acid. Proc. Nat. Rubb. Res. Conf, 1960, Kuala Lumpur. 194 pp.

Punnoose, K.T., Potty, S.N., Abdulkalam, M., Karthikakutty Amma, M. and Mathew, M. 1978. Studies on direct and residual effect of nitrogen, phosphorus and potassium on the growth and yield of rubber in the red loam soils of South India. Proc. of Plantation Crops Symposium I, 1978, Kottayam.

Pushpadas, M.V. and Karthikakutty Amma, M. 1980. Agro-ecological requirements. In: P.N. Radhakrishna Pillai (Ed.). Handbook of Natural Rubber Production in India. Rubber Research Institute of India, Kottayam. pp. 87-109.

Pushpadas, M.V. and Ahmed, M. 1980. Nutritional requirements and manurial recommendations. In: P.N. Radhakrishna Pillai (Ed.). Hand Book of Natural Rubber Production in India. Rubber Research Institute of India, Kottayam. pp. 154-184.

Pushparajah, E. 1966. Studies on effects of rock phosphate on growth and yields of Hevea brasiliensis. Thesis submitted for the degree of Master of Agricultural Science, University of Malaya.

Pushparajah, E. 1969. Response on growth and yield of Hevea brasiliensis to fertilizer application on Rengum series of soil. J. Rubb. Res. Inst. Malaya, 21: 165.

Pushparajah, E. 1977. Nutrition and fertilizer use in Hevea and associated covers in Peninsular Malaysia - A review. J. Rubb. Res. Inst. Sri Lanka, 54: 270-283.

Pushparajah, E., Sivanadhyan, K., P'Ng Tat Chin and Ng, E.K. 1972. Nutritional requirements of Hevea brasiliensis in relation to stimulation. Proc. Rubb. Res. Inst. Malaya, Planters' Conf., 1971, Kuala Lumpur. 189-200.

Pushparajah, E., Sivanadyan, K. and Yew, F.K. 1974. Efficient use of fertilizers. Proc. Rubb. Res. Inst. of Malaya. Planters Conference, Kuala Lumpur, 102 pp.

Pushparajah, E. and Yew, F.K. 1977. Management of soils. In: E. Pushparajah and L.L. Amin (Eds.). Soils under Hevea and their management in Peninsular Malaysia. Rubber Research Institute of Malaysia. pp. 94-116.

Ram Beaux, J. and Danjard, J.C. 1963. Teneronge basaltique et nutrition de l'hevea dans pes conditions ecolgiques dn Cambodge Opuse. Techq. Inst. Res. Caoutch Cambodge No. 2/63.

Rubber Research Institute of India. 1988. Annual Report 1988, Rubber Research Institute of India, Kottayam-9.

Samsidar, Abdul Aziz Mahmood, Sivanadyan, K. and Fornez, J.B. 1975. Effect of mineral deficiencies on bark anatomy of Hevea brasiliensis. Proc. Rubb. Res. Inst. Malaysia Int. Rubb. Conf., 1975, Kuala Lumpur.

Schoonncveldt, J.C. van. 1948. Optimumbem estings praef bij jange rubber in proeftuin Peweja. Arch. v.d. Rubber Cult., 26: 201-221.

Shorrocks, V.M. 1961. Leaf analysis as a guide to the nutrition of Hevea brasiliensis. I. Sampling technique with mature trees. Principles and preliminary observations on the variation in leaf nutrient compositions with position on the tree. J. Rubb. Res. Inst. Malaya, 17: 1-14.

Shorrocks, V.M. 1962b. Leaf analysis as a guide to the nutrition of Hevea brasiliensis. V. A leaf sampling technique for mature trees. J. Rubb. Res. Inst. Malaya, 17: 167-190.

Shorrocks, V.M. 1962a. Leaf analysis as a guide to the nutrition of Hevea brasiliensis. II. Sampling technique with mature trees. Variation in nutrient composition of the leaves with position to the tree. J. Rubb. Res.Inst. Malaya, 17: 91-101.

Shorrocks, V.M. 1964. Mineral deficiencies in Hevea and associated cover plants. Rubb. Res. Inst. Malaya.

Shorrocks, V.M. 1965a. Mineral nutrition growth and nutrient cycle of Hevea. I. Growth and nutrient content. J. Rubb. Res. Inst. Malaya, 19: 32-40.

Silva, C.G. 1976. Discriminatory fertilizer recommendations for rubber in Sri Lanka. In: Proc. of the International Rubber Conference, 1975, Kuala Lumpur, Vol. 3, pp. 132-144.

Singh, M.M. 1970. Exchange reactions of potassium, magnesium, aluminium in some Malayan soil. Ph.D. Thesis, University of Malaya.

Singh, M.M. and Talibudeen, O. 1969. Thermodynamic assessment of the nutrient status of rubber growing soils. J. Rubb. Res. Inst. Malaya, 21: 240.

Sivanadyan, K. 1972. Lysimeter studies on the efficiency of some potassium and nitrogen fertilizers on two common soils inwest Malaysia. Proc. 2nd ASEAN Conf., 1972, Djakarta.

Soong, N.K. and Lau, C.H. 1977. Physical and Chemical properties of soil. E. Pushparajah and L.L. Amin (Eds.). Soils under Hevea in Peninsular Malaysia and their management. Rubber Research Institute of Malaya, Kuala Lumpur. pp. 25-56.

Sys, C. 1975. Report on the adhoc expert consultation on land evaluation. FAO world soil resources report. No. 45, Rome. pp. 59-79.

Talha, M., Amberger, A. and Burkart, N. 1979. Effect of soil compaction and soil moisture level on plant's growth and potassium uptake. Z. Acker-Pflamgenban, 148: 156-164.

Watson, G.A. 1989. Nutrition. In: C.C. Webster and W.J. Baulkwil (Eds.). Rubber. Longman Scientific and Technical, England. pp. 291-348.

Webster, C.C. and Baulkwill, W.J. 1989. Rubber. Longman SCientific and Technical, England. 604 pp.

CHAPTER 12

TAPPING OF HEVEA BRASILIENSIS

P.D. ABRAHAM
Rubber Research Institute of Malaysia, Kuala Lumpur, Malaysia.

In tapping the tree for latex, a groove of the required length and slope - 25° to 30° from horizontal - is cut with a tapping knife. The cut penetrates to within 1 mm of the cambium; the precise depth varying with the skill of the different tappers. The same cut is regularly reopened by the removal at each tapping of a thin shaving of bark. The object of a well considered tapping should be to get as much latex as possible from the trees with the smallest excision of bark convenient, so as to interfere as little as can be helped, with the health of the trees and their capacity for continuing to yield latex year after year during a long life (Maclaren, 1913).

With the advent of synthetic rubber as a competitor to natural rubber, together with rising tappers' wages, the cost of tapping became a very important factor when considering which tapping systems to employ. It will be apt to define an ideal tapping system as one which gives the highest yields at the lowest tapping cost, satisfactory growth and bark renewal and the lowest incidence of Brown bast disease (Baptiste, 1962).

Unfortunately, there is not one tapping system that will give the best results on all cultivars and under all conditions. In times of low rubber prices and high wages, the yield per tapper is of greater importance than the yield per acre. Hence, tapping systems which give a high yield per unit of labour are of considerable economic importance. Tapping cost is the largest single item in the cost of production (Westgarth and Narayanan, 1964). Because of this, economic considerations are given high priority in any investigation of tapping systems, the aim being to obtain as high a yield as possible with as few as possible tappings.

TAPPING

Evolution of Tapping Systems

One of the main reasons for the successful establishment of Hevea brasiliensis on plantation scale in the Far East was the discovery of excision method of tapping for harvesting rubber from the tree. In this method, the same cut is regularly reopened by the removal at each tapping of a thin shaving of bark from a sloping cut, a principle which is in general use today (Ridley, 1890-91). On each occasion, a tree is tapped by means of a suitable knife, so that a channel is prepared along which the latex can flow. This method avoids wounding of the trees as the tissues of the tree can be recognised.

Hevea does not accumulate more than a certain amount of rubber in the latex vessels, and draining out the latex by tapping stimulates the production of latex to replace it. Thus, a tree can be trained by regular tapping to a continuous process of rubber regeneration leading to high cumulative yields. Regeneration of rubber and maintenance of yield is directly in response to regular tapping, requiring the continual application of manual labour (Wycherley, 1964). Different tapping systems modify the amount of rubber produced per unit labour or per unit capital invested in the planting and abandoning tapping means permanent loss of crop. The importance of excision tapping lay in the fact that the method was based on the specific characteristics of Hevea bark.

When the demand for rubber increased in the beginning of the twentieth century, planters became daring and began increasing the length of the tapping cut and practised intensive tapping systems to obtain greater yields. However, experience soon taught them that with lengthening the cut, the yield per unit length of the tapping cut became less and they learned that yield was not proportional to the amount of bark incised. They also noticed that though they obtained good yield responses with intensive tapping at the beginning, the yield declined after some time. Bark renewal too became poor and the planters returned to less intensive tapping systems (Dijkman, 1951). Since in those days nothing was known of the physiology of latex flow it was a matter of finding a tapping system by empirical means. This situation prevailed until 1920.

The results of Dutch scientists (De Jong, 1916; Maas, 1925; Bobilioff, 1923) who carried out many tapping experiments and studies on anatomy and physiology of Hevea brasiliensis helped to formulate tapping systems on a scientific and rational basis. Thus the early tapping systems slowly evolved into the modern systems largely by reducing the number of cuts and the frequency of tapping.

Tapping Notations

Many tapping systems were evolved through the years which included complicated techniques such as the herringbone system. Local names were given for each system and this led to a lot of confusion and difficulty. Thus, the various rubber research institutes formulated a uniform method of expressing various tapping systems commonly used. At the initiative of the International Rubber Research and Development Board (IRRDB), a standard international tapping notation for Hevea tapping systems was revised (Lukman, 1983). The notation consists of a set of symbols which should be used in regular sequence (for details please see Chapter 13. Eds.).

The first symbol describes the number and nature of cuts. Three symbols, representing three types of cuts are:

1. S - a spiral cut
2. V - a V cut
3. C - a circumferential cut; this could be a V cut or a spiral extending around the entire circumference of the tree.

The number of cuts is indicated by the numeral on the left of the letter S, C or V excepting the number 1. The numeral following the symbol and an oblique represents the fraction of the cut.

e.g.: S   - full spiral cut

S/R   - reduced spiral cut (full spiral cut less 15 cm)

2x½S - two-half spiral cuts

1/3S - one third spiral cut.

Figure 1 illustrates a typical tapping schedule for alternate day tapping on a half spiral cut i.e. ½S d/2.

The second symbol represents the frequency of tapping. The small letter 'd' is used to denote day. The numerals after the oblique indicates the interval in days between tappings.

e.g.: d/1 - daily tapping

d/2 - alternate day tapping

d/3 - third tapping

d/4 - fourth day tapping

(t,t) - two panels each in tapping for two days, alternately. The change-over system designated ¼S d/2(t,t) would be tapped on the lower quarter spiral on the first day, the adjacent quarter on the third day, and again the lower quarter on the fifth day and so on (Figure 2).

The relative intensity which is a percentage of intensity based on the standard system ½S d/2 is no longer used because of tapping rest in a week. The tapping rest of one day in a week which is widely practised is denoted by 6d/7 after the frequency notation.

266

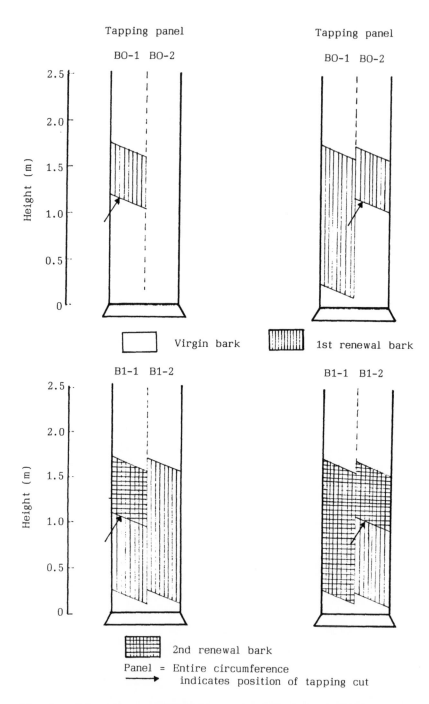

Fig. 1. Schematic representation of tapping panels. Figure shows the two-half panels arranged one beside the other.

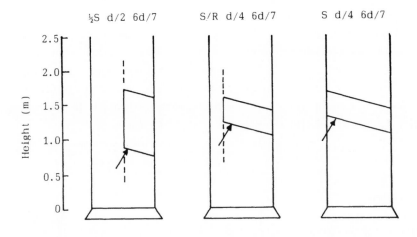

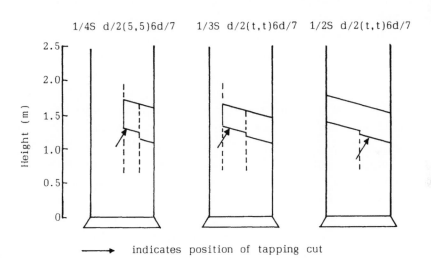

⟶ indicates position of tapping cut

Fig. 2. Schematic representation of some common tapping systems.

Stimulation Notation

The stimulation notations are not separated from the tapping notations. The two should be presented together as a complete notation with a full stop inserted in between them. The notation for stimulation are grouped into three units which are in the following order - stimulant, application and periodicity. Only the active ingredient of the stimulant is expressed in the notation with a specific code and not its trade name.

> e.g. ET5% Lam 8/y - ethephon at 5.0% concentration by lace application method for eight rounds in a year.

Tapping Techniques

Height of opening: There is a need to open trees for tapping as soon as the required minimum girth has been obtained. Budded trees have cylindrical trunks and can be opened at a height which tappers can reach without any aid. Seedling trees are conical with a bigger girth at the base of the tree; in order to take advantage of this, a lower height of opening is recommended. With conventional tapping, the recommendation is to open budded trees for tapping with a girth of 46 cm and above which reached at a height of 1.6 m from the ground when the first tapping cut is made. For seedling trees, the recommendation is to open the trees for tapping when a similar girth is reached at a height of 75 cm from the ground. Immature field can be brought into opening for tapping when 70% of the trees have attained 46 cm and above at the recommended height.

Direction and slope of cut: The latex vessels in the bark of Hevea spiral from bottom left to top right at an angle of 3.7° from the vertical in a counter-clockwise direction in most cultivars. Hence, the cut from the high left to low right will severe a greater number of latex vessels at a similar cut in the reverse direction. This has led to the current practice of sloping cut from high left to low right on all spiral cuts.

The 25° slope from the horizontal is preferred for seedling because it results in lesser bark consumption and a smaller area of bark that will be lost when the cuts reach ground level without much loss of yield. Further, the presence of a thick corky layer in bark provides a channel for the flow of latex. In the case of budded trees where total bark thickness is less than seedling trees, the latex may overflow the sides of the tapping cut with a 25° slope. Thus, an angle of slope of 30° from the horizontal is recommended for budded trees.

When the tapping cut approaches the base, a new cut on the opposite panel can be similarly opened. When a tapping cut is opened on renewed bark, care should be taken to ensure that no virgin bark island is created.

This can be done by commencing tapping from the previous height of opening.

Depth of tapping: The yield obtained from the tree is greatly influenced by the skill of the tapper. A skilled tapper will tap to optimum depth to within 1 mm of the cambium. Greatest number of latex vessels are situated near the cambium and hence, to obtain good yield, the tapper has to tap to correct depth without wounding the cambium. This is where the skill of the tapper is critical in that he is able to tap deep without wounding the trees. Generally, low intensity tapping systems benefit more from deep tapping than high intensity systems (Abraham and Tayler, 1967).

Bark consumption: Experiments have shown that beyond a minimum bark consumption yield is not enhanced with increasing thickness of bark shaving (De Jonge and Warrior, 1965). Low frequency tapping systems caused more drying of the bark tissue between tappings; hence, a thicker bark shaving per tapping is required which experienced and skilled tappers adjust automatically. Annual bark consumption from different frequencies of tapping on a half-spiral cut is given in Table 1.

TABLE 1

Annual bark consumption from different frequencies of tapping on half-spiral cut (average of seven clones).

Frequency of tapping	Annual bark consumption (cm)		
	Panels BO-1 and BO-2	Panels BI-1 and BI-2	Panels HO-1, HO-2, HO-3 and HO-4
d/2	25	30	50
d/3	18	22	40
d/4	15	20	35
d/6	12	15	28

Time of tapping: Higher yields can be obtained from early tapping than late tapping. Thus, wherever possible, tapping should commence at dawn for better yields to be obtained.

Tapping task: The number of trees the tapper is given to tap in a day is called a tapping task. The task will depend on the tapping system, stand per hectare, topography of land, position of the tapping cut and skill of the tapper. A normal task size with ½S d/2 tapping in Malaysia on young trees is between 550-600 trees. With shorter cuts e.g. ¼S, the task size may be slightly increased. Increasing task size gives substantial increase per tapper with smaller losses in yield per hectare.

Effect of Tapping on Growth

Tapping is controlled wounding and it retards the growth of all trees especially budded trees. There are several high yielding cultivars whose girthing is depressed with tapping. Trees of such clones are likely to be unbalanced in their development and will become prone to wind damage. Hence, tapping systems must be tailored to the growth habit of cultivars after tapping. Longer cuts than half spiral tend to reduce the rate of girthing and hence, long cut systems are not preferred on young rubber (Ng et al. 1969).

Micro Tapping Systems

Two micro tapping systems have been used in the planting industry viz. puncture tapping and micro-X tapping systems. Both systems only work with stimulation and hence, have their limitations. Puncture tapping has some attractions for bringing trees into early tapping (Abraham, 1981).

Micro-X system: Micro-X combines the use of puncture tapping and excision tapping (Ismail Hashim et al. 1979).

Factors Affecting Tapping Efficiency

The efficiency of tapping is determined by a host of factors. Some of these will be reviewed. Tapping is a skilled operation and hence, quality of tapping varies from person to person. Usually the best tappers are given young trees to tap where tapping should be carried out with minimum wounding.

Field supervision is necessary to ensure that the tappers complete their tasks and do not leave trees untapped. They must also ensure that latex flows along the cut into the cup and not to the ground. Effort must be made to ensure that tapping is carried out to proper depth otherwise yield loss may occur. Field maintenance is essential for proper tapping. If weeds and other obstacles are not cleared from the inter-row, tappers will find it difficult to tap their trees efficiently. The efficiency of tapping is also influenced by the sharpness of the tapping knife. If the knife is not sharp, it is likely that the cut will not be tapped cleanly and wounds may be caused.

The topography of the field is another factor affecting the efficiency of tapping. Hence, a tapper can tap more trees on flat land than on hill slope. On flat land a tapper may be able to tap 600 trees on Panel BO-1.

But on a steep hill slope he may only be able to tap 500 trees. In a large estate the management and local union usually have an understanding on the task size for the hilly areas in the estate.

Besides topography, the age of the trees also influences tapping efficiency. The girth of older trees are usually larger, and hence the tapping cuts are longer. Consequently, it requires more time to tap the older trees than the younger trees. Hence, a tapper who taps 600 trees on Panel BO-1 may only be required to tap 575 trees on Panel BO-2, and 530 trees on Panel BI-1. The length of the tapping cut is also determined by the tapping system. Tappers tapping the reduced spiral cut and the full spiral cut are required to tap only 92% and 80% respectively on the appropriate number of trees for half spiral tapping. Besides this, the height of the tapping cut also influences the efficiency of tapping. Where tapping involves the use of a ladder, the tapper should be given only 65% of the appropriate numbers of trees for low level tapping. In Control Upward Tapping, a tapper is given a slightly bigger task size than ladder tapping (Ismail Hashim et al. 1981).

It may also be necessary to state here that efficiency of tapping is also influenced by the quality of the tapping knife. A tapper who does not sharpen his knife regularly is expected to perform poorly in his task. One who continues to use an old worn-out knife may also inflict more wounds on the trees, besides being inefficient in his job.

Recommended Tapping System

Recommended tapping and stimulation procedure to follow for modern cultivars have been given (Abraham and Ismail Hashim, 1983). The schedule covers conventional tapping from opening to felling recommended separately for clones, seedlings and individual smallholders over a period of 25-28 years. This recommended schedule will make it unnecessary to tap renewed bark of high panels and bark of second renewal on the base panel. They also suggest exploitation procedure of immature trees with puncture tapping and stimulation for some clones. If this is followed, another 1½ years of tapping will be possible.

Tapping Implements

With manual tapping, two tapping knives are commonly used in the industry i.e. Jebong knife and gouge. The Jebong knife is suitable for shaving off a thin layer of bark along the tapping cut and it is the most popular knife in the country. Various modifications have been made to the Jebong particularly in the length, size and angle of the knife. But its basic structure remains the same; with a sharp edge at the point of contact on the tapping cut so that when the knife is pulled along skilfully, a thin layer of bark is shaved off. A gouge differs from a Jebong in that it is used to push along the tapping cut to shave off the bark instead of being pulled along as in the case of the Jebong. A modified gouge with a long handle is now widely used in control upward tapping. Bidirectional knives are also available and are used where upward and downward tapping systems are tapped by one person (Abraham, 1981). Figure 3 shows various types of tapping knives currently in use in Malaysia.

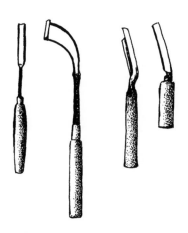

Fig. 3. Various types of tapping knives currently in use in this country.

Recently, a mechanical tapping tool (Motoray) has been introduced which may be a boon for areas experiencing shortage of skilled tappers. With this knife, tapping may be carried out with less skill and minimum effort.

Rubber cup is another essential implement for the industry. The earthenware cup is the most widely used. They are cheap but heavy. Glass cups are used in few plantations. Plastic cups have recently been introduced to the industry. Glass cups are lighter and easy to clean but are more expensive and easily broken. Further, they are easily stolen from the field. Plastic cups are lighter and easy to transport. However, rubber tends to stick to the cups permanently making cleaning difficult. Initially, smaller cups were used of the 450 ml capacity but now with higher yield per tree, bigger cup is placed in a wire hanger attached to two pieces of wire and extensible spring attached to the rubber trees. On young trees where the spring and the wire are newly fixed on to the trees, these materials perform satisfactorily but with girth increment and long period of neglect, the wire and the spring need replacement. A spout has to be fixed to the rubber trees to enable the latex to flow from the tapping cut into the cup. Spouts are made of this metal strips and are forced into the bark at the front channel about 20 cm from the end of the tapping cut. In this process, care must be taken to ensure that the spout does not damage the cambium otherwise wounding may be created.

In areas where cuts are opened higher than the tapper can reach, one-step or two-step ladders are used to enable the tappers to tap comfortably. Such ladders are specially made with a platform so that when it is placed against the rubber trees, the tapper can stand on the platform to tap the tree with a good degree of stability. The ladder should be light to make carrying it from tree to tree feasible.

Stimulation of Latex Flow

Stimulation of latex flow is principally an exogenous process to increase the yield above that normally obtained by tapping from a rubber tree. The methods that have been employed cover a wide spectrum and have progressed from simple physical techniques in the early years to more refined and effective chemical methods. Stimulation of latex flow is now an integral part of most exploitation methods. The early history of stimulation has been reviewed tracing the development to the commercial use of synthetic growth substances such as 2,4-D and 2,4,5-T (Abraham and Tayler, 1967).

A wide range of substituted phenoxyacetic acids and substituted benzoic acids were screened for stimulant activity during the period 1956-1968. Results obtained from screening of some of these compounds which were selected on the basis of observations that the yield response increased (1) with increase in chlorine atoms substituted in the benzene ring, and (2) with decrease in atomic weight of the substituted halogen. It was concluded from a number of experiments that they only compound which gave comparable or better responses than 2,4,5-T was 2,4,dichloro-5-fluoro-phenoxyacetic acid but was not considered economical because of high production costs (Blackman, 1961; Abraham et al. 1968). The results supported the continued use of 2,4-D and 2,4,5-T as yield stimulant of Hevea.

Although auxin analogues have been much studied and many are active, some substances apparently quite unrelated in structure to auxin also show stimulant effects on Hevea, as was reported for copper sulphate (Compagnon and Tixier, 1950). Several herbicides, bactericides and organomercurial compounds were later found to be active (Abraham et al. 1968). They remarked that all the active compounds were phototoxic to some extent and observed that the common factor may be a selective and limited toxicity.

The first use of a gas (ethylene oxide, which is toxic) was reported to increase latex flow (Taysum, 1961). Subsequently, acetylene was found to be a yield stimulant (Banchi, 1968; Banchi and Poliniere, 1969). Yield stimulation was obtained by ethylene gas, by several halogenoparaffins and by (2-chloroethyl)-phosphonic acid; acetylene and another herbicide of the growth regulator type (4-amino 3,5,6-trichloropicolinic acid or Picloram) were added to the list of known stimulants. All effective compounds increase the period of flow and the volume of latex harvested.

The yield stimulant action of acetylene, ethylene and ethephon was confirmed (D'Auzac and Ribaillier, 1969a,b). The activity of β-hydroxy-ethyl-hydrazine, another compound which can decompose releasing ethylene and its yield stimulation by β-hydroxy-ethyl-hydrazine has also been reported (Ribaillier and D'Auzac, 1970; Pakianathan, 1970).

All effective non-gaseous stimulants of latex flow seemed to have one common feature, namely production of ethylene (Abraham et al. 1968). A large body of circumstantial evidence for this theory was summarised. Ethylene gas is thought to act more or less directly by a fundamental mechanism till not known, by inhibiting the plugging reaction of the trees. Ethephon (Abraham et al. 1972) and β-hydroxy-ethyl-hydrazine decompose liberating ethylene, which then acts as in the direct application of the gas itself; auxin (IAA), auxin analogues, copper sulphate, herbicides, bactericides and other compounds of selective and limited toxicity stimulate Hevea tissue to produce ethylene from endogenous substrates, and this ethylene (as also ethylene produced by injury) acts similarly as applied ethylene. Subsequent report that Picloram stimulates the production of ethylene by plants is consistent with this theory (Baur and Morgan, 1969).

It has been known for some time that physical injury produces stimulant effects of Hevea (Buttery and Boatman, 1967; Baptist, 1955). Experimental evidence is now available that Hevea tissue, like those of other plants, produce ethylene after application of 2,4,5-T or 2,4-D (Audley et al. 1977). Thus, it seems that some yield stimulants act via ethylene and it is quite possible, though not proven, that they all do. The behaviour of acetylene may be one of mimicry as in the case of propylene. More details by the efforts of acetylene and ethylene on yield responses have been reported (Abraham et al. 1971a,b,c).

The success of ethephon as a yield stimulant generated an intensive search for alternative ethylene based stimulants. Numerous chemicals were screened covering a wide spectrum but with exception of two compounds found the responses to others inferior to ethephon (Pakianathan, 1971). These two stimulants coded as Edbroza and Edsia involved generating ethylene by reacting ethylene dibromide with either zinc or sodium iodide in the presence of alcohol in a test-tube placed in a bored hole above the stock-scion union. These two stimulants though cheaper than ethephon were found to be impractical for use because of the method of application.

The development of a novel technique of applying ethylene gas adsorbed in molecular sieves (Zeolites) and formulated in a carrier such as petroleum jelly known as Ethad and the synthesis of a whole range of organosilicon compounds which on hydrolysis released ethylene provided the best potential for development of alternative yield stimulants to ethephon. Preliminary results on initial screening of these stimulants showed promise (Sivakumaran et al. 1971). Further confirmation of good stimulant activity of these stimulants and their corresponding commercial preparation in large-scale trials were reported (Dickenson et al. 1975; Sivakumaran et al. 1978). The responses to selected silane compounds were found to be generally inferior to ethephon but showed potential for development as mild stimulants particularly on young rubber (Dickenson et al. 1975; Sivakumaran, 1974). Ethad was found to be as effective as ethephon on most cultivars.

Methods of Application

The need to screen chemicals of varying nature such as gases, liquids and solids for stimulant activity in Hevea has necessitated the development of different methods of application tailored for specific requirements (Abraham, 1977). However, a majority of these methods are impractical for large-scale adoption and therefore of no commercial value. Methods which are popular and widely used in the industry are those which are not labour demanding and consequently cheap, simple and very practical. These methods are described below together with the different variations (Abraham et al. 1975):

Scraped bark below cut application: A very popular and common method of application. This method entails demarcating a narrow band below the tapping cut, scraping off the outer corky tissue and then applying a thin layer of stimulant mixture over the scraped area. The width to be applied will be determined by both the tapping frequency and of stimulant reapplication. The applied portion is consumed before the next application.

This method has been used for auxins before and currently in use for ethephon. The method has been varied by positioning the strip either at mid-point of the panel or above the stock/scion union. It has been established that the most effective position is immediately below the tapping cut.

Panel or above cut application: The stimulant mixture is applied to the regenerating bark immediately above the cut with no scraping. The width of application varies with the frequency (Puddy and Warrior, 1961). This method has been effective for application of ethephon.

Groove application: The stimulant is applied to the groove of the tapping cut by means of a Chinese paint brush after removal of tree lace, on a non-tapping day. The stimulant is reapplied at monthly intervals. This method has been shown to be just as effective as the scraped bark method and suitable for most stimulants (P'Ng et al. 1973). A variation of this method called the modified groove method involves application of stimulant to a groove formed three inches below the tapping cut. This method though shown to be effective was not developed further because of its practical limitations. Application of stimulant along the front and back channels or guidelines has been studied as a variation of the groove method but the responses obtained were not superior to that obtained from groove application.

Lace application: This is a slight modification of the groove method involving application along groove of tapping cut without removal of tree lace. The responses obtained by this method of application have been shown to be comparable to that of groove application (Sivakumaran and Ismail Hashim, 1983). This is an easier method than groove application particularly for trees with very thin laces.

Effects of Stimulation

The application of yield stimulants primarily results in increased yields, with two or three-fold increases during the first few tappings following stimulation (Abraham et al. 1975). The increase in yield is largely as a result of increased flow-time, with flow extended for several hours more than that of unstimulated trees. This is very evident from percentage increase in the amount of late drip obtained in stimulated trees in contrast to unstimulated trees. The increase in flow time induced by stimulation is a consequence of two effects of stimulant namely the ability to delay plugging and extension of drainage areas in the tapping panel (Southorn, 1969; Pakianathan, 1984). The anti-plugging effect of stimulants is apparent from observations that stimulants are most effective in relation to yield increases in clones with high plugging indices and least effective in clones with low plugging indices and its effectiveness is better with short-cuts than with long-cuts where plugging is feeble (Moir, 1970; Southorn and Gomez, 1970).

The dry rubber content of latex obtained from stimulated trees is generally lower than that of unstimulated trees. This depression is particularly marked in the initial years when high yield responses are obtained (Sivakumaran et al. 1981). The extent of depression varies between cultivars and age of panel. However with repeated stimulation, the d.r.c. values stabilize with no further drop, despite a declining yield response (Abraham et al. 1975). On the contrary with prolonged stimulation and subsequent panel change over, the d.r.c. values in stimulated trees appear to recover with smaller magnitude of difference in the later years between stimulated and unstimulated trees.

The incidence of dryness in stimulated trees is generally higher than that of unstimulated trees. However, the extent of dryness during the initial years of stimulation is low but with increasing trend on continuous stimulation. The incidence of dryness increases markedly in stimulated trees when the cut approaches the union (Sivakumaran et al. 1981). Generally, trees on which stimulation was first introduced on virgin panels have higher incidence than trees which first had stimulation on renewed panels. The incidence is particularly marked in panels which have renewed over a previously tapped and stimulated panel.

Girth increment is negatively affected by stimulation (Abraham et al. 1975). However, extent of girth depression is largely influenced by age of trees and intensity of tapping in stimulated trees. Thus, stimulation on virgin panels generally has a greater adverse effect on girth increment than stimulation on renewed panel when growth rate is lower and competition for assimilates is less. Generally, the depression in girth increases in proportion to the increase in length of cut with the most depression obtained in trees that are intensively tapped.

The effect of stimulation on bark thickness and number of latex rings varied according to cultivars. Thus, in a study (Sivakumaran et al. 1981) of eight clones bark thickness of renewed bark was significantly increased in ethephon stimulated trees of five clones relative to respective controls, while in three others the increase was not significant. Similarly, for latex vessel number, there was an increase in six clones, while in two, there was a marginal decrease (P'Ng, 1982).

Stimulation tends to cause greater drainage of both macro and micro nutrients from the tree because of the increased outflow of latex. Thus, there is a need to supplement the losses with additional manuring in stimulated trees to avoid nutritional stress (P'Ng, 1982; Pushparajan et al. 1971).

Yield stimulation has been observed to cause some changes in raw rubber properties but these are not considered to be significant. At peak yield, there is a tendency for rubbers of lower average molecular weight (Po value) to be produced but are of no technological consequence (Subramaniam, 1971).

The long-term adverse effects of stimulation are most apparent in trees which first had stimulation on virgin panel and which have been subjected to intensive tapping and continuous stimulation (Sivakumaran et al. 1981). In these trees generally there is marked decline in yields as the cut approaches the union. However, a resurgence in yield response is obtained when cuts are changed to a new panel but rapid decline occurs after the initial few months. The yield responses on lower panels are generally negative.

In trees stimulated for three years with ethephon both the initial rate of latex flow after tapping and the turgor pressure in the laticiferous tissues was markedly reduced by comparison with controls (Pakianathan, 1977). A marked reduction in initial flow rates and large areas of low in situ turgor pressures on the tapping panel in long-term ethephon treated trees, relative to unstimulated trees was observed (Pakianathan et al. 1982). Anatomical examination of bark taken from these low pressure areas have shown increase in stone cells in the soft bark and partial emptiness of latex vessels stage together with changes in its nuclei. Reduced sucrose levels in latex has been observed in repeatedly stimulated trees with failing yield response (Tupy and Primot, 1976).

Mode of Action of Stimulants

All effective yield stimulants increase the period of latex flow after tapping and the volume of latex collected. They all act, at least in large measure by inhibiting the latex vessel plugging and by extending latex flow time. Thus, despite the diverse chemical nature of the numerous yield stimulants, it appears that there is strong evidence for a common chemical factor in all forms of yield stimulation. It was suggested that this common factor is ethylene (Abraham et al. 1968).

Numerous evidences and suggestions that have been advocated to explain ethylene mode of action on the basis of investigations carried out by various researchers have been reviewed (Gomez, 1983).

REFERENCES

Maclaren, W.F. de Bois. 1913. The Rubber Tree Book. London, Maclaren and Sons. p. 141.

Baptiste, E.D.C. 1962. Present possibilities of Hevea culture. Rev. Gen Caoutch., 39: 1347.

Westgarth, D.R. and Narayanan, R. 1964. The effect of rubber price and yield per acre on estate production costs. J. Rubb. Res. Inst. Malaya, 18: 51.

Ridley, H.R. 1890-91. Annual report straits settlement. Royal Botanic Gardens for 1890 and 1891.

Wycherley, P.R. 1964. The cultivation and improvement of the plantation rubber crop. Res. Archives of the Rubber Research Insitute of Malaya. Document No. 29, pp. 1-30.

Dijkman, M.J. 1951. Hevea. Thirty years of research in the Far East. University of Miami Press.

De Jong, A.W.K. 1916. Wetenschappelijke tapproeven bij Hevea brasiliensis. Meded. Agricultuur-Chemisch Laboratorium. 14. 25. quoted by Dijkman (1951).

Maas, J.G.J.A. 1925. Periodentap. Arch. Rubber Cultuur. 9: 129. quoted by Dijkman (1951).

Bobilioff, W. 1923. Anatomy and physiology of Hevea brasiliensis. Part I. Anatomy of Hevea brasiliensis. Zurich. Art. Institut Orell Fussli. quoted by Dijkman (1951).

Lukman. 1983. Revised International Notation for Exploitation Systems. J. Rubb. Res. Inst. Malaysia, 31(2): 130-140.

Abraham, P.D. and Tayler, R.S. 1967. Tapping of Hevea brasiliensis. Trop. Agric. Trin., 44(1): 1-11.

De Jonge, P. and Warriar, S.M. 1965. Influence of depth of tapping. Plrs' Bull. Rubb. Res. Inst. Malaysia, 1965, No. 80: 158.

Ng,E.K., Abraham, P.D., P'Ng, T.C. and Lee, C.K. 1969. Exploitation of modern Hevea clones. J. Rubb. Res. Inst. Malaya, 21(3): 292-329.

Abraham, P.D. 1981. Recent innovations in exploitation of Hevea. The Planter, Vol. 57, No. 668, November 1981, 631.

Ismail Hashim, P'Ng, T.C., Chew, P.K., Abraham, P.D. and Anthony, J.L. 1979. Microtapping and the development of Micro-X system. Proc. Rubb. Res. Inst. Malaysia Plrs' Conf., Kuala Lumpur, 1979, 128-159.

Ismail Hashim, Ahmad Zarin, Othman Hashim and Yoon, P.K. 1981. Further results of controlled upward tapping on high panels. Proc. Rubb. Res. Inst. Malaysia Plrs' Conf., Kuala Lumpur, 1981, 74.

Abraham, P.D. and Ismail Hashim. 1983. Exploitation procedures for modern Hevea clones. Proc. Rubb. Res. Inst. Malaysia Plrs' Conf., Kuala Lumpur, 1983, 126-156.

Abraham, P.D. and Tayler, R.S. 1967. Stimulation of latex flow of Hevea brasiliensis. Expl. Agric., 3(1): 1-12.

Blackman, G.E. 1961. The stimulation of latex flow by plant growth regulators. Proc. Nat. Rubb. Conf., Kuala Lumpur, 1960, 19.

Abraham, P.D., Boatman, S.G., Blackman, G.E. and Powell, R.G. 1968. Effects of plant growth regulators and other compounds on flow of latex in Hevea brasiliensis. Ann. Appl. Biol. 62: 159-173.

Compagnon, P. and Tixier, P. 1950. Sur ure possibilite d'ameliorer la production d'Hevea brasiliensis par l'apport d'oligo-elements. Rev. Gen Caoutch, 27: 525.

Taysum, D.H. 1961. Effects of ethylene oxide on the tapping of Hevea brasiliensis. Nature, London, 191, 1319.

Banchi, Y. 1968. Effects de l'acetylene sur la production en latex de l'Hevea brasiliensis. Arch. Inst. Rech. Caoutch. Viet-Nam No. 2/68.

Banchi, Y. and Poliniere, J.P. 1969. Effects of minerals introduced directly into the wood and of acetylene applied to the bark of Hevea. J. Rubb. Res. Inst. Malaya, 21: 192.

Abraham, P.D., Wycherley, P.R. and Pakianatha, S.W. 1968. Stimulation of latex flow in Hevea brasilinesis by 4-amino-3,5,6-trichloropicolinic acid and 2-chloroethane phosphonic acid. J. Rubb. Res. Inst. Malaya, 20(5): 291.

D'Auzac, J. and Ribaillier, D. 1969a . L'ethylene, nouvel agent stimulant de la production de latex chez l'Hevea brasilinesis. Compt. Rend., 268: 3046.

D'Auzac, J. and Ribaillier, R. 1969b. L'ethylene, nouvel agent stimulant de la production de latex chez l'Hevea brasiliensis. Revue. gen. Caoutch. Platiq., 46: 857.

Ribaillier, D. and D'Auzac, J. 1970. Nouvelles perspectives de stimulation hormonale de la production chez l'Hevea brasiliensis. Revue. gen. Caoutch. Plastiq., 47: 433.

Pakianathan, S.W. 1970. The search for new stimulants. Plrs' Bull. Rubb. Res. Inst. Malaya No. 111, 351.

Abraham, P.D. Gomez, J.B., Southorn, W.A. and Wycherley, P.R. 1972. A method of stimulation of rubber yield in Hevea brasiliensis. British Patent 1281524: appl. 22.10.68, publ. 12.7.72.

Baur, J.R. and Morgan, P.W. 1969. Effects of Picloram and Ethylene on leaf movement in Huisache and Mesquite seedlings. Plant Physiol., 44: 831.

Buttery, B.R. and Boatman, S.G. 1967. Effects of tapping, wounding and growth regulators on turgor pressure in Hevea brasiliensis Muell. Arg. J. Exp. Bot., 18: 644.

Baptist, E.D.C. 1955. Stimulation of yield in Hevea brasiliensis. I. Pre-war experiments with vegetable oils. J. Rubb. Res. Inst. Malaya, 14: 355.

Audley, B.G., Archer, B.L. and Runswitck, M.J. 1977. Ethylene production by Hevea brasiliensis tissue treated with latex yield stimulatory compounds. Ann. Bot., 42: 63.

Abraham, P.D., Blencowe, J.W., Chua, S.E., Gomez, J.B., Moir, G.F.J., Pakianathan, S.W., Sekhar, B.C., Southorn, W.A. and Wycherley, P.R. 1971a . Novel stimulants and procedures in the exploitation of Hevea: I. Introductory Review. J. Rubb. Res. Inst. Malaya, 23(2): 85-89.

Abraham, P.D., Belncowe, J.W., Chua, S.E., Gomez, J.B., Moir, G.F.J., Pakianathan, S.W., Sekhar, B.C., Southorn, W.A. and Wycherley, P.R. 1971b. Novel stimulants and procedures in the exploitation of Hevea: II. Pilot trials using (2-chloroethyl)-phosphonic acid (Ethephon) and acetylene with various tapping systems. J. Rubb. Res. Inst. Malaya, 23(2): 90-113.

Abraham, P.D., Blencowe, J.W., Chua, S.E., Gomez, J.B., Moir, G.F.J., Pakianathan, S.W., Sekhar, B.C., Southorn, W.A. and Wycherley, P.R. 1971c. Novel stimulants and procedures in the exploitation of Hevea: III. Comparison of alternative methods of applying stimulants. J. Rubb. Res. Inst. Malaya, 23(2): 114-137.

Pakianathan, S.W. 1971. Trials with some promising stimulants. Proc. Rubb. Res. Inst. Malaya Plrs' Conf., Kuala Lumpur, 1971, 72.

Sivakumaran, S., Pakianathan, S.W. and Abraham, P.D. 1971. Further trials with new yield stimulants. Rubb. Res. Inst. Malaya Yield Stimulation Symp. 1972, Preprint No. 3.

Dickenson, P.B., Sivakumaran, S. and Abraham, P.D. 1975. Ethad and other new stimulants for Hevea brasiliensis. Proc. Int. Rubb. Conf., Kuala Lumpur, 2: 315-344.

Sivakumaran, S., Abraham, P.D., P'Ng, T.C. and Stephen Thomas. 1978. Comparable performance of Ethad and Ethephon in field trials. Proc. Int. Rubb. Res. Dev. Board Symposium, Kuala Lumpur, 1978.

Sivakumaran, S. 1974. Search for economic and effective yield stimulants for Hevea brasiliensis. Project Report, RRIM, 1974.

Abraham, P.D. 1977. Stimulation of the yield of Hevea brasiliensis Muell. Arg. by ethylene releasing substances - Thesis submitted to the University of Reading for degree of Doctor of Philosophy in Agriculture and Horticulture, 1977.

Abraham, P.D., P'Ng, T.C., Lee, C.K., Sivakumaran, S., Manikam, B. and Yeoh, C.P. 1975. Ethrel stimulation of Hevea. Proc. Int. Rubb. Conf., Kuala Lumpur, 1975, II, 347, Rubb. Res. Inst. Malaysia.

Puddy, C.A. and Warrior, S.M. 1961. Yield stimulation of Hevea brasiliensis by 2,4-dichlorophenoxyacetic acid. Proc. Nat. Rubb. Res. Conf., 1960, Kuala Lumpur, 194.

P'Ng, T.C., Leong Wing and Abraham, P.D. 1973. A new method of applying novel stimulants. Proc. Rubb. Res. Inst. Malaya Plrs' Conf., Kuala Lumpur, 1973, 122.

Sivakumaran, S. and Ismail Hashim. 1983. Factors influencing responses to ethephon. Proc. Rubb. Res. Inst. Malaysia Plrs' Conf., Kuala Lumpur, 1983.

Southorn, W.A. 1969. Physiology of Hevea (latex flow). J. Rubb. Res. Inst. Malaya, 21: 494.

Pakianathan, S.W. 1984. Private Communication.

Moir, G.F.J. 1970. A radical approach to exploitation. Plrs' Bull. Rubb. Res. Inst. Malaya, No. 111, 342.

Southorn, W.A. and Gomez, J.B. 1970. Latex flow studies. VII. Influence of length of tapping cut on latex flow pattern. J. Rubb. Res. Inst. Malaya, 23: 15.

Sivakumaran, S., Pakianathan, S.W. and Gomez, J.B. 1981. Long-term ethephon stimulation. I. Effects of continuous ethephon stimulation with half spiral alternate daily tapping. J. Rubb. Res. Inst. Malaysia, 29, 57.

P'Ng, T.C. 1982. Towards rational exploitation of Hevea brasiliensis. Ph.D. Thesis, University of Ghent, Belgium.

Pushparajah, E., Sivanadyan, K., P'Ng, T.C. and Ng, E.K. 1971. Nutritional requirements of Hevea brasiliensis in relation to stimulation. Proc. Rubb. Res. Inst. Malaysia Plrs' Conf. 1971, 189.

Subramaniam, A. 1971. Effects of ethrel stimulation on raw rubber properties. Proc. Rubb. Res. Inst. Mal. Plrs' Conf. 1971, 255.

Pakianathan, S.W. 1977. Some factors affecting yield response to stimulation with 2-chloroethylphosphonic acid. J. Rubb. Res. Inst. Malaysia, 25, 50.

Pakianathan, S.W., Samsidar Hamzah, Sivakumaran, S. and Gomez, J.B. 1982. Physiological and anatomical investigations on long-term ethephon stimulated trees. J. Rubb. Res. Inst. Malaysia, 30: 63.

Tupy, T. and Primot, L. 1976. Control of carbohydrate metabolism by ethylene in latex vessels of Hevea brasiliensis Muell. Arg. in relation to rubber production. Biologia, Pl. 18: 373.

Gomez, J.B. 1983. Physiology of latex (rubber) production. MRRDB Monograph No. 8.

CHAPTER 13

INTERNATIONAL NOTATION FOR EXPLOITATION SYSTEMS

LUKMAN

Balai Penelitian Perkebunan Sungei Putih,
National Centre for Rubber Research, P.O. Box 416, Medan, Indonesia.

INTRODUCTION

An internationally accepted common exploitation notation is needed for all the rubber producing countries to ensure mutual understanding between researchers, rubber planters and students from different countries.

The first notation to describe tapping systems was suggested in 1939 (RRIM, 1939) and was revised in 1940 (RRIM, 1940) and this was recognised as the International Tapping Notation.

Since that time many new modifications in tapping methods have been developed, particularly in association with stimulation. After 1970, this notational system became increasingly inadequate to describe correctly the various systems that have been introduced. The need for a new system of notations for exploitation systems was emphasised by scientists working on rubber in different institutes.

With the introduction of highly effective stimulants, the novel technique of puncture tapping was introduced. The combinations of cut tapping and puncture tapping were also experimented upon. Short cuts, combined with latex stimulant applications, has also been introduced as a new exploitation method. During the later phases of exploitation upward tapping is now being practiced.

These developments demanded introduction of a new set of notations to adequately describe the exploitation techniques and methods. With the increased use of stimulants, a view was expressed that the term 'tapping notation' should be replaced by the form 'exploitation notation' to cover different systems of stimulation as well.

Efforts for revision

Before 1974, International Rubber Research and Development Board (IRRDB) entrusted Indonesia to make a proposal for revision. The first proposal for revision was introduced by the author in September 1974 at the IRRDB Symposium at Cochin, India. The second discussion on the revision was held at Cisarua, Indonesia, in the November 1976 IRRDB Symposium and the third discussion was held in the IRRDB Workshop on International Tapping Notation, in November 1980, in Medan, Indonesia. An agreement on the proposal emerged at the IRRDB Meeting of the Specialist Group on Physiology of Latex Production and Exploitation held at Kottayam, India in April 1982.

This agreement has since then been formally described as the 'International Notation for Exploitation Systems'. The proposal was approved by the Directors and Board members of the IRRDB in 1982.

Revised Notation

The International Notation for Exploitation Systems as a revised notation is divided into three parts: tapping method, panel notation and stimulation for both cut tapping and puncture tapping systems.

REVISED INTERNATIONAL NOTATION

CUT TAPPING

Tapping Notation

Tapping is the action of opening the latex vessels of a rubber tree. The tapping notation is a series of symbols and numbers describing this action and its frequency in a certain period of time.

Symbol of cut: Cut tapping is the operation in which a thin shaving of bark is excised for the extraction of latex.

The symbol of type of cut is denoted by a capital letter.

Examples:

S = spiral cut; V = V cut; C = circumference; Mc = Mini cut (five cm length or less).

The symbol C is used for two or more unspecified cuts (S+V) on a tree tapped on the same tapping day. The details of cuts should be given as a footnote.

Length of cut: The length of tapping cut, except for the mini cut, is interpreted as the relative proportion of the trunk circumference that is embraced by the tapping cut and does not refer to actual lengths. In the case of mini cut however, length is not expressed relatively but directly in centimetres.

The length of cut is represented by a fraction preceding the symbol of cut, except in the case of mini cut.

Examples:

S = one full-spiral cut; V = one full-V cut; C = one full-circumference (unspecified cut); 1/2S - one half-spiral cut; 1/3V = one third-V cut; 3/4S = three fourth-spiral cut; 1/2C = one half-circumference cut; Mc2 = mini cut, the length of cut being 2 cm; Mc0.5 = mini cut, the length of cut being 0.5 cm.

Note:

- For full circumference cuts, no fraction is written before the symbol of cut.

- For reduced spiral (S-15 cm), the symbol S/R can be used.

<u>Number of cuts</u>: A tapping system with a number of cuts may be applied on a rubber tree, either on the same tapping day or on alternate tapping days.

The number of cuts is represented by a figure before the length of cut notation, and a multiplication sign inserted in between.

Examples:

2x1/2S = two half-spiral cuts; 3x1/2V = three half-V cuts; 4xMc2 = four mini cuts of 2 cm length.

<u>Direction of tapping</u>: The direction of tapping is normally downward, but since the upward one is also used in some systems, it is necessary to indicate the direction of the cuts. When tapping is downward only, no symbol of direction is used. For upward tapping, the symbol is an upward arrow (↑) written immediately after the cut notation. When two directions of tapping are applied on the tree, add both upward and downward arrows (↑↓) after the concerned cut notation. In combination tapping, the downward arrow need not be indicated.

Examples:

1/2S = one half-spiral cut tapped downward; 1/3V = one third-V cut tapped downward; 1/2S↑ = one half-spiral cut tapped upward; 1/3C↑ = one third-circumference tapped upward; 2x1/2V↑↓ = two half-V cuts, one half-V cut tapped upward and the other tapped downward; 3/4C↑↓ = three fourth-circumference, one half-V cut tapped upward and the other one fourth-spiral cut tapped downward.

Note: 3/4C ↑↓ =1/2V ↑+1/4S or 1/4V↑+1/2S

2x1/2C↑ ↓ = two half-circumference, two one fourth-V cuts tapped downward and the other two one fourth-spiral cuts tapped upward.

Note: 1/2C ↑↓ = 1/4V+1/4S↑.

<u>Frequency of tapping</u>: The frequency of tapping notation describes the interval between tapping and is expressed as one fraction or a series of fractions. The first fraction is called the 'actual frequency' notation

while all following notations which are necessary to fully describe the frequency of tapping are the explanation of the actual frequency. The other notations of the tapping of the tapping frequency which may possibly follow this are for practical frequency, periodicity and change-over. One letter space is to be left in between the notations of the frequency of tapping.

Actual frequency: The notation of actual frequency is a fraction in the time unit of the day (=d). The numerator of the fraction is d without any numeral before it and denotes the tapping period (day) while the denominator denotes the tapping interval of the actual frequency in days or in fraction of day.

Examples:

d/1 = daily tapping; d/2 = alternate-daily (one day in two); d/3 = third-daily (one day in three); d/0.5 = twice a day tapping.

Practical frequency: Where continuous tapping is broken by a regular day (or days) of rest, a fraction is written after the actual frequency. This fraction showing the practical frequency has, as numerator, the number of days tapped in a period, the period being denoted by the denominator.

Examples:

d/1 2d 3 = daily tapping, two days in tapping followed by one day of rest; d/2 6d/7 = alternate daily, six days in tapping followed by one day of rest.

Periodicity: The notation of periodicity may consist of one or more fractions in the time unit of week (=w), months (=m) and years (=y). The numerator of each fraction denotes the tapping period and may be with or without numeral before the symbol while the denominator of each fraction denotes the length of the cycle (tapping period + rest). Each succeeding fraction in the periodicity notation modifies the period of operation of the previous fraction, the denominator of the final fraction giving the full cyclic period of the system.

Examples:

2w/4 = two weeks in four (two weeks in tapping followed by two weeks of rest); 6m/9 = six months in nine (six months in tapping followed by three months of rest); 2w/4 6m/9 = two weeks in four during six months in nine (two weeks in tapping followed by two weeks of rest during six months and then followed by three months of rest).

Complete example:

d/2 6d/7 3w/4 8m/12 = alternate-daily tapping, six days in seven for three weeks in four, during eight months out of twelve (alternate daily tapping for six days followed by one day of rest, for three weeks followed by one week of rest, during eight months followed by four months of rest).

In the above notation, the full cyclic period of the system is 12 months and d/2 is the actual frequency, 6d/7 the practical frequency while 3w/4 and 8m/12 denote periodicity.

This revised system of notation of tapping frequency introduces some small changes into the previous system. Examples of correction of these old notations are as follows:

Old system	New system
- 2d/1	- d/0.5
(twice daily).	(twice a day tapping).
- 2d/3	- d/1 2d/3
(two days in tapping followed by one day of rest).	(daily tapping for two days followed by one day of rest).
- d/2 with one day rest	- d/2 6d/7
(alternate-daily with one day rest).	(alternate-daily tapping for six days followed by one day of rest).

Change of cut: The type of tapping cut and frequency can be changed in one production period (eg. one year). For example, presentation of yield is yearly, but after six months of tapping, the length of cut can be changed. The old and the new notations of this change is separated with a horizontal arrow (→).

Change of tapping cut: When the tapping cut is shortened or lengthened or changed in direction, a horizontal arrow separates the notation for the old cut (on the left) and the new cut (on the right).
Examples:

1/2S → 1/3S = a half-spiral cut tapped downward changed to one third-spiral cut tapped downward (for the rest of the tapping of one production period). Shortening of cut.

1/2S → 3/4S = a half-spiral cut tapped downward, changed to three fourth-spiral cut tapped downward. Lengthening of cut.

1/2S → 1/2S ↑ = a half-spiral cut tapped downward changed to a half-spiral cut tapped upward. Change of cut direction.

Change of tapping frequency: When there is a change in the frequency of a cut tapping, a horizontal arrow separates the notations of the old system (on the left) and the new system (on the right).

Example:

1/2S d/2 6m/12 → 1/2S 2xd/2(t,t) 6m/12 = a half-spiral cut tapped down ward, alternate daily for the first six months, changed into two half spiral cuts, each cut tapped downward alternatively on every tapping day for the next six months.

Tapping intensity: Tapping intensity values can be calculated from various components of the tapping notation, to provide parameters for comparison and evaluation of tapping systems. The parameter of relative intensity was previously popular for comparing tapping systems, but is now little favoured as experience has shown that it is inadequate as an estimate of the physiological intensity of tapping systems. The value of actual intensity, which takes into account the actual number of tapping days, provides a more realistic basis for certain comparisons. Neither relative nor actual intensities are now included in the tapping notation. For each tapping system however, the total number of actual tapping days per year should be given as a footnote.

Note: Actual tapping days/year = days of tapping per year plus recovery tapping (if used).

Relative intensity: The relative intensity is expressed as percentage of the standard systems : 1/2S d/2 or 1/4S d/1 = 100%. To calculate the relative intensity, multiply the fractions in the formula, and then multiply the product by 400.

Example:

1/2S d/2 = 1/2 x 1/2 x 400 = 100%.

Actual intensity

The actual intensity is the amount of tapping actually realized, expressed in percentage. To calculate intensity, multiply four times the length of cut in the formula by the average number of tappings (tapping days $year^{-1}$) and divide by the total number of days in the given period (year).

Example:

$$1/2S \ d/2 = (\frac{4 \ x \ 1/2 \ x \ 167}{365}) \ x \ 100\% = 92\%.$$

Note: - actual tapping days $year^{-1}$: d/2 = 167.
- total days $year^{-1}$ = 365.

Panel Notation

Panel is the area of bark of the rubber tree, in which the tapping cut is located.

Panel notation is the symbol, or series of symbols, which describe the panel location and the panel renewal succession of the tapping panels.

The notation is not included in a tapping notation but it should be indicated in the tapping descriptions or treatment details.

With the introduction of short cuts, the need for changing the presently adopted panel notation, A and B to denote virgin bark and C and D to denote first renewed bark has been recognized. It has been decided to designate the base panel of virgin bark, first renewal and second renewal, by new symbols BO, BI and BII respectively, and the number of subsequent panels by a number. Therefore, it was decided that the virgin bark shall be given the symbol BO, the first renewed bark the symbol BI and the second renewed bark the symbol BII. For the high panel the letter H is used. The sequence of panels is denoted by number.

The location of the panels may be grouped into those of a circular succession and those of a vertical succession (Fig. 1).

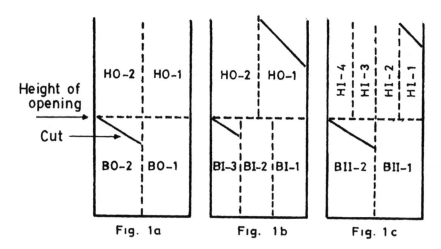

Fig. 1a   Fig. 1b   Fig. 1c

Note: The standard method of showing location of panels and cuts is a front sectional view of the tree trunk as in the above diagrams. The height of opening is indicated by a broken horizontal line. Vertical broken lines distinguish the panels.

Fig. 1.   Panel notation of cut tapping.

Vertical succession: The panels located above the height of the first opening for cut tapping of clones are called the high panels and are denoted by a capital letter H (=high).   The panels formed below this level are considered as the base panels and for them the letter B is used.
Examples:

BO-1 = base panel 1; HO-3 = high panel 3.

Panel renewal succession: The panel renewal succession in relation to the progress of tapping is considered for virgin bark and renewed bark. Renewed bark is the bark which has regrown again after tapping. Virgin bark is denoted by the letter O, the first renewed bark by the number I and the second renewed bark by the number II.

Examples:

BO-I = the first base panel on virgin bark

BI-3 = the third base panel on the first renewed bark

HO-4 = the fourth high panel on the virgin bark

Stimulation Notation

The stimulation notations are not separated from the tapping notations. The two should be presented together as a complete notation, with a full stop inserted between them. The notations of stimulation are grouped into three units and in the following order : stimulant, application and periodicity. Full stops must be inserted between these units to differentiate between them clearly, ie., stimulant unit, application unit, periodicity unit.

Stimulant - Active ingredient: The active ingredient of the stimulant must be expressed in the notation with a specific code, but for some stimulants the notations should be the same as their chemical names. The code consists of two capital letters which are taken from the technical or chemical name of the stimulant.

Examples:

ED = Ethad; ET = Ethepon; 2,4-D = 2,4-D; 2,4,5-T = 2,4,5-T; $CuSO_4$ = Copper sulphate; CaC_2 = Calcium carbide; ST = Stimulant unspecified.

Note: Carrier (in case the stimulant is diluted) and technical name are specified in footnote.

Concentration: The concentration of the active ingredient of the stimulant in the formulation used should be noted immediately after the code of the stimulant.

The notations of active ingredient and the concentration form the stimulant unit are presented consecutively.

Example:

ET 10% = Stimulated with 10% ethepon.

Method of application: The method of using a stimulant is indicated by a symbol describing the place of its application on the tree. The symbol consists of two letters.

Examples:

Pa = panel application; Ba = bark application; La = lace application; Ga = groove application; Wa = wood application (hole); Ta = tape or band application (in puncture and upward tapping); Sa = soil application.

Quantity of formulation: The quantity of formulation applied at one application is expressed by its weight in grams or by its volume in millilitre (ml) and is written in the notation without g or ml. For wood application, a liquid stimulant is always used and is measured by volume.

Width of band: The width of the band on which the stimulant is applied is measured in cm and written in the notation without cm. In groove and wood applications, the width of the groove and size of the hole in the wood are not presented: a das (−) for them is put in the notation to avoid confusion.

The method of application, the quantity of formulation and the width of band form the application unt.

Frequency of application: The frequency of stimulant application is always stated in weeks (=w) or months (=m).

Number of application per period: The total number of stimulant applications per period is denoted by a number. The period is usually expressed in years (=y).

Example:

8/y = eight applications/year.

The number of applications and frequency of application form the periodicity unit. The stimulant unit, application unit and periodicity unit are presented consecutively, and are separated by full stops.

Complete stimulation notation: Examples of complete stimulation notations are as follows:

ET5%.Pa2(1).16/y(2w) = stimulated with 5% ethepon, panel application, two g of stimulant/application on one cm of band, sixteen applications/year applied fortnightly.

ET5%.Pa2(2).3y* = stimulated with 5% ethepon, panel application, two g of stimulant/application on two cm of band, three applications/year, at irregular intervals.

Example of a complete Cut Tapping Notation

1/2S d/2 6d/7.ET5%.Pa2(1).16/y(2w) on panel BO-2 = a half spiral cut tapped alternate daily, six days in tapping followed by one day of rest. stimulated with 5% ethepon, panel application, two g of stimulant/ application on one cm of band, sixteen applications/year applied fortnightly. The panel is the second panel on virgin bark of the base panel (Fig. 1a).

PUNCTURE TAPPING

Tapping Notation

Symbol of puncture: In puncture tapping, the latex is extracted by incision on a band of bark or on the groove of the tapping cut.

The symbol of puncture tapping is given by the capital letter   P (=puncture). Various methods of puncture tapping have been studied by different countries and some of these are mentioned below:

PI = punctures on vertical band

Pg = punctures on vertical groove

PS = punctures on spiral band

PG = punctures on groove of tapping cut

PC = punctures on a channel

PB = punctures on a scraped bark

Note: The most common puncture tapping is made on a band, so the puncture tapping notation has been designed for this system. For the other types of puncture tapping, use this notation if it is possible - if not, explain the puncture tapping systems as a foot note.

Number of punctures: In puncture tapping, a number of punctures may be distributed in a band or groove of bark.

The number of punctures on one band is designated by a number before the symbol for puncture (=P). The length and width of band is given in centimeters (=cm), but the symbol cm is not present in the notation. The length of the band is written in brackets after the notation for puncture, while the width of the band is shown after the bracket.

The length of the band is relatively unlimited, and could be as long as is convenient to tap. Hence the total punctures on one band should be expressed together with the length of the band concerned.

Examples:

3PI(45)2 = three punctures on a vertical band of forty five cm length and two cm width.

4Pg(80)1 = four punctures on a vertical groove of eight cm length and one cm width.

6PG(1/2S) = six punctures on a half spiral groove.

5PB(1/2S)2 = five punctures on a scraped bark. The length of band/scraped bark is a half spiral, and its width is two cm.

4PS(1/2S)2 = four punctures on a half spiral band, two cm wide.

It is assumed that the distance between punctures on a band must be always the same unless otherwise indicated.

Number of bands: A tapping system with several bands may be applied on a rubber tree, and are either tapped on the same tapping day or on alternate days.

The number of bands is denoted by the figure preceding the number of punctures notation and a multiplication sign inserted in between.

Note: for one band, no numeral is required before the number of puncture notation.

Examples:

2x3PI(60)2 = two three-puncture vertical bands of sixty cm length and two cm width (there are two bands, three punctures are made on each band and the band is of sixty cm length and two cm width).

2x6PG(1/2S) = two six-puncture grooves of a half spiral length (there are two grooves, six punctures are made on each groove and the groove is a half spiral).

2x5PB(1/2S)2 = two five-puncture scraped barks of a half spiral length and two cm width (there are two bands of scraped barks, five punctures are made on each band and the band is a half spiral two cm wide).

2x4PS(1/2S)2 = two four-puncture spiral bands of a half spiral length and two cm width (there are two spiral bands, four punctures are made on each band and the band is a half spiral two cm wide).

When the bands are more than one, the number of punctures on each band should be the same.

Frequency of tapping: The notation of puncture tapping frequency is the same as that for cut tapping frequency.

Change of puncture tapping: The puncture or band number and the tapping frequency can be changed in one production period (eg. one year). For example, presentation of yield is yearly, but after six months of tapping, the number of punctures or the number of bands is changed. The old and the new notations of this change are separated with a horizontal arrow (→).

Band Notation

The band is a strip of bark which is punctured with the same number of punctures on each tapping day. The band notation is the symbol or series of symbols which describe the puncture band of a puncture tapping concerning the aspects of the band location and the band renewal succession.

The band notation is not included in a tapping notation but should be indicated in the tapping descriptions and in band diagrams.

The pattern of bands can be made with unlimited variations and combinations. The pattern and location of the band may be illustrated in band diagrams as the frontal views and the cross-cut views of the band concerned. They are not designated with special symbols.

The location of the bands on the trunk are on the panels of cut tapping. The panels are classified into those of a circular succession and those of a vertical succession and the notations used are similar to the ones used for cut tapping.

Stimulation Notation

Stimulation notation for puncture tapping is the same as stimulation notation for cut tapping except for the method of application. In cut tapping, there are six methods of application, ie. panel application (=Pa), bark application (=Ba), groove application (=Ga), wood application (=Wa), tape application (=Ta) and lace application (=La). However, in puncture tapping there are only five application methods: Ta, Ba, Ga, La and Sa (soil application).

Example:

ET5%.Ta2(2).12/y(m) = Stimulated with 5% ethepon, stimulant applied on tape or band of bark after scraping. The weight of stimulant is 2 g on 2 cm width of band. Number of applications is 12/year, applied at monthly intervals.

Example of complete Puncture Tapping Notation

3PI(45)2 d/2 6d/7.ET5%.Ta2(2).12/y(m) on band BI(1) : three punctures on a vertical band of fortyfive cm length and two cm width, tapped alternate daily, six days in tapping followed by one day of rest. Stimulated with 5% ethepon, stimulant applied on tape or band of bark after scraping. The weight of stimulant is 2 g on 2 cm width of band. Number of applications is 12/year, applied at monthly intervals. The band is the first band on the first renewed bark of the base panel (Fig. 2a).

COMBINATION OF TAPPING SYSTEMS

There can be situations where different systems of tapping are combined. The different cuts may be tapped on the same day or on alternate tapping days.

294

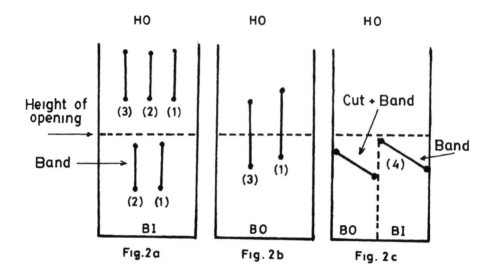

Note: The standard method of showing location of bands is a front sectional
        view of the tree trunk as in the above diagrams. The height of
        opening of cut tapping is indicated by a broken horizontal line.
        Vertical broken lines distinguish the panels (for cut tapping).

Fig. 2.   Band notation of puncture tapping.

Tapping on the same tapping day: When different cuts are on the
same tapping day, the notations for the system are joined by a plus sign (+).
Example:

1/2S↑+1/4S = a half-spiral cut tapped upward, tapped on the same tapping
        day with one fourth spiral cut tapped downward.

Tapped on alternate tapping days: When different cuts are tapped,
separately on alternate tapping days, the notations for the systems are
separated by a comma (,).
Example:

1/4S,1/8S↑ = one fourth-spiral cut tapped downward, and one eight-spiral
        cut tapped upward alternating the system on every tapping day.

Frequency of tapping: The tapping of a tree may be done continuously
on one panel or one group of panels tapped on the same tapping day. On
the other hand, it also can be done on several panels or on several groups
of panels, each tapped on alternate tapping day or in alternate tapping
periods. The second method, called change-over system, is denoted by
the cycle of changes of each tapping panel given in brackets.

In change-over system, the first figure (in brackets) indicates the
cycle of change of the first tapping panel and the second figure indicates
the cycle of change of the second tapping panel. A comma is inserted between

the cycle of changes of tapping panels. The cycle of changes of tapping
is denoted by t (=tapping), w (=week), m (=month) and y (=year).

Examples:

(t,t) = two cuts, each tapped alternately.

(6m,6m) = two cuts, each tapped alternately every six months.

(w,2w) = two cuts, the first cut tapped for one week followed by the second
cut tapped for two weeks.

(10t,m) = two cuts, the first cut tapped for ten tappings followed by the
second cut tapped for one month.

These are called the change-over symbols, and follow immediately
after the statement of actual frequency.

Examples:

d/2(t,t) = alternate-daily tapping, two cuts, each tapped alternatively
on every tapping day.

d/0.5(t,t) = twice a day tapping, two cuts, each tapped alternatively once
a day.

d/2 (t,t) 6m/9 = alternate-daily tapping, two cuts, each tapped alternatively
on every tapping day for six months followed by three months of
rest.

Note: figures in brackets not to be considered for the calculation
of the relative intensity.

1/3S+1/2S d/3 6d/7 = one-third spiral cut tapped on the same tapping day
with a half spiral cut, both tapped third-daily, six days out of seven.

1/2S,5PI(45)2 d/2(t,t) = a half spiral cut tapped alternately with five
punctures on a vertical band of forty five cm length and two cm width,
tapped alternate-daily, each tapped alternatively on every tapping day.

Examples of complete Notation for Combination of Tapping Systems

1/2S+5PI(45)2 d/3 6d/7.ET5%.Ba2(2).8/y(m)+Ta2(2).12/y(m)   on   panels

BI-2+HO(1) = a half spiral cut tapped on the same tapping day with five
punctures on a vertical band of fortyfive cm length and two cm width,
both tapped third-daily, six days out of seven. Stimulated with 5%
ethepon, for cut tapping the stimulant applied on bark and for the
puncture tapping is on a band of bark, both after scraping. The weight
of stimulant per application is two grams, the width of scraped bark
is two cm, eight applications/year (for cut tapping) and for frequency
of application is at monthly intervals. These details are the same
for both systems of applications except that the number of applications/
year for band is twelve. Both tapping cut and band are tapped on
the same tapping day. The cut is the second panel on first renewed

bark of base panel, meanwhile the band is the first band on virgin bark of high panel (Fig. 3a).

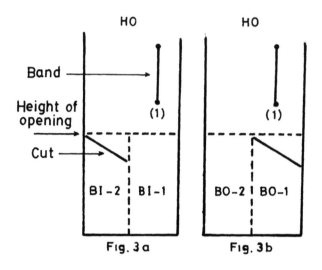

Fig. 3. Panel and band notation of a combination tapping.

1/2S,5PI(45)2 d/2(t,t).ET5%.Ba2(2).8/y(m),Ta2(2).12/y(m) on panels

BO-1,HO(1) = a half spiral cut tapped alternately with five punctures on vertical band of fortyfive cm length and two cm width. Alternate daily tapping, with each method being used alternately on every tapping day. Stimulated with 5% ethepon, for cut tapping the stimulant applied on bark and for puncture tapping on a band of bark, both after scraping. The weight of stimulant per application is two grams, the width of scraped bark is two cm, eight applications/year (for cut tapping) and the frequency of application is at monthly intervals. These details are the same for both systems of application except that the number of applications/year for band is twelve. Both tapping cut and band are tapped alternately, ie. in a change-over system. The cut is the first of base panel, virgin bark, while the band is the first band on virgin bark of high panel (Fig. 3b).

ACKNOWLEDGEMENT

The author is very grateful to the IRRDB members who were involved actively in discussion in IRRDB Symposium at Cochin, India in September 1974 and at Cisarua, Indonesia in November 1976, followed by the discussion in Workshop on International Tapping Notation in Medan, in November 1980 and in the meeting of specialist group on Physiology of Latex Production and Exploitation at Kottayam, India, in April 1982. Many thanks for them who have offered their critical remarks for formulation of the new international exploitation notations.

REFERENCES

Rubber Research Institute of Malaya, 1939. Communication 240. J. Rubb. Res. Inst. Malaya, 9: 164.
Rubber Research Institute of Malaya, 1940. Communication 247. J. Rubb. Res. Inst. Malaya, 10: 26.

CHAPTER 14

CERTAIN ASPECTS OF PHYSIOLOGY AND BIOCHEMISTRY OF LATEX PRODUCTION

S.W. PAKIANATHAN, S.J. TATA, LOW FEE CHON
Formerly of: Rubber Research Institute of Malaysia, Kuala Lumpur, Malaysia.
Present address: 95, Jalan Terasek 8, Bangsar Baru, Kuala Lumpur, Malaysia.
and
M.R. SETHURAJ
Rubber Research Institute of India, Kottayam-686009, Kerala, India.

The production of rubber is largely dependent upon (a) volume and type of laticiferous tissues in which latex is stored, (b) capacity of storage vessels (c) physiological and biochemical processes controlling latex flow and (d) capacity of the tree to resynthesise latex and other organic constituents within the drained area (for a detailed treatment of biosynthesis of rubber, please see Chapter 6, Ed.).

The organic non-rubber constituents such as proteins, carbohydrates, lipids, phospholipids, nucleic acids play a significant role in the re-synthesis of latex to replace that which had been drained off through tapping.

NATURE AND COMPOSITION OF LATEX

Organic non-rubber constituents of latex

The _Hevea_ latex, as it flows out of the tree, is a complex cytoplasm containing a suspension of rubber and non-rubber particles in an aqueous medium (Frey-Wyssling, 1929; Southorn, 1961; Archer et al. 1969). Using high speed centrifugation (59,000 g), Cook and Sekhar (1953) separated latex into four fractions. These were: an upper white fraction of rubber cream, an orange or yellow layer containing Frey-Wyssling complexes, a colourless serum named C-serum and a greyish yellow gelatinous sediment the 'bottom fraction' consisting mainly of lutoids (Dickenson, 1969; Southorn, 1966).

Moir (1959) using differential staining and high speed centrifugation techniques, showed that the sedimentable material in latex did not consist wholly of one species of particle. By treating the latex with trace amounts of Janus Green B or neutral red before centrifugation he obtained eleven zones. Zone 1 corresponded to the 'top whitish fraction' of Cook and Sekhar (1953) which consists mainly of hydrocarbon particles. Zone 2 was a much smaller, translucent layer situated under the lowest end of Zone 1. Zone 3 was a suspension of rubber particles in the serum. Zone 4 was the yellow and orange layer of Cook and Sekhar (1953). The aqueous Zone 5 corresponded to C-serum and Zones 6 - 11 together were broadly equivalent to the 'bottom fraction'.

Pujarniscle (1968) later extended Moir's results and showed by isopycnic centrifugation of fresh latex on sucrose gradients that latex could be distinguished by thirteen fractions. However, the lutoids, as obtained by the above technique were found to be either damaged or seriously aggregated (Low, 1976). Ficoll was later proved to be a more superior gradient medium than sucrose. Diluted fresh latex was successfully separated on linear Ficoll gradients with minimal damage to the lutoids (Low and Wiemken, 1982; 1984).

Though 'bottom fraction' contains particles other than lutoids, it was observed that the main protein constituents of the 'bottom fraction' originated from the lutoids (Southern and Edwin, 1968). By repeated freezing and thawing or by ultrasonication of the bottom fraction, the lutoids can be disrupted. A further centrifugation of this disrupted bottom fraction, yields another serum, the B-serum. This represents the original fluid within the lutoids with perhaps minor contributions from the entrapped C-serum and from particles other than the lutoids (Southern and Edwin, 1968).

The organic non-rubber constituents of fresh latex are presented in Table 1. Bearing in mind the great variability of latex, these organic non-rubber constituents may vary both in composition and concentration, depending on various physiological and physical parameters.

Proteins

Of the several non-rubber constituents of latex, the one which has received most attention is proteins. The early literature on the proteins of Hevea has been reviewed recently (Tata, 1975; 1980b). The earliest report of the presence of proteins in Hevea latex was by Spencer (1908) who detected peroxidase and catalase activities in dialysed aqueous extracts of rubber sheets, and subsequently, in dialysed latex. Even though enzymes play a major role in the biological function in Hevea latex, discussions on the enzymology of Hevea latex are kept to a minimum in this chapter.

The total protein content in latex has been estimated to be about 1% (Archer and McMullen, 1961; Archer et al. 1963b; Tata, 1980a). However,

TABLE 1

Organic non-rubber constituents of latex.

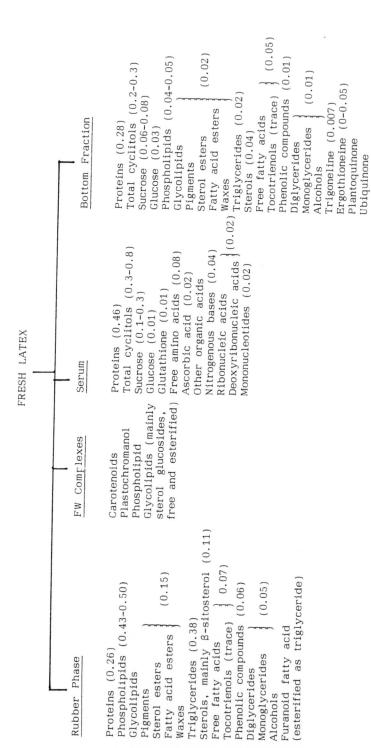

FRESH LATEX

Rubber Phase	FW Complexes	Serum	Bottom Fraction
Proteins (0.26)	Carotenoids	Proteins (0.46)	Proteins (0.28)
Phospholipids (0.43-0.50)	Plastochromanol	Total cyclitols (0.3-0.8)	Total cyclitols (0.2-0.3)
Glycolipids	Phospholipid	Sucrose (0.1-0.3)	Sucrose (0.06-0.08)
Pigments	Glycolipids (mainly	Glucose (0.01)	Glucose (0.03)
Sterol esters } (0.15)	sterol glucosides,	Glutathione (0.01)	Phospholipids (0.04-0.05)
Fatty acid esters	free and esterified)	Free amino acids (0.08)	Glycolipids
Waxes		Ascorbic acid (0.02)	Pigments
Triglycerides (0.38)		Other organic acids	Sterol esters } (0.02)
Sterols, mainly β-sitosterol (0.11)		Nitrogenous bases (0.04)	Fatty acid esters
Free fatty acids } 0.07		Ribonucleic acids	Waxes
Tocotrienols (trace)		Deoxyribonucleic acids } (0.02)	Triglycerides (0.02)
Phenolic compounds (0.06)		Mononucleotides (0.02)	Sterols (0.04)
Diglycerides } (0.05)			Free fatty acids
Monoglycerides			Tocotrienols (trace) } (0.05)
Alcohols			Phenolic compounds (0.01)
Furanoid fatty acid			Diglycerides
(esterified as triglyceride)			Monoglycerides } (0.01)
			Alcohols
			Trigoneline (0.007)
			Ergothioneine (0-0.05)
			Plantoquinone
			Ubiquinone

Figures within brackets indicate their approximate concentration in g/100 g latex.

discrepancies in the distribution of the proteins between the major phases of latex exist. Archer and McMullen (1961) reported that 20% of the total proteins was absorbed on the rubber surface, 66% in the C-serum and 14% in the bottom fraction. Later, reports variously described the distribution as 20%, 60% and 20% (Archer et al. 1963) and 27.2%, 47.5% and 25.3% (Tata, 1980) for the rubber phase, C-serum and bottom fraction respectively.

Proteins on the surface of the rubber particles: The existence of proteins in association with phospholipids on the surface of rubber particles was recognised as early as 1953 by Bowler (1953). He attributed that this protein-phospholipid layer imparted a net negative charge to the rubber particle, thereby contributing to the colloidal stability of these particles (Bowler, 1953). By measuring the iso-electric points of various latex samples, he concluded that there was more than one protein adsorbed on the rubber surface and that the relative proportions of the adsorbed proteins varied with clones.

Apart from the estimation that the protein adsorbed on rubber surface accounted for about 1% of the weight of rubber (Cockbain and Philpott, 1963) the proteins on the rubber surface remained unknown until recently. The major protein on the rubber surface has been shown to be negatively charged and has a molecular weight of approx. 65,000 (RRIM, 1982). It migrates towards the anode at a higher rate than the major C-serum protein α-globulin and contrary to the earlier suggestion, is therefore not identical with the latter.

Of the numerous enzymes reported in Hevea latex, only two have been found to be associated with the rubber surface. These isopentenyl pyro-phosphate polymerase (Lynen, 1967; Archer et al. 1963a) and rubber transferase (Lynen, 1967; Archer et al. 1963; Archer and Cockbain, 1969; McMullen and McSweeney, 1966; Archer et al. 1966). Their presence on the rubber surface is not surprising, since they are involved in rubber biosynthesis. What is perhaps more surprising is that so few have been detected on the rubber surface.

Proteins in the serum: Nearly half the enzymes examined in Hevea latex appeared to be located in the C-serum of latex. These include enzymes for the glycolytic pathway (Bealing, 1969; d'Auzac and Jacob, 1969) as well as many of the enzymes for rubber biosynthesis (Archer and Audley, 1967). Recently, twenty-seven enzymes were separated by electrophoresis by Jacob and co-workers, of which, seventeen were shown to exist in multiple forms (Jacob et al. 1978).

Until the development of the ultracentrifugal separation of fresh latex (Cook and Sekhar, 1953; Moir, 1959) the proteins investigated in the serum of _Hevea_ were mainly those which remained in the serum after acid coagulation or other treatments of latex (Bishop, 1927; Kemp and Twiss, 1936; Bondy and Freundlich, 1938; Kemp and Straitiff, 1940; Roe and Ewart, 1942). However, with the development of the ultracentrifugation techniques, seven protein components were demonstrated in the C-serum by Archer and Sekhar (1955) using paper electrophoresis. The same workers also confirmed the presence of seven protein components in the B-serum, ie. serum which is obtained by prolonged freezing and thawing of latex.

The first protein to be isolated from _Hevea_ latex was from C-serum. It was named α-globulin by Archer and Cockbain (1955). This protein is the major protein component of C-serum. It is readily adsorbed at a water-air or oil-water interface with a resulting fall in the interfacial tension. This led to the suggestion that α-globulin was one of the proteins on the surface of rubber particles and that it contributed to the colloidal stability of fresh latex (Archer and Cockbain, 1955). However, as mentioned earlier, α-globulin was later found not to be present on the surface of the rubber particles (RRIM, 1982).

With the introduction of more sensitive techniques, further discoveries on the proteins of C-serum were made. Using starch gel electrophoresis, Tata and Moir (1964) reported the presence of twenty-two protein bands in C-serum. Seventeen of these were anionic at pH 8.2, whilst five were cationic and existed in much lower concentrations. A comparative study on the proteins in the C sera from four clones viz. RRIM 501, GT 1, Tjir 1 and Pil A44, revealed very little differences between their general electro-phoretic patterns (RRIM, 1963). There also no significant difference in the proteins with seasonal variation within a single clone. Later, the list of proteins in C-serum was enlarged to twenty-four (Tata and Edwin, 1970), using the same starch gel electrophoretic technique. Using poly-acrylamide gel electrophoresis, Yeang et al. (1977) reported 26 protein bands from C-serum at alkaline pH and 15 bands at acid pH. These workers also did not observe significant differences in the protein patterns of C sera between clones (Tjir 1, PR 107, GT 1, PB 86 and BR 2), in agreement with the earlier conclusions from starch gel electrophoresis (RRIM, 1963).

Proteins in the bottom fraction: Proteins in the bottom fraction are essentially studied as the soluble proteins in B-serum. These have been examined with various techniques, including paper electrophoresis (Moir and Tata, 1960), starch gel electrophoresis (Tata, 1975; Tata and Edwin, 1969) and polyacrylamide gel electrophoresis (Yeang et al. 1977). Irrespective of the technique used, the proteins of B-serum were found to be markedly different from those of C-serum. Upon electrophoresis, the B-serum proteins were usually separated into two major protein bands at the extreme anionic and cationic ends, with several minor bands in between.

Hevein: The major protein in B-serum is hevein, which accounts for about 70% of the water soluble proteins in the bottom fraction (Archer et al. 1969). Hevein is a low molecular weight anionic protein (Approx. 5,000 daltons) with a higher (5%) sulphur content (Tata, 1975; Archer, 1960; Tata, 1976). All the sulphur in hevein exists as eight disulphide (S-S) bridges of cystine (Archer, 1960; Tata, 1976). Because of its low molecular weight and the large number of S-S bridges, hevein is heat stable, and is not precipitated by the common reagents for precipitating proteins eg. trichloroacetic acid (Tata, 1975; Tata, 1976). The molecular weight of hevein was first estimated to be about 10,000 ± 500 daltons by Archer (1960). Subsequent analysis showed that earlier preparations (Archer, 1960; Karunakaran et al. 1961) of hevein were mixtures containing hevein, traces of esterase and a protein with slightly less anionic mobility than hevein, termed pseudo-hevein (Tata, 1975; Tata, 1976). When pure hevein (free of pesudo-hevein) was isolated and characterised, it was found to be a single peptide chain with glutamic acid as the N-terminus and a molecular weight of approximately 5,000 daltons (Tata, 1975; Tata, 1976). (The molecular weight of pseudo-hevein was also 5,000 daltons). Later, an almost complete amino acid sequence of hevein was reported (Walujono et al. 1976). It contained 43 amino acid residues in a single polypeptide chain and an estimated molecular weight of 4729 daltons. A comparison of the three-dimensional structure of hevein with that of wheat germ agglutinin and ragweed pollen allergen (Drenth et al. 1980) showed that there were similarities in the position of the disulphide bridges (Walujono et al. 1976). The biological functions of hevein and pseudo-hevein are unknown. Although hevein resembles other proteinase inhibitors in general structural characteristics, it did not exhibit inhibitory activity against trypsin, chymotrypsin or carboxypeptidase (Walujono et al. 1976).

The microfibrillar protein: Dickenson (Dickenson, 1965; 1969; 1963) in his ultrastructural studies and electron microscopic investigations of lutoids, first described some fibrillar components having a tightly coiled helical structure, which he named microfibrils. These structures were observed within lutoids of young latex vessels but were absent from mature vessels. These microfibrils were later shown to be proteins containing upto 4% carbohydrate, and having an isoelectric pH of about 4 (Audley, 1965; 1966). At ambient temperature (20°C), the microfibrils break up into smaller segments which reassemble on freezing (Audley, 1965; 1966).

The microhelices: These structures were first observed by Dickenson (1963) in lutoids from mature trees. However, Dickenson described them as stretched microfibrils. Later, Southorn and Yip (1968) and Gomez and Yip (1974; 1975; 1976) carried out detailed investigations and reported that these zig-zag structures differed from microfibrils in that they were larger in dimensions and were open helcies (not lightly coiled helices of the microfibrils). They were called 'Microhelices' by Gomez and Yip (1975). Lowering of the ionic concentration of B-serum by dialysis against water or by dilution with water resulted in the formation of microhelices (Tata, 1975; Gomez and Yip, 1974, 1975, 1976). Furthermore, their formation required the combination of two glycoproteins in a certain ratio. These are an acidic 'assembly factor' (molecular weight 160,000) and a slightly basic 'pro-helical protein' (molecular weight 22,000) (Tata, 1975). A third glycoprotein termed the 'bundling factor' (molecular weight 5,000) appeared to promote the combination of single microhelices into bundles (Tata, 1980). The 'pro-helical protein' has some flocculating activity on suspensions of rubber particles in vitro. Microhelices are rarely seen in lutoids from young trees.

The basic proteins: The presence of basic proteins in B-serum was first demonstrated when B-serum or an aqueous extract of freeze-dried bottom fraction was electrophoresed (Tata and Edwin, 1970; Moir and Tata, 1960; Karunakaran et al. 1961). Two basic proteins - a major and a minor basic protein - which account for about 4% of the total proteins in latex were found to have lysozyme and chitinase activities (Tata, 1980; Tata et al. 1983). The Hevea lysozymes were found to have similar pH optima and molecular weight as lysozymes from papaya and fig, but their activities were different from avian lysozymes (Tata et al. 1983). The major basic protein has been crystallised and its molecular weight (approx. 26,000) determined. Its first 21 amino acid residues were elucidated, and found to differ significantly from those of hen egg, duck egg, baboon milk and T4 phage lysozymes (Tata et al. 1983). The major basic protein, also referred to earlier as 'band (i) first peak protein' (Tata, 1980) was found to be identical with heveamine A, a cationic protein described by Archer (Archer, 1976), another basic protein in B-serum.

Carbohydrates: The major soluble carbohydrates in Hevea latex are the total cyclitols, sucrose and glucose in that order (Low, 1978). It had been reported earlier that latex contained mainly sucrose and a smaller amount of raffinose (Tupy and Resing, 1969). However, Bealing (1969) later found that glucose, fructose and sucrose were the only free sugars present in latex in significant quantity and that other unidentified substrates (probably pentoses) sometimes detected in paper chromatograms had no quantitative significance. The particularly low fructose concentration in latex sera is a result of rapid metabolism of this sugar in preference to glucose (Bealing, 1969; d'Auzac and Jacob, 1967) by a specific hexokinase present in latex (d'Auzac and Jacob, 1969; Jacob and d'Auzac, 1967).

Quebrachitol is the predominant cyclitol in latex with smaller amounts of ℓ-and myo-inositols (Bealing, 1969). The distribution and concentration of the major soluble carbohydrates in latex have been described (Low, 1978). The concentration of total cyclitols ie. quebrachitol, ℓ- and m-inositols appear to vary with clones and are in the range of 13.0-32.0 mg per ml of C-serum. Like total cyclitols, the concentration of sucrose in C-serum also varies with clones, and is usually between 4.0-10.5 mg per ml serum in the five clones examined. Total cyclitols and sucrose are confined mainly to C-serum whilst glucose is located mainly in the lutoids. The finding that sucrose and total cyclitols are located mainly in C-serum is hardly surprising since the enzymes for carbohydrate metabolism are present in C-serum (Bealing, 1969; d'Auzac and Jacob, 1969).

Lipids and phospholipids: Lipids and phospholipids associated with the rubber and non-rubber particles in latex play a vital role in the stability and colloidal behaviour of latex. Earlier studies (Cockbain and Philpott, 1963; Blackley, 1966) demonstrated that the rubber particles are strongly protected by a complex film of protein and lipid material. It is believed that some of the lipids are present within the rubber particles. The concentration and distribution of lipids between the rubber cream and the bottom fraction had been studied (Ho et al. 1976). These lipids were isolated and divided into neutral lipids and phospholipids for further analysis. There appeared to be distinct clonal variation in the total amount of neutral lipids extractable from rubber cream and from bottom fraction. Colloidal stability of latex was found related to the natural lipid content of rubber particles (Sheriff and Sethuraj, 1978). Lipids from different clones, however, were qualitatively similar. Triglycerides and sterols were the main components of the neutral lipids of rubber particles, whilst sterols and long-chain free fatty acids mainly made up the neutral lipids of the bottom fraction. More recently, a furanoid fatty acid containing a methylfuran group was found mainly in the triglyceride fraction of the neutral lipids (Hasma and Subramaniam, 1978). It constituted about 90% of the total esterified acids. It was suggested that the main triglyceride in Hevea latex contained three furanoid fatty acids, hence making it a rare triglyceride known in nature. The phospholipid content of the rubber particles (approx. 1% on the dry weight of rubber) was similar between different clones. The total phospholipid content of the bottom fraction was much less (only about 10%) than that in the rubber cream. It was suggested that the amount of neutral lipid (especially triglycerides) associated with the rubber particles was inversely related to the plugging index of the clone which the latex originated from (Ho et al. 1976). Lutoid stability, as indicated by bursting index, was found to be negatively correlated with the phospholipid content of the bottom fraction of latex (Sheriff and Sethuraj, 1978).

A systematic study of the glycolipids from natural rubber was reported (RRIM, 1980). The glycolipid fraction was found to consist mainly of esterified sterol glucoside (ESG), monogalactosyldiglyceride (MGDG), sterol glucoside (SG) and digalactosyldiglyceride (DGDG). The sterol attached to ESG and SG was mainly β-sitosterol, while the acid components of ESG, MGDG and DGDG were of 14:0; 16:1; 18:0; 18:1; 18:2; 18:3 and furanoic acids. The constituents of the phospholipids are mainly phosphatidyl ethanolamine (PE), phosphatidyl choline (PC) and phosphatidyl inositol (PI).

Nucleic acids and polysomes

The presence of nucleic acids in latex was discovered by McMullen (McMullen, 1962) and confirmed by Tupy (1969a). According to Tupy, Hevea latex contains both ribosomal RNA and soluble RNA, DNA and messenger RNA. These are all present in the serum fraction of latex. Later, functional polysomes were also discovered in the serum phase of latex (Coupe and d'Auzac, 1972). More recently, ribonucleic acid (Marin and Trouslot, 1975) and ribosomes (Marin, 1978) have been found to be located in lutoids. The lutoid ribosomes represent 11.9% of the total ribosomal content of Hevea latex. Two high molecular weight RNA components have also been identified and their nucleotide base composition determined (Tupy, 1969a). The presence of these membrane-bound ribosomes in lutoids led to the speculation that these ribosomes were transported from the groundplasm to the lutoids (which are also lysosomes) where they are rapidly destroyed (Marin, 1978).

PHYSIOLOGY OF LATEX FLOW

Response to tapping: When a tree is brought into tapping for the first time, the latex extruded from the vessels is viscous and contains a high density of rubber particles. The flow of latex is curtailed after a short time. Subsequent tappings at regular intervals result in increased yield due to longer duration of flow and more dilute latex until it reaches more or less a steady equilibrium. The increase in yield before reaching a state of equilibrium was termed by early workers (Pakianathan, 1967;

Pakianathan and Milford, 1977) as wound response. Thus, regular and controlled tapping not only increases the time of flow but also enhances the biosynthesis of rubber in the drained vessels below the tapping cut.

Water relations of latex vessels: Turgor pressures within latex vessels in the early morning hours range from 7.9 to 15 atmospheres. Pakianathan and Milford (1977) using a vapour pressure osmometer, obtained values of 10-12 atmospheres on drop samples of latex. Diurnal turgor and osmotic pressure measurements taken at various intervals from 0530 to 1900 h showed maximum turgor values at 0530 h whereas, maximum osmotic pressure values were recorded between 1300 to 1600 h (Buttery and Boatman, 1966). The extent of dilutions, five minutes after the tapping had commenced, were 24.7, 18.8 and 12.1%, for trees tapped at 0400 h, 0800 h and 1230 h respectively. The diffusion pressure deficit was highest in trees tapped at 1230 h. Trees tapped at 0400 h yielded more latex than those tapped at 0830 h or 1230 h (Buttery and Boatman, 1966). Thus, it appeared that latex production was largely influenced by the internal water relations of the tapping panel. These observations showed that latex vessels behaved as a relatively simple osmotic system. Turgor pressure falls during the day as a result of withdrawal of water under transpirational stress (Pakianathan, 1967; Buttery and Boatman, 1964).

Seasonal variations in yield, yield components and components of water relations were studied in some Hevea clones by Gururaja Rao et al. (1988) and Devakumar et al. (1988). Summer yield drops were found to be low in clones like RRII 105 and Gl 1. High latex vessel turgor and low solute potentials in the dry season in clone RRII 105 indicate the presence of osmotic adjustment. Higher plant water status and lowered transpiration rates in this clone might help in maintaining better turgor. Their studies also indicate that low transpiration coefficients are associated with high yields and drought tolerance in RRII 105 and Gl 1.

Events following tapping: On tapping, release of pressure occurs to a greater extent in the latex vessels than in the surrounding tissues. This results in a rapid elastic expulsion of latex flow through the vessels along the pressure gradient. The gradient is highest near the cut and becomes smaller with increasing distance away from the tapping cut. Frey-Wyssling (1952) and Riches and Gooding (1952) made extensive studies on the mechanism of latex flow and cessation of flow. Further work by Boatman (1966) and Buttery and Boatman (1967) demonstrated that flow is rapidly restricted by plugging of the vessels at or near the cut surface and this was usually the major factor causing a decline in the flow rate.

Plugging index

Paardekooper and Samosorn (1969) showed that the latex flow characteristics can be empirically defined by the expression $y = b.e^{-a.t}$ in which the flow rate   y at a given time t is a function of the initial flow rate b (at zero time) and a time flow constant a. The time flow constant may be regarded as an index of plugging (Milford et al. 1969). This was expressed as the ratio of the flow for the initial five minutes to the total yield. For ease of handling, this ratio was multiplied by 100 (Paardekooper and Samosorn, 1969).

Mechanism of vessel plugging: It is clear that latex contains de-stabilising factors normally located in the lutoid particles. Consequently, any physiological or biochemical factor which affects the stability of the lutoids would undoubtedly affect the latex flow and plugging of the vessels. By repeated reopening of the tapping cut, Boatman (1966) demonstrated that flow was restricted rather rapidly by some process occuring at or near the surface of the cut. Pakianathan et al. (1966) observed flocs of damaged lutoids in tapped latex and suggested that dilution of latex during flow might damage the osmotically sensitive lutoids and provide a possible mechanism of latex vessel plugging. Electron microscopical observation of the ends of the tapping cut revealed both a cap of coagulum on the surface of the cut and internal plugs within the latex vessels (Southorn, 1968a). Lutoid counts taken before tapping and at various intervals during flow showed a rapid loss during the initial thirty minutes of flow indicating that lutoids were trapped on the cut surface and initial cap formation during the early stages of flow. Shear may play an important part in lutoid damage. Internal plugging occurs mainly during the fast initial flow whereas coagulation on the surface of the cut is effective when the flow is slow. It seems that there is no substantial reason to suppose that the two types of sealing processes are separated in time (Southorn, 1968b). (See Chapter 7. Eds.).

TAPPING PANEL DRYNESS (BROWN BAST)

Brown bast or Tapping Panel Dryness (TPD) is a syndrome encountered in rubber plantations, characterised by spontaneous drying up of the tapping cut resulting in abnormally low yield or stoppage of latex production. The disease was reported for the first time in Brazil in 1887 in Hevea in the Amazon forest and at the beginning of the century in plantations in Asia (Rutgers and Dammerman, 1914).

Symptoms

The symptoms range from partial dryness with no browning of the tapping cut, browning and thickening of the bark and cracking and deformation of the bark in some instances. The syndrome is characterised by the appearance of tylosoids and the coagulation of latex in situ (de Fay, 1981; de Fay and Hebant, 1980; Paranjothy et al. 1976). abnormal behaviour of the parenchyma cells adjoining the laticifers and general increase in synthesis of polyphenols (Rands, 1921). A detailed review of the histological, histochemical and cytological study of the diseased bark was presented by de Fay and Jacob (1989).

Investigations on causative organisms

The involvement of a causative organism was doubted by early workers (Keuchenius, 1924; Rands, 1921; Sharples, 1922). But these workers were unable to demonstrate the existence of an agent responsible for causing tapping panel dryness. Later the possibility of certain types of cortical necrosis which leads to stopping of flow through some pathogenic causes was reported by Nandris et al. (1984), Peries and Brohier (1965) and Zheng Guanbiao et al. (1982). Though rickettsia-like organisms (RLO) was implicated by Zheng Guanbiao et al. (1988), no confirmatory evidence could so far be made available.

Soil, climatic and clonal characters in relation to TPD

Influence of climate and growth period on the incidence of brown bast disease was reported by early workers (Harmsen, 1919; Vollema, 1949; Compagnon et al. 1953; Bealing and Chua, 1972). Through the analysis of soil, leaves and latex, the effect of unbalanced nutrition favouring the incidence of disease was reported by Pushpadas et al. (1975). Clonal sensitivity to tapping panel dryness was observed by many workers (Bangham and d'Agremond, 1939; Dijkman, 1951; Heusser and Holder, 1930; Ostendorf, 1941; Vollema and Dijkman, 1939).

Biochemical and physiological studies

The biochemical and biophysical changes take place at the later stage of this syndrome. The most common symptoms include a phase of excessive late dripping of latex and a simultaneous fall in the rubber content and after a period of time, the volume per tapping gradually declines. The colloidal stability of the latex will also be reduced resulting in particle damage, flocculation of rubber particles in situ, and early plugging of latex vessels (Chrestin et al. 1985). A reduction in turgor pressure

(Sethuraj et al. 1977), change in latex flow pattern (Sethuraj, 1968) and a sharp increase in bursting index (Eschbach et al. 1983) were also reported.

According to Chua (1967) the reserves of starch and other soluble carbohydrates are not depleted. Recent investigations by de Fay (1981) reported abundance of starch grains in the wood of affected trees and the vascular rays were reported to function normally.

The existence of an endogenous NAD(P)H oxidase in lutoids which generates toxic forms of oxygen ($O_2^{\bullet}$, H_2O_2, $OH^{\bullet}$) responsible for the per-oxidase degradation of organelle membranes in the latex from diseased trees was reported by many workers (Chrestin, 1984; Chrestin et al. 1984; Chrestin, 1985; Cretin and Bangratz, 1983). Simultaneously, decrease in concentrations of latex cytosol scavengers (reduced thiols and ascrobate) (Chrestin, 1984) as well as virtual disappearance of scavenging enzyme activities (SOD and catalase) (Chrestin, 1984, 1985) was reported. The combination of increased peroxidative activities and considerably diminished quantities of scavengers in latex from affected trees result in destabilisation and lysis of lutoids leading to coagulation (Chrestin, 1989). The possible damage to all the membrane structures in latex cells and resulting impairment of nutrient supply and water exchange at plasmalemma was suggested by Chai Kim Chun et al. (1969) and Pushpadas et al. (1975).

High intensity of exploitation is known to promote incidence of tapping panel dryness in plantations; the proportion of dry trees increases with tapping intensity and particularly with tapping frequency (Bealing and Chua, 1972; Chua, 1967; Paranjothy et al. 1977). The intensive exploitation is reported to result in excessive outflow of latex and consequent nutritional stress (Chua, 1967; Schweizer, 1949; Sharples and Lambourne, 1924; Taylor, 1926), inadequate organic resources (Chua, 1966; Tupy, 1984), and Cu and K deficiency (Compagnon et al. 1953). Changes in mineral ratios, especially of K_2O/CaO and Mg/P was reported by Beaufils (1957). Also, an increase in K content and K/Ca, K/P ratios in latex was observed by Pushpadas et al. (1975).

Certain forms of bark dryness are transitory and do not display the characteristic symptoms of the formation of tylosoids or activation of the phenolic metabolism (de Fay and Jacob, 1989). Numerous traumatism (mechanical such as tapping, chemical or pathological infection) cause formation of ethylene (Yang and Pratt, 1978) and its influence in bio-chemical, anatomical and histological phenomena is proved (Liebermann, 1973). Over stimulation (dose and frequency) or over tapping can lead to excessive endogenous ethylene production and deleterious effect on cellular systems (Chrestin, 1984, 1985). Induction of bark dryness through deliberate over stimulation with ethrel results in imbalance in peroxidase activities and consequently disorganisation of the membrane structures. This may lead to the onset of bark dryness.

According to Eschbach et al. (1986) a reduction in sucrose, thiol and Mg contents and increase in redox potential (RP) are connected with a higher rate of bark dryness. The reduced availability of assimilates and the essential enzyme systems may be the principal cause of more frequent occurrence of the disease.

Incidence of TPD can be reduced by reducing the exploitation intensity. Tapping rest imposed for varying periods may revive certain trees, but in majority of cases reoccurrance of the syndrome is encountered.

Recent thinking centres round the question why only certain percentage of trees in a monoclonal population get affected. The involvement of the genetics of root stock has been implicated and this aspect will receive adequate attention in the international network research programme envisaged by the International Rubber Research and Development Board (Sethuraj, 1989).

PROMOTION OF LATEX FLOW WITH CHEMICALS

A wide variety of chemicals ranging from simple inorganic salts to complex organic compounds bring about stimulation of latex yield in Hevea (Blackman, 1961). Periodic scraping of the outer layers of bark caused stimulation resulting in higher yields of latex (Kamerun, 1912). Synthetic auxins such as 2,4-dichlorophenoxyacetic acid (2,4-D) and 2,4,5-trichloro-phenoxyacetic acid (2,4,5-T) applied to scraped areas of bark below the tapping cut resulted in marked increase of latex flow (Chapman, 1951; Baptiste and de Jong, 1953). Copper sulphate, when injected into trees, also caused stimulation (Mainstone and Tan, 1964). A number of workers have tested a variety of chemicals which included growth regulators, herbicides, carbonates, nitrides, bactericides, cyclic and acyclic olefines and halogenoparaffins to study their effects (Pakianathan, 1970; Taysum, 1961).

Increase in yields have also been obtained by the application of gases to the trunk of the tree (Banchi, 1968; Taysum, 1961). How is it that such a wide range of chemicals with completely different molecular structures could bring about yield stimulation activity? The common feature of all these chemicals which induce yield stimulation activity is that they produce ethylene either directly, or through injury of tissues. Some other chemicals have the ability to induce the production of ethylene in the tissues. The ethylene-releasing compound commercially known as 'Ethrel' or 'Ethephon' has been found to be very effective in prolonging flow which resulted in marked increases in yields.

Mechanism of action of yield stimulants

Boatman (1966) and Buttery and Boatman (1967) showed that 2,4,5-T delayed the plugging. Pakianathan et al. 1966) observed that the proportion of damaged lutoids in latex samples collected during flow increased rather than decreased following 2,4,5-T treatment. In trees treated with ethephon or 2,4,5-T the loss of lutoids did not commence until 45 minutes after tapping, but thereafter increased until by the end of flow nearly 50% of the lutoids were lost compared with unstimulated trees. These observations in some way indicate that yield stimulants tend to delay the retention of lutoids by the cut vessel ends and hence delay plugging and prolong flow (Pakianathan and Milford, 1977).

Application of 2,4,5-T and ethephon result in an increase in yield and decrease in plugging index. The initial flow rates were not affected, and the yield difference was a consequence of extension of flow time (Abraham et al. 1968). Treatment of Hevea bark with 2,4-D or Naphthalene acetic acid (NAA) increased sucrose level, invertase activity, and sucrose utilisation in the latex.

Stimulation results in an increase in the displacement area (Lustinec and Resing, 1965; Pakianathan et al. 1975). This suggests a localised reduction in the vicinity of the tapping cut to form plugs.

Osborne and Sargent (1974) suggested a probable mechanism of promotion of flow by ethylene. Ethylene treatment possibly results in wider latex vessels with thicker, more rigid walls. This in turn leads to less constriction of severed vessels after tapping, reduced shearing forces at the orifice, a lower proportion of damaged lutoid particles, and less enzymic coagulation of rubber particles. The delayed occlusion of the vessel orifice results in prolonged flow.

THE ROLE OF CARBOHYDRATES IN LATEX FLOW

Relatively little is known about the role of carbohydrates in latex flow. Because of its high concentration in latex (1-2%) (Rhodes and Wiltshire, 1931) quebrachitol has been suggested to contribute as much as over 30% of the total osmotic pressure of latex serum (Sheldrake, 1973). The osmotic role of quebrachitol was further emphasised when quebrachitol (measured as total cyclitols, together with lesser amounts of ℓ- and m-inositols) was found to be present, in high concentrations, in a high plugging clone viz.

Tjir 1 and in low concentrations in a low plugging clone viz. RRIM 501 (Low, 1978). A high total cyclitols concentration would lead to a faster and higher dilution reaction during latex flow, resulting in possibly greater lutoid damage, a faster plugging reaction and an earlier cessation of flow. Similarly, clones with low total cyclitols content probably experience a slower dilution reaction, less lutoid damage, a slower plugging reaction and consequently, a more prolonged flow. The concentration of total cyclitols in latex was shown to decrease as the flow proceeded (Low, 1981). This lowering of total cyclitols concentration was similar to the decrease in total solids content in latex in the same latex samples, implying the dilution of total cyclitols during latex flow. Since the osmotic concentration of latex decreases in the progressively collected drop fractions (Pakianathan, 1967) after tapping (and hence flow time), the fall in total cyclitols concentration with flow must be a reflection of the osmotic role of total cyclitols in latex, as suggested earlier (Low, 1978; Sheldrake, 1973).

Sucrose is the next major soluble carbohydrate in latex, after quebrachitol (Low, 1978) and it is plausible that sucrose also contributes to the osmolarity of latex. However, the concentration of sucrose is small compared to that of total cyclitols. Therefore, the osmotic role of sucrose must be minor compared to that of total cyclitols in general and quebrachitol in particular (Low, 1978).

Recently, cell sap has been shown to exhibit strong destabilisation activity (Gomez, 1977; Yip and Gomez, 1978, 1984) on latex particles. This may be important in the cap coagulum formation over the vessel ends, during the later stages of flow (Yip and Gomez, 1984) resulting in the final cessation of flow. There is experimental evidence to suggest that simple sugars such as monosaccharides may delay the destabilisation effect of the cell sap (RRIM, 1980). Though the role of monosaccharides in the destabilisation activity of cell sap towards latex particles is not clearly understood at present, it may nonetheless be important in the final cessation of latex flow after tapping.

CARBOHYDRATES AND RUBBER BIOSYNTHESIS

Carbohydrates, particularly the hexoses in latex, are probably the primary source of acetate and acetyl-CoA, essential for the biosynthesis of rubber in Hevea. Hence, a discussion on the role of carbohydrates in rubber production is included in this chapter. (For detailed account on biosynthesis, please see Chapter 6. Eds.).

The importance of carbohydrates in rubber production had been studied in relation to various aspects. These include changes in the levels of carbohydrates in latex (Low, 1968; Low and Gomez, 1982) and bark (Low and Gomez, 1984) in response to yield stimulation by ethephon, the effects of stimulation and exploitation on invertase activity in latex and the correlation of latex invertase activity with latex vessel plugging in Hevea (Yeang et al. 1984) Ethephon stimulation was shown to result in a decrease in sucrose and total cyclitols concentrations in latex (Low, 1968; Low and Gomez, 1982). The decline in sucrose concentration in latex after ethephon stimulation usually preceeded a similar decline in total cyclitols concentration in latex (Low and Gomez, 1982). In the unexploited tree, gentle gradients of starch and total sugars appeared to exist in opposite directions in the bark (Low and Gomez, 1984). Initiation of exploitation on previously un-exploited trees resulted in a significant lowering of bark starch at the tapping panel, creating a 'source-sink' situation in the tree. The depression of bark starch with continued exploitation was however, not continuous. After a certain period, a lowered level of bark starch was maintained (Low and Gomez, 1984). Ethephon stimulation enhanced this depression of bark starch. The depletion of bark starch was confined to the tapped panel and did not extend to the opposite untapped panel or to an untapped region above the tapped panel (Low and Gomez, 1984). Hence, the benefits of panel changing, in relation to the exhaustion of bark carbohydrates on the tapped panel, are immediately obvious.

The importance of invertase in latex physiology and production has been widely studied (Yeang et al. 1984; Tupy, 1969b, 1973b, 1973c). Invertase, by virtue of its extremely low activity (d'Auzac and Jacob, 1969), has been suggested as a possible pacemaker enzyme in the glycolytic pathway of Hevea latex. Since the utilisation of sucrose, the primary carbo-hydrate substrate in latex, is controlled by invertase (Tupy, 1969b), it is plausible that invertase also controls the overall metabolic activity of latex (Tupy, 1973b). Ethephon stimulation on previously unstimulated trees, resulted in an increased invertase activity in latex. However, repeated ethephon stimulation caused an eventual decline in the invertase activity, which at times, was lower than that of the unstimulated controls, even though the yield response was positive (Yeang et al. 1984). Contrary to the findings of Tupy (1973) invertase activity was not always correlated with latex production (Yeang et al. 1984). Invertase activity was, however, found to be negatively correlated with the inclination to latex vessel plugging (Yeang et al. 1984).

The feasibility of using latex sucrose levels as an early warning signal to indicate overexploitation and subsequent dryness of the Hevea trees was examined. This investigation was based on the equivocal assumption that latex with a higher sucrose content was a reflection of the better

315

physiological status of the tree (Leong and Tan, 1978). Based on this criterion, the superiority of microtapping as compared to conventional tapping was assumed since microtapped latex was reported to contain a higher sucrose content than the latex from conventional tapping (Leong and Tan, 1978; Tupy, 1973a; Gener et al. 1977; Tupy and Primot, 1974; Primot and Tupy, 1976; Leong et al. 1976; Ramachandran, 1978; Ramachandran and Lee, 1980). However, other workers had demonstrated that the microtapped latex did not contain more sucrose than in the latex latex from conventional tapping (Samosorn et al. 1978; Low et al. 1983). Possible reasons for this discrepancy had been discussed (Low et al. 1983) and in the present uncertainty, it may not be prudent to evaluate the physiological health of the Hevea tree solely on the basis of its sucrose levels in latex.

Many non-rubber compounds emerge along with rubber in the latex when a tree is tapped, and the tree has to replace these compounds by synthesis. Among these compounds the very important ones are the proteins (enzymes) and the nucleic acids. The latex vessels have the machinery for synthesizing specific proteins. The synthesis is dependent upon the presence of the nucleic acids in the cells: DNA, transfer RNA (t-RNA), ribosomal RNA (r-RNA) and messenger RNA (m-RNA). Except DNA, all these nucleic acids have been found in latex (McMullen, 1959, 1962; Dikenson, 1965; Tupy, 1969). Most of the nuclei are left behind and do not come out with latex on tapping and this may be the reason for the lack of DNA in tapped latex.

Some correlation between the yield of rubber and the amount of nucleic acid in latex has been reported (Tupy, 1969) and it has been claimed that latex from high yielding trees synthesises nucleic acids much faster in vitro than latex from low yielding trees of the same clone. Furthermore, when a tree is intensively tapped, the nucleic acids at first decline in amount but are then resynthesised and quickly rise to a concentration well above that in trees tapped at normal frequencies. Nucleic acids and proteins, therefore, play an important role in the production of rubber and thus in the final yield.

ACKNOWLEDGEMENT

We wish to acknowledge the assistance of Encik-Encik Aw Kim Fatt, C. Tharmalingam, Choo Gee Tiem and Puan Paramaswari for Checking and preparation of the manuscript.

Abraham, P.D., Wycherley, P.R., Pakianathan, S.W. 1968. Stimulation of latex flow in Hevea brasiliensis of 4-amino-3,5,6-trichloropicolimic acid and 2-chloroethanephosphoric acid. J. Rubb. Res. Inst. Malaya, 20: 291-305.

Archer, B.L. 1960. The proteins of Hevea brasiliensis latex. 4. Isolation and characterisation of crystalline 'hevein'. Biochem. J., 75: 236.

Archer, B.L. 1976. Hevamine: a crystalline basic protein from Hevea brasiliensis latex. Phytochemistry, 15: 297.

Archer, B.L. and Audley, B.G. 1967. Biosynthesis of rubber. Adv. in Enzymol and related areas of mol. bil., 29: 221.

Archer, B.L., Audlye, B.G., Cockbain, E.G. and McSweeney, G.P. 1963a. The biosynthesis of rubber: Incorporation of mevalonate and isopentenyl pyrophosphate into rubber by Hevea brasiliensis latex fractions. Biochem. J., 89: 565.

Archer, B.L., Audley, B.O.G., McSweeney, G.P. and Tan, G.H. 1969. Studies on composition of latex serum and bottom fraction particles. J. Rubb. Res. Inst. Malaya, 21: 560.

Archer, B.L., Barnard, D., Cockbain, E.G., Cornforth, J.W., Cornforth, R.H. and Popjak, G. 1966. The stereochemistry of rubber biosynthesis. Proc. Roy. Soc. B., 163: 519.

Archer, B.L., Barnard, D., Cockbain, E.G., Dickenson, P.B. and McMullen, A.I. 1963b. Structure, composition and biochemistry of Hevea latex In: L. Bateman (Ed.). The Chemistry and Physics of Rubber Like Substances. London, Maclaren and Sons Ltd. p. 41.

Archer, B.L. and Cockbain, E.G. 1955. The proteins of Hevea brasiliensis latex (2) Isolation of the α-globulin of fresh latex serum. Biochem. J., 61: 508.

Archer, B.L. and Cockbain, E.G. 1969. Rubber transferase from Hevea brasiliensis latex. Meth. Enzymol., 15: 476.

Archer, B.L. and McMullen, A.I. 1961. Some recent studies of the non-rubber constituents of natural rubber latex. In: Proc. Nat. Rubb. Res. Conf. Kuala Lumpur, 1960, 787 pp.

Archer, B.L. and Sekhar, B.C. 1955. The proteins of Hevea brasiliensis latex. I. Protein constituents of fresh latex serum. Biochem. J., 61 503.

Audley, B.G. 1965. Studies of an organelle in Hevea latex containing helical protein microfibrils. In: L. Mullins (Ed.). Proc. Nat. Rubb. Prod. Res. ASsoc. Silver Jubilee Conf. Cambridge, 1964. Maclaren and Sons Ltd., London. p. 67.

Audley, B.G. 1966. The isolation and composition of helical protein microfibrils from Hevea brasiliensis latex. Biochem. J., 98: 335.

Banchi, Y. 1968. Effects de l acetylene sur la production on latex de l' Hevea brasiliensis. Arch. Inst. Resh. Caoutch. Vietnam No. 2/68.

Bangham, W.N. and d'Agremond, A. 1939. Tapping results on some new AVROS Hevea clones which originated in cross-pollination. Arch. Rubbercult., 23: 191.

Baptiste, E.D.C. and de Jonge, P. 1953. Stimulation of yield in Hevea brasiliensis. II. Effect of synthetic growth substances on yield and bark renewal. J. Rubb. Res. Inst. Malaya, 14: 362.

Bealing, F.J. 1969. Carbohydrate metabolism in Hevea latex - availability and utilisation of substrate. J. Rubb. Res. Inst. Malaysia, 21: 445.

Bealing, F.J. and Chua, S.E. 1972. Composition and metabolic activity of Hevea latex in relation to tapping intensity and the onset of brown bast. J. Rubb. Res. Inst. Maalya, 23: 204.

Blackley, D.C. 1966. High polymer latices, Vol. 1. Chapter IV, Maclaren and Sons Ltd., London. pp. 214-237.

Blackman, G.E. 1961. The stimulation of latex flow by plant growth regulators. In: Proc. Nat. Rubb. Res. Conf. 1960, Kuala Lumpur, 16 pp.

Bishop, R.O. 1927. Studies of Hevea latex. VI. The proteins in serum from frozen latex. Malay agric. J., 15: 27.

Boatman, S.G. 1966. Preliminary physiological studies on the promotion of latex flow by plant growth substances. J. Rubb. Res. Inst. Malaya, 19: 243-258.

Bondy, C. and Freundlich, H. 1938. The proteins in preserved Hevea latex. The Rubber Age, 42: 377.

Bowler, W.W. 1953. Electrophoretic mobility of fresh Hevea latex. Ind. Eng. Chem., 45: 1790.

Buttery, B.R. and Boatman, S.G. 1964. Turgor pressure in phloem: measurement on Hevea latex. Science, 145: 285-286.

Buttery, B.R. and Boatman, S.G. 1966. Manometric measurement of turgor pressures in laticiferous phloem tissues. J. Exp. Bot., 17: 283-296.

Buttery, B.R. and Boatman, S.G. 1967. Effect of tapping, wounding and growth regulators on turgor pressure in Hevea brasiliensis. Mull. Arg. J. Exp. Bot., 18: 644-659.

Bryce, G. and Campbell, L.E. 1917. On the mode of occurrence of latex vessels in Hevea brasiliensis. Dept. Agric. Ceylon Bull., 30: 1.

Chai, Kim Chun, J., Tupy, J. and Resing, W.L. 1969. Changes in organo-mineral composition and respiratory activity of Hevea latex, associated with intense tapping. J. Rubb. Res. Inst. Malaya, 2: 184.

Chapman, C.W. 1951. Plant hormones and yield in Hevea brasiliensis. J. Rubb. Res. Inst. Malaya, 13: 167.

Chrestin, H. 1984a. Le compartiment Vacuo-Lysosomal (les Lutoides) du latex d'Hevea brasiliensis, son role dans le maintein de l'homeostasie et dans les processus de senescence des cellules laticiferes, These Doct. Etat, Universite Montpellier 2, Montpellier, France (USTL).

Chrestin, H. 1984b. Biochemical basis of bark dryness, C.R. Coll. Exp. Physiol. Amel. Hevea, IRCA-GERDAT, Montpellier, France, 273 p.

Chrestin, H. 1985. La vacuole dans l'homeo stasie et la senescence des cellules laticifers d'Hevea, Etudes et Theses, ORSTOM, Ed., Paris.

Chrestin, H. 1989. Biochemical aspects of bark dryness induced by over-stimulation of rubber trees with ethrel. In: J. d'Auzac, J.L. Jacob and H. Chrestin (Eds.). Physiology of Rubber Tree Latex. CRC Press, Boca Raton, Florida. pp. 432-439.

Chua, S.E. 1966. Physiological changes in Hevea brasiliensis tapping panels during the induction of dryness by interruption of phloem transport. 1. Changes in latex. II. Changes in bark. J. Rubb. Res. Inst. Malaya, 19: 277.

Chua, S.E. 1967. Physiological changes in Hevea trees under intensive tapping. J. Rubb. Res. Inst. Malaya, 20: 100.

Cockbain, E.G. and Philpott, M.W. 1963. Colloidal properties of latex in L. Bateman (Ed.). The Chemistry and Physics of Rubber-like Substances. Maclaren and Sons Ltd., London. p. 73.

Compagnon, P., Tixier, P. and Roujansky, G. 1953. Contribution a' l'eude des accidents physiologiques de saignee. Arch. Rubbercult. Ext. Numb., 54: 153.

Cook, A.S. and Sekhar, B.C. 1953. Fractions from Hevea brasiliensis latex centrifugation at 59,000 g. J. Rubb. Res. Inst. Malaya, 14: 163.

Coupe, M. and d'Auzac, J. 1972. Demonstration of functional polysomes in the latex of Hevea brasiliensis (Kunth) Mull. Arg. Comptes rendues de 1 Academic des Science de Paris, 274, series D. 1031.

Cretin, H. and Bangratz, J. 1983. Une activite enzymatique endogene (NAD(P)H dependante), responsible de la degradation peroxidative des organites membranaires et de la coagulation prococe, on in situ, due latex d'Hevea brasiliensis, C.R. Acad. Sci. Paris, Ser. III, 296, 101 pp.

d'Auzac, J. and Jacob, J.L. 1967. Sur le catabolisme glucidique du latex d'Hevea brasiliensis. Rapp. ech. N 60, Inst. for Caoutch.

d'Auzac, J. and Jacob, J.L. 1969. Regulation of glycolysis in latex of Hevea brasiliensis. J. Rubb. Res. Inst. Malaya, 21: 417.

Devakumar, A.S., Gururaja Rao, G., Rajagopal, R., Sanjeeva Rao, P., George, M.J., Vijayakumar, K.R. and Sethuraj, M.R. 1988. Studies on soil-plant-atmosphere system in Hevea. II. Seasonal effects on water relations and yield. Indian J. Nat. Rubb. Res., 1: 45-60.

Dickenson, P.B. 1963. In: L. Bateman (Ed.). The Chemistry and Physics of the Rubber-like Substances. Maclaren and Sons Ltd., London. 43 pp.

Dickenson, P.B. 1965. The ultrastructure of the latex vessel of Hevea brasiliensis. In: L. Mullins (Ed.). Proc. Nat. Rubb. Prod. Res. Assn. Jubilies Conf. Cambridge, 1964, London. Maclaren and Sons Ltd. 52 pp.

Dickenson, P.B. 1969. Electron microscopical studies of latex vessel system of Hevea brasiliensis. J. Rubb. Res. Inst. Malaya, 21: 543.

de Fay, E. 1981. Histophysiologie comparee des Ecorces saines et pathologiques (Maladie des Encoches Seches) de l'Hevea brasiliensis. These Doc. Third cycle. Universite Montpellier, 11 (USTL), Montpellier, France.

de Fay, E. and Hebant, C. 1980. Etude histologique des ecorces d'Hevea brasiliensis atteintes de la maladie des encoches seches, C.R. Acad. Sci. Paris Ser. D. 291, 867.

de Fay, E. and Jacob, J.L. 1989. Symptomatology, histological and cyto-logical aspects. In: J. d'Auzac, J.L. Jacob and H. Chrestin (Eds.). Physiology of Rubber Tree Latex. CRC Press, Boca Raton, Florida, pp. 408-428.

Dijkman, J. 1951. Thirty years of research in the Far East. University of Miami Press, Miami.

Drenth, J., Low, B.W., Richardson, J.S. and Wright, C.S. 1980. The toxin agglutinin organised around a four disulphide core. J. Biol. Chem., 255: 2652.

Eschbach, J.M., Lacrotte, R. and Serres, E. 1989. Conditions which favour the onset of Brown bast. In: J. d'Auzac, J.L. Jacob and H. Chrestin (Eds.). Physiology of Rubber Tree Latex. CRC Press, Boca Raton, Florida. pp. 444-458.

Eschbach, J.M. van de Sype, H., Roussel, D. and Jacob, J.L. 1983. The study of several physiological parameters of latex and their relation-ship with production mechanisms. Int. Rubb. Res. Dev. Bd. Symp., Peking, 1983.

Eschbach, J.M., Tupy, J. and Lacrotte, R. 1986. Photosynthetate allocation and productivity of latex vessels in Hevea brasiliensis. Biol. Plant., 28: 321.

Frey-Wyssling, A. 1929. Microscopic investigations on the occurrence of resins in Hevea latex. Archf. Rubbercult. Ned-Indie., 13: 371.

Frey-Wyssling, A. 1952. Latex flow. In: A. Frey-Wyssling (Ed.). Deformation and flow in biological systems. North-Holland Publ., Amsterdam. pp. 322-349.

Gener, P., Primot, L. and Tupy, J. 1977. Recent progress de saignee par piqures. Caoutch. et. Plastiq., 54: 141.

Gomez, J.B. 1966. Electron Microscopic studies on the development of latex vessels in Hevea brasiliensis Muell. Arg. Thesis submitted for the degree of Doctor of Philosophy, University of Leeds.

Gomez, J.B. 1977. Demonstration of latex coagulants in bark extracts of Hevea and their possible role in latex flow. J. Rubb. Res. Inst. Malaysia, 25: 109.

Gomez, J.B., Narayanan, R. and Chen, K.T. 1972. Some structural factors affecting the productivity of Hevea brasiliensis. I. Quantitative deter-mination of the laticiferous tissue. J. Rubb. Res. Inst. Malaya., 23: 193-203.

Gomez, J.B. and Yip, E. 1974. Microhelices in Hevea latex: their isolation and electron microscopy. Proc. Symp. Int. Rubb. Res. Dev. Board, Part I, Cochin, 1974.

Gomez, J.B. and Yip, E. 1975. Microhelices in Hevea latex. J. Ultrastructure Res., 52: 76.

Gomez, J.B. and Yip, E. 1976. Microhelices in Hevea latex: their isolation and electron microscopy. Rubber Board Bull. India, 13: 14.

Gururaja Rao, G., Devakumar, A.S., Rajagopal, R., Annamma, Y., Vijaya-kumar, K.R. and Sethuraj, M.R. 1988. Clonal variation in leaf epicuti-cular waxes and reflectance: possible role in drought tolerance in Hevea. Indian J. Nat. Rubb. Res., 1: 84-87.

Harmsen, J.R. 1919. Bruine Binnenbastziekte, Ruygrok Batavia.

Hasma, H. and Subramaniam, A. 1978. The occurrence of a furanoid fatty acid in Hevea brasiliensis latex. Lipids, 13: 905.

Heusser, C. and Holder, H.J.V.S. 1930. Tapresultaten met let nieuwe twee-sneden tapsysteam Bij Hevea Oculaties, Arch. Rubbercult., 14: 275.

Ho, C.C., Subramaniam, A. and Yong, Y.M. 1976. Lipids associated with particles in Hevea latex. Proc. Int. Rubb. Conf. Kuala Lumpur, 1975, 2: 441.

Jacob, J.L. and d'Auzac, J. 1967. Sur I existence conjointe d'une hexokinase et d'une fucrokinase au sein du latex d'Hevea brasiliensis. C-r Acad. Sci. Paris, Ser. D., 265: 260.

Jacob, J.L., Nouvel, A. and Prevot, J.C. 1978. Electrophorese etmise en evidence d'activities enzymatiques dans le latex d'Hevea brasiliensis. Rev. gen. Caoutch. Plastiq., 582: 87.

Kamerun, P.S. 1912. British Patent No. 19615. Cited by Baptiste, 1955.

Karunakaran, A., Moir, G.F.J. and Tata, S.J. 1961. The proteins of Hevea latex: Ion exchange chromatography and starch gel electrophoresis. Proc. Nat. Rubb. Res. Conf. Kuala Lumpur, 1960, 798.

Kemp, A.R. and Straitiff, W.G. 1940. Hevea latex. Effect of proteins and electrolytes on colloidal behaviour. J. Phy. Chem., 44: 788.

Kemp, I. and Twiss, D.F. 1936. The surface composition of the rubber globules in Hevea latex. Trans. Farad. Soc., 32: 890.

Keuchenius, P.E. 1924. Consideration on Brown bast disease of rubber. Arch. Rubbercult., 8: 810.

Leong, T.T., RAvoof, A.A. and Tan, H.T. 1976. A preliminary report on exploitaing rubber through puncture tapping. Planter, Kuala Lumpur, 52: 209.

Liebermann, M. 1973. Biosynthesis and action of ethylene. Annu. Rev. Plant. Physiol., 30: 533.

Leong, T.T. and Tan, H.T. 1978. Results of Chemara Puncture Tapping trials. Proc. Rubb. Res. Inst. Malaysia Plrs' Conf. Kuala Lumpur, 1977. III.

Low, F.C. 1981. The role of carbohydrates in the exploitation and latex flow of Hevea. Thesis for the degree of D.Sc. University of Gent (Belgium).

Low, F.C. and Gomez, J.B. 1982. Carbohydrate status of exploited Hevea. I. The effect of different exploitation systems on the concentration of the major soluble carbohydrates in latex. J. Rubb. Res. Inst. Malaysia, 30: 1.

Low, F.C. and Gomez, J.B. 1984. Carbohydrate status of exploited Hevea in the bark. J. Rubb. Res. Inst. Malaysia, 32: 82.

Low, F.C., Gomez, J.B. and Ismail bin Hashim. 1983. Carbohydrate status of exploited Hevea. II. Effect of microtapping on the carbohydrate content of latex. J. Rubb. Res. Inst. Malaysia, 31: 27.

Low, F.C. and Wiemken, A.A. 1982. The separation of Hevea latex by density gradient centrifugation. Proceedings of the Eighth Malaysian Biochemical Society Conference, 1982, Kuala Lumpur, 37-41.

Low, F.C. and Wiemken, A. 1984. Fractionation of Hevea brasiliensis latex on Ficoll density gradient. Phytochem., 23: 747.

Low, F.C. 1976. Unpublished results. Rubb. Res. Inst. Malaysia.

Low, F.C. 1978. Distribution and concentration of major soluble carbohydrates in Hevea latex, the effects of ethephon stimulation and the possible role of these carbohydrates in latex flow. J. Rubb. Res. Inst. Malaysia, 26: 21.

Lynen, F. 1969. Biochemical problems of rubber synthesis. J. Rubb. Res. Inst. Malaya, 21: 389-406.

Lynen, F. 1967. Biosynthesis pathways from acetate to natural products. Pure Appl. Chem., 14: 137.

Lustinec, J. and REsing, W.L. 1965. Methodes pow la detimetation de laire drainee a l aide microsaignees et des radioisotopes. Rev. Gen. Caout. Plast. 42: 1161-1165.

Marin, B. and Trouslot, P. 1975. The occurrence of ribonucleic acid in the lutoid fraction (lysosomal compartment) for Hevea brasiliensis Kunth (mull Arg) latex. Planta (Bul)., 124: 31.

Marin, B. 1978. Ribosomes in the lutoid fraction (lysosomal compartment) from Hevea brasiliensis Kunth (Mull Arg). latex. Planta., 138: 14.

Mainstone, B.J. and Tan, K.S. 1964. Copper sulphate as an yield stimulant for Hevea brasiliensis. I. Experimental stimulation of 1931 budded rubber with 2,4-D or 2,4,5-T in the presence and absence of copper sulphate injection. J. Rubb. Res. Inst. Malaya, 18: 253.

McMullen, A.I. 1959. Nucleotides of Hevea brasiliensis latex: a ribonucleoprotein component. Biochem. J., 72: 545-549.

McMullen, A.I. 1960. Thiols of low molecular weight in Hevea brasiliensis latex. Biochem. Biophys. Acta., 41: 152.

McMullen, A.I. 1962. Particulate ribonucleoprotein components of Hevea brasiliensis latex. Biochem. J., 85: 491.

McMullen, A.I. and McSweeney, G.P. 1966. The biosynthesis of rubber: incorporation of insopentenyl pyrophosphate into purified rubber particles by a soluble latex enzyme. Biochem. J., 101: 42.

Milford, G.F.J., Paardekooper, E.C. and Ho, C.Y. 1969. Latex vessel plugging: its importance to yield and clonal behaviour. J. Rubb. Res. Inst. Malaya, 21: 274-282.

Moir, G.F.J. 1959. Ultracentrifugation and staining of Hevea brasiliensis latex. Nature, London, 184: 1626.

Moir, G.F.J. and Tata, S.J. 1960. The proteins of Hevea brasiliensis latex. 3. The soluble proteins of bottom fraction. J. Rubb. Res. Inst. Malaya, 16: 155.

Nandris, D., Chrestin, H., Geiger, J.P., Nicole, M. and Thouvenel, J.C. 1984. Occurrence of a phloem necrosis on the trunk of rubber tree. In: Phytopathol. Meet. Ceylon, Sri Lanka, 1984.

Ng, T.S. 1961. Isolation and identification of the free amino acids in fresh unammoniated Hevea latex. Proc. Nat. Rubb. Conf. Kuala Lumpur, 1960, 809.

Osborne, D.J. and Sargent, J.A. 1974. A model for the mechanism of stimulation of latex flow in Hevea brasiliensis by ethylene. Ann. Appl. Biol., 78: 83-88.

Ostendorf, F.W. 1941. Rubber plant material 1936-1941. De Bergcult., 15: 852.

Paardekooper, E.C. and Samosorn, S. 1969. Clonal variation in latex flow patterns. J. Rubb. Res. Inst. Malaya, 21: 264-273.

Pakianathan, S.W. 1970. The search for new stimulants. Plrs' Bull. Rubb. Res. Inst. Malaya, No. 111-351.

Pakianathan, S.W. 1967. Determination of osmolarity of small latex samples by vapour pressure osmometer. J. Rubb. Res. Inst. Malaya, 20: 23.

Pakianathan, S.W., Wain, R.L. and Ng, E.K. 1975. Studies on displacement area on tapping in mature Hevea trees. Proc. Int. Rubb. Conf. Vol. 2, p. 255.

Pakianathan, S.W. and Milford, G.F.J. 1977. Changes in the bottom fraction contents of latex during flow in Hevea brasiliensis. J. Rubb. Res. Inst. Malaya, 23: 391-400.

Pakianathan, S.W., Botman, S.G. and Taysum, D.H. 1966. Particle aggregation following dilution of Hevea latex: A possible mechanism for the closure of latex vessels after tapping. J. Rubb. Res.Inst. Malaya, 19: 259-271.

Paranjothy, K., Gomez, J.B. and Yeang, H.Y. 1976. Physiological aspects of Brown bast development. In: Proc. Int. Rubb. Conf. 1975, Rubber Research Institute of Malaysia, Kuala Lumpur, 181 p.

Peries, O.S. and Brohier, Y.E.M. 1965. A virus as the causal agent of bark cracking in Hevea brasiliensis. Nature, 205 p. 624 p.

Primot, L. and Tupy, J. 1976. Sur l'exploitation de l'Hevea par micro-saignee. Rev. gen. Caoutch. Plastiq., 53: 588.

Pujarniscle, S. 1968. Caractere lysosomal des lutoides du latex d'Hevea brasiliensis Mull. Arg. Physiol. Veg., 6: 27.

Pushpadas, M.V., Kochappan Nair, K., Krishnakumari, M. and Karthikakutty Amma, M. 1975. Brown bast and nutrition. Rubber Board Bull., 12: 83.

Ramachandran, P. 1978. Proc. Rubb. Re. Inst. Malaysia Plrs' Conf. Kuala Lumpur, 1977. p. 130.

Ramachandran, P. and Lee, T.P. 1980. Preliminary results of Socfin puncture tapping trials. Proc. Rubb. Res. Inst. Malaysia plrs' Conf. Kuala Lumpur 1979. p. 166.

Rands, R.D. 1921. Histological studies on Brown bast disease of plantation rubber. Inst. Plantenziekten Meded., 49: 26.

Rubber Research Institute of Malaysia. 1982. Rep. Rubb. Res. Inst. Malaysia, p. 43.

Rubber Research Institute of Malaysia. 1980. Rep. Rubb. Res. Inst. Malaysia, 1980. p. 231.

Rubber Research Institute of Malaysia. 1980. Rep. Rubb. Res. Inst. Malaysia, 1979. pp. 76.

Rutgers, A.A.I. and Dammerman, 1914. Disenes of Hevea brasiliensis, Java Med. v. H. Lab. v. Plant No. 10.

Riches, J.P. and Gooding, E.G.N. 1952. Studies in the physiology of latex. I. Latex flow on tapping - theoretical consideration. New. Phytol., 51: 1-10.

Rhodes, E. and Wiltshire, J.L. 1931. Quebrachitol - a possible by-product from latex. J. Rubb. Res. Inst. Malaya, 3: 160.

Rodwell, V.W., Nordstrom, J.L. and Mitscheleu, J.J. 1976. Regulation of HMG-CoA reductase. Adv. Lipid Res. 14: 1-74.

Roe, C.P. and Ewart, R.H. 1942. An electrophoretic study of the proteins in rubber latex serum. J. Amer. Chem. Soc., 64: 2628.

Rubber Research Institute of Malaya. 1963. Rep. Rubb. Res. Inst. Malaya, 1962. p. 78.

Samosorn, S., Creencia, R.P. and Wasuwat, S. 1978. Study on yield, sucrose level in latex and other important characteristics of Hevea brasiliensis Mull. Arg. III. As influenced by microtapping system. Thai. J. Agric. Sci., 11: 193.

Sanderson and Sutcliff. 1929. Vegetative characters and yield of Hevea. Quar. J. Rubb. Res. Inst. Malaya., 1: 75.

Schweizer, J. 1949. Brown bast disease. Arch. Rubbercult., 26: 385.

Sethuraj, M.R. 1968. Studies on the physiological aspects of rubber pro-duction. 1. Theoretical considerations and preliminary observations. Rubber Board Bull., 9: 47-62.

Sethuraj, M.R. 1989. Present status of investigations in the Rubber Research Institute of India on panel dryness syndrome. In: Foo Kah Yoon and P.G. Chuah (Eds.). Proc. IRRDB Workshop on Tree Dryness, 1989, Penang, Rubber Research Institute of Malaya, pp. 37-40.

Sethuraj, M.R., George, M.J., Usha Nair, N. and Mani, K.T. 1977. Physiology of latex flow as influenced by intensive tapping. J. Rubb. Res. Inst. Sri Lanka, 54: 221-226.

Sharples, A. 1922. Consideration of recent work on Brown bast problem. Malaya. Agric. J., 10: 155.

322

Sharples, A. and Lambourne, J. 1924. Field experiments relating to Brown bast disease of Hevea brasiliensis. Malay. Agric. J., 12: 290.

Sheldrake, A.R. 1973. Yearbook of the Royal Society 1973. The Royal Society, London. pp. 371-372.

Sherief, P.M. and Sethuraj, M.R. 1978. The role of lipids and proteins in the mechanism of latex vessel plugging in Hevea brasiliensis. Physiol. Plant., 42: 351-353.

Sipat, A.B. 1982. The occurrence of 3-Hydroxy-3-Methyl Glutaryl CoA Reductase (NADPH) in the latex of regularly tapped Hevea brasiliensis. Pertanipra, 5: 246-254.

Spencer, D. 1908. On the presence of oxidases in India rubber, with a theory in regard to their function in the latex. Biochem. J., 3: 165.

Southorn, W.A. 1961. Microscopy of Hevea latex. Proc. Nat. Rubb. Res. Conf. Kuala Lumpur, 1960. p. 766.

Southorn, W.A. 1966. Electron microscope studies on the latex on Hevea brasiliensis. Proc. 6th Int. Cong. Electron Microsc. Kyoto 1966, Maruzen Company Ltd., Tokyo. p. 385.

Southorn, W.A. 1968. Latex flow studies. I. Electron microscopy of Hevea brasiliensis in the region of the tapping cut. J. Rubb. Res. Inst. Malaya, 20: 176-186.

Southorn, W.A. 1968. Latex flow studies. IV. Thixotrophy due to lutoids in fresh latex demonstrated by a microviscometer of new design. J. Rubb. Res. Inst. Malaya, 20: 226-235.

Southorn, W.A. and Edwin, E.E. 1968. Latex flow studies. II. Influence of lutoids on the solubility and flow of Hevea latex. J. Rubb. Re. Inst. Malaya, 20: 187.

Southorn, W.A. and Yip, E. 1968. Latex flow studies. III. Electrostatic considerations in the colloidal stability of fresh Hevea latex. J. Rubb. Res. Inst. Malaya, 20: 201.

Tan, C.H. and Audley, B.G. 1968. Egothioneine and hercynine in Hevea brasiliensis latex. Phytochem, 7: 109.

Tata, S.J. 1975. A study of the proteins in the heavy fraction of Hevea brasiliensis latex and their possible role in the destabilisation of rubber particles. Thesis for a degree of Masters of Science, University Malaya, Kuala Lumpur.

Tata, S.J. 1976. Hevein. Its isolation, purification and some structural aspects. Proc. Int. Rubb. Conf. 1975, Kuala LUmpur, Rubb. Res. Inst. Malaysia. p. 499.

Tata, S.J. 1980a. Distribution of proteins between the fractions of Hevea latex separated by ultracentrifugation. J. Rubb. Res. Inst. Malaysia, 28: 77.

Tata, S.J. 1980b. Studies on the lysozyme and components of microhelices of Hevea brasiliensis latex. Thesis for a degree of Doctor of Philosophy. Universiti Malaya. Kuala Lumpur.

Tata, S.J., Beintema, J.J. and Balabaskaran, S. 1983. The lysozyme of Hevea brasiliensis latex. Isolation, purification, enzyme kinetics and a partial amino acid sequence. J. Rubb. Res. Inst. Malaysia, 31: 35.

Tata, S.J. and Edwin, E.E. 1969. Significance of nonstaining white zones in starch gel electrophoresis. J. Rubb. Res. Inst. Malaya, 21: 477.

Tata, S.J. and Edwin, E.E. 1970. Hevea latex enzyms detected by zymogram technique after starch gel electrophoresis. J. Rubb. Res. Inst. Malaya, 23: 1.

Tata, S.J. and Moir, G.F.J. 1964. The proteins of Hevea brasiliensis latex. Starch gel electrophoresis of C-serum proteins. J. Rubb. Inst. Malaya, 18: 97.

Taylor, R.A. 1926. A note on Brown bast. Trop. Agric., 72: 323.

Taysum, D.H. 1961. Effect of ethylene oxide on the tapping of Hevea brasiliensis. Nature, London, 191: 1319.

Taysum, D.H. 1961. Yield increase by treatment of Hevea brasiliensis with antibiotics. Proc. Nat. Rubb. Conf. Kuala Lumpur, 1960. p. 224.

Tupy, J. 1969a. Nucleic acids in latex and production of rubber in Hevea brasiliensis. J. Rubb. Res. Inst. Malaya, 21: 468.

Tupy, J. 1969b. Stimulatory effects of 2,4-dichlorophenoxyacetic acid and 1-Naphthylacetic acid on sucrose level, invertase activity and sucrose utilisation in the latex of Hevea brasiliensis. Planta (Berl.)., 88: 144.

Tupy, J. 1973a. Possiblite d'exploitation de l'Hevea par microsaignee. Rev. gen. Caoutch. Plastiq., 50: 620.

Tupy, J. 1973b. The activity of latex invertase and latex production in Hevea brasiliensis Mull. Arg. Physiol. Veg., 11: 633.

Tupy, J. 1973c. The regulation of invertase activity in latex of Hevea brasiliensis Mull. Arg. The effects of growth regulators, bark wounding and latex tapping. J. Expt. Bot., 24: 516.

Tupy, J. 1984. Translocation, utilisation and availability of sucrose for latex production of Hevea. In: C.R. Coll. Exp. Physiol. Amel. Hevea, IRCA-GERDAT, Montpellier, France. 249 p.

Tupy, J. and Primot, L. 1974. Physiology of latex production. Int. Rubb. Res. Dev. Bd. Scient. Symp. Cochin, India, 1974, Part I, Session C1.

Tupy, J. and Resing, W.L. 1969. Substrate and metabolism of carbon dioxide formation in Hevea latex in vivo. J. Rubb. Res. Inst. Malaya, 21: 456.

Vollema, J.S. 1949. Enige Waarnemngen oven het optreden der BBB, Berg-cultures, 243.

Vollema, J.S. and Dijkman, M.J. 1939. Results of the testing of Hevea clones in experimental garden Tjiomass-II. Arch. Rubbercult., 23: 47.

Walujono, K., Schloma, R.A. and Beintema, J.J. 1976. Amino acid sequence of hevein. Proc. Int. Rubb. Conf. 1975, Kuala Lumpur. Rubb. Res. Inst. Malaysia, Vol. 2: 518.

Yeang, H.Y., Ghandimathi, H. and Paranjothy, K. 1977. Protein and enzyme variation in some Hevea cultivars. J. Rubb. Res. Inst. Malaysia, 25: 9.

Yeang, H.Y., Low, F.C., Gomez, J.B., Paranjothy, K. and Sivakumaran, S. 1984. A preliminary investigation into the relationship between latex invertase and latex vessel plugging in Hevea brasiliensis. J. Rubb. Res. Inst. Malaysia, 32: 50.

Yeang, S.F. and Pratt, M.K. 1978. The physiology of ethylene in wounded plant tissues. In: G. Kalh and A. de Gruyter (Eds.). Biochemistry of wounded plant tissues, Springer Verlag, Berlin, 595 pp.

Yip, E. and Gomez, J.B. 1978. Destabilisation of rubber particles by bark extracts: A possible mechanism for cessation of latex flow after tapping. Int. Rubb. Res. Dev. Bd. Symp. Kuala Lumpur, 1978.

Yip, E. and Gomez, J.B. 1984. Characterisation of cell sap of Hevea and its influence on cessation of latex flow. J. Rubb. Res. Inst. Malaysia, 32: 1.

Zheng Guanbiao, Chen Murong, Chen Zuoyl and Shen Juying. 1982. A preliminary report on the study of causative agents of Brown bast. Chinese J. Trop. Crops., 3: 62.

Zheng Guanbiao, Chen Murong, Yung Shivwa, Chen Zuoji and Shen Juying, 1988. A further report on the study of causative agents and control of brown bast disease of rubber. J. South China Agric. 9(2): 22-33.

CHAPTER 15

DISEASES OF ECONOMIC IMPORTANCE IN RUBBER

A. de S. LIYANAGE
Rubber Research Institute of Sri Lanka, Dartonfield, Agalawatta, Sri Lanka.

C. KURUVILLA JACOB
Rubber Research Institute of India, Kottayam-686009, Kerala, India.

The para rubber tree, Hevea brasiliensis Muell. Arg., is cultivated in many countries in the world, as a monocrop, in over six million hectares of land. Rubber makes a significant contribution to the export earnings and agricultural employment especially in countries of south and south-east Asia, where it has traditionally occupied a predominant position. The world's production of natural rubber exceeds five million tonnes and is still increasing and most of it is produced by the smallholders who are often subsistence farmers. The losses caused by diseases are a key factor affecting the productivity of their holdings and eventually the quality of their life.

The rubber tree is susceptible to several diseases but their economic importance and severity vary with the climatic conditions and the cultural practices in each country. This chapter discusses the major diseases of economic importance, which can be divided, for convenience, into four categories: leaf, stem and branch, panel and root diseases.

LEAF DISEASES

Leaf diseases had assumed greater importance during the post-war years, owing to the extensive cultivation of clones primarily selected for high yield at the expense of resistance to diseases. There are five major leaf diseases that can cause damage of economic importance to Hevea in different countries. These are caused by Oidium heveae Steinm., Colletotrichum gloeosporioides (Penz.) Sacc., Phytophthora spp., Corynespora cassiicola (Berk. and Curt.) Wei., and Microcyclus ulei (P. Henn.) V. Arx. The diseases caused are secondary or abnormal leaf fall of immature and mature leaflets.

Oidium leaf fall

General: Oidium leaf disease (OLD) caused by the fungus Oidium heveae Steinm., attacks the immature leaflets when trees refoliate after the annual wintering, causing secondary leaf fall. It is widespread in many countries (Stoughton Harris, 1925; Beeley, 1930; Cramer, 1956; Ramakrishnan and Radhakrishna Pillai, 1962a; Kaiming, 1987), but in a few it is of minor importance (Peries, 1966a). The severity of the disease varies with the pattern of wintering (Liyanage, 1977a), leaf age (Lim and Sripathi Rao, 1976), elevation (Murray, 1929; Liyanage et al., 1971), clonal susceptibility (Peries, 1966b; Van Emden, 1954; Lim, 1973b) and weather conditions prevailing at the time of refoliation (Beeley, 1932; Fernando, 1971; Lim, 1972).

Symptoms: The copper-brown and apple green leaflets are most susceptible to the infection. Symptoms appear as white powdery patches on both leaf surfaces, especially on the lower leaf surface near the veins as the fungal hyphae grow radially to form extensive circular colonies, sometimes covering the entire leaf surface with the fungus. When the tender leaflets are affected, they shrivel and fall off, leaving the petioles on the stem for sometime. When semi-mature leaves are attacked leaflets become distorted and characteristic translucent brownish spots develop which later become necrotic and persist throughout the life of the leaflets.

A severe attack of Oidium leads to extensive defoliation, resulting in poor canopies being retained on the trees and often with loss of yield. It can also result in a serious retardation of the rate of growth and bark renewal. Repeated defoliation in areas where the disease is not controlled, particularly at high elevations leads to the depletion of food reserves in the trees resulting in dieback of twigs and branches, assisted by secondary parasites, mainly Botryodiploidia theobromae Pat.

O. heveae also affects the inflorescences causing them to wither and shed prematurely, with consequent loss in seed production.

Biology: Studies in vitro indicated that the temperature and humidity are the most important environmental factors that affect the viability, germination, sporulation and infectivity of the fungus (Lim, 1972; Peries, 1974; Liyanage et al., 1985; Chua, 1970). The optimum temperature for germination, infection and sporulation ranged from 23-25°C, while the humidity requirement exceeded 90% (Lim, 1972; Peries, 1974; Liyanage et al. 1985). Spore viability was adversely affected by temperatures exceeding 32-35°C, exposure to ultra-violet light and direct sun light for a short period and prolonged immersion in water (Chua, 1970). Spore germination and conidophone formation were significantly better in artificial light than in the dark (Chua, 1970). It was possible to evolve short-term methods of forecasting oidium leaf disease incidence using weather parameters, especially temperature and humidity (Lim, 1972; Liyanage et al. 1985).

Epidemiology: Although wintering is a physiological process (Chua, 1970), its commencement, duration and completion are largely influenced by weather conditions. It has been observed that PB 86 which is a clone susceptible to Oidium escapes the infection due to its early wintering habit. Dry weather conditions encourage early and rapid wintering and this often helps the refoliating leaves to mature rapidly and escape an attack of Oidium. Similarly, wet weather at the time of wintering delays it, causing the refoliating leaves to be exposed to weather factors conducive for disease development, resulting in a high incidence of the disease (Liyanage, 1977a).

The fungus survives from one season to the next on young leaves which emerge periodically within the canopy or on tender foliage in rubber nurseries. Although it has been suggested that Euphorbia pilulifera is an alternative host of O. heveae (Young, 1952), it is not readily transferable from one species to the other (Ramakrishnan and Radhakrishna Pillai, 1962a; Peries, 1966b).

The fungus is disseminated by means of air borne conidia. O. heveae has a typical afternoon pattern of spore release reaching a peak around 1300 h. The spore production and spread are influenced by weather conditions, especially low humidity, high temperature and high air turbulence prevalent during the period of spore liberation (Fernando, 1971; Chua, 1970; Peries, 1965a).

General field observations between weather and disease outbreaks have revealed that overcast humid and relatively cool weather with misty mornings and intermittent light rain provide ideal conditions for spread of the fungus (Peries, 1965a; Wastie and Mainstone, 1969; Liyanage, 1976). Bright sunny periods militate against the development of the fungus and have a lethal effect on detached spores (Liyanage et al. 1985; Peries, 1965a). Heavy rains, as well as leaf wetness for long periods, were inimical to the development of the fungus, as the conidia get washed off by heavy rains (Fernando, 1971) and the presence of free water kills the fungus (Liyanage et al. 1985), thus preventing the build up of inoculum.

Control: Young (1951) considered that manipulation of the time of refoliation to coincide with a period less favourable to disease development appeared to be an efficient means of indirect control of Oidium and used calcium cyanide in the initial defoliation trials. Aerial spraying has been tried out with good effect to artificially defoliate trees with 2,4,5-T (Hutchison, 1958). Subsequently, organo-arsenical dessicants were areially sprayed to advance and accelerate wintering of susceptible clones to coincide refoliation with the dry period thus favouring disease avoidance. However, several practical difficulties have made it difficult for the industry to adopt wider use of this technique of indirect control (Lim, 1982).

The application of a supplementary dose of nitrogenous fertilizer towards the end of the wintering season helped to minimise the adverse effects of repeated defoliation (Beeley, 1935; Murray, 1936), by enhancing the leaf maturity before the onset of conditions favourable for Oidium leaf disease. Thus, a judicious application of nitrogen, up to twice the currently recommended rate at the onset of refoliation, was considered as another means of disease avoidance (Lim, 1974). This method yielded additional benefits from improved bark renewal and growth as well as reduced weeding costs.

An alternative to growing resistant clones is to crown-bud a high yielding clone with a mildew resistant crown. This method was adopted especially to control oidium leaf disease of susceptible clones with the Oidium tolerant clone LCB 870 (Van Emden, 1954). However, the delay in reaching maturity of crown budded trees, their generally lower yields and changes in the properties of latex have prevented its wider use. Nonetheless, it still remains a useful technique.

Protective dusting with sulphur with portable or tractor-mounted machinery has been the standard method of economically controlling O. heveae for many years (Murray, 1921; Wastie and Mainstone, 1969). At the inception, 10-12 rounds of sulphur dusting were used at 5-7 days intervals as a blanket cover throughout the refoliation period (Murray, 1921). Later, the number of rounds was reduced to 3-4, at weekly intervals, if accurately timed, based on the pattern of refoliation and weather parameters (Lim, 1972; Liyanage et al. 1985; Peries, 1965a). Currently, with better knowledge of these factors, the routine control of the disease has been dispensed with and dusting is restricted to areas where the disease is severe (Liyanage et al. 1985). Sulphur dust has the disadvantage of being easily washed off by rain, resulting in poor control of the disease, necessitating repeated application thereby increasing the cost. To overcome these difficulties oil-based systemic fungicides, such as tridemorph, which have anti-sporulant action have been used. Further, low-volume spraying of new rain-fast fungicidal formulations with shoulder mounted mini micron sprayers provide cost effective and better control of the disease (Lim, 1982). More recently, an efficient ground based mechanised fogging has been used to dispense oil-based fungicides at an ultra low rate. Fogging leaves adequate quick-drying and rain-fast residues on leaflets giving a fairly lasting protecting against fungal infestations (Lim, 1982).

Phytophthora leaf fall

General: Abnormal leaf fall caused by Phytophthora species occurs during periods of prolonged rain and generally originate from sporangia and zoo-spores produced on infected green rubber pods. Several species of Phytophthora have been reported as the causal organism of various diseases affecting the rubber tree. These are Phytophthora meadii Mc Rae (Mc Rae, 1918), P. palmivora (Butl.) Butl. (Tucker, 1931), P. heveae Thompson (Thompson, 1929), P. botryosa Chee (Chee, 1969), P. citrophthora (R.E. Sm., & E.H. Sm.) Leonian (Ho et al. 1984), P. nicotianae Van Breda de Haan Var. parasitica (Dast.) Waterhouse (Thomson and George, 1976), P. phaseoli Thaxter (Ho et al. 1984), P. citricola Swada (Liyanage, 1989) and P. cactorum (Leb & Cohn) Schroet (Ho et al. 1984). A species currently designated P. palmivora Morphological Form 4(MF_4) (=P. capsici), has been identified and the presence of P. megakarya Brasier & Griffus is suspected to be present in Brazil. The causal organism is separated into different species on the basis of several diagnostic features which include colony characteristics as well as sporangial, oospore and chlamydospore production and morphology.

Symptoms: The most conspicuous symptom is seen on the leaf petiole. A greyish to black lesion develops along the petiole with white globules of coagulated latex, at the point of entry of the pathogen (Fig. 1). Petiolar infection results in leaf fall due to the formation of a premature abscission layer. At this stage the leaflets are still attached to the petiole but they

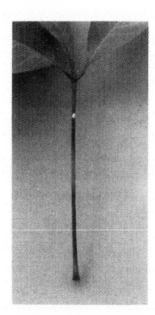

Fig. 1.   Coagulation of latex on the petiole at the point of entry of Phytophthora.

often become reddish brown while some remain green. On leaflets, circular brownish water soaked lesions appear on the lamina. These lesions eventually coalesce to form large irregular areas and the leaflets are easily shed on vigorous shaking. Under favourable conditions trees of susceptible clones defoliate rapidly and they remain bare until the subsequent refoliation season. The vigour and yield of such trees are considerably reduced.

Biology: Cardinal temperatures for growth of P. meadii (Peries and Fernando, 1966; Dantanarayana et al. 1984), P. palmivora (Chee, 1969), P. botryosa (Chee, 1969), P. citrophthora (Ho et al. 1984) and P. cactorum (Ho et al. 1984) ranged between 5-35°C with an optimum between 25-28°C. The fungus sporulated freely at room temperature but germination of sporangia was favoured at temperatures slightly below room temperature. Free water is essential for the propagation of the fungus, which is rapidly killed on exposure to conditions of low humidity, direct sun light or ultra-violet radiations (Peries and Fernando, 1966).

The occurrence of both compatibility types of P. meadii, with compatibility type A_2 predominating, was initially reported (Satchuthanantha-vale, 1963). Subsequently, the existence of homothallic P. meadii was shown (Peries and Dantanarayana, 1965). Recent studies provided evidence for a shift in the P. meadii population towards a predominance of A_1 compatibility type (Liyanage and Wheeler, 1989; Dantanarayana et al. 1984). In P. cactorum abundant sex organs were produced readily in single cultures but none were produced in P. citrophthora and P. palmivora single cultures or by pairing among isolates of the same species or any other species. However, Chee and Turner claim to have seen homothallic isolates of P. palmivora in India and Sarwak, respectively. Rubber, cacao and atypical (non-complementary) strains of P. palmivora, and (+), (-) and non-complementary strains were recognised for P. botryosa (Chee, 1969).

Oospores and chlamydospores are important propagules for long term survival. Incubation of paired cultures of P. meadii in the dark at temperatures between 18°-22°C proved most favourable for the formation of oospores. At higher temperatures (25-35°C) the number of oogonia were considerably reduced but moderate numbers were formed at 30°C (Liyanage, 1986). P. citrocola, on the other hand was well adapted and produced oospores in abundance at 20°, 25° and 30°C both in the light and in the dark but their germination was poor (Liyanage, 1986). Production of a few oospores of P. botryosa was stimulated by light but none were induced in P. palmivora (Chee, 1969).

Higher temperatures favoured chlamydospore production of P. palmivora in deionized water and there was a fourfold increase between 17 and 32°C.

The optimum pH was about 5.8 and the optimum temperature for germination was 27°C although it germinated over a wide range of temperatures (Chee, 1973). On the otherhand, chlamydospores of P. meadii are formed abundantly in the dark and at temperatures below 20°C, even as low as 10°C (Liyanage, 1986). In P. botryosa, Chlamyolospore production was infrequent (Chee, 1969) and none were formed in P. cactorum and P. citrophthora (Ho et al. 1984).

Epidemiology: The fungus survived in mummified rubber pods and pod stalks as chlamydospores under Sri Lankan conditions (Peries, 1965b) and as oospores in infected plant parts in India (George and Edathil, 1975), respectively. With the advent of favourable weather conditions, the dormant propagules germinated to act as primary inoculum. A selective antibiotic medium was successfully used to isolate both P. palmivora and P. botryosa, from old lesions in different plant parts in Thailand but the dormant structures from which the infection occured was not identified (Tsao, 1976).

In Sri Lanka and India, inoculum of P. meadii for leaf fall was derived mainly from sporangia produced on mature green rubber pods where sporulation was most prolific (Peries, 1969; Liyanage et al. 1983). Infected pods provided inoculum for periods ranging from three to five weeks depending on the interim weather conditions after infection (Liyanage et al. 1983). The inoculum disseminate in water droplets and repeatedly inoculate other plant parts to cause infection. However, P. botryosa is more freely sporulating and has a lower requirement for free moisture for infection and is capable of causing direct petiolar infection rather than via pods (Wastie, 1973). Water is essential for both dispersal of sporangia and germination of zoospores (Peries and Fernando, 1966; Peries, 1969). Little germ tube growth occurs in the absence of free water even in a saturated atmosphere. The minimum period of surface wetness for in vitro field respectively (Peries, 1969), but for P. botryosa it was 30 min (Wastie, field respectively (Peries, 1969), but for P. botryosa was 30 min (Wastie, 1973). An attempt was made to forecast outbreaks of Phytophthora disease using the formula: if the temperature was not above 29°C, relative humidity above 80%, atleast 25 mm of rain per day, less than 3 h sunshine per day prevailing for four consecutive days when mature green pods are present, leaf fall was likely to occur within fourteen days (Peries, 1969). The work in Malaysia showed that the severity of defoliation is closely correlated with the duration of surface wetness and 100% relative humidity seven days previously. Rainfall, temperature and solar radiation do not influence defoliation directly once it has begun, but are important in determining its onset (Wastie, 1973). Later, it was shown that rainfall is the main climatic factor governing the onset and severity of the disease.

Control: Phytophthora leaf fall is an annually recurring disease of rubber in India, causing severe yield losses ranging from 38-56% (Satchuthananthavale and Dantanarayana, 1973; Ramakrishnan, 1960; Radhakrishna Pillai, 1977). Recent experiments indicate an yield loss of 9 to 16% in susceptible clones (RRIM 600 and PB 86) of 10 to 25 years age when prophylactic spraying against the disease is skipped for one season (Jacob et al. 1989). However, in Malaysia and Sri Lanka the maximum yield reduction reported was about 8% (Tan and John, 1985; Tan et al. 1977) and 2% (Lloyd, 1963), respectively.

In India, where the disease occurs in an epiphytotic scale almost every year, spraying with 1% Bordeaux mixture as a premonsoon application has been for many years, the most effective method of control of Phytophthora leaf fall, since it was first recommended. Although it is a laborious process and uneconomical for large scale use, high volume spraying of Bordeaux mixture with rocker sprayers is still widely practised (Radhakrishna Pillai and George, 1973). Dusting copper fungicides as a prophylactic measure did not provide adequate protection, as dusts were easily washed-off by heavy rains. Further, dusting was as expensive as spraying, owing to the high cost of fungicides, offsetting the savings in labour (Ramakrishnan, 1960; Lloyd, 1963). Pre-monsoon low volume application of copper-in-oil, using mini micron sprayers from the ground or aerial application with helicopters (Fig. 2), confers complete protection

Fig. 2. Aerial spraying for protection against abnormal leaf fall disease.

against the disease (Radhakrishna Pillai and George, 1973). More recently, fogging copper-in-oil or captafol-in-oil, significantly reduced leaf fall caused by Phytophthora (Lim et al. 1981; Tand and John, 1985). Fortunately, the severity of the disease in all rubber growing countries, except India is not of enough economic importance to warrant expenditure on its routine control.

Good control of Phytophthora leaf fall has been achieved in Sri Lanka by the planned restriction of sulphur dusting to control oidium leaf fall, where the causal fungus affects the flowers too. This reduces pod set and consequently the inoculum for causing severe leaf fall (Peries et al. 1985).

Although crown budding has not become widely accepted as a method of disease control, due to some adverse effects of the crown on the growth, yield and properties of latex, this technique is being used in India for more than three decades to control Phytophthora leaf fall successfully (Sarma and Ramakrishnan, 1985).

Colletotrichum leaf fall

General: Colletotrichum leaf disease caused by the fungus Colletotrichum gloeosporioides (Penz.) Sacc., occurs in many rubber growing countries with different degrees of severity. This disease attacks tender leaves of immature plants and also leaves developing towards the latter part of the refoliation season of mature rubber trees. However, this disease occurs throughout the year but becomes more dominant with the onset of wet weather conditions to cause extensive defoliation, in most susceptible clones. The severity of secondary leaf fall depends on both the suscept-ibility of the clone to the pathogen and to the weather conditions at the time of refoliation.

Symptoms: Tender leaves produced soon after bud burst are most susceptible to infection. When the immature leaves are affected, the infection begins at the tip of the leaf and spreads towards the base causing it to produce a necrotic area. If the damage is extensive, the leaves become distorted, shrivel and fall off leaving the petioles on the stem for a short period. Sometimes, the portions affected by the disease drop away leaving behind unaffected area of the lamina on the shoot. When semi-mature or mature leaves are infected, the natural resistance of the host usually prevents extensive damage. Such leaves are covered with numerous spots having a brown margin surrounded by a yellow halo. The spots become raised and prominent as the leaf gets older.

Repeated defoliation due to Colletotrichum could result in dieback of succulent shoots of young buddings. Sometimes the fungus grows down

Affecting the bud patch and killing the entire plant. It can also cause gradual death of the twigs and branches and may even kill the entire tree, when it is highly susceptible to the disease, especially at higher elevations or in areas where wet weather is experienced continuously.

Biology: The germination of the fungus was poor when the spore concentrations were above 7×10^6 per ml (Wimalajeewa, 1967). The optimum temperature for growth and sporulation was between 26-32°C (Wastie, 1972a), with a maximum around 28°C (Wimalajeewa, 1967). Germination decreased on exposure of spores to sunlight (Wimalajeewa, 1967; Wastie, 1972a) and exposure to ultra-violet radiation for short periods (Wimalajeewa, 1967). Continuous light favoured spore production in vitro, but spores produced in the dark had a higher percentage germination (Wastie, 1972a). Spore viability and germination were sensitive to atmospheric conditions. At 99% relative humidity spore viability and germination were reduced by 50% when compared to 100% relative humidity (Wimalajeewa, 1967). Germination decreased by up to 30% after 3 h storage at 80% relative humidity (Wastie, 1972a). Loss of spore viability was also observed when they were stored at temperatures of 50°C and above for more than 6 h. In contrast, spores were found to remain viable for relatively long periods at very low humidity (Wimalajeewa, 1967). Complete inhibition of germination was observed in the absence of oxygen in the ambient atmosphere (Wimalajeewa, 1967). No differences were detected between the numbers of spores germinating on leaves of different ages (Wastie, 1972a), but on slightly older leaflets spores tended to form longer germ tubes (Senechal et al. 1987). Similar germination occured on both resistant and susceptible clones (Liyanage and de Alwis, 1978; Samarajeewa and Liyanage, 1986; Zainuddin and Omar, 1988), but more appressoria were formed on resistant clones. Leaf diffusates obtained from susceptible and resistant cultivars stimulated and inhibited spore germination, respectively, indicating the possibility of using this as a technique for screening clones for disease resistance in the laboratory (Liyanage and de Alwis, 1978). However, there was no correlation between thickness of the cuticle and clonal susceptibility (Wastie and Sankar, 1970). In susceptible clones, extensive ramification of the pathogen was observed within the leaf tissues and acervuli were formed 72 h after inoculation, but in resistant cultivars disorganisation and necrosis of epidermal and mesophyll cells were observed (Senechal et al. 1987; Liyanage and de Alwis, 1978; Samarajeewa and Liyanage, 1986).

C. gloeosporioides obtained from Hevea vary considerably in their growth, morphology, intensity of sporulation and relative infectivity (Wastie and Janardhanan, 1970). Studies carried out in Sri Lanka showed that there were differences in the pathogenicity of isolates obtained from different agro-climatic regions and also that an inverse relationship exists between the leaf infection and leaf area damaged (Liyanage, 1976; Liyanage, 1977b).

Epidemiology: It has been shown that free water is necessary for optimum germination of the fungus and it accounted for the close relationship between secondary leaf fall and rainfall. However, disease establishment can occur in a few hours at 100% relative humidity (Wastie, 1967), or longer at humidities down to 96% (Wimalajeewa, 1965), even in the absence of free water. The severity of leaf fall was correlated to the higher total amount and frequency of rainfall (Wastie, 1972b). The spore discharge followed a regular diurnal pattern and peak spore catch usually occurred late at night but declined to low concentrations as the humidity dropped during the day time and also during rainy periods (Wastie, 1972a). The overall incidence and severity varied in different environments, but the relative degree of resistance of the clones was not markedly affected by the environmental differences. The relationship between the incidence and severity was linear when the plants were less affected by the disease while a curvilinear trend was evident in susceptible clones when they were heavily infected (Samarajeewa et al. 1985).

Control: This disease was controlled by the application of Bordeaux mixture (RRIM, 1968) and colloidal copper (Wimalajeewa and Shanmuganathan, 1963), with the former being more effective than the latter as it provided better initial coverage and good rainfastness. The use of carbamate fungicides such as Zineb (zinc ethylenebisdithiocarbamate) and Ferbam (ferric di-methyldithiocarbamate) and organic formulations, Daconil (chlorothalonil) and Difolatan [Cis-N(1,1,2,2-tetrachloroethylthio) cyclohex-4-ene-1,2-dicarboxymide] gave good control of the disease (RRIM, 1968). However application of organo-mercurials was more effective and advantageous due to their systemic nature (Peries and Wimalajeewa, 1970) but later it was found that Daconil and Benlate (Benomyl) were more effective and could replace antimucin (phenyl mercuric acetate) (RRIM, 1977). The application of water miscible fungicides was done with knapsack sprayers or power driven mist blowers, but the fungicides were easily washed off during rains and did not give the desired results in controlling the disease. To overcome this problem low volume application of Daconil as an oil-water emulsion with a mist blower was undertaken to reduce the severity of the disease (RRIM, 1977). Subsequently, mechanised fogging of captafol-in-oil thrice, at weekly intervals, during the refoliation period gave good control of the disease in Malaysia (Tan and John, 1985). Mechanised spraying or fogging with chlorthalonil in water and in oil, respectively, gave better results than with captafol (Lim, 1987). In Cameroon, artificial defoliation using ethrel was also used as a means of induced defoliation and refoliation to effect avoidance of secondary leaf fall (Senechal and Gohel, 1988).

Corynespora leaf fall

General: Corynespora leaf fall disease caused by Corynespora cassiicola (Berk & Curt.) Wei., was first observed in seedling nurseries in India (Ramakrishnan and Pillai, 1961) and later in Malaysia on iron deficient nursery plants (Newsam, 1961). Subsequently, the disease was reported from other countries in Asia (Situmorang and Budiman, 1985; Liyanage et al. 1986; Kajornchaikul, 1987), Africa (Awoderu, 1969) and South America (Liayanage, 1986). In many rubber growing countries the disease has caused severe damage only in a few introduced and indigenous clones but in Sri Lanka it has become the most devastating disease, affecting more than 3000 ha of susceptible clones, mainly the clone RRIC 103, both in immature and mature plantations (Liyanage, 1987a). Several alternate hosts are known to harbour the fungus (Situmorang and Budiman, 1985; Liyanage et al. 1986; Liyanage, 1987b).

Symptoms: The fungus affects the immature and mature leaves, the former being more susceptible. The symptoms first appear as greyish brown spots which enlarge into conspicuous circular or irregular lesions of varying sizes and shapes. Several spots may coalesce to produce extensive crisp brown areas on the leaf, some of which may become irregular papery lesions giving a scorched, shrivelled appearance. On mature leaves the characteristic feature of the disease is the browning or blackening of the veins adjacent to the lesions giving a 'fish bone' appearance. The area around the lesions gradually become chlorotic due to the destruction of chloroplasts. Even a single lesion on a leaflet could result in defoliation. Greyish black lesions may also be seen on some petioles causing defoliation even without lesions on the leaf blade (Liyanage et al. 1986; Liyanage, 1987b). Repeated defoliation results in the severe retardation of growth, extending the period of immaturity and eventually causing die back of shoots and branches (Liyanage et al. 1986), bark splitting along the trunk (Kajornchaikul, 1987) and even death of trees (Liyanage et al. 1986; Kajornchaikul, 1987).

Biology: The fungus is very variable in cultural morphology, growth rate and sporulation (Liyanage, 1987b; Chee, 1987; Soekirman and Purwantara, 1987). Abundant sporulation was observed with two hours daily exposure to ultra-violet (UV) light and progressively less with continuous light, alternate light and dark, two hours daily fluorescent light and continuous dark (Chee, 1987). Exposure to near UV light for varying periods did not encourage heavy sporulation (Liyanage, 1988). Sporulation was enhanced by washing the culture with running water for 24 h and drying it at 25±2ºC for a period of two weeks (Liyanage, 1988), but daily scraping of the aerial mycelium of the culture did not increase sporulation (Chee, 1987). When cultures were incubated in the dark for three days followed by incubation in the light for a further three, six or nine days, there was significantly more sporulation after six days than after three or nine days (Chee, 1987). Most conidia germinated within 3-4 h at 25-30ºC producing one or more germ tubes which arose from both ends of the spore (Liyanage et al. 1986; Liyanage, 1987b; Chee, 1987; Soekirman and Purwantara, 1987). At lower temperatures of 15ºC and 20ºC a period of 12 h was required for spore germination to occur. At 40ºC conidia germinated but germ tube growth was limited (Liyanage, 1987b; Soekirman and Purwantara, 1987). Free water stimulated spore germination (Liyanage, 1987a; Soekirman and Purwantara, 1987) but it was not essential as the spores germinated when the relative humidity was close to dew point. At lower humidities it required longer periods of incubation for spore germination (Soekirman and Purwantara, 1987; Liyanage, 1988). Spore germination was best at 100% relative humidity with a progressive reduction at lower humidities and none germinated after 24 h exposure to 50% relative humidity (Soekirman and Purwantara, 1987; Liyanage, 1988). The lower leaf surface supports better germination of spores than the upper surface. The spores germinated in 3-4 h and entered pallisade cells in between epidermal cells and ramified inter-cellularly in the mesophyll cells, producing conidia at the opposite leaf surface 96 h after inoculation (Liyanage, 1987b). In many instances a single lesion around the vein was sufficient to cause defoliation, though mechanical removal of such lesions prevented defoliation of leaves (Liyanage, 1987b; Chee, 1987). The fungus produced a toxin in a synthetic medium and the maximum amount of toxin was produced after 10 days of incubation at 28±2ºC (Liyanage and Liyanage, 1986). A technique was developed for rapid screening of clones in the laboratory using the culture filtrate containing the toxin (Liyanage and Liyanage, 1986).

Epidemiology: The conidia are wind dispersed. The pattern of spore release followed a diurnal rhythm. Spore release began around 0500 h and tailed off by 1800 h. The maximum spore release occured between 0730 and 1130 h with a peak production at 0930 h and very few spores or none were released during the night (Liyanage, 1987b). However, in Malaysia the spore release started at 0800 h and reached a peak around noon and declined to a low level until sunrise the next day (Chee, 1987). Fewer spores were caught during wet weather but a high spore count was recorded on five days succeeding a day of wet weather (Chee, 1987). Conidia on leaves remained viable for a period of 30 days after their exposure to natural conditions.

Over 80 alternate hosts of the fungus have been reported (Situmorang and Budiman, 1985; Liyanage et al. 1986; Chee, 1987). Under Malaysian conditions Hevea isolates did not cross infect other hosts nor did the Carica papaya (pawpaw) isolate infect Hevea (Chee, 1987) but in Sri Lanka they infected both C. papaya and Mikania scandens, but the leaves were only mildly infected (Liyanage, 1987a).

Control: In India two rounds of prophylactic spraying with 1% Bordeaux mixture or 0.2% Dithane Z-78 (zineb) was recommended for budwood and seedling nurseries and immature clearings, as a prophylactic measure (Ramakrishnan and Pillay, 1961). In Sri Lanka, application of 0.3% Benomy, mancozeb and Orthocide and 0.4% Propineb at 4-5 day intervals during wet weather and seven day intervals during dry period was recommended. In Thailand 1% Bordeaux mixture was recommended as a prophylactic treatment on untapped trees, and application of 0.75% a.i. was suggested as an alternative spray (Kajornchaikul, 1987). In Malaysia, benomyl was found to be the most effective fungicide under laboratory conditions, but under field conditions, only partial control was achieved with the application of six rounds of 0.15% benomyl and thiram, at weekly intervals, during the refoliation period (RRIM, 1975). More recently, several chemicals such as chlorothalanil, benomyl, triademtan and tridemorph have been used as protective sprays (Liyanage, 1986). In Sri Lanka spraying mature trees with vehicle mounted thermal foggers using fungicides such as benomyl and mancozeb did not give adequate control. Field trials using portable thermal foggers and mist blowers indicated that neither is suitable for use in mature trees but were adequate for immature clearings and nurseries (Liyanage, 1987b). It was also observed that if spraying was done at five day intervals, commencing at refoliation, it prevented the establishment of the disease. When spraying was stopped after six months, the leaves succumbed to the disease and within a further period of six months dieback was observed (Liyanage, 1987b). Thus, chemical control of corynespora leaf fall is likely to be much more difficult than that of secondary leaf fall caused by other foliar diseases as the disease occurs throughout the year and on leaves of all ages. Further, the fungus produces a toxin which induces leaf abscission, so that one lesion on the petiole or main vein is sufficient to cause leaf fall.

In view of the seriousness of the disease and the extended period of fungicide application, many susceptible clones, especially the immature trees were successfully crown budded and base budded (Liyanage, 1987a). Alternatively, planting of resistant clones appears to be the only practical method of combating the disease.

South American leaf blight

General: South American leaf blight (SALB) caused by Microcyclus ulei (P. Henn.) Arx is indigenous to South America. It occurs in the region of the Amazon river system, the Guianas, the upper Orinoco and the Matto Grosso. SALB is presently confined to South America and is absent from Africa and Asia where most of the world's rubber is planted. This disease has caused many plantations of rubber in South America to be abandoned and is a serious impediment for any future expansion of the industry.

M. ulei is known only from Hevea species. Of the nine species of the host which are not uniformly distributed throughout the geographical range of the genus, five are known to be susceptible, ie. H. brasiliensis, H. benthamiana, H. guianensis and H. spruceana (Chee, 1976a) and H. camporum (Junqueria, 1988). The rest were not infected naturally. The disease spread throughout the South American countries from infections arising from wild trees (Alandia and Bell, 1957), but spread to Trinidad and Central America and to Bahia and Sao Paulo areas in Brazil, presumably through planting material when attempts were made to grow rubber in these regions.

Symptoms: The symptoms of the disease are most conspicuous on the leaves. When immature leaves are infected conidia appear in grey black lesions, usually on the lower surface, accompanied by leaf distortion. If the infection is severe, the leaves blacken, shrivel and fall off. When older leaves are infected distortions may be absent or slight, but a few lesions which may develop assure a ragged shothole appearance. The leaves that survive the primary infection show characteristic black pycnidia mostly on the upper surface around the edges of shotholes. At full maturity of the leaves, the stroma increase in size and become rough to touch, having well developed ascospores. These symptoms can also occur on petioles, green stems, inflorescences and young fruit leading to distortions, suberisation and splitting.

Biology: The obligate nature of the causal fungus M. ulei makes it difficult to manipulate in vitro. Growth of the fungus in culture is correlated with conidial viability, which depends on seasonal changes (Chee, 1978a). The medium most suitable for growth and sporulation was potato sucrose agar (Chee, 1978a; Holliday, 1970), but enriching this medium with purified vitamins and growth substances often resulted in inhibition of growth (Chee, 1978a). Fresh rubber leaf extract stimulated growth and sporulation (Langford, 1943a) but subsequent studies showed that air dried rubber leaves gave better results (Chee, 1978a). Two morphological strains and three distinct colony types were formed amongst field isolates and laboratory produced conidia, respectively (Chee, 1978a). The optimum temperature for germination and germ tube growth on various media was around 24-28°C (Holliday, 1970; Langford, 1943a; Blasquez and Owen, 1957) but later investigations showed that the fungus grew best at 23°C (Chee, 1978a). The optimum temperature for lesion development on inoculated leaf discs was between 24-26°C under a 16/8 h light/dark period of six days (Chee, 1976b). The growth of germ tubes was slower in distilled water than on susceptible leaves (Holliday, 1970). Exposure of the fungal culture to near ultra violet light daily, for a period of 45 min increased sporulation (Chee, 1978a; Holliday, 1970). Good conidial germination was observed after three days when the cultures were exposed to 70% relative humidity at 27°C (Chee, 1976b). Conidia stored under normal laboratory conditions survived for two weeks (Brookson, 1963; de Jonge, 1962). Similar observations were made when conidia in leaf lesions germinated (50%) after leaves were stored in dessicators for 15 weeks or frozen (-20°C) for two weeks or kept at 40°C for several days (Chee, 1976c). Leaf infection occured more readily with dry conidia at 100% relative humidity than at 65% relative humidity. However, sporulation occured at both 65% and 100% relative humidity after lesion formation, but was mot abundant at 100% relative humidity (Chee, 1976c), especially at temperatures between 23-25°C (Kajornchaiyakul et al. 1984). Dry conidia require 6 h of high humidity immediately after deposition and the disease intensity was higher when plants were incubated at 19-25°C than at 26-32°C with the optimum around 23-25°C (Kajornchaiyakul, et al. 1984). Exposure to ultra violet light (2537 °A) for 4 min, killed 90% of the conidia (Brookson, 1963; de Jonge, 1962), and light reduces viability with time, compared with dark and natural indoor light (Holliday, 1970).

The temperature for ascospore germination was 24°C, but none germinated at 12°C or 32°C (Chee, 1976c). Ascospores completed germination in 2.5 h in darkness and took 6 h in the light (Chee, 1976c). Germ tubes were three times longer in the dark than in the light after 6 h (Chee, 1976b). No germination occurred after nine days at relative humidities above 80% or after 15 days under desiccation (Chee, 1976c). The germination rate of ascospores was consistently high irrespective of the conditions under which they were produced (Chee, 1976b). The germination of ascospores was higher in floating than in submerged spores (Chee, 1976c). Ascospores were killed by a 4 min exposure to ultra violet light (Chee, 1976c).

Variation in cultural characteristics such as colony appearance, growth rate and sporulation have been observed (Chee, 1978a; Holliday, 1970; Junqueira et al. 1984). Experimentally, races 1 and 2 (Langdon, 1965) and races 3 and 4 (Miller, 1968) were differentiated. The existence of nine races with eight of them present in the state of Bahia, Brazil was demonstrated (Chee et al. 1986) by using field inocula introduced on leaf discs of various Hevea clones from which the differentiating clones were selected. Three main groups of isolates were identified in Brazil from cultural characteristics but no attempt was made to classify them into races (Junqueria et al. 1985). However, they were designated 4a, 4b and 4c (Sudhevea, 1970). These were later renamed as races 4, 5, 6, respectively (Chee et al. 1986). However race 4c was similar to race 7 of Chee et al. (1986). Later, physiologic races of isolates were identified by inoculating leaf discs and differential plants grown as polybags. More recent work in the state of Bahia indicated that isolates studied could be classified into more predominant races 2, 4, 5 and 6 and morphologically into two groups based on differences in their growth and sporulation in medium (Hashim and de Almeida, 1987). It was also observed that a clone could be infected by more than one race in the field simultaneously. It has also been shown that a race may consist of strains of different virulence which can account for the differences in pathogenicity from locality to locality (Langdon, 1966; Liyanage and Chee, 1981).

Epidemiology: Both conidia and ascospores play an important role in the spread of the disease. The diurnal periodicity of conidial production with a maximum around 1000 h and decreasing towards the evening, with a minor peak between 2000 h and 2100 h, reaching a minimum in the early morning was reported (Holliday, 1969). Subsequent investigations showed that peak production of conidia occurred a little later than 1000 h. However, similar number of conidia were trapped between 1000-1200 h on both cloudless and cloudy days (Chee, 1976d). No conidia were trapped during the dry season although a few were found on old lesions. The maximum conidial production coincided with the time of widespread field infection (Holliday, 1969; Chee, 1976d). On the otherhand, ascospores were present throughout the year. Ascospore production on dry days followed a marked diurnal pattern, with low catches during the day but increasing to reach a peak at 0600 h (Chee, 1976d). However, in wet weather the heavy ascospore production also occured during dry weather with a day time maximum coinciding with a rainfall peak which is usually higher than the peak production on dry days. Wetting of perithecia is a prerequisite for ascospore discharge (Holliday, 1969) and the presence of ascospores in the air coincided with rise in relative humidity, temperature and dew than with rain (Chee, 1976d).

Epidemics of the disease occur when the daily temperature is under 22°C for longer than 13 h, relative humidity over 92% for a period over 10 h and rainfall exceeding 1 mm per day the preceeding seven days (Holliday, 1969; Chee, 1976d). Continuous heavy rain is inimical to the spread of the disease as it can wash off the spores, but light intermittent showers distributed throughout the year is favourable for disease development (Holliday, 1969; Chee, 1976d; Stahel, 1917). There is evidence to suggest that the spread of the disease to Haiti is from the spores brought over by wind and rain from Guiana and Trinidad (Compagnon, 1976). Similarly, there is strong circumstantial evidence that the spread of the disease from the Amazon basin to the surrounding areas was caused by long distance dissemination and deposition of spores (Liyanage, 1981a).

Control: The beneficial effect of Bordeaux mixture to control the disease was first demonstrated in 1913 (Bancroft, 1913) and later the use of insoluble copper and wettable sulphur gave good results (Langford, 1943b). Subsequently, ferbam (ferric dimethyldithiocarbamate) (Hilton, 1955) and zineb (zinc ethylenebisdithiocarbamate) were found to be more effective (Langford and Echeverri, 1953; Langford and Townsend, 1954). More recent experiments in Brazil have shown that both mancozeb (Dithane M 45) and benomyl (Benlate) were effective (Rogers and Peterson, 1976). In field trials thiophanate methyl (0.07% a.i.) and benomyl (0.025% a.i.) were most effective in controlling leaf infection, followed by chlorothalonil (0.15% a.i.) and mancozeb (0.32% a.i.). Benomyl suppressed conidial sporulation, whereas one application of thiophanate methyl (0.14% a.i.) to perithecia, inhibited ascospore release; half of this concentration applied to conidial lesions or pycnidia caused the perithecia formed subsequently to abort (Chee, 1978b). These fungicides have been used widely in plantations in Brazil (Chee and Wastie, 1980). More recently, triadimefon (0.015% a.i.), triforine (0.038% a.i.) and bitertanol (0.03% a.i.) have also given promising results in nursery trials. It was also shown that mixtures of systemic and protective fungicides were more effective than the application of individual fungicides alone (Santos et al. 1984).

The difficulty of spraying fungicides to reach the canopy of mature rubber trees was recognised early. The use of portable mist-blower to spray several fungicides was unsuccessful (Rocha, 1972). Aerial spraying was soon therefore considered the only feasible means of spraying mature trees. The first attempt to control the disease by aerial spraying was made in 1971. Since then, numerous fungicides have been used involving different dilutions, intervals and diluents (Rogers and Peterson, 1976; Mainstone et al. 1977), and also various types of aircrafts were tried for spraying in Brazil (Mathews, 1976). An alternative to aerial spraying was fogging with portable thermal foggers (Tifa vehicle mounted 'TART' and wheel drawn 'TIGA', vehicle mounted 'LECO' and Dynafog). These machines can cover 200-600 ha in a day and use either 200 g thiophanate methyl or 1 kg mancozeb per hectare (Chee and Wastie, 1980). These two fungicides are used on a large scale in Brazil starting before refoliation and continuing at four day intervals initially and later at seven day intervals to complete 12 rounds. Thiophanate methyl was applied at a rate of 175 g ha^{-1} suspended in 6 l of an 80:20 mixture of shell spray oil and diesel fuel (Chee and Wastie, 1980).

The technique of artificial defoliation to hasten refoliation while the weather is still dry so as to enable the trees to escape infection as practised in Malaysia (Rao, 1972) was also tried in Brazil. Two defoliants Folex (merphos 2.2 kg ha^{-1}) and Dropp (thidiazuron 0.6 kg ha^{-1}) were tested by aerial application and fogging but results were variable (Romano et al. 1982) and this method was not widely adopted in Brazil.

Crown budding of high yielding clones of eastern origin with crowns of resistant clones was first attempted at Ford's Beltera in the Amazon basin. This method was used on a large scale for many years (Tollenaar, 1959). Several clones of Fx and IAN origin as well as H. pauciflora clone PA 31 were recommended for crown budding (Rands, 1946; Ostendorf, 1948). This technique was not popular because of the adverse effects on depressing growth and yield due to the interaction of the crown on high yielding scion clone. To minimise the unfavourable interactions, crown budding at a height of 2.5 m has been recommended (Rands, 1946). Later, when H. brasiliensis was crown budded with H. guianensis, H. spruceana, H. collina and H. confusa it yielded better than the latter clones when used alone (Ostendorf, 1948).

Breeding programmes to incorporate SALB resistance with high yield were done mostly in Brazil at Belterra and Belem (IPEAN) and also on a small scale in Costa Rica and Guatemala (Langford, 1943). Although a large number of resistant clones were bred in Brazil, only six appear to be acceptable for commercial planting. They are Fx 3810, Fx 3899, Fx 3925, IAN 717, IAN 710 and Fx 25. Of these, the first four have H. benthamiana, clone F4542 as the resistant parent while the last two derived their resistance from F409 and F351 (also H. benthamiana clones) respectively. However, variation in susceptibility between localities (Chee and Wastie, 1980) and also breakdown of resistance due to occurrence of physiologic races (Langdon,

1965) are two problems which affect the progress of the breeding programme. Breeding for SALB resistance was started in Malaysia and Sri Lanka with 25 clones and 42 clones, respectively (Brookson, 1956; Baptiste, 1958), obtained on an exchange basis. These clones were used as in a hybridization programme and also tested for their vigour and yield under Malaysian and Sri Lankan conditions. Out of these clones, an illegitimate progeny of Fx 25 and IAN 873 gave yields comparable to PB 86 in Malaysia (Subramaniam, 1969). Several clones with resistance to SALB together with all desirable secondary attributes have been obtained from a breeding programme initiated in 1961, some of which having a combination of different sources of resistance, have been recommended for large scale planting in Sri Lanka (Fernando, 1962; Fernando and Liyanage, 1980).

More recently biological control methods have been successfully carried out using the hyperparasite Hansfordia pulvinata (Berk & Curt Huges), which grows well on conidial lesions as well as on stromatic layers of M. ulei. Under high humidity conditions, the parasite colonized 93% of the conidial lesions, and 86% of the stromatic areas were destroyed (Lieberel et al. 1989).

STEM AND BRANCH DISEASES

Stem and branch diseases are present in many rubber growing countries but they usually do not cause serious damage except in certain localities where conditions favourable for disease development are present. Pink disease, Ustulina stem rot, Phellinus stem rot and Botryodiploida die back caused by Corticium salmonicolor Berk. & Br., Ustulina deusta (Hoffm. ex Fr.) Lind., Phellinus noxius (Corner) G.H. Cunn. and Botryodiploida theobromae Pat., respectively are the more common diseases. However, pink disease can cause extensive damage and the remaining diseases are essentially caused by weak parasites, which gain entry in to trees through wounds, mainly caused by natural or accidental injuries.

Pink disease

General: Pink disease caused by Corticium salmonicolor Berk. & Br., is the only important stem disease of rubber. The fungus is widely distributed in all the countries and has a wide host range. The fungus attacks the bark of the main stem and branches of 3-7 year old immature trees, particularly at the fork region during periods of wet weather. Pink disease also occurs on mature trees and on such trees, stems and branches are somewhat slower to develop disease symptoms.

Symptoms: The first indications of an attack are exuding drops of latex from the region of the fork. This is followed by the appearance of white silky threads on the bark surface, which later give a cob-web appearance (Fig. 3). Under favourable conditions the disease spreads and a pink mass

Fig. 3.   Pink disease affected tree showing exudation of latex and cob web like mycelium.

of sterile mycelium begins to appear, when the outer bark dies. At this stage pink pustules erupt in lines through cracks in the bark, which disappear as the dead bark sloughs off. The 'corticium' stage that produces basidiospores consists of a smooth, pinkish-white surface covering the pink crust, while the nector stage consists of organge-red pustules composed of a tight mass of spores which are scattered over the upper surface of side branches. When the disease spreads it causes ring-barking causing the dormant buds below the injured portion to produce numerous side shoots. In a more advanced stage of infection, death of branches occurs, exposing bare shoots.

Biology: Both basidiospores and necator spores can be grown readily in culture on a variety of media (Hilton, 1958) and on distilled water at room temperature giving rise to sterile mycelia. At times mycelial aggregates formed in culture have to be regarded as analogous with the necator stage (Brooks and Sharples, 1951) but later investigations showed that such aggregations are structureless and are comparable with the pustular stage (Hilton, 1958).

Epidemiology: The disease occurs in wet weather and also in humid locations but it does not occur in the coastal areas of Malaysia (Hilton, 1958) and in soils where boron toxicity occurs. The former indicates the effect of drying winds in reducing humidity and thereby the disease incidence, and the latter is caused by a nutritional effect. The severity of the attack varies from one locality to another according to the rainfall pattern (Yeoh and Tan, 1974).

Unlike leaf and panel diseases which have well defined patterns of severity varying from clone to clone, pink disease shows more uniformity. A few clones are known to be of above average susceptibility but most cultivars are prone to the disease.

Control: Since the beginning of this century Bordeaux mixture was used widely to control this disease (Anstead, 1914). Eventhough copper was only mildly toxic to the fungus, it was preferred for its high tenacity under heavy rainfall conditions. With the refinements in the application methods it was used either as a brush-on 10% paste or as 1% liquid sprayed as a jet from the ground using a knapsack sprayer with a long lance with its swirl plate removed (Ramakrishnan and Radhakrishna Pillay,1 962b). In wet weather the chemical gets washed off an becomes less effective necessitating repeated applications. Further, trees in tapping cannot be treated with copper fungicides because of the risk of contaminating latex with copper (Wastie and Yeoh, 1972). Later, application of 0.5% solution of Fylomac 90 (tetradecyl pyridinium bromide) as a spray or brush-on application gave satisfactory control and was used to control the disease in mature trees (Alston, 1953; Yeoh and Tan, 1974). A number of non-copper fungicides are highly effective against the fungus in laboratory tests but under field conditions, most are ineffective since they rapidly get leached by rain (Wastie and Yeoh, 1972). This difficulty was overcome by in-corporating 5% Calixin (75% N-tridecyl 2,6 dimethyl morpholene) with natural rubber latex as a binder. One brush-on application of it gave good control for periods upto three months (Wastie and Yeoh, 1972). Subsequently, a formulation based on prevulcanised natural rubber (50%) as a carrier in 1.5% tridemorph as a fungicide was more effective than Bordeaux mixture (Yeoh and Tan, 1974). A polyvinyl acetate based formulation using either propiconazole or tridemorph was also reported to be effective (Jacob and Edathil, 1986). A formulation of 1.5% MK 23 [n-(P-fluorophenyl)-2,3 dichloromaleimide] in diluted field latex, applied as two sprays at six weekly intervals have given good control upto three months, and the costs are comparable to the brush-on calixin application (Tan and Yeon, 1976).

DRY ROT

General: Dry rot disease caused by Ustulina deusta (Hoffm. ex Fr.) Lind. causes complete loss of attacked trees as the trees break at the point of infection (Radhakrishna Pillai and George, 1980). This disease is rare in young rubber plantations. The pathogen is known to cause collar rot and root rot besides stem rot.

Symptoms: The initial symptom is copious exudation of latex from the point of attack. The attacked wood turns pale brown and is readily fragmented. Irregular flat grey fructifications of the fungus appear on the bark surface which later turn black and brittle. Infected wood shows a network of black double lines (Sripathi Rao, 1975).

Biology: The fungus grows well in vitro at 23°C, the mycelium being white to grey initially but becoming greyish brown with formation of conidia and black after 2 to 3 weeks (Hawksworth, 1972).

Epdidemiology: Ustulina deusta is essentially a wound parasite (Petch, 1921). But the pathogen also gains entry through lenticels and moribund root initials. The pathogen is known to show selectivity to live hosts rather than dead trees (Varghese, 1971). The incidence of the disease is observed to be more after heavy winds during the rainy season.

Control: Field sanitation and cutting and removal of infected branches followed by painting the cut surfaces with tar or asphalt mixutre has been recommended by early workers (Petch, 1921; Sharples, 1936). Application of a organomercurial fungicide solution followed by wound dressing was found to contain the disease (Radhakrishna Pillai and George, 1980). Direct application of any of the fungicides viz. methoxyethyl mercurychloride, thiram, oxycarboxin, carbendazim or thiophanate methyl, incorporated in a petroleum wound dressing compound after removal of affected tissues, was as effective as fungicide wash followed by painting with wound dressing compounds. Bordeaux paste was ineffective (Idicula et al. 1990).

PATCH CANKER

General: The occurrence of patch canker was reported first from Sri Lanka in 1903 and subsequently from other rubber growing countries. The disease is caused by infection of untapped bark by Phytophthora palmivora or Pythium vexans (Radhakrishna Pillai and George, 1980).

Symptoms: Exudation of latex is observed from the point of infection. Accumulation and coagulation of latex under the bark at these points forms a pad which give a foul odour and leads to cracking of bark. Internal tissues show discolouration and rotting. The purplish discolouration (Sharples, 1936) is absent in some cases (Chee, 1968).

Biology and Epidemiology: The biology of the pathogen has already been discussed under Phytophthora leaf fall. Pythium vexans is associated only when the canker is on the collar or on roots. Infection is through wounds. Wet weather favours the spread of the disease.

Control: Excision of the disease affected tissues followed by painting of the wound with organomercurial fungicides is effective (Sharples, 1936; Radhakrishna Pillay and George, 1980). This should be followed by application of a wound dressing compound. Application of Bordeaux paste also is effective (Ramakrishnan, 1964).

PANEL DISEASES

Besides the stem diseases described, panel diseases like black stripe, mouldy rot and panel necrosis also are reported.

Black stripe

General: Among the panel diseases, black stripe disease caused by Phytophthora palmivora (Butl) Butl., P. meadii Mc Rae or P botryosa Chee, is more important (Chee, 1990).

Symptoms: The disease appeares as a vertical linear depression in the tapping panel which when scraped shows black lines on the wood beneath. The pathogen destroys the bark tissue leaving large wounds which make subsequent tapping on the same panel difficult (Radhakrishna Pillay and George, 1980).

Biology and Epidemiology: The biology of the pathogen has already been discussed. The disease is severe when tapping is continued during rainy season unless regular panel protection measures are undertaken (Ramakrishnan and Radhakrishna Pillay, 1963). Humidity over 90% and frequent wetting of tapping panel favour the spread of the disease.

Control: Black stripe can be controlled by application of organo-mercurial fungicides at frequent intervals on the tapping panel during rainy season (Ramakrishnan and Radhakrishna Pillay, 1963a). Other fungicides like captafol (Yeoh and Tan, 1980), oxadixyl (Tran, 1986) and mancozeb (Thomson et al. 1988) are also reported to be effective. Application of a panel dressing compound on renewed bark before the onset of monsoons helps in the prevention of disease incidence.

ROOT DISEASES

Among the root diseases of rubber, three are considered to be of significance - white root disease, brown root disease and red root disease (Chee, 1976). As the infection of roots and collars of trees leads to death of trees, these diseases cause reduction in the mature stand and yield per hectare.

White root disease

General: White root disease caused by Rigidoporus lignosus (Kl.) Imazeki is the most serious root disease of rubber due to its fast spreading nature (Sharples, 1936) and its early appearance in the field. This disease was first reported from Singapore (Ridley, 1904) and from Sri Lanka in 1905 (Petch, 1921).

Symptoms: In affected trees general discolouration of the foliage turning off-treen in colour and giving a ripened appearance are the earliest above ground symptoms. Leaves later turn yellow and drop. Premature flowering and die back of the branches are also common. On the affected roots white rhizomorphs, which turn pale orange red when old, are seen. The fruiting bodies are firm, flesh and usually tiered (Hilton, 1959). Wood newly affected by the fungus is brown and hard but in later stages it is white or cream and firm (Chee, 1976a).

Biology: Rigidoporus lignosus is a basidiomycete which produces bracket like sporophores on the collar of naturally infected trees. The fungus spreads through rhizomorphs. The isolates of the fungus could be identified by its characteristic penetration pattern on Jenson's agar medium. Fruitification of the fungus was induced in inoculum containing pathogen on malt extract agar (600 ml) when field conditions were simulated in the laboratory (Fox, 1960).

Twenty isolates collected from a range of geographical origins could be grouped according to their origin on the basis of soluble protein banding patterns obtained by isoelectric focussing of isozyme profiles (Louanchi et al. 1992).

Epidemiology: Freshly felled rubber stumps get infected by airborne spores (Sharples, 1936) but the chief method of spread is by root contact (Hilton, 1959). The mycelium can travel in soil for some distance but the need for a food base for the rhizomorph to remain viable has been demonstrated (John, 1961). The fructifications appear near the collar of affected trees during wet weather. The influence of planting distance on the incidence of the disease also has been established (Liyanage, 1981b).

Control: Field sanitation and removal of all affected root materials have been suggested as the most important preventive measures (Petch, 1921; Sharples, 1936: Hilton, 1959). Complete mechanical removal of roots of trees while clearing before planting was preferred to stump poisoning method (Newsam, 1967). Leguminous cover crops grown in rubber plantations act as decoy hosts and help in reduction of inoculum potential (Fox, 1965). Application of sulphur to the plant bases has been of advantage in reducing soil pH and stimulating soil antagonists which cause lysis of the pathogen (Peries and Liyanage, 1983). In vitro antagonism of several soil fungi like Trichoderma viride, T. harzianum, Gliocladium roseum has been demonstrated (Jollands, 1983). Several fungicides have been reported to be useful as collar drench. These include tridemorph (Tran, 1986), triademefon, triademenol, propiconazole (Tan, 1990). An integrated approach using both fungicides (Triademofon and Tridemorph) and the antagonistic fungus Trichoderma has been observed to improve the disease control. Introduction of the antagonist two months after drenching with fungicides ensured that a high population of the antagonist was maintained in the soil (Hashim, 1990).

BROWN ROOT DISEASE

General: Brown root disease caused by Phellinus noxius (Corner.) was first reported from Sri Lanka (Petch, 1921). It is of greater significance in Sri Lanka and India (Rajalekshmy, 1980). Higher disease incidence is observed in light soil.

Symptoms: The incidence of the disease can be detected by the discolouration of the foliage along with cessation of growth (Ramakrishnan and Radhakrishna Pillay, 1963b). The trees showing such external symptoms can not often be saved as the infection might have progressed considerably killing the root tissue. However, prophylactic measures can be undertaken if roots are only partially infected. The rhizomorphs form a continuous fungal mat over infected roots, brown in colour turning black with age (Chee, 1976d). A layer of soil mixed with fungal mycelium forms a hard brittle mass which is difficult to be washed off (Petch, 1921). Wood also shows brownish discolouration and in advanced stages honey combing is seen (Ramakrishnan and Radhakrishna Pillay, 1962c).

Biology and Epidemiology: Phellinus noxius is a basidiomycete forming bracket shaped sporophores. The fruiting bodies are brownish purple or black on the upper surface with concentric growth rings and grey on the under surface. Abortive sporophores are often seen at the collar region. These are tawny brown and water soaked. The fungus colonises the stumps of trees left in the plantations which form a source of infection. Underground root contact between healthy and diseased roots is the chief method of spread. The progress of disease from tree to tree is slow when compared to that of white root disease (Petch, 1921). Several cultivated tree crops and forest trees are alternate hosts of the pathogen (Sripathi Rao, 1975).

Control: Curative treatments are effective only if the disease is detected early. The root systems of affected trees are excavated and portions of dead roots removed. The partially affected roots are scraped, washed with an organomercurial fungicide and then painted with a petroleum wound dressing compound. The tree bases are then refilled and packed firmly. Prophylactic dressing with a bitumin or grease compound in which 10% tridemorph is incorporated is reported to be effective (RRIM, 1974b).

RED ROOT DISEASE

General: Red root disease caused by Ganoderma philippii (Bres. & P. Henn.) Bres is a less common disease which develops slowly and is more frequently encountered in mature plantations (Hilton, 1956). However recent reports from China indicate that even young plantations are affected and that the spread is relatively fast (Tan and Fan, 1990).

Symptoms: The symptoms of brown root disease and red root disease are similar except that in the case of red root disease, the mycelium which covers the root surface is red or reddish brown with creamy white growing margins. The affected wood is pale brown and hard at first becoming pale buff, wet and spongy and breaks easily in layers when dry.

Biology and Epidemiology: Ganoderma philippii is a basidomycete which forms bracket like fruiting bodies which are hard and woody with dark reddish brown wrinkled upper surfaces and ashy white lower surfaces. Mature fructification bear abundant spores (RRIM, 1974a). Although infection from spores is likely, the spread is largely by root contact. Larvae of certain flies which breed within the fructifications are capable of spreading the disease as the spores are viable even after passing through their gut (Lim, 1971b).

Control: The control measures include preplanting eradication of the sources of infection and prevention of the spread in the stand. Protective dressing of roots with fungicides (10% Drazoxolon or 10% Iridemorph) is recommended (Tan and Lim, 1971). Soil fumigation along with mulching and application of fertilizers and drenching with Drazoxolon is also reported to be effective as it helps in stimulation of native anatagonists (Varghese et al. 1975). An integrated control system involving preplanting inspection and removal of source of infection, post planting inspection and treatment prior to and after opening for tapping both by eliminating infected material and by drenching 0.75% Tridemorph at the rate of 200 ml per tree at 6 months interval for two years was found useful in China (Tan and Fan, 1990).

REFERENCES

Alandia, S. and Bell, F.H. 1957. Diseases of warm climate crops in Bolivia. Pl. Prot. Bull. F.A.O. 5: pp. 172-173.

Alston, R.A. 1953. Report of Rubber Research Institute of Malaya. Pathological Division Report for the year 1950-1951.

Anstead, R.D. 1914. Pink disease of para rubber and Bordeaux mixture. Plant. Chron., 6: 98.

Ashplant, H.T. 1928. Bordeaux and burgundy spraying mixtures. Scientific Dept. Bull. United Planters of South India, United Planters Association.

Awoderu, V.A. 1969. A new leaf spot of para rubber Hevea brasiliensis in Nigeria. Pl. Dist. Reptr., 53: 406-408.

Azaldin, M.Y. and Rao, B.S. 1974. Practicability and economics of large scale artificial defoliation for avoiding secondary leaf fall. In: Proc. Rubb. Res. Inst. Malaysia Plrs' Conf. 1974, Kuala Lumpur, 161 pp.

Bancroft, C.K. 1913. A leaf disease of para rubber. Bd. Agric. Brit. Guiana, 7: 37-38.

Baptiste, E.D.C. 1958. Director's Report. Ann. Report. Rubb. Res. Inst. Cey. 1957. pp. 1-16.

Beeley, F. 1930. A recent outbreak of secondary leaf fall due to Oidium heveae. J. Rubb. Res. Inst. Malaya, 2: 61.

Beeley, F. 1932. Effect of meteorological factors on the virulence of Oidium heveae in Malaya. J. Rubb. Res. Inst. Malaya, 4: 104.

Beeley, F. 1935. Diseases and pests of the rubber tree. Planter, Kuala Lumpur, 16: 520.

Blazquez, C.H. and Owen, J.H. 1957. Physiological studies on Dothidella ulei. Phytopathology, 47: 727-732.

Brooks, F.T. and Sharples, A. 1915. Pink disease of plantation rubber. Ann. App. Biol., 2: 58.

Brookson, C.W. 1956. Importation and development of new strains of Hevea brasiliensis. J. Rubb. Res. Inst. Malaya, 14: 423-447.

Brookson, C.W. 1963. Botanical Division Report. Rubb. Res. Inst. Malaya, 1962. pp. 34-35, 64-66.

Chee, K.H. 1968. Patch canker of Hevea brasiliensis caused by Phytophthora palmivora. Plant Disease Reporter, 52: 132-133.

Chee, K.H. 1969. Variability of Phytophthora species from Hevea brasiliensis. Trans. Br. Mycol. Soc., 52: 425-436.

Chee, K.H. 1973. Production, germination and survival of chlamydospores of Phytophthora palmivora from Hevea brasiliensis. Trans. Br. Mycol. Soc., 61: 21-26.

Chee, K.H. 1976a. Assessing susceptibility of Hevea clones to Microcyclus ulei. Ann. App. Biol., 84: 135-145.

Chee, K.H. 1976b. South American Leaf Blight of Hevea brasiliensis: spore behaviour and screening for disease resistance. In: Proc. Int. Rubber Conf. 1975, Kuala Lumpur, Vol. 3, pp. 228-235.

Chee, K.H. 1976c. Factors affecting discharge, germination and viability of spores of Microcyclus ulei. Trans. Br. Mycol. Soc., 66: 449-504.

Chee, K.H. 1976d. Microorganisms associated with rubber (Hevea brasiliensis Muell. Arg.). Rubb. Res. Inst. Malaysia. 78 p.

Chee, K.C. 1978a. South American leaf blight of Hevea brasiliensis: Culture of Microcyclus ulei. Trans. Br. Mycol. Soc., 70: 341-344.

Chee, K.H. 1978b. Evaluation of fungicides for control of South American leaf blight of Hevea brasiliensis. Ann. App. Biol. 90: 51-58.

Chee, K.H. 1987. Studies on sporulation, pathogenic and epidemiology of Corynespora cassiicola on Hevea rubber. In: Proc. Int. Rubb. Res. Dev. Bd., pp. 6-17.

Chee, K.H. 1990. Present status of rubber diseases and their control. Review of Plant Pathology, 69: 423-430.

Chee, K.H. and Wastie, R.L. 1980. The status and future prospects of rubber disease in Tropical America. Rev. Plant Pathol., 59: 541-548.

352

Chee, K.H. and Kai-Ming, Z. 1985. Diseases of Hevea in South Bahia, Brazil, caused by Phytophthora spp. Planter, 61: 299-305.

Chee, K.H., Zhang, K. and Darmono, T.W. 1986. The occurrence of eight races of Microcyclus ulei on Hevea rubber in Bahia, Brazil. Trans. Br. Mycol. Soc., 87: 15.

Chua, S.E. 1970. The physiology of foliage senescence and abscission in Hevea brasiliensis Muell. Arg. Thesis submitted for the degree of Doctor of Philosophy, University of Singapore.

Compagnon, M.P. 1976. Note on the influence of climatic conditions on the spread of SALB. 8 pp.

Cramer, P.J.S. 1956. Half a century of the rubber planting industry in Indonesia. Archs. Rubb. Cult., 33: 345.

Dantanarayana, D.M., Peries, O.S. and Liyanage, A. de S. 1984. Taxonomy of Phytophthora species isolated from rubber in Sri Lanka. Trans. Br. Mycol. Soc., 82: 113-126.

FErnando, D.M. 1962. Review of the Plant Breeding Section. Ann. Rev. Rubb. Res. Inst. Cey., 1961. pp. 38-65.

Fernando, D.M. and Liyanage, A. de S. 1980. South American leaf blight resistance studies on Hevea brasiliensis selections in Sri Lanka. J. Rubb. Res. Inst. Sri Lanka, 57: 41-47.

Fernando, T.M. 1971. Oidium leaf diseases - the effect of environment and control measures on incidence of disease and atmospheric spore concentration. Q. J. Rubb. Res. Inst. Ceylon, 48: 100.

Fox, R.A. 1960. White root disease of Hevea brasilienis. The identity of the pathogen. Proceedings of the Nat. Rubb. Res. Conf., 1960. Kuala Lumpur, pp. 473-482.

Fox. R.A. 1965. The role of biological eradication in root disease control in replantings of Hevea brasiliensis. Ecology of soil borne diseases - a prelude to biological control. pp. 348-362.

George, M.K. and Edathil, T.T. 1975. Over summering of Phytophthora causing abnormal leaf fall disease of rubber. Rubb. Bd. Bull. 12: 112-114.

Hashim, I. 1990. Possible integration of Trichoderma with fungicides for the control of white root disease of rubber. Root disease of Hevea brasiliensis. Proc. of IRRDB Symposium, Kunming, Chia, pp. 1-8.

Hashim, I. and De Almeida, L.C.C. 1987. Identification of races and in vitro sporulation of Microcyclus ulei. J. Nat. Rubb. Res., 2: 111.

Hawksworth, D.L. 1972. Ustulina deusta. CMI descriptions of pathogenic fungi and bacteria, Commonwealth Agricultural Bureau, No. 360.

Hilton, R.N. 1955. South American leaf blight: A review of literature relating to its depredations in South America, its thread to the Far East and the methods available for its control. J. Rubb. Res. Inst. Malaya, 14: 287-377.

Hilton, R.N. 1958. Pink disease of Hevea caused by Corticium salmonicolor Berk. et Br. J. Rubb. Res. Inst. Malaysia, 15: 275.

Hilton, R.N. 1959. Maladies of Hevea in Malaya. Rubb. Res. Inst. Malaya, 101 pp.

Ho, H.H., Liang, Z.R. and Yu, Y.N. 1984. Phytophthora spp., from rubber tree plantations in Yunnan Province of China. Mycopathologia, 86: 121-124.

Holliday, P. 1969. Dispersal of coinidia of Dothidella ulei from Hevea brasiliensis. Ann. Appl. Biol., 63: 435-447.

Holliday, P. 1970. South American leaf blight Microcyclus ulei of Hevea brasiliensis. No. 12, 31 pp.

Hutchison, F.W. 1958. Defoliation of Hevea brasiliensis by aerial spraying. J. Rubb. Res. Inst. Malaya, 15: 241.

Idicula, S.P., Edathil, T.T., Jayarathnam, K. and Jacob, C.K. 1990. Dry rot disease management in Hevea brasiliensis. Indian J. Nat. Rubb. Res., 3: 35-39.

Jacob, C.K. and Edathil, T.T. 1986. New Approaches to pink disease management in Hevea. Planter, 62: 463-467.

Jacob, C.K., Edathil, T.T., Idicula, S.P., Jayarathnam, K. and Sethuraj, M.R. 1989. Effect of abnormal leaf fall occured by Phytophthora spp. on the yield of rubber tree. Indian J. Nat. Rubb. Res., 2: 77-80.

John, K.P. 1958. Inoculation experiments with Fomes lignosus Klotzsch. J. Rubb. Res. Inst. Malaya, 15: 223-240.

Jollands, P. 1983. Laboratory investigations on fungicides and biological agents to control three diseases of rubber and oil palm and their potential applications. Tropical Pest Management, 29: 33-38.

Junqueira, N.T.V. 1988. Unpublished data.

Junqueira, N.T.V., Chaves, C.M., Zambolim, L. and Gasporotto, L. 1984. Variabilide Fisiologica de Microcyclus ulei. Agente Etiologico do Mal Das Pothas de Seringueira. Fitopathologia Brasiliera, 9, 385 p.

Junqueria, N.T.V., Zambolim, L. and Chaves, G.M. 1985. Resistance de clones de seringueira ao Mal. Das Folhas. Infme Agropecuario, Belo Horizonte, 11, 42 p.

Kaiming, Z. 1987. Important diseases of rubber trees in China with special reference to Oidium and Phytophthora. In: Proc. Int. Rubb. Res. Dev. Bd., Chiang Mai, 1987, pp. 40-51.

Kajornchaikul, P. 1987. Corynespora disease of Hevea in Thailand. In: Proc. Int. Rubb. Res. Dev. Bd., pp. 1-5.

Kajornchaikul, P., Chee, K.H., Darmono, T.W. and De Almeida, L.C.C. 1984. Effect of humidity and temperature on the development of South American Leaf Blight (Microcyclus ulei) of Hevea brasiliensis. J. Rubb. Res. Inst. Malaysia, 32: 217-223.

Langdon, K.R. 1965. Relative resistance or susceptibility of several clones of Hevea brasiliensis x H. benthamiana to two races of Dothidella ulei. Pl. Dis. Rep., 49: 12 p.

Langdon, K.R. 1966. Development of a new medium for culturing Dothidella ulei in quantity. Phytopathology, 56: 564-565.

Langford, M.H. 1943a. South American Leaf Blight of Hevea rubber trees. Tech. Bull. U.S. Dep. Agric., 882, 31 pp.

Langford, M.H. 1943b. Fungicidal control of South American Leaf Blight of Hevea. (U.S. Dept. of Agric. Circular) No. 686, 20 pp.

Langford, M.H. and Echeverri, H. 1953. Control of South American Leaf Blight by use of a new fungicide. Revista Turrialba (Costa Rica) 3: 102-105.

Langford, M.H. and Townsend, Jr. C.H.T. 1954. Control of South American Leaf Blight of Hevea rubber tree. Pl. Dis. Rept. Supplement 225, pp. 42-48.

Lieberel, R., Junqueira, N.T.V. and Feldman, F. 1989. Integrated disease control in rubber plantations in South America. In: Proc. Int. Pest mang. in Tropical and Sub-Tropical Cropping Systems, pp. 445-456.

Lim, T.M. 1971a. Dispersal of conidia of Oidium heveae and Colletotrichum gloeosporioides in a field of Hevea in West Malaysia. Epidemiology of Plant Diseases, Vol. III, Wageningen: Advanced Study Institute, 237 pp.

Lim, T.M. 1971b. The role of a tipulid fly in the germination of and dispersal of basidiospores of Ganoderma pseudoferrum. Abstr. 2nd International Symposium on Plant Pathology, New Delhi, 1971. p. 173.

Lim, T.M. 1972. A forecasting system for use in the chemical control of Oidium secondary leaf fall on Hevea. In: Proc. Rubb. Res. Inst. Malaya. Plrs' Conf., Kuala Lumpur, 1972, 169 p.

Lim, T.M. 1973. A rapid laboratory method of assessing susceptibility of Hevea clones to Oidium heveae. Expl. Agric., 9: 275.

Lim, T.M. 1974. Enhancing post wintering tree vigour for avoiding Oidium secondary leaf fall. In: Proc. Rubb. Res. Inst. Malaya, Plrs' Conf. 1974, Kuala Lumpur, 178 p.

Lim, T.M. 1982. Recent developments in the chemical control of rubber leaf diseases in Malaysia. In: Proc. Int. Conf. Pl. Prot. in Tropics, Kuala Lumpur, 219 p.

354

Lim, T.M., Radziah, N.Z. and Abdul Aziz bin S.A. Kadir. 1981. Rubber Leaf disease control - a case for mechanisation. In: Proc. Rubb. Res. Inst. Malaya, Plrs' Conf., pp. 19-21.

Lim, T.M. and Sripathi Rao, B. 1976. An epidemiological approach to the control of Oidium secondary leaf fall of Hevea. In: Proc. Int. Rubb. Conf. Kuala Lumpur, 1975, 3: 293 p.

Liyanage, A. de S. 1976. Review of the Plant Pathology Department. Ann. Rev. Rubb. Res. Inst. Sri Lanka, pp. 67-114.

Liyanage, A. de S. 1977a. Influence of some factors on the pattern of wintering and on the incidence of Oidium leaf fall in clone PB 86. J. Rubb. Res. Inst. Sri Lanka, 53: 31.

Liyanage, A. de S. 1977b. Review of the Plant Pathology Department. Ann. Rev. Rubb. Res. Inst. pp. 66-98.

Liyanage, A. de S. 1981a. Long distance transport and deposition of spores of Microcyclus ulei in tropical America - a possibility. Bull. Rubb. Res. Inst. Sri Lanka, 16: 3-8.

Liyanage, A. de S. 1981b. Review of Plant Pathology Department. In. Annual Review of Rubber Research Institute of Sri Lanka, 53-68.

Liyanage, A. de S. 1986. Unpublished data.

Liyanage, A. de S. 1987a. Management of corynespora leaf spot disease under Sri Lankan conditions. In: Proc. Int. Rubb. Res. Dev. Bd., pp. 21-23.

Liyanage, A. de S. 1987b. Investigations on corynespora leaf spot disease in Sri Lanka. In: Proc. Int. Rubb. Res. Dev. Bd., pp. 18-20.

Liyanage, A. de S. 1988. Review of the Plant Pathology Department. Ann. Rev. Rubb. Res. Inst. Sri Lanka, pp. 52-67.

Liyanage, A. de S. and Chee, K.H. 1981. The occurrence of a virulent strain of Microcyclus ulei on Hevea rubber in Trinidad. J. Rubb. Res. Inst. Sri Lanka, 58: 73-78.

Lim, T.M. 1987. Recent developments in the chemical control of rubber leaf diseases in Malaysia. In: Proc. Int. Conf. Pl. Prot. in Tropics. pp. 219-231.

Liyanage, A. de S. and De Alwis, K. 1978. Review of the Plant Pathology department. Ann. Rev. Rubb. Res. Inst. Sri Lanka, 54 pp.

Liyanage, A. de S., Jayasinghe, C.K., Liyanage, N.I.S. and Jayaratne, A.H.R. 1986. Corynespora leaf spot disease of rubber (Hevea brasiliensis) - A new record. J. Rubb. Res. Inst. Sri Lanka, 65: 47-50.

Liyanage, A. de S., Peries, O.S., Dharmaratne, A. 1983. Effect of weather factors on disease establishment and sporulation of Phytophthora meadii on rubber pods. J. Rubb. Res. Inst. Sri Lanka, 61: 41-48.

Liyanage, A. de S., Peries, O.S., Dharmaratne, A., Fernando, B., Irugalbandara, Z.E., Wettasinghe, S. and Wettasinghe, P.C. 1985. Biology of Oidium heveae, the powdery mildew fungus of Hevea brasiliensis. In: Proc. Int. Rubb. Conf. Kuala Lumpur, 3, pp. 291-313.

Liyanage, A. de S., Peries, O.S. and Sebastian, R.D. 1971. Assessment of the incidence of Oidium leaf fall and economics of its control in the smallholdings. Q. J. Rubb. Res. Inst. Ceylon, 48: 112.

Liyanage, N.I.S. 1986. Phytophthoras on rubber (Hevea brasiliensis): Pathogen taxonomy, survival and disease control. Ph.D. thesis, University of London.

Liyanage, N.I.S. 1989. Phytophthora citricola on rubber in Sri Lanka. Plant Pathology, 38: 438-439.

Liyanage, N.I.S. and Liyanage, A. de S. 1986. A study of the production of a toxin in Corynespora cassiicola. J. Rubb. Res. Inst. Sri Lanka, 65: 51-53.

Liyanage, N.I.S. and Wheeler, B.E.J. 1989. Comparative morphology of Phytophthora species on rubber. Plant Pathology, 38: 592-597.

Lloyd, J.H. 1963. The control of abnormal leaf fall disease (Phytophthora palmivora Butler) of Hevea in Ceylon. Bull. Rubb. Res. Inst. Ceylon, 57: 87 pp.

Louanchi, M., Fofana, A., Rohin, P., Balesdent, M.H. and Despreaux, P. 1992. Intraspecific variation in soluble protein and isozyme patterns in Rigidoporus lignosus the causative agent of white root disease on Hevea brasiliensis. Abstracts of papers presented in International Natural Rubber Conference, 1992, Bangalore, India. p. 35.

Mathews, G.A. 1976. Fungicide application for the control of South American Leaf Blight of Hevea brasiliensis. Mimeograph, 29 pp.

Mainstone, B.J., McManaman, G. and Begeer, J.J. 1977. Aerial spraying against South American Leaf Blight of rubber. Plrs' Bull. Rubb. Res. Inst. Malaysia, 148: 15-26.

McRae, W. 1918. Pytophthora meadii sp on Hevea brasiliensis. Mem. Dep. Agr. India Bot. Ser., 9: 219-273.

Miller, J.W. 1968. Differential clones of Hevea for identifying races of Dothidella ulei. Pl. Dis. Rep. 50: 187 p.

Murray, R.K.S. 1921. On the occurrence and signifiance of Oidium leaf disease in Ceylon. Q. Cir. Ceylon Rubb. Res. Sch., 2: 7.

Murray, R.K.S. 1936. Report on the Botanist and Mycologist for 1935. Ann. Rep. Rubb. Res. Sch., 36.

Newsam, A. 1961. Pathology Division Report. Rubb. Res. Inst. Malaya, pp. 63-70.

Newsam, A. 1967. Clearing methods for root disease control. Planters Bull., 92: 178-182.

Ostendorf, F.W. 1948. Two experiments with multiple grafts on Hevea. Archief. 26: 1.

Peries, O.S. 1965a. Review of the Plant Pathology Division. Ann. Rev. Rubb. Res. Inst. Ceylon, 1964, pp. 48-74.

Peries, O.S. 1965b. Review of the Plant Pathology Division. Ann. Rev. Rubb. Res. Inst. Ceylon, 1964, 48 p.

Peries, O.S. 1966a. Present status and methods of control of leaf and panel diseases of Hevea in South East Asian and African countries. Q. Jl. Rubb. Res. Inst. Ceylon, 42: 35-47.

Peries, O.S. 1966b. Host induced changes in the morphology of a powdery mildew fungus. Nature, 212(50610: 540-541.

Peries, O.S. 1969. Studies on epidemiology of Phytophthora leaf disease of Hevea brasiliensis in Ceylon. J. Rubb. Res. Inst. Malaya, 21: 73-78.

Peries, O.S. 1974. A study of the factors affecting spore germination in Oidium heveae. Q. Jl. Rubb. Res. Inst. Sri Lanka, 51: 57.

Peries, O.S. and Dantanarayana, D.M. 1965. Compatibility and variation in Phytophthora cultures isolated from Hevea brasiliensis in Ceylon. Trans. Br. Mycol. Soc., 48: 631-637.

Peries, O.S. and Fernando, T.M. 1966. Studies on the biology of Phytophthora meadii. Trans. Br. Mycol. Soc., 48: 311-325.

Peries, O.S. and Liyanage, N.I.S. 1983. The use of sulphur for the control of white root disease caused by Rigidoporus lignosus. J. Rubb. Res. Inst. Sri Lanka, 61: 35-40.

Peries, O.S. and Liyanage, A. de S. 1985. Hevea diseases of economic imprtance and integrated methods of control. Proc. Int. Rubb. Conf., 1985, Kuala Lumpur, Vol. 3, pp. 255-269.

Peries, O.S. and Wimalajeewa, D.L.S. 1970. Control of Gloeosporium leaf disease of Hevea in Ceylon. Trop. Agric. Trinidad, 47: 221 pp.

Petch, 1921. Diseases and pests of the rubber tree. MacMillan & Co., London, 278 p.

Radhakrishna Pillay, P.N. 1977. Aerial spraying against abnormal leaf fall disease of rubber in India. Plrs' Bull. Rubb. Res. Inst. Malaysia, 148: 10 pp.

356

Radhakrishna Pillai, P.N. and George, M.K. 1973. Recent experiments on the control of abnormal leaf fall disease of rubber in India. Q. J. Rubb. Res. Inst. Sri Lanka, 50: 223-227.

Radhakrishna Pillay, P.N. and George, M.K. 1980. Stem diseases. In: P.N. Radhakrishna Pillay (Ed.). Handbook of Natural Rubber Production in India. Rubber Research Institute of India, Kottayam. 281-292.

Rajalekshmi, V.K. 1980. Root diseases. In: P.N. Radhakrishna Pillay (Ed.). Handbook of Natural Rubber Production in India. Rubber Research Institute of India, Kottayam. 295-304.

Ramakrishnan, T.S. 1960. Experiments on the control of abnormal leaf fall of Hevea caused by Phytophthora palmivora in South India. Proc. Nat. Rubb. Res. Conf., Kuala Lumpur, 1960, pp. 454-466.

Ramakrishnan, T.S. 1964. Patch canker or bark canker caused by Phytophthora palmivora Butl. and Pythium vexans De Bary. Rubber Board Bull., 7: 11-13.

Ramakrishnan, T.S. and Radhakrishna Pillay, P.N. 1962a. Powdery mildew of rubber. Rubber Board Bull., 5: 187.

Ramakrishnan, T.S. and Pillay, P.N.R. 1961. Leaf spot of rubber caused by Corynespora cassiicola Berk. & Curt. Wei. Rubber Board Bull., 5: 32-35.

Ramakrishnan, T.S., Radhakrishna Pillay, P.N. 1962a. Pink disease of rubber caused by Pellicularia salmonicolor (Berk & Br.) Dastur (Corticium salmonicolour Berk & Br.). Rubber Board Bull., 5: 120-126.

Ramakrishnan, T.S. and Radhakrishna Pillay, P.N. 1962b. Brown root disease, Fomes nodius Corner, Fomes lamaoensis Murr. Rubber Board Bull., 6: 8-11.

Ramakrishnan, T.S. and Radhakrishna Pillai, P.N. 1963a. Black stripe black thread and bark rot. Rubber Board Bull., 6: 110-112.

Ramakrishnan, T.S. and Radhakrishna Pillay, P.N. 1963b. Brown root disease in nurseries. Rubber Board Bull., 7: 67-69.

Rands, R.D. 1946. Progress on tropical American rubber planting through disease control. Phytopathologia, 36: 688 (Abstract).

Rao, B.S. 1972. Chemical defoliation of Hevea brasiliensis for avoiding secondary leaf fall. J. Rubb. Res. Inst. Malaysia, 23: 248.

Rao, B.S. and Azaldin, M.Y. 1973. Progress towards recommending artificial defoliation for avoiding secondary leaf fall. Proc. Rubb. Res. Inst. Malaya. Plrs' Conf. Kuala Lumpur, 1973,2 67 p.

Ridley, H.N. 1904. Parasitic fungi on Hevea brasiliensis. AGriculture Bulletin Straits & Federated Malay Station, 3: 173-175.

Rocha, H.M. 1972. Problems de enfermidades nos seringais da Bahia. In Seminario Nacional da Seringueira, 10, Cuiba, Mato Grosso, 1972, Anais. Cuiaba, Superintedencia da Borracha, 1972, pp. 99-100.

Rogers, T.H. and Peterson, A.L. 1976. Control of South American Leaf Blight on a plantation scale in Brazil. Proc. Int. Rubb. Conf. 1995, Kuala Lumpur, 3: pp. 266-277.

Romano, R., Rao, S., Souza, A.R. and Castro, A.M.G. 1982. Defolhamento quimi co da seringueira por termonebulizacao. Pesq. Agropec. Bras. Brasilia, 17: 1621-1626.

Rubber Research Institute of Malaysia. 1974a. Root diseases, Part I. Detection and Recognition. Planters' Bulletin, 133: 111-120.

Rubber Research Institute of Malaysia, 1974b. Root diseases, Part 2. Control. Planters Bulletin, 134: 157-164.

Rubber Research Institute of Malaysia, 1968. Gloeosporium. Planters Bulletin, 97: 110-113.

Rubber Research Institute of Malaysia, 1975. Corynespora leaf spot. Planters Bulletin, pp. 84-86.

Rubber Research Institute of Malaysia. 1977. Director's Report, Crop Protection, 149.

Santos, A.F. Dos., Perera, J.C.R. and Almeida, L.C.C. de. 1984. Chemical control on South American Leaf Blight (Microcyclus ulei) da Seringueira (Hevea sp.). Abstract in iv Seminario Nacional da Seringueiria. Junbo, 1984, Salvador, 148.

Samarajeewa, P.K. and Liyanage, A. de S. 1986. Reactions of resistant and susceptible Hevea clones to Colletotrichum gloeosporioides. J. Nat. Rubb. Res., 1: 187-194.

Samarajeewa, P.K., Liyanage, A. de S. and Wickremasingha, W.N. 1985. Relationship between the incidence and severity of Colletotrichum gloeosporioides leaf disease in Hevea brasiliensis. J. Rubb. Res. Inst. Sri Lanka, 63: 1-8.

Sarma, S. and Ramakrishna, 1985. Some aspects of abnormal leaf fall of Hevea caused by Phytophthora palmivora in South India. Proc. INt. Rubb. Conf., Rubber Research Institute of Malaysia, Kuala Lumpur, 1985, Vol. 3, pp. 238-251.

Rao, B., Romano, R., De Souza, A.R. and De Castro, A.G.M. 1980. Surtos de requeima de Phytophthora nas seringueiras do sul da Bahia em. Sudhevea, 26 pp.

Satchuthananthavale, V. 1963. Complementary strains of Phytophthora plamivora from Ceylon rubber. Phytopathology, 53: 729.

Satchuthananthavale, V. and Dantanarayana, D.M. 1973. Observations on Phytophthora diseases of Hevea. J. Rubb. Res. Inst. Sri Lanka. 50: 228-243.

Senechal, Y. and Gohel, E. 1988. Influence d' une matadie de feulle (Colletotrichum gloeosporioides) su da composition minerale foliaire de l'Hevea brasiliensis. Effect d' um traitment defoliant preventif. Physiologie Vegetale. 307, pp. 445-450.

Senechal, Y., Sanier, C., Gohet, E. and d'Auzac, J. 1987. Differents modes de penetration du Colletotrichum gloeosporioides dans res Feuilles d'Hevea brasiliensis. Physiologie Vegetale, 305: 537-542.

Sharples, A. 1936. Diseases and pests of the rubber tree. MacMillan and Co. London. 480 p.

Situmorang, A. and Budiman, A. 1985. Corynespora cassiicola (Berk. & Curt.) Wei Penyebab Penyakit.

Soekirman, P. and Purwantara, A. 1987. Sporulation and spore germination of Corynespora cassiicola. Proc. Int. Rubb. Res. Dev. Bd., 24-33.

Staahel, G. 1917. De-Zuid-Amerikaansche Hevea bladekte veroorzaakt door Melanopsammopsis ulei Hov. en Dothidella ulei P. Hennings Paramaribo, Department Vanden Landbouw, 34 pp.

Sripathi Rao, B. 1975. Maladies of Hevea in Malaysia. Rubb. Res. Inst. Malaysia, Kuala Lumpur, 108 p.

Stoughton-Harris, R.H. 1925. Oidium leaf disease of rubber. Q. Circ. Ceylon Rubb. Res. Sch., 8-11.

Subramaniam, S. 1969. Performance of recent introductions of Hevea in Malaya. J. Rubb. Res. Inst. Malaya, 21: 11-18.

Sudhevea. 1970. Plana Nacional de Borracha. Anexo XI - Pesquisas e Experimentacao com Seringueira.

Tan, A.M. 1990. The present status of fungicide drenching on the control of white root disease of rubber. Root disease of Hevea brasiliensis. Proceedings of IRRDB Symposium, Kunming, China. 47-54.

Tan, A.M. and John, C.K. 1985. Economic benefits of disease control in rubber. Malaysian experience. Proc. Int. Rubb. Conf. Kuala Lumpur, 1985, Vol. 3, pp. 270-279.

Tan, A.M., Leong, M.W., John, C.K. and Tan, K.J. 1977. Current status of Phytophthora diseases of rubber in Peninsular Malaysia. Proc. Rubb. Res. Inst. Malaysia. Plrs' Conf. Kuala Lumpur, 1977, pp. 65-73.

358

Tan, A.M. and Lim, C.M. 1971. Development of a collar protectant dressing against Ganoderma pseudoferrum. Proceedings of RRIM Planters Conference Kuala Lumpur, 1972. 172-177.

Tan, A.M. and Yeoh, C.S. 1976. A spray on natural rubber latex formulation for controlling pink disease. Proc. Plrs' Con. Rubb. Res. Inst. Malaysia, 1976, Kuala Lumpur, pp. 243-249.

Tan, X. and Fan, H. 1990. Development and control of red root disease in rubber plantations. Root disease of Hevea brasiliensis. Proceedings of IRRDB Symposium, Kunming, China. 31-36.

Thompson, A. 1929. Phytophthora species in Malaya. Malay. Agric. J., 17: 53-100.

Thomson, T.E. and George, M.K. 1976. Phytophthora nicotianae var parasitica (Dastur) Water house on Hevea brasiliensis in South India. Rubb. Bd. Bull.,13: 3-4.

Thomson, T.E., Idicula, S.P. and Jacob, C.K. 1988. Field evaluation of fungicides to identify a substitute for organo mercurials in the control of black stripe disease of rubber in India. Indian J. Nat. Rubb. Res., 1: 42-47.

Tran, V. 1985. Use of Calixin and Sandofan against white root disease and black stripe of Hevea brasiliensis. International Rubber Conference, Kuala Lumpur, 1985, 222-236.

Tran, V.C. 1986. Use of Calixin and Sandofan F against white root disease and black stripe of Hevea brasiliensis. Proceedings of International Rubber Conference, Kuala LUmpur, 1985, 222-237.

Tsao, P.H. 1976. Recovery of Phytophthora species from old, badly decayed, infected tissues of Hevea brasiliensis, 66: 557-558.

Tucker, C.M. 1931. Taxonomy of the genus Phytophthora de Bary. Research Bulletin of the University of Missouri Agricultural Experimental Station, 1953, 208 pp.

Varghese, G. 1971. Infection of Hevea brasiliensis by Ustulina zonata (Lev.) Sacc. J. Rubb. Res. Inst. Malaya, 23: 157-163.

Vargese, G., Chew, P.S. and Lim, J.K. 1975. Biology and chemically assisted biological control of Gonoderma. Proceedings of International Rubber Conference, Kuala Lumpur, Vol. 3, 278-292.

Wastie, R.L. 1967. Gloeosporium leaf disease of rubber in West Malaysia. Planter, 43: 553 pp.

Wastie, R.L. 1972a. Secondary leaf fall of Hevea brasiliensis: factors affecting the production, germination and viability of spores of Colletotrichum gloeosporioides. Ann. App. Biol. 72: 273-282.

Wastie, R.L. 1972b. Factors affecting secondary leaf fall of Hevea in Malaya. J. Rubb. Res. Malaya, 23: 232-247.

Wastie, R.L. 1973. Influence of weather on the incidence of Phytophthora leaf fall of Hevea brasiliensis in Malaysia. J. Rubb. Res. Inst. Malaya, 23: 381-390.

Wastie, R.L. and Janardhanan, P.S. 1970. Pathogenicity of Colletotrichum gloeosporioides, C. dematium and C. crassipes to leaves of Hevea brasiliensis. Trans. Br. Mycol. Soc., 54: 150-152.

Wastie, R.L. and Mainstone, B.J. 1969. Economics of controlling secondary leaf fall of Hevea caused by Oidium heveae Steinn. J. Rubb. Res. Inst. Malaya, 21: 64.

Wastie, R.L. and Sankar, G. 1970. Variability and pathogenicity of isolates of Colletotrichum gloeosporioides from Hevea brasiliensis. Trans. Br. Mycol. Soc., 54: 117-121.

Wastie, R.L., Yeoh, C.S. 1972. New fungicides and formulations for controlling pink disease. Proc. Rubb. Res. Inst. Malaysia Plrs' Conf. 1972, Kuala Lumpur, pp. 163-168.

Van Emden, J.H. 1954. Notes on the incidence and control of Oidium hevea in Ceylon. Q. Cir. Rubb. Res. Inst. Ceylon, 30: 20.

Wimalajeewa, D.L.S. 1967. Studies on the physiology of spore germination in Gloeosporium alborubrum. Q. J. Rubb. Res. Inst. Ceylon, 43: 4-12.

Wimalajeewa, D.L.S. and Shanmughanathan, T. 1963. A preliminary report of physical assessments of products for Gloeosporium control trials. Q. J. Rubb. Res. Inst., 39: 25.

Wimalajeewa, D.L.S. 1965. The significance of the factors affecting spore germination in the spread of Gloeosporium leaf disease in Hevea. Q. J. Rubb. Res. Inst. Ceylon, 41: 63.

Yeoh, C.S. and Tan, A.M. 1974. National Rubber latex for controlling pink disease. Proc. Rubb. Res. Inst. Malaysia Plrs' Conf. Rubb. Res. Inst. Malaya, 171.

Yeoh, E.S. and Tan, A.M. 1980. A new formulation for controlling Black stripe. Proceedings of RRIM Planters Conference, Kuala Lumpur, 1979, pp. 400.

Young, H.E. 1951. Report of the Oidium Research Officer and Mycologist for the year 1951. Ann. Rep. Rubb. Res. Board, 152, 38.

Young, H.E. 1952. Leaf mildew of rubber - a review. Q. Cir. Rubb. Res. Inst. Ceylon, 27: 3.

Zainuddin, R.N. and Omar, M. 1988. Influence of the leaf surface of Hevea on activity of Colletotrichum gloeosporioides. Trans. Br. Mycol. Soc., 91: 427-432.

CHAPTER 16

PESTS

K. JAYARATHNAM
Rubber Research Intitute of India, Kottayam-686009, Kerala, India.

Among the agricultural crops, rubber is an exception with regard to pests, as the attack is usually mild, mostly sporadic and localized. However, a few pests become serious at times and cause considerable damage in almost all rubber growing countries. One reason attributed for the un-attractiveness of rubber to pests is the presence of latex all over the plant from root tip to shoot tip, which may coagulate and block the mouth parts of insects and other animals. But a few insects and non-insect pests do surmount this problem and feed on rubber plants. The cover crops grown as intercrops during immaturity period and rubber wood are attacked by many pests. There are also pests which cause health hazard and in-conveniences to the personnel residing in rubber plantations. Detailed accounts of pests in rubber plantations were furnished by Sharples, 1936; Edgar, 1958; Sripathy Rao, 1965. The pests attacking rubber can be classified into three major groups such as insects, non-insects (invertebrates) and vertebrates.

INSECT PESTS

Root grubs

Root grubs are larvae of Cockchafer beetles (Coleoptera: Melolonthidae) which feed on roots of rubber. The pest is polyphagous and attacks rubber in nurseries and planted fields, especially in areas adjacent to virgin forests. The pest has been reported mainly in nurseries in India (Ramakrishnan and Radhakrishna Pillai, 1963) and also in main fields in Malaysia and Papua New Guinea (Smee, 1964). Some of the species are Lachnosterna [Holotrichia bidentata (Burn)] reported to be most serious in Malaysia (Sripathi Rao, 1965; RRIM, 1968) and Ceylon, and Psilopholis vestita (Sharp) and Leucopholis rorida in Malaysia. Holotrichia serrata (F.) is the most common in India, and the other species present are H. rufoflava F., H. fissa Brenske and Anomala variance O. (Nehru and

Jayarathnam, 1988). The first instar grubs feed on humus and very tender roots. The third and final instar grubs can feed on tap roots and collars of hardened plants. At this stage the grubs feed voraciously. The whole root system is fed on, in upto 6 month old nursery plants, which sometimes sway in the morning due to the active feeding of grubs at the collar region. In such cases the leaves turn yellow, shed off and then the whole plant dries up. Larger plants can survive the attack, but will lose their vitality. The grubs are fleshy with wrinkled 'c' shaped bodies (Fig. 1). Third and final instar grubs measure about 48 mm.

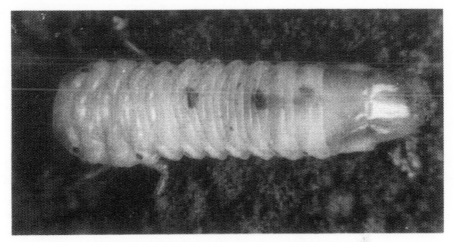

Fig. 1.   Third instar grub of H. serrata.

The life cycle of white grubs lasts for one year. Adults emerge from soil with the first rains after summer and are active mostly between 7.30-8.30 pm. Adult beetles are sturdy and females are much larger than males. In India , adults are noticed to feed on leaves of Acacia, neem and tapioca and some specific forest plants, though not on rubber. They mate and lay eggs singly around the host plants at depths of 8-10 cm, enclosed in earthen cells. Adults live for about two months. The grubs emerge in 10-12 days. The three larval instars occupy about 30, 35 and 102 days. The population of grubs can be as high as 500,000 per hectare. Pupation takes place in the month of November-December and lasts for 12 days. The adults lie quiescent until first summer rains and then emerge. The adults are highly attracted to light.

There is good scope for adopting integrated pest management. Control of the pest is achieved to some extent by natural enemies like wasps, vertebrate predators and fungal and bacterial pathogens. But they are not very effective. In Malaysia, black light traps were found to be useful for attracting the adults and killing them (Sripathi Rao, 1964; Van Iddekinge and Gill, 1969). Spraying DDT 0.4% or endosulphan 0.1% on adult host plants was found effective in reducing the adult population. In nurseries, incorporation of insecticides like sevidol 4:4 G or phorate 10 G at the rate of 25 kg ha^{-1} or BHC 10 D at 100 kg ha^{-1}, in the soil at the time of preparation of nursery beds proved very effective (Jayarathnam and Nehru, 1984; Nehru and Jayarathnam, 1988). Drenching insecticides like aldrin 0.1% solution is recommended for field plants, though it is found to be less effective. Application of insecticide granules in main field also is recommended in Malaysia for better protection.

Bark feeding caterpillar

At present, this pest is reported in rubber from India only (Nehru et al. 1983). Caterpillars of two species of moths Aetherastis circulata Meyr. (Yponomeutidae) and Ptochoryctis rosaria Meyr (Xyloryctidae) feed on brown corky bark of mostly mature trees, the most common being A. circulata. Field incidence of bark feeding caterpillar, A. cirulata, on different alternative host plants and its control was published by Nehru et al. (1987). They build galleries with silk and chewed bark and live within, damaging brown bark all over the tree. They thrive only during dry months and in the wet season they disappear, as the galleries get wet and damaged. The caterpillars feed deeper at certain points in the prepupal stage and pupate. Since they feed as close to the green bark as possible, the latex vessels break at such points due to bending and twisting of trees in strong wind and latex oozes out. This damage is more in clones with stem bleeding character. Latex oozes out continuously and forms a thick pad on the trunk and branches. During rainy months these latex oozing points help easy penetration of wood rotting fungi like Phytophthora, Botryo-diploidea, Pythium, etc. With the rotting of tissues upto wood region, there can be branch snap or trunk snap.

The caterpillar is brick red in colour, flat and with a broad head and thorax and tapering abdomen (Fig. 2). At prepupal stage the caterpillar measures about 1.5 cm in length. When disturbed it moves briskly inside the gallery. The adult moths are white in colour with black dots in the fore wing. Each female moth can lay about 400 eggs on the bark. Larva emerges in about 4 days. Larval period lasts for 25-30 days. Pupa is enclosed in a dome shaped cocoon made of thick gallery.

363

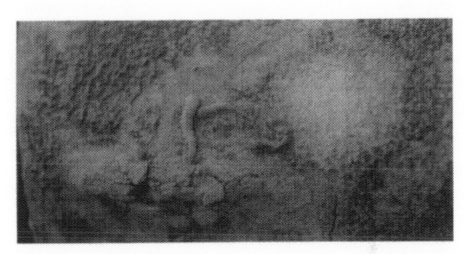

Fig. 2.   Bark feeding caterpillar A. circulata on rubber tree.

Satisfactory control is achieved by dusting insecticides fenval 0.4%, methyl parathion 2%, quinalphos 1.5% or carbaryl 5% with a power duster at the rate of 10-15 kg ha^{-1} (Jayarathnam et al. 1989). The duster need be carried only at every fourth row and dusting operation has to be done in the early morning hours, when there is no wind.

Termies

In Malaysia, termites are a big menace to rubber cultivation. A species of termite Coptotermes curvignathus Holmgren (Isoptera: Rhinotermitidae), which can feed on green plant materials was reported in Malaysia (Newsam and Sripathi Rao, 1958; RRIM, 1966). These penetrate rubber trees through tap root, main laterals or collar and build galleries in the trunk (Sripathy Rao, 1965). In many cases, no sign of attack is seen above ground before the tree is killed or uprooted by strong wind. When such trees are examined it can be seen that the central wood portion is consumed by termies. This termite is also reported from Indonesia. C. elisae (Resn.) occurs in Papua New Guinea. Species present in other countries are C. testaceus (L.) in Brazil, Neotermes greeni (Desn.) and N. militaris (Desn.) in Sri Lanka and Odentotermes obesus (Ramb) in India. The species occurring in India damages only dead or partially dead trees. Termite colonies normally originate from old logs buried in soil. They damage the dry mulch, shade baskets, etc.

Control of termites for 23 years could be achieved in India (Jayarathnam, 1968) by drenching aldrin 0.1% solution in the soil around the affected plants or objects. In Malaysia, drenching with a few soil insecticides is recommended. The earlier practice of tracing buried old logs was discarded due to high costs involved (Sharples, 1936).

Scale insects and mealy bugs

These insects come under the category of plant bugs (Hemiptera: Coccidae) which suck the sap from terminal green parts of rubber plant and trees. The species of scale insects involved in India is Saissetia nigra Nietin and mealy bug Ferrisiana virgata (Ckll.) (Ramakrishnan and Radhakrishna Pillai, 1961; Jayarathnam, 1980). In other countries, scale insects Pulvinaria maxima (Green), Lepidosaphes cocculi (Green), Laccifer greeni (Chamberlain) and mealy bug Planococcus citri (Risso) occasionally attack rubber (RRIM, 1968). Symptoms of attack are yellowing and shedding of leaves and die back of twigs. These insects secrete a sugary secretion known as honey dew, which spreads on the stem and leaves. A black fungus, sooty mould (Capnodium sp.), grows on it. This gives a black and dirty appearance to the leaves and stem. Since these insects are sedentary in nature a number of natural enemies like parasites and predators and a fungal pathogen Hypocrelia reineckiana attack them and keep them under check. However, at times due to absence of natural enemies, population explosion occurs. In such cases malathion 0.05% solution along with a wetting agent, sprayed on infested parts gives effective control.

Leaf eating caterpillar

The insect causing serious damage in Brazil and Guyana (Winder, 1976) is the larva of the moth Erinnys ello L. (Lepidoptera: Sphingidae). This pest also attacks other Euphorbiaceous plants, as well as crops like papaya, tobacco and cotton. Young larvae can feed only on tender leaves while the older ones feed on mature leaves and bark of green twigs. Hence outbreaks normally occur during refoliation. The attacked trees produce less latex. Completely defoliated trees refoliate only the following year.

Eggs are laid on lower surface of leaves. The larvae grow to a length of 8-9 cm and are green or greyish green in colour. There are five instars and larval period lasts for two weeks. Pupation takes place in the soil. Adult moths have a wing span of 34-48 mm and are grey in colour. Even though as many as 30 parasites were identified, for control, spraying of insecticides such as carbaryl, lindane, etc. is practiced (de Abreu, 1982).

The other lepidopterous pest noticed in Indonesia, Sri Lanka, Malaysia and Papua New Guinea is Tiracola plagiata Wlk. (Noctuidae).

Other insect pests

Occasional damage occurs on tender leaves due to the weevil, Hypomyces squamosus (F.) (Coleoptera: Curculionidae). The grass hopper, Valanga nigricornis (Burn.) (Acrididae) appears in swarms and consumes rubber and cover crop leaves. The cricket Brachytrypes portentosus Licht. (Gryllidae) cuts the stem of young nursery plants a few centimeters above ground. The tip wilt bug, Amblypelta lutescens papuensis Brown (Coreidae) damages the young leaves at the tip. Thrips also damage tender leaves 2-3 days from bud burst in Malaysia, though they are not noticed in India (RRIM, 1962).

Wood boring beetles

Some species of powder post beetles attack partially or fully dead rubber wood such as Mynthea rugicollis, Heterobostrychus acqualis and Sinoxylon anale and Platypus sp. The borer beetles invade rubber wood after the attack of diseases, sun scorch and fire. Severe attack was noticed in Malaysia (RRIM, 1959; Tan et al. 1979) subsequent to fire damage. Their presence reduces the timber value. In live plants they are not a problem and can be controlled by swabbing BHC 10 D. Timber preservation methods are detailed by Sekhar (1989).

Pests of cover crops

In India the popular cover crop of rubber Pueraria phaseoloides is attacked by many pests and the important ones are leaf lacerating flea beetle, Pagria signata, stem borer Eucomatocera vittata and flower and pod borer Maruca testulalis. Pests are noticed in cover crops in Malaysia also (RRIM, 1963, 1969). But their attack is significant only in initial stage of growth of the cover crop and they can be managed by dusting BHC 10 D. Other cover crops such as Calopogonium mucunoides, Centrosema pubescens, Mimosa invisa and Mucuna bracteata are rarely attacked by pests.

Pests causing health hazards and inconveniences to personnel

Mosquitoes are a real menace in rubber plantations and can cause health hazard by transmitting diseases (Edgar, 1958). All stagnant waters in plant-ations other than used for drinking should be covered with a thin film of insecticide mixed diesel oil. The latex collecting cups should be kept upside down when not in use to prevent collection of rain water inside. The beetles Lyprops corticollis Frm. invade the dwelling places in rubber estates in very large numbers and cause considerable inconvenience by their presence and secretion of stain. These can be collected by putting up light traps with very bright light, or killed by spraying or dusting DDT.

NON-INSECT (INVERTEBRATE) PESTS

Slugs and Snails

Some terrestrial molluscs can considerably retard the growth of young rubber plants and also feed on latex from tapped mature trees. The species of slugs attacking rubber in Malaysia, Ceylon and India are Mariaella dussumieri Grey and Semperula maculata (Templeton) (Ramakrishnan and Radhakrishna Pillai, 1962; RRIM, 1967). The snails present in Malaysia are Parmarion martensi Simroth, Xestina striata and Achatina fulica while Cryptozona (Xestina) bistrialis Beck is reported in India (Jayarathnam and Rajendran, 1979). They are nocturnal in habit and in the day time lie concealed under natural mulch, rocks and crevices of soil. Only by dusk do they become active and climb on rubber plants and trees, returning to their hideouts by day break. On young plants, they move upto the terminal part and rasp the terminal bud with their denticulated tongue. Due to the damage to the tissues, latex oozes out and they feed on latex. With repeated damage, the terminal bud does not grow and get calloused. Similarly the axillary buds also get damaged. Thus the growth of the plant is arrested and the terminal part gives a clubbed appearance. In some cases fasciation and die back occur. In trees, they feed on tender parts and on latex from tapped trees. They are not affected by the latex they feed on and excrete coagulated latex. After feeding they cross the tapping cut to move downwards. When they get habituated to feeding on latex, they are seen to collect at the tapping panel and collecting cup as if waiting for the tappers to begin tapping. Due to this, the latex gets contaminated with slime. During day time, their presence can be identified by the watch spring like excreta and glistening lines of dried slime present all over the shoot. Only a very small portion of the total population attack rubber, as most of them feed on tender parts of cover crops and weeds.

Slugs and snails have an unsegmented, fleshy, wormlike body and continuously secrete slime. They glide on a sole like foot, leaving a track of slime. Snails bear a spiral shell, whereas slugs have a small concealed shell. They are hermaphrodites, but mating is required to lay eggs. Upto 400 leathery, pearly white eggs are laid in groups. Life cycle is completed in a year.

The well known molluscicide metaldehyde is used as 2-5% bait (RRIM, 1964, 1969) or 0.1% spray (Nair et al. 1968). Metabait is prepared by mixing metaldehyde, cement, lime and wheat bran (rice bran) in the ratio

2:3:5:16 by weight and making it into pellets or briquettes with water (Edgar, 1958). They can be effectively controlled by sprinkling the insecticide granule temik 10 G on the surface of soil around the collar region of plants. Painting 0.1% slurry of temik in wheat flour about 30 cm wide at the base of stems is more effective and economical. Similar application of 10% Bordeaux paste repelled the slugs and snails for a period of 30-40 days (Jose et al. 1989).

Mites

Mites infest rubber only during summer months and mostly, the attack is confined to nursery. The species Hemitarsonemus latus Banks (Acaria: Tarsonemidae) suck sap from the underside of tender leaves in India and also in Malaysia (RRIM, 1988). On the upper surface, numerous minute white chlorotic spots appear. Severely affected leaves turn yellow and drop.

VERTEBRATE PESTS

A variety of vertebrate pests ranging from rats to elephants cause various types of damage to rubber plants from nursery to very old plantations (RRIM, 1964; Sripathy Rao, 1965; Jayarathnam, 1980). As these are intelligent animals, controlling with poison baits, scaring devices and traps is difficult. The damage due to large animals is mostly confined to areas near virgin forests. The burrowing rats and mole rats cause considerable damage to nursery plants and upto three year old field plants. They feed on the roots and collar from below or above the ground and the plants dry up. The species of rats attacking rubber in India are Rattus meltada Gray and Bandicota bengalensis Gray. In Malaysia, Rattus jalorensis Bonhote and Bandicota spp. are reported. Various types of traps are available for catching and killing rats. Poison baiting with zinc phosphide or temik 10 G is effective, but results in bait shyness. Mass baiting is required to achieve success. Single dose blood anti-coagulant rodenticides like brodifacoum (Nehru and Jayarathnam, 1985) and bromadiolone are very effective bait poisons.

Porcupines

They feed on the bark of mature trees at the base and when fed all around, the trees dry up. They pull out young plants and chew the roots. The common species found are Hystrix indica Kerr in India and H. brachyura, Atherurus macrourus and Trichys lipura in Malayasia. Porcupines are much more intelligent than rats. Killing them by shooting is the only practicable method. Baiting with zinc phosphide or temik 10 G in salt meats can be tried.

Wild pigs, elephants, monkeys, deer and sambar

Wild pigs ransack nursery and field plants by pushing out soil with their snouts and chew the roots of toppled plants. The species found in Malaysia are Sus scroba and Sus barbatus. Elephants damage the young plants by trampling and pulling out the plants. They rub their body on the trunk and damage the bark. They also chew roots of young plants. Monkeys break the branches and tear off the leaves. Deer and sambar feed on the bark of the trees. In Malaysia, sambar deer Cervus unicolor and barking deer, Muntiacus muntjak are recorded. These animals can be driven off by scaring devices or prevented from entering the plantation by putting up ordinary or electric fences or by digging deep trenches.

REFERENCES

RRIM 1959. Fire damage and borers. Plrs' Bull., Rubb. Res. Inst. Malaya. 41: 36-37.
RRIM 1962. Thrips. Plrs' Bull., Rubb. Res. Inst. Malaya, 59: 47-49.
RRIM 1963. Pests of leguminous covers in Malaya. Plrs' Bull., Rubb. Res. Inst. Malaya, 68: 182-186.
RRIM 1964. Warm blooded animals. Plrs' Bull., Rubb. Res. Inst. Malaya, 70: 10-18.
RRIM 1964. Slug bait pellets. Plrs' Bull., Rubb. Res. Inst. Malaya, 72: 53.
RRIM 1966. Termites. Plrs' Bull., Rubb. Res. Inst. Malaya, 84: 67-71.
RRIM 1967. Snails and slugs. Plrs' Bull., Rubb. Res. Inst. Malaya, 93: 284-287.
RRIM 1968. Cock chafer. Plrs' Bull., Rubb. Res. Inst. Malaya, 97: 102-106.
RRIM 1968. Scale insects, Mealy bugs and lac insects. Plrs' Bull., Rubb. Res. Inst. Malaya, 98: 146-152.
RRIM 1969. Pests of legume covers. Plrs' Bull., Rubb. Res. Inst. Malaya, 102: 105-112.
RRIM 1988. Mites. Plrs' Bull. Rubb. Res. Inst. Malaysia, 194: 8-12.
De Abreu, J.M. 1982. Investigations on rubber leaf caterpillar Erinnys ello in Bahia, Brazil. Revista Theobroma, 12: 85-199.
Edgar, A.T. 1958. Manual of rubber planting. The Incorporated Society of Planters, Kuala Lumpur.
Jayarathnam, K. 1968. Termite control in rubber plantations. Rubb. Bd. Bull., 9: 34-38.
Jayarathnam, K. 1980. Pests in rubber plantations. In: Radhakrishna Pillai etal. (Eds). Handbook of natural rubber production in India, Rubber Research Institute of India, Kottayam. pp. 315-323.
Jayarathnam, K. and Nehru, C.R. 1984. White grubs and their management in rubber nursery. Pesticides, 18: 27-29.
Jayarathnam, K., Nehru, C.R. and Jose, V.T. 1989. Field evaluation of some newer insecticides against the bark feeding caterpillar Aetherastis circulata Meyr. infesting rubber. Paper presented at the International Conference on biology and control of pests of medical and agricultural importance held at American College, Madurai, India, 1989.
Jayarathnam, K. and Rajendran, T.P. 1979. Control of the snail Cryptosona bistrialis attacking Hevea brasiliensis. Proceedings of Second Symposium on Plantation Crops. pp. 4237-4242.

Jose, V.T., Nehru, C.R. and Jayarathnam, K. 1989. Effect of Bordeaux paste as a repellant of slugs (Mariaella dussumieri Gray) infesting rubber plants. Indian J. Nat. Rubb. Res., 2: 70-71.

Nair, M.R.G.K., Mohandas, N. and Abraham Jacob, 1968. Use of metaldehyde as dusts and sprays to control giant African snail, Achatina fulica Bowdich. Indian J. Ent., 30: 58-60.

Nehru, C.R. and Jayarathnam, K. 1985. Field evaluation of some granular insecticides against the Indian mole rat, Bandicota bengalensis Gray. Planter, 61: 172-175.

Nehru, C.R., Jayarathnam, K. and Radhakrishna Pillai, P.N. 1983. Incidence of bark feeding caterpillar Aetherastis circulata (Meyr.) on rubber (Hevea brasiliensis Muell. Arg.). Indian J. Plant Protection, 11: 150.

Nehru, C.R., Jayarathnam, K. and Thankamony, S. 1987. Field incidence of bark-feeding caterpillar, Aetherastis circulata (Meyr.) on different alternative host plants and its control. Pesticides, 21: 39.

Nehru, C.R. and Jayarathnam, K. 1988. Control of white grub (Holotrichia serrata F.) attacking rubber at the nursery stage in India. Indian J. Nat. Rubb. Res., 1: 38-41.

Newsam, A. and Sripathi Rao, B. 1958. Control of Coptotermes curvignathus Holmgrer with chlorinated hydrocarbons. J. Rubb. Res. Inst. Malaya, 15: 209-215.

Ramakrishnan, T.S. and Radhakrishna Pillai, P.N. 1961. Scale insects (Saissetia nigra Nietn and Pulvinaria maxima Green). Rubb. Bd. Bull., 5: 209-212.

Ramakrishnan, T.S. and Radhakrishna Pillay, P.N. 1962. Snails and slugs. Rubb. Bd. Bull., 6: 26-28.

Ramakrishnan, T.S. and Radhakrishna Pillay, P.N. 1963. Cock chafer grubs. Rubb. Bd. Bull., 7: 76-78.

Sekhar, K.C. 1989. Rubber Wood Production and Utilisation. Rubber Research Institute of India, Kottayam-9.

Smee, L. 1964. Insect pests of Hevea brasiliensis in the territory of Papua New Guinea. Agricultural Journal, 17: 21-28.

Sharples, A. 1936. Diseases and Pests of Rubber Trees, McMillan Co., London, pp. 423-426.

Sripathi Rao, B. 1964. The use of light traps to control the cockchafer Lachnosterna bidentata Burm. in Malayan rubber plantations. J. Rubb. Res. Inst. Malaya, 18: 243-252.

Sripathy Rao, B. 1965. Pests of Hevea plantations in Malaya. Rubber Research Institute of Malaya, Kuala Lumpur.

Tan, A.G., Ali Sujan, Chong, K.F. and Tam, M.K. 1979. Biodeterioration of rubber wood and control measures. Plrs' Bull., Rubb. Res. Inst. Malaya, 160: 106-117.

Van Iddekinge, F.E.W.P. and Gill, D.S. 1969. Light traps for control of cockchafers. Plrs' Bull. Rubb. Res. Inst. Malaya, 104: 183-189.

Winder, J.A. 1976. Ecology and control of Erinnys ello and E. alope: important insect pests in the New World. PANS, 22: 449-466.

CHAPTER 17

PRIMARY PROCESSING

BABY KURIAKOSE
Rubber Research Institute of India, Kottayam-686009, Kerala, India.

The main crop from rubber plantation consists of latex and field coagulum. The proportion of these materials varies depending upon the tapping system followed, climatic conditions, stability of latex, stimulant application, age of the tree etc. On an average, it is observed that the proportion of latex and field coagulum is in the range of 80 to 85% and 15 to 20%, respectively. While latex can be processed and marketed in any one of the forms such as latex concentrate, ribbed smoked sheet, technically specified rubber, crepe rubber or chemically modified rubber, field coagulum is processed and marketed as either technically specified rubber or crepe rubber. Whichever be the method adopted for processing the latex, it is to be preserved adequately so as to prevent precoagulation in the field and also during various processing operations.

PRESERVATION OF LATEX

For effective preservation of latex against precoagulation, factors affecting the stability of latex are to be understood and taken care of. There are several theories postulated for the auto-coagulation of latex, out of which the following assume importance.

Natural rubber latex contains about 4 to 5% non-rubber constituents consisting mainly of carbohydrates, proteins and lipids. Bacterial proliferation occurs at the expense of these non-rubber substances, resulting in the formation of acids, particularly volatile fatty acids (VFA). As the VFA content increases coagulation of latex occurs (John, 1974). A second theory postulates that hydrolysis of various lipids present in latex liberates higher fatty acid anions which get adsorbed onto the surface of the latex particles, possibly by replacing the adsorbed proteins. The free fatty acid anions interact with metallic ions such as calcium and magnesium, which are either present in latex initially or are gradually released by the action of enzymes. The formation of insoluble higher fatty acid soaps of calcium/ magnesium draws the latex particles together resulting in coagulation (Madge et al. 1950). Yet another theory suggests that a proteolytic enzyme, coagulase, hydrolyses the protective layer of proteins surrounding latex particles, thereby exposing the surface of the particles which leads to coagulation (Woo, 1973). Possibly all the factors described in these theories may be having a combined role in the autocoagulation of fresh natural rubber latex, as indicated by an increase in VFA, change in pH and formation of insoluble salts of calcium and magnesium during the process of auto-coagulation of natural rubber latex.

Any chemical or combination of chemicals to work as an effective preservative for NR latex should have the following basic features:

act as a bactericide - ie. it should be able to destroy micro-organisms or supress their activity.

preferably have an alkaline nature so that it will increase the colloidal stability of the rubber particles by increasing the negative electric charge.

have some chelating or precipitting effect on the divalent metallic ions present in latex.

be effective as an enzyme poison so that the effect of enzymes such as coagulase can be minimised.

have some substrate-complexing activity so that non-rubber materials such as carbohydrates may not be decomposed by bacteria.

When it is required to preserve latex for a few hours, chemicals such as formalin, sodium sulphate and ammonia at very low dosages of 0.02, 0.05 and 0.01% respectively on latex are employed. These are termed anti-coagulants and are used to prevent precoagulation in the field. Dilute solutions of these chemicals are added in latex collection cups and buckets. A detailed account on the uses of anticoagulants has been given by Cook (1960).

Since 1853, ammonia has been recognised as the most effective and popular preservative for NR latex as it fulfils most of the requirements for an ideal preservative. However, this chemical has certain drawbacks as well. To be effective for longer periods, a higher dosage is to be used. This necessitates a higher quantity of coagulant for subsequent processing operations. Higher levels of ammonia in latex also leads to atmospheric pollution. Because of these problems, preservation systems comprising of low levels of ammonia, in combination with other chemicals were introduced (Cheong and Ong, 1974; John et al. 1984). A review of the various types of chemicals used as preservatives for NR latex has been presented by Ng and Lau (1978). Mathew et atl. (1976) reported the use of monomethyl amine as an effective alternative for ammonia in preserving concentrated latex. The commercially available low ammonia systems are those containing sodium pentachlorophenate (SPP), zinc diethyldithiocarbamate (ZDC) or boric acid along with 0.2% by weight of ammonia. The main drawbacks of these systems are high toxicity, lower mechanical and storage stability, poor chemical stability and slow rate of cure respectively for the SPP, ZDC and boric acid preserved latex. A composite preservation system consisting of tetramethyl thiuram disulphide, zinc oxide and ammonia, popularly known as the LA-TZ system was introduced during 1975 (John et al. 1975) and subsequently commercialised. But this system has the drawback that the chemicals TMTD and zinc oxide, are to be prepared as dispersions and in many cases they sediment during long term storage. This system is unsuitable as a preservative for latex to be processed as block rubber and sheet rubber as it affects plasticity retention index (PRI). Another preservation system, consisting of 0.3% ammonia, and 0.05% of a biocide containing triazine/benzotriazole derivative, was reported to be capable of preserving field latex for at least one week and the latex thus preserved could be processed into concentrate, solid block or sheet rubber without any difficulty (John et al. 1986). A non-toxic system of preservative comprising of 0.4% ammonia and 0.05% Dowicil [cis-1-(3-chloroally)-3,5,7-triaza-1-azonia-adamantane chloride] was found to be as good as the LA-TZ system for field latex as well as for latex concentrate (John et al. 1985). The search for still better preservatives for latex is continuing and identification of an ammonia-loving bacteria by Shum and Wren in 1977 has intensified this search for a preservative system devoid of ammonia.

CONCENTRATED NATURAL RUBBER LATEX

The dry rubber content (DRC) of NR latex, as obtained from the tree, varies from 30 to 40% by weight. Clonal characteristics, age of the tree, length of tapping cut, frequency of tapping, stimulant application, time of tapping, environmental conditions etc. are some of the factors that affect DRC of latex (Kang and Hashim, 1982). On the average, the DRC of field latex is taken as 33%. For economic transportation of latex and for product manufacture, such a low DRC is undesirable and it has to be raised to about 60%. Increasing the DRC has the additional benefit that the latex becomes purer, since some of the nonrubber materials will be removed during the process of concentration, except in the case of evaporation. Even though several processes have been developed for concentration of latex, only two of them namely, creaming and centrifuging, are widely practised.

Creaming

Latex is a two-phase colloidal system consisting of rubber particles as the dispersed phase and the aqueous phase as the dispersion medium. Since the dispersed particles carry a negative charge, they are in constant Brownian movement which maintains the colloidal nature. Dispersed particles cream or sediment depending upon the difference between the densities of the dispersed particles and the dispersion medium. Applying Stokes's law, the velocity of creaming, v, is given by the relationship.

$$v = \frac{2gr^2 (d_s - d_r)}{9\eta}$$

where r = average radius of the rubber particle

g = acceleration due to gravity

η = coefficient of viscosity of latex serum

d_s = density of latex serum

d_r = density of rubber particles.

As is evident from the above relationship, larger particle size, higher difference between the densities of the serum and the dispersed rubber particles and lower viscosity of the serum favour higher velocity of creaming. Several materials have been reported to function as creaming agents which facilitate faster creaming of NR latex. Edgar and Sekar (1938a,b) reported the use of sodium salt of sulphonated lauryl alcohol and tamarind seed powder as creaming agents. Davey and Sekar (1947) evaluated methyl cellulose for this purpose and found that 3 g of methyl cellulose per litre of field latex was sufficient to effect the process. Baker (1937) identified a large number of naturally occurring materials such as gum arabic, pectin, gum tragacanth, gum karaya, tragon seed gum, alginic acid, alginates etc. and synthetic materials like polyacrylic acids and their salts, polyvinyl alcohol and their ethers, and derivatives of poly-ethylene oxide, which can be used for creaming of NR latex. Microscopic observations made by Baker revealed that, in the presence of creaming agents, rubber particles in latex formed large agglomerates and the Brownian movement of the particles was arrested within a few minutes of addition of the creaming agent. These clusters were redispersible on diluting the cream. This observation led to several theoretical explanations for the creaming process.

Creaming agents, in general, are hydrophilic colloids, swelling in water and forming highly viscous solutions at very low concentrations. It has been suggested that the creaming agent gets adsorbed onto the surface of the rubber particles thus favouring the agglomeration of the particles into relatively large groups of 10 to 20 microns diameter having reduced movement. Reduced Brownian movement and larger size of the agglomerates favour higher rate of creaming since there is a fairly significant difference between the density of the rubber particles (0.92 g cc^{-1}) and that of the serum (1.02 g cc^{-1}). Another theory postulated by Duckwork (1964) suggests that a definite structure is formed by the gum in solution and that adsorbed on the particles, which develops into a kind of molecular net restricting the movement of the rubber particles. The clusters thus formed, grow by continued entrapment of the particles. This structure gradually shrinks and rises to the surface, squeezing out the serum. This theory is supported by the fact that the line of separation between the serum and the cream starts at the bottom of the creaming vessel and moves upwards. However, there are several observations which remain unexplained due to the lack of proper theoretical support. It is observed that addition of a small portion of serum to latex helps in increasing the rate of creaming. Another observation is that with certain types of creaming agents, repeated creaming operation is not possible for more than three or four times, whereas the some other types unlimited repeated creaming is possible. It is also noticed that a small percentage of rubber present in the serum does not undergo further creaming eventhough the serum contains part of the creaming agent, whereas the cream always undergoes further creaming on storage.

In practice creaming operation is done in the following way. Field latex after sieving and bulking is ammoniated to about 1.0 to 1.2% by weight of latex and desludged by adding calculated quantity of diammonium hydrogen phosphate. The dosage of the creaming agent depends on the type, age and DRC of latex. Normally, depending on cost and availability, either ammonium alginate or tamarind seed powder is used for this purpose and the dosage varies from 0.2 to 0.3% of the dry material calculated on the aqueous phase of the latex. About 3% solution of tamarind seed powder is prepared by cooking the dry material with water for about one hour. It is filtered to remove the uncooked material. Addition of a small quantity of higher fatty acid soap such as potassium oleate, at the level of about 0.15% by weight of latex, increases the efficiency of the creaming agent. If the latex is sufficiently aged, the above dosage could be reduced further. Maintaining a temperature of 40°C in the creaming tank helps in reducing the induction time. The ammoniated and desludged latex is thoroughly stirred with the solutions of creaming agent and soap for about one hour and then kept undisturbed. Initially there is an induction period of several hours during which no visible creaming takes place. Between 24 and 48 h, the rate of creaming is maximum and thereafter the process is rather slow. Normally it takes about 72 to 96 h to complete the process. After this period, the serum layer is drained out and the cream collected. It is then tested for ammonia content and the same made up to the required level. The DRC of the cream depends on several factors such as concentrations of creaming agent and soap, creaming period and the quality of the field latex used. The DRC of the serum also varies from 1 to 3% depending upon the conditions of the creaming process.

The creaming process has several advantages over other methods of concentration of latex. It requires only simple equipments to operate, involves very low power consumption and labour and has the flexibility to adjust the production capacity from very low to very high levels. The disadvantages include slowness of the process, its dependence on the quality of field latex and type of creaming agent and the after creaming that invariably occurs during storage of the cream.

Centrifuging

The centrifugal method is the most important one among the various commercial methods of concentration of NR latex. Theory of centrifugal concentration of latex is basically the same as that of creaming. In this method, separation into latex concentrate and skim is effected by means of centrifugal force rather than by gravitation. The process consists of subjecting the latex to a centrifugal force many times greater than gravity, in a centrifuging machine which rotates at high speed. The centrifugal acceleration 'f' on a particle rotating with an angular velocity 'ω' and having radius of rotation 'R' is equal to $R\omega^2$. If 'n' is the speed of rotation in rpm, then:

$$f = (2\pi n)^2 R$$

The centrifugal acceleration can be expressed as a multiple of gravitational acceleration 'g' and the separating power 'c' of the machine is often described as

$$c = \frac{f}{g} = \frac{(2\pi n)^2 R}{g}$$

At a point about 15 cm away from the axis of the machine rotating at 6000 rpm, the separating power will be about 6160. Thus the velocity of separation of the particles, under the above conditions, will be over 6000 times greater than that under the action of gravity alone. Since the suspended rubber particles are lighter than the serum, the concentrate accumulates at the centre of the bowl and the serum at the rim. The separating force on the particle increases with increasing distance from the centre of rotation of the bowl. As the serum flows outwards from the centre, the rubber particles tend to be removed in the order of their size and at the same time, the particles in the serum are subjected to an increasing force. The viscosity of the cream increases very rapidly as the dry rubber content increases, until it reaches a point at which the latex will not flow. Thus in the concentrate the particle velocity falls rapidly and the tendency for further concentration is reduced by the increase in viscosity and diminishing separating force, as the concentrate flows towards the centre of the bowl.

Different types of centrifuges have been commercially available. The basic design of these machines is similar and it consists of a rotating bowl in which a set of concentric conical metallic separator discs are enclosed. Latex enters the bowl through a central feed tube and passes to the bottom of the bowl through a distributor. A series of small holes on the separator discs, positioned at definite distance from the centre, allow the latex to get distributed and broken up into a number of thin conical shells within the bowl which rotates at high speed. At steady state running of the machine, the DRC of the latex at the periphery of the bowl will be much lower than that of the latex at the centre. The latex concentrate which has above 60% DRC flows towards the axis of rotation and is collected through a galley at the top. The skim latex which contains about 6 to 10% rubber is collected through a separate galley. The DRC of the latex concentrate and that of the skim depends on several factors such as speed of rotation of the bowl, pressure head in the feed cup, diameter of feed tube, length and diameter of the skim screw and DRC of the feed latex. Shorter skim screw, low feed rate of latex and higher DRC of field latex favours a higher DRC for the concentrate. Factors affecting efficiency of different types of machines have been studied in detail and operating conditions to get the maximum bowl efficiency have been worked out (Piddlesden, 1940 and Sum et al. 1982). When the DRC of the latex is above 66%, it is known as high DRC latex, which can be produced by proper adjustment of the skim screw and feed rate (Zachariassen et al. 1972).

Latex for centrifuging is to be adequately preserved against bacterial growth and development of volatile fatty acids. The dosage of ammonia required for preserving the field latex depends on the period of storage before centrifuging operation and varies from 0.3% to 1.0%. Only the minimum essential quantity of ammonia is to be used for preserving the latex since a substantial quantity will be lost in the skim during centrifuging. The preserved latex is treated with calculated quantity of diammonium hydrogen phosphate to remove magnesium ions as magnesium ammonium phosphate. On storage, this material settles down at the bottom of the tank as sludge. Desludged latex is fed into the centrifuging machine at a constant rate. The machine can be run at standardised conditions of speed of rotation, rate of feed and skim screw settings, continuously for about 3 to 4 h. By this time, small clots of rubber and sludge get accumulated between the separator discs and the efficiency drops. At this stage, the machine is stopped and the bowl cleaned. It is possible to run the machine continuously for more than 4 h if the feed latex is properly desludged and passed through a clarifier initially. By proper setting of the skim screw and by adjusting the rate of feed, DRC and the total rubber recovered in the concentrate can be controlled. Normally it is observed that attempts to increase DRC of the concentrate beyond a certain level causes more loss of rubber in the skim. The DRC of the concentrate is usually kept slightly above 60% and finally adjusted to 60% by dilution.

Since most of the preservatives used in the feed latex would be lost in the aqueous phase of skim, it is necessary to supplement the concentrate with the required quantity of preservative depending upon the type of preservation system adopted. The concentrate may be preserved with ammonia alone or with a combination of ammonia with chemicals such as boric acid, sodium pentachlorophenate, zinc diethyl dithiocarbamate, triadine 10, zinc oxide, tetramethyl thiuramdisulphide etc. A small quantity of a higher fatty acid soap such as ammonium laurate is also normally used to boost the mechanical stability of the latex. Each of the above mentioned preservative systems has been studied in detail and the conditions for effective preservation of the concentrate standardised (Bloomfield and Mumford, 1970; Angove and Pillai, 1965; Poh, 1983; John et al. 1986). Recent legislations in some of the developed countries resulted in objections to the use of high ammonia type latex. Among the low ammonia preservative systems, the one consisting of 0.2% ammonia, 0.013% each of tetramethyl thiuramdisulpide and zinc oxide (LA-TZ system) is the most popular. Table 1 gives a list of various types of commercially available preservative systems for latex and their compositions. Details of some of the speciality natural rubber latex concentrates are given in Table 2 (Wahab, 1983).

There are about thirteen characteristics of natural rubber latex concentrate based on which the quality is assessed as per ISO specification 2004. The requirements of ISO 2004 and the test methods are listed in Table 3.

. TABLE 1

Preserviative systems for latex concentrate

Type of systems	Composition
1. High ammonia (HA)	0.6% ammonia minimum.
2. Low ammonia - sodium pentachloro-phenate (LA-SPP)	0.2% ammonia + 0.2% sodium pentachlorophenate.
3. Low ammonia - Boric acid (LA-BA)	0.2% ammonia + 0.24% boric acid + 0.05% lauric acid.
4. Low ammonia - Zinc diethyl-dithiocarbamate (LA-ZDC)	0.2% ammonia + 0.10% ZDC + 0.05% lauric acid.
5. Low ammonia - tetramethylthiuram disulphide zinc oxide (LA-TZ).	0.2% ammonia + 0.013% TMTD + 0.013% ZnO + 0.05% lauric acid.

TABLE 2

Speciality natural rubber latex concentrates

Types	Description
1. High DRC latex	Centrifuged latex having DRC in the range of 64% to 67%.
2. Purified or multiple centrifuged latex	Prepared by diluting centrifuged latex to 30% DRC with water containing ammonia and then recentrifuging to 60% DRC. It has a low non-rubber content.
3. Prevulcanized latex concentrate.	Prepared by heating a stabilised latex concentrate with dispersions of sulphur, zinc oxide and an ultrafast accelerator at temperatures of about 70°C for 2 h.
4. Methyl methacrylate grafted (MG) latex.	Prepared by grafting methyl methacrylate onto NR in latex form. MG latex having 30% and 49% by weight of polymethyl-methacrylate are available. This type of latex has a self reinforcing effect and provides a means of substantially improving the tear and puncture resistance of dipped goods.
5. Low constant viscosity (LCV) latex.	Prepared from specially selected clonal latices by treatment with 0.15% of hydroxylamine salts immediately after centrifuging. This type of low viscosity latex is advantageous in low pressure tack adhesive formulations.

TABLE 3

ISO 2004 (1974) requirements for centrifuged natural rubber latex concentrate.

Parameter	HA	LA	ISO test method No.
Total solids content, min; %	61.5	61.5	124
Dry rubber content, min; %	60.0	60.0	126
Non-rubber solids, max; %	2.0	2.0	-
Alkalinity, as ammonia, on latex weight, %	0.6 (min)	0.29 (max)	125
Mechanical stability time, min; s	650.0	650.0	35
Coagulum content, max; %	0.05	0.05	706
Volatile fatty acid number, max.	0.20	0.20	506
Potassium hydroxide numbr, max.	1.0	1.0	127
Copper content, max; mg/kg solids	8.0	8.0	1654
Manganese content, max; mg/kg solids	8.0	8.0	1655
Sludge content, max; %	0.1	0.1	2005
Colour	No blue or grey		
Odour	No putrefactive odour after neutralization with boric acid.		

While most of the parameters are stable and do not change during storage, three important characteristics namely volatile fatty acid (VFA) number, potassium hydroxide (KOH) number and mechanical stability time (MST) increase with time of storage of latex. These parameters are also affected by handling of latex, seasonal effects, pumping, exposure to air etc. The changes in VFA, KOH number and MST of latex are the result of two effects - changes associated with bacterial action in latex and those associated with hydrolytic action. Proper preservation can control the first effect whereas no control over the second can be exercised.

The volatile fatty acids in latex consist primarily of acetic and formic acids. They are produced by bacterial activity in latex utilising a glucose-amino acid complex as substrate (Lowe, 1960). The breakdown of this complex occurs via triose compounds, 3 phosphoglyceraldehyde and dihydroxy acetone phosphate, to pyruvic acid. The volatile acids can be produced from pyruvic acid and alanine through many metabolic pathways. VFA of latex concentrate is found to be between one third and one half of the VFA number of the field latex from which it is prepared (Cook and Sekhar,

1955). Hence field latex having a VFA number greater than 0.3 is considered to be unsuitable for latex concentrate production. In the event of serious bacterial contamination, the whole plant is to be disinfected by using a 1% solution of any one of the chemicals such as 5, 5'dichloro-2,2'dihydroxy diphenyl methane, alkyl-dimethyl-benzyl ammonium chloride or sodium hypochlorite, to control VFA formation. Work conducted by Lowe (1959) revealed that formation of VFA is a limited enzymic process which is influenced by the following factors:

* the quality of field latex, level of bacterial population in latex and length of time they have been allowed to remain viable in the latex
* temperature variations of latex before and after processing
* amount of serum substrate available for the enzymic process
* degree of inhibition exerted by the level of ammoniation
* the redox potential of latex

Well preserved latex concentrate may have a VFA level of 0.01 to 0.02 at the producers' end and may reach 0.08 to 0.10 at the consumers' end. At VFA levels of 0.25 to 0.35, putrefactive smell is evident and the latex concentrate is difficult/impossible to process.

Potassium hydroxide number is a measure of the ionic strength of the serum in the presence of ammonia. This represents most of the ions present in the latex but not all, since a significant amount of potassium ions is also present. KOH number of latex concentrate rises slowly during storage. Exposure to air tends to enhance the natural increase, presumably by absorption of carbon dioxide and the formation of carbonate/bicarbonate ions. The increment in KOH number during storage arises primarily from increase in the ions adsobred at the particle surface as the serum ion concentration is found to be unchanged during storage. A poorly preserved latex will have a high KOH number, but a high KOH number need not be an indication of poor preservation conditions since the presence of ions such as phosphates, amino acids and many others increases the KOH number, yet the preservation level may be quite satisfactory. Commercial latex concentrates on production may have a value of 0.5 which rises to 0.6 or more during storage.

The mechanical stability time of concentrated latex on the day of production will be low, in the range of 80 to 100 seconds, but will rise steadily within the first four to six weeks to give a value of 600-1500 seconds. External addition of soaps such as ammonium laurate will instantly increase MST. The spontaneous increase in MST during storage of latex is attributed to ammoniacal hydrolysis of lipids which produce fatty acid soaps which get adsorbed on the particle surface. Increase in temperature of latex is found to have the same effect on increasing the MST of fresh latex concentrate. The exact reason for the temperature effect is not understood but may involve the temperature/solubility dependence of the naturally occurring soaps. Recent work on MST of latex revealed that increase in MST is not related to electrical potential at the particle surface, indicating that the application of the electrostatic stabilization theory and the relation of particle charge to MST of natural rubber latex is questionable (Pendle, 1990).

Even though the production of concentrated NR latex by the centrifugal method is well established and a lion's share of the concentrate is produced by this method, further work on the following lines will be of great interest.

1. To develop a process by which the desludging process could be accelerated. At present it takes about 48 h for settling of the sludge, necessitating large storage capacities to be built up.

2. To develop a more effective and water soluble preservation system, as ammonia at higher levels is objectionable and none of the low ammonia systems is fully satisfactory. The LA-TZ system has the disadvantage of giving low PRI for skim rubber, inconsistency in quality of dispersion and settling of the added chemicals.

3. To develop a centrifuging machine having better efficiency and capability of self cleaning. The average efficiency with respect to concentrate recovery of the currently available machines varies from 86% to 90%. There is a need to re-examine the design of the machine so as to provide a higher feed rate with an accompanying high rate of concentrate recovery. Another point which requires attention is the facility to run the machine continuously. Since all the commercially available centrifuges are to be stopped for cleaning after about 3-4 h running, considerable time is lost even if a spare bowl is kept ready. A self cleaning machine will be advantageous in this respect.

Use of a latex clarifier is reported to have solved some of these problems to a certain extent (Kumaran, 1990). Passing preserved field latex through the clarifier removes most of the sludge present in it, which enables the centrifuge to be run continuously for about 8-9 h.

Treatment of skim latex

Skim latex may contain about 6-10% rubber depending upon the efficiency of separation. The rubber particles in skim are relatively smaller in size and have most of the proteins adsorbed onto them. The serum contains dissolved nonrubber solids and a major portion of the preservative used in field latex. All these factors make skim latex a highly stable one and render separation of rubber from it a difficult task. However, different methods have been developed for recovering the rubber. Normally the latex is first deammoniated by aeration and then coagulated with dilute sulphuric acid. The coagulum is thoroughly washed and processed into crepe rubber. Enzymic deproteinisation of skim latex using trypsin, followed by coagulation was reported to improve the quality of skim rubber (Morris, 1954). Creaming of skim latex using tamarind seed powder followed by coagulation also improves the quality of skim rubber (Thomas and Jacob, 1967). Skim rubber obtained by microbial fermentation is having better properties compared to those of rubber recovered by calcium chloride treatment (Resing, 1960). Quaternary ammonium surfactant as an alternative coagulant was reported (Sum, 1983). Rubber having properties equivalent to those of SMR 5L could be produced from skim latex by following the process described by Ong (1974). In this process the skim latex is treated with sodium metabisulphite (0.04% on DRC) before acid coagulation. The coagulum is soaked in 3% sodium hydroxide solution, washed and then soaked in 0.15% sulphuric acid solution. It is then washed and treated with a composite solution consisting of thiourea and oxalic acid (1% w/w). The rubber thus recovered is reported to have very high PRI and constant viscosity characteristics.

RIBBED SMOKED SHEETS

Converting natural rubber latex into ribbed smoked sheets (RSS) is the oldest method of processing. This method is widely adopted by small and medium scale rubber growers because of the following reasons.

1. This is the most convenient method when the quantity of latex is small.
2. It requires only simple equipments.
3. It does not require great technical expertise to adopt the process.
4. Overall cost of processing is less compared with other methods.

Latex intended for processing into RSS is treated with any one of the short-term preservatives such as sodium sulphite, formalin or ammonia, if it shows a tendency for precoagulation. A combination of formalin and sodium metasilicate is reported to be an ideal anticoagulant system for latex to be processed into RSS (Cook, 1960). Latex in reception at the collection centre/processing factory is sieved using 40 and 60 mesh sieves to remove suspended impurities. Since the quantity of chemicals required for processing latex into RSS is based on its DRC, a quick estimate of the same is made, usually by the hydrometric method. A detailed account of estimation of DRC was given by Chin and Singh (1980) and Chin (1981a,b). Latex from different fields/estates is pooled to make rubber of uniform quality. The bulked latex is diluted to a standard DRC of 12.5%, by adding water, to improve the quality of the sheets produced. Sheets from diluted latex have better colour, transparency and show less tendency for mould growth. It also makes the sheeting operation easy and helps faster drying of the sheet. On allowing the diluted latex to remain undisturbed for about 10-15 min, the denser impurities settle fast and the latex is transferred to another tank, without disturbing the sediments. Chemicals such as sodium bisulphite and paranitrophenol are added to the latex at this stage, if found necessary.

Sodium bisulphite is added to prevent discolouration of the coagulum, caused by enzymes of the polyphenol oxidase type. Phenols and aminophenols present in latex combine with oxygen from the air to form orthoquinones, which react with naturally occurring amino acids and proteins in latex, to give coloured products resembling melanin. Reduced melanin is tan coloured and the oxidised one is black. The enzymes are believed to catalyse the above reactions and sodium bisulphite prevents the discolouration by preferentially getting reacted with atmospheric oxygen. The dosage of sodium bisulphite depends on the extent of enzymic activity and concentrations of amino phenols, amino acids, proteins etc. in latex. About 1.2 g of sodium bisulphite is normally found to give satisfactory protection for latex containing 1 kg dry rubber. It is added as 2% solution. Use of excess quantity of this chemical may lead to slow drying of the sheets. To prevent mould growth on the surface of the sheets, the wet sheets are dipped in 1% solution of paranitrophenol or a 1% solution of the chemical is added to the latex before coagulation, at the rate of 1 g per kg DRC. The second procedure is reported to be more effective (Hastings and Piddlesden, 1938).

Diluted latex of about 12.5% DRC is coagulated in pans, troughs or tanks. Coagulating pans are usually made of aluminium whereas the tanks may be of aluminium or masonry which is lined inside with aluminium sheets or glazed tiles. When the number of sheets to be prepared is small, pan or trough coagulation is done. For larger scale operation coagulation tanks provided with partition plates are used. Solutions of volatile acids such as acetic and formic acids at low concentrations are used for coagulating latex. Such weak acids provide uniform coagulation and excess acid if any, gets volatalized off during drying of the sheets. The pH of coagulation is kept at about 4.6 to get complete recovery of rubber, avoid fermentation of latex and to get a soft coagulum. Attempts to use strong acids such as sulphuric acid for coagulating NR latex were made as early as in 1932 (Wiltshire, 1932; Martin and Davey, 1934). It was reported that properly diluted sulphuric acid, when used in correct proportions did not cause any deterioration in properties of the rubber. Corrosion to machinery and utensils also did not occur if they were properly washed after the use. Later work by Baker and Philpott (1950) and that by Best and Morrel (1955) also confirmed the earlier reports. However Neef (1950) reported that rubber coagulated with excess of sulphuric acid showed considerable softening when it was heated at 140°C for 3 h. Renewed interest in the use of sulphuric acid started during early 1980's since the price difference between the volatile acids and sulphuric acid became attractive. Othman and Lye (1980) reported that pH of coagulation is very important when sulphuric acid is used as coagulant for NR latex and recommended a pH around 5.0 as more suitable for sheet rubber production. More recent studies indicated that sulphuric acid can be used as coagulant for NR latex in sheet rubber production, if all factors such as dosage, dilution and washing of sheets and machinery are taken care of (George et al. 1990). Another nonvolatile acid that is recommended for latex coagulation is sulphamic acid. Use of optimum dosage and proper washing of the coagulum during sheeting to remove residual acid were reported to yield good quality sheets when this chemical is used (Sebastian et al. 1982).

Latex coagulum is sheeted after maturation. Sheeting operation squeezes out serum present in the coagulum and reduces its thickness to about 3.0 mm. This is done by passing the coagulum through a set of hand operated plane rolls, several times, followed by one pass through another set of grooved rolls on which the grooves are spirally cut at an angle of 45° and at 3 to 5 mm width and depth. The ribbed design helps to increase the surface area of the coagulum for faster drying and also prevents sticking of the dried sheets when these are stacked in bundles. A sheeting battery consists of four or five sets of plane rolls and one set of grooved rolls arranged in a row with their nip gap progressively reduced. It is used for sheeting the coagulum prepared in large tanks. Uniformity and softness of the coagulum are the two important parameters that affect the efficiency of the sheeting battery (Bishop and Wiltshire, 1932). The coagulum is washed thoroughly during sheeting to remove the serum and residual acid. Sheeted coagulum is soaked in 1% solution of paranitrophenol if it has not been added to latex before coagulation.

Drying of sheets can be done in a smoke house or hot air chamber and the dried sheets thus obtained are termed ribbed smoked sheets (RSS) or air dried sheets (ADS), respectively. Partial drying in sunlight followed by smoke drying is also widely practiced. A survey of the factors involved in the drying of sheet rubber (Gale, 1959) revealed that upto a moisture level of 10%, syneresis is the main process of water removal and drying below this level is diffusion controlled. Method of preparation of the wet sheet (pH of coagulation, softness of coagulum, extent of machining etc) and drying conditions such as temperature, relative humidity and speed of air circulation in the chamber etc. have profound influence on the rate of drying. The sheeted coagulum is allowed to drip off water emerging due to syneresis, before it is fed into the smoke house or drying chamber. Several types of smoke houses such as those with furnace inside or outside the drying chamber, batch type or continuous type are in operation (peries, 1970a). The RRIM tunnel type smoke houses described by Graham (1964) meet almost all the requirements of an ideal smoke house. In this type of smoke house, the furnace is outside the drying chamber and smoke and hot air are led into the chamber through underground flue inlets. By controlling the rate of burning of firewood and regulating the openings for flue inlet, ventilator and exhaust, the temperature inside the chamber at different regions can be controlled and maintained in the range of about 45°C to 60°C. The sheets are put on reapers placed on trolleys which moved on a central rail track inside the chamber. The capacities of the trolley and the chamber are designed in such a way that each trolley can accommodate one day's crop and the chamber, four days' crop. The sheets take about four days to get dried.

Since firewood has become scarce and its price increased several fold, efforts to reduce its consumption have been made. Use of solar energy to dry sheet rubber has been attempted in many ways. Direct exposure of wet sheets to sun light did not affect the dynamic properties of the sheets. However, exposure of dried sheets even for a few hours, adversely affected these properties (Tan et al. 1977). Experiments using solar power-boosted smoke houses (Rama Rao et al. 1986; Nair et al. 1988) show that there could be a saving of about 50 to 60% firewood by using solar energy for drying. In such smoke houses, the main drying process occurs with the help of hot air generated by solar panels and firewood is used only as a subsidiary heat source to maintain the temperature of the drying chamber from dusk to dawn. While the initial expenses of the solar boosted smoke house is higher than that of a conventional smoke house, its recurring expense is lower.

The dried sheets are visually examined and graded, adopting the norms prescribed by the International Rubber Quality and Packing Committee under the Secretariat of the Rubber Manufacturers Association Incorporated, USA and which are described in the 'Green Book'. The important parameters considered in grading the sheets into RSS 1X to RSS 5 are degree of dryness, presence of foreign matter, virgin rubber, oxidised spot, blisters, bubbles, resinous matter, transparency, colour, tackiness, over smoking, mould growth, dark or coloured spots etc.

TECHNICALLY SPECIFIED NATURAL RUBBER

Visual grading followed for quality assessment of sheet and crepe rubbers, has many drawbacks. Hence attempts were made as early as the 1950s to study the variability in sheet rubber and to evolve a better method of grading, based on technical parameters (Fletcher, 1950; Newton et al. 1951; Baker, 1954). By early 1960s, parameters that could provide a satisfactory indication of the quality of NR were identified (Baker et al. 1967). NR in technically specified form was first introduced in 1962 by the SOCFIN group of companies in Ivory Coast. However, only with the launching of the Standard Malaysian Rubber (SMR) Scheme in 1965, this new form got wider consumer acceptance. Now, a lion's share of NR processed in major NR producing countries is in technically specified form.

Advantages of technically specified rubber (TSR) include assurance of quality with respect to important technical parameters, consistency in quality, minimum space for storage and clean and easy to handle packing (Bekema, 1969; Pike and Ramage, 1969). This new method enabled NR producers to process both latex and field coagulum, using almost the same set of machinery and to reduce the processing time to less than 24 h. The competitive position of their produce with respect to cleanliness, presentation and appearance also improved. Even though different methods were developed to produce TSR, all these processes involve certain common steps such as coagulation of latex/precleaning of field coagulum, size reduction, drying, baling, testing, grading and packing. Differences among commercial processes lie in the method of coagulation or in the machinery used for crumbling the coagulum. Crumbs of coagulum may be prepared by purely mechanical means (Smith, 1969) or by a mechano-chemical process (Muthukuda, 1967).

The Dynat method of processing involves four basic machines - a rotary cutter, pelletiser, drier and baling press. Coagulation of latex is done at field DRC. Assisted biological coagulation produces a coagulum which is porous in nature and hence easy to process and dry (Thompson and Howorth, 1964). A slitting machine splits the slab into strips which are then fed to the pelletising machine. When field coagulum is processed, a rotary cutter with a perforated screen is employed (Thompson et al. 1966) for initial size reduction. The pelletiser operates by a process of extrusion and cutting. Extrusion takes place with minimum mechanical working of the coagulum, and cutting knives placed both inside and outside the die plate, cut the coagulum into small pieces of about 3 to 4 mm diameter. The face of the extruder is sprayed with water to lubricate the knives and for effective washing (Howorth, 1966). These machines are reported to have lower power requirements (Heyneker, 1967). The washed crumbs are dried in a vertical semicontinuous drier. Drying in the stack employs a relatively low temperature (60-65°C) with high humidity at the top or feeding point and a high temperature (90-93°C) and low humidity at the bottom or discharge point. The dried crumbs are weighed and baled using a 60 tonne self contained hydraulic unit giving 3 to 4 min dwell time under pressure (Shaw, 1968).

The Decan remill process, described by Gyss and Fleurot (1969) for preparing Nat-rubber, consists of coagulating latex in cylindrical tubes, veneering, calendering and granulating the coagulum to produce fine crumbs. Coagulation is done by assisted biological method. Macerators and scrap washers are replaced by granulators and hammermills for preparing crumbs from estate scrap and smallholders' coagulum. Rapid drying is achieved by adjusting flow rate, humidity and temperature of the air passed through a bed of crumbs. Use of zinc pentachlorothiophenate for preparing TSR (Peptorub) from field coagulum grade rubber was described by Hastings (1964). Scrap washed coagulum was put into a rotating cement mixer into which a measured quality of the peptising agent is sprayed as dispersion. It is further processed using a creper and a granulating machine. The granules are kept in open steam for 20 min and then dried in an air circulated oven maintained at about 63°C. It takes about three days for drying of the crumbs.

The Heveacrumb process introduced by the RRIM makes use of castor oil as the crumbling agent. A full account of the process has been described by its originators (Sekhar et al. 1965) and also by several others (Bateman et al. 1965; Graham and Morris, 1966). Addition of a small quantity of castor oil either in the latex stage or during processing of the coagulum prevents reagglomeration of the crumbs. The advantage of this mechano-chemical process over the purely mechanical process is that even conventional machinery such as crepers can be used for producing crumbs. For processing small holders latex coagulum and field coagulum several combinations of machinery such as crepers with hammermill, shredder or granulator may be employed. Use of prebreaker enables processing of even dried field coagulum into fine crumbs which can be blended more effectively and processed further. Polypropylene glycol of average molecular weight 2000, was reported to be an effective alternative to castor oil (Chin, 1974). Use of this chemical along with zinc stearate can reduce the dosage of the former from 0.7% to about 0.35% on DRC of latex. Methods adopted for processing latex/field coagulum have a profound influence on the quality of TSR produced. DRC and method of coagulation, type of coagulant, maturation of coagulum etc. are reported to influence the quality to a great extent (Graham, 1969; Morris, 1969; Sekhar, 1971). Coagulation at original DRC using formic acid gives a higher plasticity retention index (PRI). Assisted biological coagulation is better than the natural one to get a higher PRI.

Further efforts were made to modify the process and machinery to achieve better properties and higher outputs in TSR production. Continuous coagulation of latex was attempted by Harris et al (1974) based on the principle of matched gravitational flow of diluted acid and latex. More cost effective methods for continuous coagulation of latex using dilute sulphuric acid (Mcintosh and Wilkinson, 1975) and by heat gelation technique (Fah and Peng, 1975) were reported to produce more uniform coagulum within a very short time. Factors affecting power consumption and output of machinery used for SMR production were identified for improving the efficiency. Speed of rotation in hammermills and that of the faster roll in crepers are reported to be the deciding factors for peak load on electric motor, rather than friction ratio and nip gap of the rolls (Lim and Sethu, 1974). Higher speeds and deeper grooves on creper rolls increase the output, but require more power (Subbiah et al. 1976). Drying of the crumbs also consumes much energy. Three stages involved in drying of wet crumbs are the constant rate period and the first and second falling rate periods. Influence of parameters such as humidity gradient, air velocity and temperature during each stage of drying has been identified (Sethu, 1967). Later work by Yushan et al. (1985) revealed that the maximum temperature for drying latex grade crumbs is 125°C and the ideal bed thickness is 25-28 cm, whereas for field coagulum grade crumbs, these will be 115°C and 17-20cm respectively. The drying time in both cases will be 2.5 h. However, Roudeix (1985) reported that latex crumbs could be dried at 120°C within 2 h, if the coagulum is made porous by adding yeast (0.5 g kg^{-1} DRC) and sugar (6-8 g kg^{-1} DRC) before coagulating the latex. Since higher baling temperature influences the rate of storage hardening of NR, temperature below 60°C is recommended for baling the crumbs (Fah, 1977).

Technically specified form of NR is produced in constant viscosity (CV), oil extended (OENR), superior processing (SP) and deproteinised (DPNR) varieties also. Addition of 0.02% on DRC of xylyl mercaptan to latex along with 0.15% on DRC of latex of hydroxylamine hydrochloride is reported to yield SMR 5 CV from high viscosity rubbers (Ong and Lim, 1978). CV rubber having Mooney viscosity in the range of 50±5 can also be produced by using 0.15% by weight on DRC of latex of hydroxylamine neutral sulphate and 0.0037 to 0.014% by weight on DRC of latex of Renacit VII (Tillakaretne et al. 1981). Methods for producing OENR, CV rubber and SP rubber by Heveacrumb process was reported by Sung (1966). The preparation and properties of OENR have been described by Sung and O'Connell (1969) also. Chin et al. (1974) and Cheang et al. (1987) have given a detailed account of the production of deproteinised form of NR.

Since its introduction in 1965, the SMR scheme has been under constant monitoring and revision. There were only three grades (SMR 5, SMR 20 and SMR 50) initially, and six parameters (dirt, ash, nitrogen, volatile matter, copper and manganese) based on which the grading was done (Rubber Research Institute of Malaysia, 1965). Methods to improve parameters such as PRI of the raw material (Watson, 1969) and the significance of the specification parameters on the technological properties have been established (Bateman and Sekhar, 1966; Bristow, 1990). The first major revision of the SMR scheme was made during 1970. Copper and Manganese were replaced by PRI. Wallace plasticity (Po) was introduced as one of the specification parameters. Nitrogen and ash limits of SMR 5 were changed and two new grades, SMR 10 and SMR EQ, were introduced. The second revision of the scheme was made in 1979. The major changes involved in these revisions are described by Tong and Kamaruddin (1984). Work conducted by Ong et al. (1987) pointed out the need for introducing parameters other than PRI in predicting the mastication and mixing behaviour of CV rubber. The main complaints against TSR are the presence of wet rubber and contamination by plastic materials (Rao and Tong, 1984). The scheme is being continuously updated to meet consumer demands within the framework of producer capabilities. The latest SMR grades and their specification are given in Table 4. Other TSR producing countries also update the specifications of their produce from time to time.

Introduction of the technical specification shceme was an important step in the development of NR processing industry. It enabled the consumers to have a better idea of the quality of their raw material. However,

TABLE 4

Standard Malaysian Rubber Specification Scheme Mandatory from 1 October 1991.

Parameter	SMR CV60	SMR CV50	SMR L	SMR 5	SMR GP	SMR 10CV	SMR 10	SMR 20CV	SMR 20
	LATEX		SHEET MATERIAL[a]		BLEND	FIELD GRADE MATERIAL			
Dirt retained on 44 μ aperture (max, % wt)	0.02	0.02	0.02	0.05	0.08	0.08	0.08	0.16	0.16
Ash content (max, % wt)	0.50	0.50	0.50	0.60	0.75	0.75	0.75	1.00	1.00
Nitrogen (max, % wt)	0.60	0.60	0.60	0.60	0.60	0.60	0.60	0.60	0.60
Volatile matter (max, % wt)	0.80	0.80	0.80	0.80	0.80	0.80	0.80	0.80	0.80
Wallace rapid plasticity (Po) (min)	–	–	35	30	–	–	30	–	30
Plasticity retention index (PRI) (min, %)[b]	60	60	60	60	50	50	50	40	40
Lovibond Colour: individual value (max)	–	–	6.0	–	–	–	–	–	–
range (max)	–	–	2.0	–	–	–	–	–	–
Mooney viscosity ML(1'+4')100°C[b]	60(+5,-5)	50(+5,-5)	–	–	65(+7,-5)	c	–	c	–
Cure[d]	R	R	R	–	R	R	–	R	–
Colour coding marker	Black	Black	Light Green	Light Green	Blue	MAGENTA	Brown	Yellow	Red
Plastic wrap colour	Transparent	Transparent	Transparent	Transparent	Transparent	Transparent	Transparent	Transparent	Transparent
Plastic strip colour	Orange	Orange	Transparent	Opaque White	Opaque White	Opaque White	Opaque White	Opaque White	Opaque White

a Two sub-grades of SMR 5 are SMR 5RSS and SMR 5ADS which are prepared by direct baling of ribbed smoked sheet and air-dried sheet (ADS), respectively.

b Special producer limits and related controls are also imposed by the RRIM to provide additional safeguard.

c The Mooney viscosities of SMR10CV and SMR20CV are, at present, not of specification status. They are, however, controlled at the producer end to 60(+7,-5) for SMR 10CV and 65(+7,-5) for SMR 20CV.

d Rheograph and cure test data (delta torque, optimum cure time and scorch) are provided.

advancement in the rubber product manufacturing industry necessitated a re-examination of the specification parameters and the possibility of introducing new parameters which may truely reflect the processability and technological properties of NR. To have automated processing and computer controlled machinery in product manufacture, strict batch to batch consistency in raw rubber has become very essential. Even though TSR meets most of the requirements of the consumer, the quality specified is to be consistent also. In addition to this, for better control of manu-facturing processes such as extrusion, calendering, injection moulding etc. more information on viscosity, physico-chemical and rheological properties of the raw material is necessary. Baker and Bristow (1991) have identified the important parameters and their limits, which are to be controlled for producing TSR having the quality and consistency acceptable to the consumers. However, the inherent variability and that introduced due to difference in processing techniques make it a challenge to produce TSR of really consistent quality (Livonniere, 1991). Quality has to be designed and built into the rubber from the very beginning so that the entire produce conforms to the requirements. For meeting the ISO 9000 series of quality systems, strict control on basic source material and the production process is to be effected. The system introduced should be such that it is economically viable to the producer and the quality accept-able to the consumer at an affordable price.

CREPE RUBBERS

There are 29 grades of crepe rubber, contained in six different types which are described in the Green Book. However, based on the raw material used, crepe rubbers can be grouped into two, namely latex crepes and those produced from field coagulum. For low quality crepes field coagula such as cup lump, tree lace, shell scrap, earth scrap, bark scrap etc. and cuttings and rejections of pale latex crepe and smoked and un-smoked sheets are made use of. The manufacturing procedure depends on the type of raw material used and the grade of the crepe to be produced. However, for all the grades, steps such as sorting of the coagula, pre-cleaning/power-washing, maceration, remilling, refining etc. are involved. The quality of the final product depends, to a large extent, on the number of passes through each machine and on the extent of washing given at each stage. The lower grade crepes are dried in sheds at atmospheric temperature. Crepe rubber produced from tree lace is of inferior quality compared with that from cup lump since tree lace undergoes severe degradation on storage (Arumugam and Morris, 1964). With the advent of technically specified rubber, production of lower grade crepes has come down drastically. But, the latex grade crepes, namely pale latex crepe (PLC) and sole crepe are preferred to even the best quality TSR, in some applications such as pharmaceutical and food contact products, electrical insulation, rubber solution and cements, white and bright coloured products etc. because of their high purity and light colour (Karunaratne, 1977).

PALE LATEX CREPE AND SOLE CREPE

These are manufactured from latex, under strictly controlled conditions. Detailed procedures for their manufacture are described by Morris (1964) and Peries (1970b). Latex which contains a lower concentration of yellow colouring pigment, which is less susceptible to enzymic darkening and which yields rubber having a higher Mooney viscosity is found to be the ideal one for PLC and sole crepe. Considering the above factors, latex from PB 86 is the most suited one, even though that from other clones can also be used, either alone or in blends with PB 86 latex. The best anticoagulant for the latex to be processed into PLC is sodium sulphite. To prevent enzymic darkening sodium bisulphite is recommended.

For removing the yellow colouring materials, fractional coagulation or bleaching method or a combination of these two, is usually followed. For fractional coagulation diluted latex is treated with a small quantity of 2% acetic or oxalic acid solution and stirred well until small clots of rubber containing the yellow pigment are formed. The clotted fraction is removed by filtration and the filtered latex coagulated using 1% formic or 2% oxalic acid solutions. The bleaching process consists of treating the latex with thiols such as xylyl mercaptan which preferentially reacts with the colouring matter. Due to high toxicity, xylyl mercaptan is being replaced by other chemicals such as sodium/potassium salt of tolyl mercaptan and that of para-tertiary butyl thiophenol (Karunaratne, 1983). Quantity of the bleaching agent to be used depends on the extent of colouring matter present and is determined by trials. The bleaching action of the above chemicals is reported to be effected through oxidised carotene of xanthophyll and other oxygenated forms of carotene and not through carotene itself (Tillakarente et al. 1984). The bleached latex is coagulated using either 1% solution of formic acid or 2% solution of oxalic acid. Use of oxalic acid gives better colour and retention of colour during storage of the crepe. The coagulum is passed through a set of machinery consisting of macerators, crepers and smooth rolls. Thorough washing during machining is given to remove serum and excess coagulant but the final pass through the plane rolls is done without spraying water to facilitate quick removal of surface moisture. The thickness of the crepe coming from the plane rolls is adjusted between 0.8 to 1.0 mm. Drying of the crepe is done in sheds maintained at a temperature of 32-35°C. Several modifications to the conventional boiler-radiator system for heating the air in the sheds have been done from time to time. Use of finned air heater-forced draft blower arrangement, is reported to be more economical than the conventional system (Tharmalingam et al. 1977). A simple air heater arrangement which extracts heat from the flue gases prior to discharge up the chimney has been reported to improve the efficiency (Walpita et al. 1984a). Use of solar energy and application of partial vacuum are found to reduce the drying time of latex crepes considerably (Walpita et al. 1984b). Dried latex crepes are examined for defects and the selected grades are laminated to produce sole crepe of the required dimensions.

Discolouration of the latex crepe during manufacture and on storage is a serious problem in many manufacturing units. Nadarajah and Muthukuda (1974) suggested the use of oxalic acid as coagulant and to avoid addition of sodium bisulphite for eliminating discolouration. Nadarajah and De Silva (1983) have shown that use of boric acid in the range of 0.05 to 1.0% on DRC of latex in place of sodium bisulphite can control discolouration to some extent. Various factors causing discolouration of latex crepe and its remedial measures have been described by Nadarajah (1983). Nadarajah and Perera (1983) have shown that processing conditions and volatile matter present in latex crepe also affect mould growth and discolouration.

EFFLUENT FROM PROCESSING FACTORIES

A large quantity of effluent emerges from all types of rubber processing factories. It is to be treated properly before being discharged into the normal water ways. Different types of treatments for the effluent are described in literature (Ahmed, 1978, 1980; Ahmed et al. 1979; Nordin and Mohamed, 1989; Zaid and Sing, 1980).

REFERENCES

Ahmed bin Ibrahim. 1978. Treatment of effluent from SMR block rubber factories. Planters' Bull., 157: 133-139.
Ahmed bin Ibrahim. 1980. Starting of anaerobic/facultative ponds for treatment of rubber processing effluent. Planters' Bull., 165: 153-155.
Ahmed Ibraham, Sethu, S., Mohd. zin A. Karim and Zaid Isa. 1979. Anerobic/facultative ponding system for treatment of latex concentrate effluent. Proceedings of the Rubber Research Institute of Malaysia. Planters' Conference, Kuala Lumpur. pp. 419-435.
Angove, S.N. and Pillai, N.M. 1965. Preservation of natural rubber latex concentrate, Part III - Evaluation of various organo zinc compounds as secondary preservatives. Trans. Inst. Rubb. Ind., 41: T41-52.

Arumugam, G. and Morris, J.E. 1964. The effect of tree lace and cuplump storage conditions on the properties of brown crepe. Planters' Bull., 74: 144-154.

Baker, C.S.L. and Bristow, G.M. 1991. NR consistency - What the consumer is seeking. Proceedings of the IRRDB Technology Symposium on Quality and Consistency of Natural Rubber. Manila, Philippines. pp 33-37.

Baker, H.C. 1937. The concentration of latex by creaming. Trans. Inst. Rubb. Ind., 13(1): 70-82.

Baker, H.C. 1954. Vulcanising variability of natural rubber in pure gum and carbon black compounds. Trans. Inst. Rubb. Ind., 30: T162-179.

Baker, H.C., Barker, L.R., Chambers, W.T. and Greensmith, H.W. 1967. The properties of market grades of natural rubber. Rubb. Journal, 149(1): 10-21.

Baker, H.C. and Philpott, M.W. 1950. Coagulation with sulphuric acid. J. Rubb. Res. Inst. Malaya, 12: 265-268.

Bateman, L. and Sekhar, B.C. 1966. Significance of PRI in raw and vulcanised natural rubber. J. Rubb. Res. Inst. Malaya, 19(3): 133-140.

Bateman, L., Sekhar, B.C. and Webster, C.C. 1965. Mechano-chemical granulation of natural rubber. Rubb. Journal, 147(4): 184.

Bekema, N.P. 1969. Consumer appraisals of natural rubber. J. Rubb. Res. Inst. Malaya, 22(1): 1-13.

Best, L.L. and Morrell, S.H. 1955. Effect of sulphuric acid coagulation on properties of natural rubber. Trans. Inst. Rubb. Ind., 31(3): 133-140.

Bishop, R.O. and Wiltshire, J.L. 1932. Further notes on sheeting batteries. J. Rubb. Res. Inst. Malaya, 4(2): 85-93.

Bloomfield, G.F. and Mumford, R.B. 1960. Low ammonia latices. Trans. Inst. Rubb. Ind., 36(6): 251-262.

Bristow, G.M. 1990. The assessment of quality in natural rubber. Rubber Developments, 43(1&2): 23-26.

Cheang, K.T., Fong, C.S. and Lian, L.C. 1987. A new and improved method for deproteinised natural rubber production. Proceedings of the Rubber Growers' Conference, Desaru, Johore, Malaysia. pp. 456-473.

Cheong, S.F. and Ong, C.O. 1974. New preservation systems for field latex. J. Rubb. Res. Inst. Malaysia, 24(2): 118-124.

Chin, P.S. 1974. Alternative crumbling agent for Heveacrumb manufacture. Proceedings of the Rubber Research Institute of Malaysia Planters' Conference, Kuala Lumpur. pp. 219-228.

Chin, H.C. 1981a. Matrication and changes to the Chee method of determining the dry rubber content of field latex. Planters' Bull., 169: 136-149.

Chin, H.C. 1981b. The change over to latex hydrometer calibrated in matric units. Planters' Bull., 169: 159-162.

Chin, P.S., Chang, W.P., Lau, C.M. and Pong, K.S. 1974. Deproteinised natural rubber. Proceedings of the Rubber Research Institute of Malaysia Planters' Conference, Kuala Lumpur. pp. 252-262.

Chin, H.C. and Singh, M.M. 1980. Determination of dry rubber content of Hevea field latex. Planters' Bull., 163: 56-69.

Cook, A.S. 1960. The short-term preservation of natural latex. J. Rubb. Res. Inst. Malaya, 16(2): 65-75.

Cook, A.S. and Sekar, K.C. 1955. Volatile acids and the quality of concentrated natural latex. J. Rubb. Res. Inst. Malaya, 14: 407-422.

Davey, W.S. and Sekar, K.C. 1947. The evaluation of creaming agents. J. Rubb. Res. Inst. Malaya, 12: 62-77.

Duckwork, I.H. 1964. Creamed latex concentrate. Planters' Bull., 77: 111-122.

Edgar, R. and Sekar, K.C. 1938a. Creaming of latex with synthetic creaming agents. J. Rubb. Res. Inst. Malaya, 9: 343-345.

Edgar, R. and Sekar, K.C. 1938b. Creaming of latex with tamarind seed power. J. Rubb. Res. Inst. Malaya, 9: 346-349.

Fah, C.S. 1977. Storage behaviour of crated tyre rubber - Effect of baling temperature and viscosity stabilizers. J. Rubb. Res. Inst. Malaysia, 25(2): 81-92.

Fah, C.S. and Peng, L.F. 1975. Continuous heat gelation of latex - A potentially cost effective processing technique for SMR. Proceedings of the International Rubber Conference, Kuala Lumpur. Vol. IV. pp 295-305.

Fletcher, W.P. 1950. Some problems involved in the grading and testing of natural rubber - A progress report. Rubber Chem. Technol., 23(1): 107-116.

Gale, R.S. 1959. A survey of the factors involved in an experimental study of the drying of sheet rubber. J. Rubb. Res. Inst. Malaya, 16: 38-64.

George, K.M., Varghese, L. and Mathew, N.M. 1990. Use of sulphuric acid as coagulant for natural rubber Latex. Ninth Symposium on Plantation Crops PLACROSYM IX, Bangalore.

Graham, D.J. 1964. New tunnel type smoke houses. Planters' Bull., 74: 123-130.

Graham, D.J. 1969. New presentation process and SMR scheme. J. Rubb. Res. Inst. Malaya, 22(1): 14-25.

Graham, D.J. and Morris, J.E. 1966. Manufacture of Heveacrumb. Planters' Bull., 86 130-147.

Gyss, P.R. and Fleurot, M. 1969. Five years of Nat-rubbers. J. Rubb. Res. Inst. Malaya, 22(1): 70-77.

Harris, E.M., Graham, D.J. and Chang, W.P. 1974. A method for continuous addition of acid for coagulation of latex. Proceedings of the Rubber Research Institute of Malaya. Planters' Conference, Kuala Lumpur. pp. 208-218.

Hastings, J.D. 1964. Peptorub - An improved natural rubber. Planters' Bull., 74: 176-186.

Hastings, J.D. and Piddlesden, J.H. 1938. Prevention of mould growth on sheet rubber. J. Rubb. Res. Inst. Malaya, 8(3): 250-257.

Heyneker, W.G. 1967. Technological Development in natural rubber manufacture - The Dynat Process. Rubber Research Institute of Ceylon Bulletin, 2(3&4): 72-73.

Howorth, H. 1966. Preparation of Dynat rubber - A new form of natural rubber. Planters' Bull., 86: 126-129.

John, C.K. 1974. A novel method for stabilizing Hevea latex. J. Rubb. Res. Inst. Malaysia, 24(2): 111-117.

John, C.K., Nadarajah, M. and Lau, C.M. 1974. Microbiological degradation of Hevea latex and its control. J. Rubb. Res. Inst. Malaysia, 24(5): 261-271.

John, C.K., Nadarajah, M., Rama Rao, P.S., Lau, C.M. and Ng, C.S. 1975. A composite preservation system for Hevea latex. Proceedings of the International Rubber Conference, Kuala Lumpur. Vol. IV, pp. 339-357.

John, C.K., Wong, N.P., Chin, H.C., Latiff, A. and Lim, H.S. 1986. Recent developments in natural rubber latex preservation. Proceedings of the Rubber Growers' Conference, Ipoh, Perak, Malaysia, pp. 320-341.

John, C.K., Wong, N.P., Chin, H.C., Rama Rao, P.S. and Latiff, A. 1985. Further development in Hevea latex preservation. Proceedings of the International Rubber Conference. Kuala Lumpur, Malaysia. Vol. II. pp. 451-467.

Kang, L.C. and Hashim, Ismail. 1982. Factors influencing the dry rubber content of Hevea latex. Planters' Bull., 172: 89-98.

Karunaratne, S.W. 1977. Technical aspects of crepe rubber. J. Rubb. Res. Inst. Sri Lanka, 54: 637-639.

Karunaratne, S.W. 1983. Review of the Chemistry Department. Annual review of the Rubber Research Institute of Sri Lanka. pp. 83-100.

Kumaran, M.G. 1990. Latex clarifier - A boon to processors. Rubber Asia, 4(1&2): 47-49.

Lim, F.P. and Sethu, S. 1974. Performance improvement of machinery used for block rubber production. Proceedings of the Rubber Research Institute of Malaysia Planters' Conference, Kuala Lumpur. pp. 189-207.

Livonniere, H.de. 1991. Natural rubber consistency: A challenge for a bio-product. Proceedings of the IRRDB Technology Symposium on Quality and Consistency of Natural Rubber. Manila, Philippines. pp. 38-44.

Lowe, J.S. 1959. Formation of volatile fatty acids in ammonia preserved natural rubber latex concentrate. Trans. Inst. Rubb. Ind., 35(1): 10-18.

Lowe, J.S. 1960. Substrate for VFA formation of natural rubber latex. Trans. Inst. Rubb. Ind., 36(1): 202-210.

Madge, E.W., Collier, H.M. and Peel, J.D. 1950. Treatment of abnormal latexes. Trans. Inst. Rubb. Ind., 26(4): 305-312.

Martin, G. and Davey, W.S. 1934. Rubber from latex coagulated with sulphuric acid. J. Rubb. Res. Inst. Malaya, 5: 282-294.

Mathew, N.M., Varghese, L., Kothandaraman, R. and Thomas, E.V. 1976. Preservation of concentrated natural rubber latex with methylamine. Rubber Board Bull., 13(3): 58-66.

Mcintosh, J.B. and Wilkinson, B.C. 1975. Continuous coagulation of latex. Proceedings of the International Rubber Conference, Kuala Lumpur. Vol. IV. pp. 283-294.

Morris, J.E. 1954. Improved rubbers by enzymatic deproteinisation of skim latex. Proceedings of the 3rd Rubber Technology Conference, London, pp. 13-37.

Morris, J.E. 1964. Sole crepe. Planters' Bull., 74: 155-175.

Morris, J.E. 1969. Heveacrumb process. J. Rubb. Res. Inst. Malaya, 22(1): 39-55.

Muthukuda, D.S. 1967. Technological developments in the rubber industry - New forms of natural rubber. Rubb. Res. Inst. of Ceylon Bull., 2(3&4): 69-71.

Nadarajah, M. 1983. Principal factors causing discolouration of bleached crepe rubber. Rubb. Res. Inst. Sri Lanka Bull., 17: 15-17.

Nadarajah, M. and De Silva, A. 1983. Prevention of discolouration of pale crepe during storage. Rubb. Res. Inst. Sri Lanka Bull., 18: 23-27.

Nadarajah, M. and Muthukuda, D.S. 1974. Standardisation of production methods of latex crepes to suit to end-use requirements. Rubb. Res. Inst. of Sri Lanka Bull., 9(1&2): 45-49.

Nadarajah, M. and Perera, D.C.R. 1983. Influence of processing conditions on volatile matter content of pale crepe. Rubb. Res. Inst. of Sri Lanka Bull., 17: 12-14.

Nair, N.R., Thomas, K.T., Varghese, L. and Mathew, N.M. 1988. Solar-cum-smoke drier for raw sheet rubber. Indian J. Nat. Rubb. Res., 1(2): 13-21.

Neef, De, J.C. 1950. Identification of rubber coagulated with too much sulphuric acid. J. Rubb. Res. Inst. of Malaya, 12: 263-264.

Newton, R.G., Philpott, M.W., Smith, H.F. and Wren, W.G. 1951. Variability of Malayan Rubber. Ind. Eng. Chem., 43: 329-334.

Ng, C.S., Cheng, S.F. and Ong, C.T. 1982. Evaluation of West Lake centrifugal separater. Planters' Bull., 173: 134-137.

Ng, C.S. and Lau, C.M. 1978. Review and classification of natural rubber latex preservation. Technology series report No. 8. Rubber Research Institute of Malaysia.

Nordin, AB Kadir Bakti and Mohd. Zin AB Karim. 1989. Treatment of rubber effluent with rate algal pond. J. Nat. Rubb. Res., 4(3): 179-185.

Ong, C.O. 1974. High quality rubber from skim latex. Proceedings of the Rubber Research Institute of Malaysia Planters' Conference, Kuala Lumpur. pp. 243-251.

Ong, C.O. and Lim, H.S. 1978. Production of SMR 5CV from high viscosity rubbers by chemical peptisation. J. Rubb. Res. Inst. Malaysia. 26(1): 6-12.

Ong, E.L., Lim, H.S., Ong, C.O., Lok, K.M., Chen, K.Y. and Aziz, A. 1987. Further efforts towards achieving consistency of natural rubber. Proceedings of the Rubber Growers' Conference, Desaru, Johore, Malaysia. pp. 440-455.

Othman, A.B. and Lye, C.B. 1980. Effect of pH of coagulation and sulphuric acid as a coagulant on natural rubber properties. J. Rubb. Res. Inst. Malaysia, 38(3): 109-118.

Pendle, T.D. 1990. Production, properties and stability of NR latices. Rubber Chem. Technol., 63: 234-243.

Peries, O.S. 1970a. Ribbed smoked sheets; factory operations. In: A Handbook of rubber culture and processing. Rubber Research Institute of Ceylon, Agalawatta. Chapter 17, pp. 119-141.

Peries, O.S. 1970b. Pale crepe. In: A Handbook of rubber culture and processing. Rubber Research Institute of Ceylon, Agalawatta. Chapter 18. pp. 142-164.

Piddlesden, J.H. 1940. The concentration of latex by centrifugal machines. J. Rubb. Res. Inst. Malaya, 10: 78-107.

Pike, M. and Ramage, J. 1969. Uniformity and processability of new type natural rubbers. J. Rubb. Res. Inst. Malaya, 22(1): 26-38.

Poh, W.N. 1983. Developments in Malaysian latex concentrate. Planters' Bull., 177: 133-144.

Rao, R. and Tong, N.Y. 1984. Consumer complaints on SMR and remedial procedures. Planters' Bull., 180: 81-85.

Rama Rao, P.S., Graham, D.J. and Muniandy, V. 1986. Solar power-boosted smoke houses. Planters'Bull., 187: 52-56.

Resing, W.L. 1960. Observations on the properties of skim rubber. Proceedings of the natural rubber research conference, Kuala Lumpur. pp 686-696.

Roudeix, H. 1985. Improvement of NR processing and drying conditions by getting a more controlled structure of latex coagulum. Proceedings of the International Rubber Conference, Kuala Lumpur, Malaysia. Vol. II, pp. 423-433.

Rubber Research Institute of Malaya, 1965. Planters' Bull., 78: 81-88.

Sebastian, M.S., Balan, M.V. and Thomas, E.V. 1982. A new coagulant for natural rubber latex. Proceedings of the Fifth Annual Symposium on Plantation Crops - PLACROSYM V. pp. 316-323.

Sekhar, B.C., Sung, C.P., Graham, D.J., Sethu, S. and O'Connel, J. 1965. Heveacrumb. Rubber Developments, 18(3): 78-84.

Sekhar, B.C. 1971. New presentation process - An e ssential feature of modernisation of the natural rubber industry. Quarterly Journal of the Rubber Research Institute of Ceylon, 48(3&4): 212-238.

Sethu, S. 1967. Through-circulation drying of particulate natural rubber. I. Heveacrumb. J. Rubb. Res. Inst. Malaya, 20(2): 65-79.

Shaw, S. 1968. Dynat - The process and machinery. Rubber Journal, 150(2): 42-43.

Shum, K.C. and Wren, W.G. 1977. Observations on bacterial activity in natural rubber latex - Plate counts of latex bacteria on a supplemented medium. J. Rubb. Res. Inst. Malaysia, 25(2): 69-80.

Smith, M.G. 1969. Recent aspects of block natural rubber production by mechanical methods. J. Rubb. Res. Inst. Malaya, 22(1): 78-86.

Subbiah, R.M., Sethu, S., Pong, C.W. and Nambiar, J. 1976. Some factors influencing the design and selection of crepers for SMR production. Proceedings of the Rubber Research Institute of Malaysia Planters'; Conference, Kuala Lumpur. pp. 275-278.

Sum, Ng, C. 1983. Quaternary ammonium surfactants as alternative coagulants of skim latex - A laboratory study. J. Rubb. Res. Inst. Malaysia, 31(1): 49-59.

398

Sung, C.P. 1966. Versatality of the Heveacrumb process - Applications to oil extended and constant viscosity natural rubber. Planters' Bull., 86: 111-125.

Sung, C.P. and O'Connell, J. 1969. Oil extension of natural rubber at latex stage. J. Rubb. Res. Inst. Malaya, 22(1): 91-103.

Tan, A.S., Lim, C.L. and Chan, B.L. 1977. Effect on the dynamic properties of rubber exposed to sunlight. J. Rubb. Res. Inst. Malaysia, 25(3): 127-134.

Tharmalingam, R., Ponniah, W.T., Koelmeyer, C. and De Silva, K.P.N. 1977. Some improvements in crepe rubber drying. J. Rubb. Res. Inst. Sri Lanka, 54: 640-648.

Thomas, E.V. and John Jacob, P. 1967. Improved skim rubber: creaming of skim latex. Rubber Board Bulletin, 9: 33-38.

Thompson, C.W. and Howorth, H. 1964. A new form of technically specified natural rubber. Rubber Developments, 17(3): 62-68.

Thomposon, C.W., Howorth, H. and Smith, M.G. 1966. Production and preparation of pelletised natural rubber to technical specifications. Rubber Journal, 148(1): 24-30.

Tillakeratne, L.M.K., Sarathkumara, P.H., Weeraman, S., Mahanama, M. and Nandadewa, R. 1984. A study of mechanism of action of alkaline metal salts of aromatic thiols on carotenoid compounds present in natural rubber latex. Proceedings of the International Rubber Conference. Rubber Research Institute of Sri Lanka. Part 1, Vol. 2, pp. 89-95.

Tillakeratne, L.M.K., Vimalasiri, P.A.D.T. and De Silva, G.A. 1981. The manufacture of constant viscosity natural rubber from clones producing high viscosity rubbers. J. Rubb. Res. Inst. Sri Lanka, 59: 7-12.

Tong, N.Y. and Kamaruddin, A.A.B.Y. 1984. Patterns of standard Malaysian rubber expansion. Rubber Research Institute of Malaysia. Technology Series Report No. 12. pp. 2-3.

Walpitta, N.C.C., Fernando, T.L.G. and Nandadewa, R. 1984a. The efficiency of hot water heating system in a crepe drying tower. Proceedings of the International Rubber Conference. Rubber Research Institute of Sri Lanka, Agalawatta. Part 1, Vol. 2. pp. 51-64.

Walpitta, N.C.C., Goonatilleka, M.D.R.J. and Weerasinghe, S. 1984b. Use of solar energy for the drying of crepe rubber. Part 1: Model solar collector and drying tower. J. Rubb. Res. Inst. Sri Lanka, 62: 1-17.

Watson, A.A. 1969. Improved ageing of natural rubber by chemical treatment. J. Rubb. Res. Inst. Malaya, 22(1): 104-119.

Wiltshire, J.L. 1932. Sulphuric acid as a latex coagulant. J. Rubb. Res. Inst. Malaya, 4(2): 94-103.

Woo, C.H. 1973. Rubber coagulation by enzymes of Hevea brasiliensis latex. J. Rubb. Res. Inst. Malaysia, 23(5): 323-331.

Yushan, W., Jiahan, H., Lingzhan, Y., Peiming, L., Mongquan, L. Zanxing, L. Guoren, T. and Qingai, M. 1985. A production test for continuous drying and baling of standard rubber. Proceedings of the International Rubber Conference, Kuala Lumpur, Vol. 2. pp. 440-447.

Zachariassen, B., Looi, Y.S., Pillai, N.M., Pong, K.S., Wong, N.P. and Gorton, A.D.T. 1972. Further developments in high DRC centrifuged latex using the 410 centrifuge. Proceedings of the Rubber Research Institute of Malayasia Planters' Conference. pp. 287-298.

Zaid bin Isa and Mohinder Sing, M. 1980. Chemical analysis of rubber effluent. Planters' Bull., 163: 70-80.

CHAPTER 18

PHYSICAL AND TECHNOLOGICAL PROPERTIES OF NATURAL RUBBER

N.M. MATHEW
Rubber Research Institute of India, Kottayam-686009, Kerala, India.

Natural rubber (NR) is a high molecular weight polymer, whose chemical structure is cis-1,4-polyisoprene. The crude rubber as obtained from the tree contains, in addition to the pure rubber hydrocarbon, various other substances like proteins, fats and fatty acids, carbohydrates, mineral matter etc. The hydrocarbon content is reported to be about 94% (Allen and Bloomfield, 1963). The non-rubber substances, although only present in low concentrations, influence the chemical and physical properties of the hydrocarbon polymer. The cis content of the polymer was reported to be almost 100%. But a recent work (Tanaka, 1985) indicated the presence of about three trans units per chain. The properties of NR depend very much upon the state of crosslinking. Therefore, the important properties of the raw unvulcanized rubber and those of the vulcanized rubber are discussed separately.

UNVULCANIZED RUBBER

Molecular weight

As natural rubber is a linear long chain polymer, it is composed of molecules of different size. Therefore, the numerical value of its molecular weight depends on how the heterogeneity is averaged. Thus it can be expressed either as number average molecular weight, $\bar{M}n$ or as weight average molecular weight, $\bar{M}w$. The relation between $\bar{M}n$ and $\bar{M}w$ depends on the molecular weight distribution. The two averages are equal only when the polymer is homogeneous. In other cases, it is found that $\bar{M}w > \bar{M}n$. The weight average molecular weight of natural rubber ranges from 30000 to about 10 million. A random blend would have a weight average molecular weight of about 2×10^6 and a number average molecular weight of about 5×10^5. Determination of $\bar{M}n$ of natural rubber involves measurement of colligative properties, like osmotic pressure (Bristow and Place, 1962). Measurement of $\bar{M}w$ is possible by the light-scattering method which is particularly useful for polymers with molecular weights in the region of 10^6 (Schulz et al. 1956).

Intrinsic viscosity measurements have been widely used for the determination of molecular weight of natural rubber using toluene as solvent (Onyon, 1959).

Molecular weight distribution

In addition to the measurements of $\bar{M}n$ and $\bar{M}w$, a full characterization of a polymer like natural rubber requires determination of its molecular weight distribution (MWD). Using direct visual measurement of sizes of natural rubber molecules, Schulz and Mula (1960) determined its molecular weight distribution. This method is restricted to molecular weights in excess of approximately 5×10^5. Later, using more advanced techniques like gel permeation chromatography (GPC), Subramaniam (1972) demonstrated that the molecular weight distribution of unmasticated natural rubber is distinctly bimodal. The MWD curve shows that the peak at lower molecular weight is less pronounced than that of the higher molecular weight. The districtuion is wide, the ratio $\bar{M}/\bar{M}n$ being in the range of 2.5 to 10. The various commercial grades of NR show differences in molecular weight and its distribution. Storage hardening of NR tends to change the shape of its MWD curve from bimodal to unimodal and to raise $\bar{M}w$ slightly.

It has long been known that properties of raw NR are characteristic of the clones from which the rubber is obtained. Subramaniam (1975) studied clonal variation in molecular weight and its distribution using GPC. Though the range of molecular weight is nearly the same for all the clones, the mean values and the shapes of the MWD curves are different. While the low and average molecular weight clonal rubbers show distinct bimodal distribution, the high molecular weight clonal rubbers usually show a unimodal distribution with a shoulder of shallow plateau in the low molecular weight region. The effect of the yield stimulant 'ethrel' on the molecular weight and its distribution was studied by Subramaniam (1971) and it was shown that the average molecular weight decreases quite sharply a few days after stimulation. As the effect of the stimulant wears off, the molecular weight recovers to reach normal values, MWD studies indicated that the additional latex obtained from the stimulated trees contains a greater proportion of lower molecular weight material.

Macro and micro gel

When NR is immersed in a solvent, the rubber first swells. On prolonged standing some soluble rubber is extracted from the swollen gel. This phenomenon gives rise to a two-phase theory of rubber, comprising 'sol' and 'gel' phases and various factors are responsible for increasing the proportion of either of these phases at the expense of the other. The gel phase consists of the more highly branched and lightly crosslinked components of the rubber closely intertwined with insoluble high-molecular non-rubber substances like proteins. The gel phase thus observed in solid or latex rubber which underwent prolonged storage is known as macrogel. The increase in the macrogel content during storage is responsible for the hardening of rubber stored in bulk (Wood, 1953). Factors such as mechanical shear or oxidative degradation etc. are known to disaggregate the macro gel and to make it soluble.

Fresh NR latex contains crosslinked particles of colloidal dimensions. Bloomfield (1951) coined the name microgel for this crosslinked fraction in Hevea latex considering their similarity with the microgel in SBR latex. The usual concentration of microgel in normal Hevea latex is of the order of 7-30%, but in long rested, and particularly newly opened trees, it may be as high as 60-80%. Both microgel and macrogel in rubber have technological implications. While macrogel is responsible for storage hardening, formation of microgel in latex affects only the original level melt viscosity of the resulting rubber. Unlike macrogel, which is formed in dry rubber on storage, microgel is formed in the latex present within the vessels of the tree. Sekhar (1962) reported that microgel formation in latex is initiated by aldehyde condensing groups, numbering between 100 and 420 per polyisoprene molecule.

Chain branching

The presence of abnormal chemical groups on the rubber chain is believed to cause formation of branched chains. Bristow (1962) showed the existence of branched chains in natural rubber through dilute solution viscometry. Chain branching is responsible for the lower values of molecular weight determined by GPC techniques. The rheological properties of NR are strongly influenced by long chain branching. The slow rate of stress relaxation of Hevea rubber compared to guayule and synthetic polyisoprene rubber has also been attributed to chain branching (Montes and White,1982).

Storage hardening

Natural rubber, either in the form of latex or solid rubber, when stored for long periods, develops higher hardness, as measured by Mooney viscosity or Wallace plasticity. It is also known that the change in viscosity is greater if the initial viscosity is lower. The hardening process is accelerated by low relative humidity and higher temperature of storage.

The increase in hardness occurring when ammoniated latex is stored is believed to be the result of intra-particle crosslinking and microgel formation. However, with solid rubber the crosslinking process is not confined to the original latex particle.

The mechanism of storage hardening is known to involve carbonyl groups in rubber, since hardening is almost fully suppressed by the addition to latex of reagents that could block carbonyl groups (Sekhar, 1958). It is possible to estimate the number of such carbonyl groups per rubber molecule by measuring the concentration of hydroxylamine required to fully inhibit storage hardening, and values of 9-29 were found for a number of clonal rubbers. The amino acids present among the non-rubber constituents are also believed to be playing a role in the hardening reaction (Gregory and Tan, 1975). The change in hardness could be quite large. Nair (1970) compared melt viscosity of rubbers prepared from ordinary and viscosity-stabilised latices of different clones. For the former, Wallace plasticity values were found to be 10-45% higher, depending upon the clonal source.

Low temperature crystallization

In the raw form, natural rubber freezes even at 0°C if the exposure time is sufficiently long (say, one week). This also causes stiffening of raw rubber during storage (Bristow and Sears, 1982). The maximum rate of crystallization occurs at about -24°C when crystallization is virtually complete in about eight hours. Such frozen rubber can be thawed to its original amorphous condition in several hours at 70-100°C. Any significant stiffening due to crystallization can be avoided by ensuring storage above 15°C. It may be noted that laboratory measurements at room temperature can be influenced by very small amounts of pre-existing crystallization.

Melting temperature, Tm

The temperature at which the last traces of crystallinity disappear is usually described as melting temperature, Tm. The experimental procedures used for the determination of Tm present difficulties and the observed values are influenced by factors like chemical modification, presence of diluents, degree of deformation of the amorphous material etc. The observed melting temperature also depends markedly on the temperature at which crystallization occurs. It is probable that in the case of NR, the low melting temperatures observed when the crystallization takes place at a relatively low temperature, reflect the improbability of forming large crystals under these conditions. Although a value of 28°C has been assigned for the Tm of NR, a value of 30°C and higher have been occasionally reported. However, they have been attributed to the presence of some degree of orientation in the amorphous rubber prior to crystallization (Andrews and Gent, 1963).

Transition temperatures

One of the most fundamental measurements on any polymer is the measurement of the temperature(s) at which solid state transitions occur, since specific properties and the manner of usage depend to a large extent on the relation of these transition temperatures to the temperature at which the material is to be used. All polymeric materials will, at some temperature, undergo a glass transition (Tg) change from a plastic to a rubbery state. In addition, many polymeric materials exhibit a first order transition at a temperature (Tm), resulting from the melting of the polymer crystals to form an amorphous rubber. In general, a useful rubber should have a Tg considerably below the temperature of application and have such a structure that crystallization with its associated increase in hardness does not take place on long term standing of the finished product.

Specific volume measurements on NR have established a Tg of -72°C. Boyer (1963) reported many of the complexities associated with solid state transitions. Since such transitions are associated with the allowance of some hitherto restricted chain segmental motion, it has been found to change with degree of crosslinking in NR stocks (Wood et al. 1972). Presence of carbon black has relatively little effect on Tg. If low operating temperatures are desired, plasticisers may be incorporated and this is found to depress the Tg of NR, the degree of depression being dependent on the viscosity of the plasticiser. Of all the elastomeric materials showing first order transition, none has been studied as extensively as NR. This situation arises from a number of factors, particularly the fact that NR was found to undergo crystallization at a readily measurable rate at convenient temperatures and hence served as a model for the development of experimental and theoretical treatments of crystallization of polymers in general. A detailed summary of a variety of experimental work on the crystallization of NR was made by Andrews and Gent (1963).

Rubber solvent interactions

Information on the interactions of a polymer with a given solvent can be obtained from measurements of properties of dilute solutions such as viscosity, osmotic pressure, light scattering etc. However, measurements made over a much broader concentration range provide data on solvent resistance of polymers and on the characteristics of vulcanizates prepared therefrom. The most generally applied technique for measuring such inter-actions involves combining equilibrium swelling measurements on vulcanized rubber with some parameters characterising the vulcanizate network by means of the following equation (Flory and Rehner, 1943).

$$- \ln(1-V_r) - V_r - \chi V_r^2 = \frac{2VoC_1}{RT} V_r^{1/3} \qquad (1)$$

where V_r is the volume fraction of rubber at equilibrium swelling, Vo is the molar volume of the swelling liquid, C_1 is a network parameter from elasticity measurements and χ is the solvent-polymer interaction parameter. A modification of this equation includes the term involving the functionality (f) of the crosslink points (Flory, 1950).

$$- \ln(1-V_r) - V_r - \chi V_r^2 = \frac{2VoC_1}{RT} (V_r^{1/3} - \frac{2V_r}{f}) \qquad (2)$$

The application of gas liquid chromatography to solvent-polymer interactions resulted in the rapid generation of data on many solvents and into high polymer concentration ranges (Summers et al. 1972).

Among all the measurements that have been made combining stress-strain and swelling measurements on NR, the most extensive appear to be those of Bristow (1965) with the following results for NR.

$$\chi = 0.411 \text{ (decane, } 25°C, \text{ equation [1])}$$

$$\chi = 0.40 + 0.20 V_r \text{ (decane, } 25°C, \text{ equation [2])}$$

Much less extensive data give the following results.

$$\chi = 0.42 \text{ (benzene, } 25°C, \text{ equation [1])}$$

$$\chi = 0.41 + 0.20 V_r \text{ (benzene, } 25°C, \text{ equation [2])}$$

$$\chi = 0.425 + 0.20 V_r \text{ (heptane, } 25°C, \text{ equation [1])}$$

$$\chi = 0.415 + 0.35 V_r \text{ (heptane, } 25°C, \text{ equation [2])}$$

Cohesive energy density

The polymer-solvent interactions, as described in the previous section cover specific polymer-solvent systems. However, it is highly desirable to be able to consider a more general case where it is possible to predict the solvent-polymer behaviour of a particular combination from a knowledge of two parameters each representing a given polymer and a given solvent. Such parameters can be derived from measurements of heat of mixing. The parameter usually employed is the cohesive energy density (CED) or more commonly its square root δ, the solubility parameter. The values reported for natural rubber are 8.08 and 8.10 (Trick, 1977).

Flow properties

Studies on flow properties of rubbers started with the advent of the parallel-plate plastimeters like the Wallace rapid plastimeter, and the rotating disc viscometers like the Mooney viscometer. The former provides a measure of the flow produced during a simple compression at 100°C, while the latter a measure of the shear viscosity at a particular strain rate also at 100°C. Both instruments provide a measure of the flow behaviour at low strain rates and there is a broad correlation between the results of the two tests (Anon, 1981; Subramaniam, 1975). Freshly prepared natural rubber is variable in plasticity and viscosity. Normal modifications of the method of preparation have only minor effects. It is now clear that variations in plasticity and viscosity of raw NR are due to the differences in the molecular size and structural arrangements of the rubber hydrocarbon, and these characteristics are specific to the clone. Rubber with a low Mooney viscosity normally has a low intrinsic viscosity in solution and vice versa, but this correlation is not as good as might be expected, due to the presence of microgel, as discussed earlier in this chapter. The presence of microgel causes increase in Mooney viscosity without contributing much to solution viscosity. Subramaniam (1975) studied clonal variation in Mooney viscosity and reported that the value normally ranges from 63 to 94, seasonal variation being apparent in a few cases. The Wallace rapid plasticity values for the same set of clones ranges from 42 to 68.

Ong and Subramaniam (1975) derived a linear correlation between the initial maximum torque $V_{i\ max}$ and the Mooney viscosity V_r of raw NR. Mastication, storage hardening and method of coagulation did not alter the linear relationship. A hypothesis involving the mechanism of disentanglement orientation of polymer chains with chain slippage was proposed to explain the variation of torque with time of shearing. It was also suggested that the initial maximum increase in torque in Mooney viscosity determination and the subsequent thixotropic phenomenon are of the same origin.

The inherent disadvantages of flow measurements at low shear rates as in Wallace rapid plastimeter and Mooney viscometer, led to the development of capillary rheometers which are being extensively used to extend measurements to high shear rates (10^3 s^{-1}). Investigations on the capillary flow behaviour of raw and filled NR stocks have been reported (Ong and Lim, 1983; Bristow, 1985; Gupta, 1989). The relation between viscosity and shear stress depended upon the natural rubber grade and samples of different grades with similar Mooney viscosities exhibited significantly different flow behaviour at high shear rates (Bristow and Sears, 1988; 1989). This again suggests that useful additional information could be gained by high shear rate tests.

Viscoelastic behaviour

Natural rubber shows viscoelastic behaviour indicating that its physical properties are partly liquid-like (viscous) and partly solid-like (elastic). This in fact, is the case for all rubbers. Rubbers behave in many ways like highly viscous liquids before they are crosslinked. The vulcanization process which introduces crosslinks reduces the flow properties and makes a rubber more elastic. Nevertheless, there is still evidence of flow behaviour even in a crosslinked rubber. This is demonstrated in the creep and stress-relaxation behaviour of rubbers.

Tack and green strength

The term processability refers to the way in which a rubber behaves through a series of processing operations or to the reproducibility of that behaviour. Easy processability of NR has been regarded as one of its main attributes. The most important aspects of processability of NR are its high inherent tack and good green strength. These two characteristics are of utmost importance in the manufacture of products like tyres. Tack is important so that the components of a green tyre will hold together until moulding. Green strength is needed so that the uncured tyres will not creep and hence distort excessively before moulding, or tear during the expansion that occurs upon moulding. A practical definition of tack is the ability for two similar materials to resist separation after they are brought into contact for a short time under a light pressure. The tack of NR and NR/SBR blends have been compared with that of SBR by Hamed (1981). The higher tack of NR compared to the SBR stock has been attributed to its greater ability to flow under compressive load and its higher green strength. NR is an ideal material for developing high tack. It can be processed to a low viscosity and still maintain high green strength. Furthermore, the mechanism responsible for high green strength (strain crystallization) is not active in the bond formation step, and hence does not interfere with contact and inter-diffusion, but rather develops upon stressing. The high level of molecular inter-diffusion is also responsible for the high tack of NR (Skewis, 1966). Tack and stickiness, although inter-related, are not the same. Juve (1944) was one of the first to try to differentiate between tack and stickiness. According to him, if two pieces of a rubber compound are pressed firmly together and form a joint so strong that attempts to separate them cause a failure at another point, that is excellent tack. If only partial tearing at the former interfaces occurs. tack is fair. If separation occurs at the interface, it is stickiness, and the degree of stickiness is dependent on the force required to separate the interface. Tack is also denoted as autoadhesion or autohesion, green strength as cohesion and stickiness as adhesion. All measurements of tack involve green strength or cohesive strength of the rubber compound, and all rubber compounds with high tack values have good green strength and this applies well to NR.

VULCANIZED RUBBER

Strength properties

As in the case of other engineering materials, strength properties are of great importance in most of the practical applications of rubber. A number of strength properties can be defined and measured. The most important among these are tensile strength, tear strength and resistance to fatigue.

Tensile strength

Tensile strength, in which the material is subjected to a uniform uniaxial tensile stress, is the simplest strength property as far as measurement is concerned. Perhaps the most striking characteristic of natural rubber, compared with most synthetic elastomers is its very high tensile strength even without the help of reinforcing agents. This is undoubtedly due to its ability to crystallize considerably on extension at normal temperatures. Strain induced crystallization in rubbers has been directly investigated by X-ray, density and other methods. Tensile strength of NR vulcanizates frequently exceeds 30 MPa which is almost ten times the values reported for gum vulcanizates of noncrystallizing rubbers such as SBR, under similar test conditions (Thomas, 1960; Greensmith et al. 1963). The tensile strength of NR gum vulcanizates depends on various factors (Hofmann, 1967) including the type and extent of crosslinking; a peroxide-cured vulcanizate showing a maximum tensile strength of 15 Mpa, a TMTD-cured rubber, containing mostly monosulphidic crosslinks, having tensile strength values upto 25 MPa and accelerated sulphur vulcanizates giving values above 30 MPa. The effect of reinforcing fillers on the tensile strength of NR is not as significant as in the case of noncrystallizing rubbers. Temperature is found to influence significantly the tensile strength of gum NR vulcanizates and there is a critical temperature around 100°C, above which the strength falls abruptly, crystallization being suppressed at that temperature. However, tensile strength of reinforced NR vulcanizates is found to be less temperature dependent.

Tear resistance

The tear resistance of elastomers reflects their tensile strength characteristics. As the tip of the tear sustains high strains, crystallization occurs and high tear resistance is observed in NR. The energy parameter for tear fracture is termed tearing energy, T, and is defined mathematically as:

$$T = \frac{-(\delta U)}{(\delta A)\,l} \tag{3}$$

where U is the total strain energy stored in the specimen containing a crack, A the area of one fracture surface, and the partial derivative indicates that the specimen is considered to be held at constant length, l, so that the external forces do no work. From measurement of tearing forces the values of T at which tearing occurs can be calculated. For noncrystallizing rubbers tearing energy shows a strong dependence on rate of tearing. Tearing in such rubbers often proceeds in a 'steady' manner in the sense that the force, when a 'trousers' type test piece is tested at a constant rate of separation of legs, remains relatively constant. However, in the case of crystallizing rubbers like NR, tearing generally proceeds in a 'stick-slip' manner with the force increasing during the 'stick' periods until a catastrophic failure point is reached at which the tear jumps forward. Over wide ranges, the catastrophic tearing energy is insensitive to rate and temperature for a crystallizing rubber like NR (Greensmith et al. 1963). It appears that in such materials the effect of crystallization which can induce substantial hysteresis at high strains, overshadows viscoelastic effects. Another factor promoting high tear resistance is roughening or branching of the tear tip. In extreme cases roughening can lead to 'knotty' tearing in which the tear tip circles around on itself under increasing force until finally a new tear breaks ahead. Although knotty tearing is not exclusive to filled rubbers, the tendency for it to occur can be greatly increased by re-inforcing fillers (Gent, 1978). It seems probable that a strength anisotropy arising from orientation effects is at least partly responsible for tear deviation (Gent and Kim, 1978). Another factor that may influence tear deviation is cavitation ahead of the crack tip due to the hydrostatic component of the tensile stresses.

Crack growth and fatigue

Fatigue failure of rubber under cyclic deformation has been shown to be a crack growth process initiating from small pre-existing flaws, usually of size 2×10^{-3} cm (Gent et al. 1964; Lake and Lindley, 1964a; Mathew and De, 1983a). Thus crack growth behaviour and fatigue are intimately related. The strain dependence of fatigue life of different elastomers vary widely. Natural rubber is very good at high strains, compared with noncrystallizing elastomers. The difference is more pronounced under nonrelaxing conditions, that is, when the deformation cycle is repeated, stress does not return to zero (Fielding, 1943; Lake and Lindley, 1964b). This effect could be utilized to advantage in certain engineering applications such as springs. In these applications, suitable designs ensuring incomplete relaxation could exploit this desirable feature. Even for non-crystallizing elastomers, fatigue life under nonrelaxing condition is found to be longer which is essentially attributable to the reduction in the strain energy of the cycle (Lindley, 1974). For a crystallizing rubber, the much larger enhancement is attributed to two additional factors: an effective increase in the threshold energy (To), required for the initiation of mechanical crack growth, and a reduction in the rate of growth once the new To is exceeded. The minimum tearing energy needs to be a small but definite fraction of the maximum for these effects to become apparent. It has been shown that To can be approximately calculated from the molecular structure of the vulcanizate and from the strength of the chemical bonds (Lake and Thomas, 1967). It has also been shown that To for NR increases substantially if atomospheric oxygen is excluded and/or if certain anti-oxidants are incorporated in the vulcanizate. This behaviour is reflected in enhanced life, particularly in the region of the 'fatigue limit', that is a strain below which the fatigue life of elastomers is very long.

Frequency of deformation is found to have very little influence on the fatigue life of NR vulcanizates. The effect of temperature on crack growth and fatigue is also found to be much less for NR than for noncrystallizing rubbers. This difference is believed to be associated with the origin of mechanical hysteresis. In NR, hysteresis due to strain-induced crystallization, which does not vary greatly with temperature, far outweighs the viscoelastic contribution. Carbon black and other fillers provide an additional source of hysteresis and their inclusion can greatly modify the temperature dependence of crack growth and fatigue of noncrystallizing rubbers. For both crystallizing and noncrystallizing rubbers, resistance to crack growth could be increased by fine particle size fillers, the effect being attributable to blunting of the crack tip due to branching.

ELASTIC PROPERTIES

Modulus and hardness

According to the statistical theory of elasticity (Trelor, 1975), in simple shear the stress, τ, is proportional to the strain, γ, even for large deformations. Thus

$$\tau = G\gamma \tag{4}$$

where G is the shear modulus. In simple tension or compression the nominal stress, σ, is related as

$$\sigma = G(\lambda - 1/\lambda^2) \tag{5}$$

where λ is the extension ratio. At a low elongation, e, this becomes

$$\sigma = 3\,Ge = Eoe$$

where Eo is the Young's modulus. The two moduli at low strain obey the relationship Eo = 3G for an incompressible solid. The statistical theory explains reasonably well the stress-strain behaviour of unfilled rubber upto strains of a few hundred per cent when the stress rises much more steeply than the theory predicts. This is either due to the molecular chains between crosslinks approaching their limiting extension or, in the case of crystallizing elastomers like natural rubber, to the onset of strain-induced crystallization. However, the statistical theory of elasticity is not obeyed by filled rubbers. The values of G derived from both shear and tension/

compression tests decrease with increasing strain, though the values obtained are similar when e equals γ up to strains of about 50%. At higher strains the deformation within the rubber matrix becomes sufficient for strain-crystallization or limiting chain extensibility, to steepen the stress-strain curve.

Generally hardness measurements are used to characterize vulcanized rubbers approximately. In the case of rubbers, hardness is essentially a measurement of the reversible, elastic deformation produced by a specially shaped indentor under a specified load and is therefore related to the low strain modulus of the rubber. Readings are usually in International Rubber Hardness Degree (IRHD). Hardness is relatively simple and easy to measure, but is subject to some uncertainty in measurement and hence ±2 degrees tolerance is given. Shear modulus values are much more accurate, but are more difficult to measure. However, they are preferred as a basis for design calculations, particularly for filled rubbers as the modulus is dependent on strain.

The bulk modulus of rubber, β, is many times larger than its Young's modulus, Eo. For most purposes the Poisson's ratio can be taken as ½. The much larger bulk modulus indicates that rubber hardly changes in volume even under high loads, so that for most types of deformation, there must be space into which rubber can deform. The more restriction that is made on its freedom to deform, the stiffer it will become, a characteristic used in the design of compression springs.

Resilience, hysteresis and heat build-up

Resilience is a basic form of dynamic test on rubber in which the strain is applied by impacting the test piece with an indentor which is free to rebound after the impact. Rebound resilience is defined as the ratio of the energy given up on recovery from deformation to the energy required to produce the deformation and is usually expressed in percentage. Resilience is not an arbitrary parameter, but is approximately related to the loss tangent:

$$R = \frac{EA}{ER} = \pi \tan \delta \tag{6}$$

where ER = reflected energy,

EA = absorbed energy = E_T - ER

where ET = incident energy

The relationship is not particularly accurate because $\tan \delta$ is strain dependent and in an impact test, the form of applied strain is complex and its magnitude not controlled.

Hysteresis is the energy lost per cycle of deformation. It is the result of internal friction and is manifested by the conversion of mechanical energy into heat. Heat build-up is the temperature rise in a rubber body resulting from hysteresis. As the heat generated is not easily conducted away in a material of low thermal conductivity such as rubber, the rise in temperature may assume so much magnitude in products like heavy duty truck tyres as to cause failure through tread lift, blow-out and other delamination or crack growth processes. It is because of such risks that heat build-up influences the design, compounding and use of large tyres. Natural rubber has been the preferred polymer in such applications, considering its outstanding resilience, low hysteresis and heat build-up characteristics. Unlike in SBR, hysteresis in NR is contributed mostly by strain induced crystallization, which does not vary much with temperature and therefore the effect of temperature on crack growth and fatigue tends to be much less in the case of NR.

Creep, stress relaxation and set

When a vulcanized rubber is held under constant strain, the stress is found to decrease gradually with time as the crosslinked network approaches an equilibrium condition. This phenomenon is called stress relaxation. The same process leads to creep, which is defined as the additional strain occurring, after a lapse of time, beyond the immediate elastic deformation. Although all but a few per cent of the original deformation is recovered immediately on removal of the load, further recovery takes much longer and may never be complete. The extent of deformation not recovered is known as permanent set. If the time scale and the temperature are such that chemical effects are negligible, creep and stress relaxation are approximately proportional to the logarithm of the time after loading.

One of the most notable features of natural rubber, compared with most other elastomers, is its good elastic behaviour. This results in low creep and a lower stress relaxation rate. If the stress relaxation rate is expressed as per cent stress relaxation per decade of time, a typical NR gum vulcanizate may give a value of about two per cent per decade. If carbon black is present, the rate will he higher, about seven per cent per decade for a 70 IRHD rubber containing 50 phr of a nonreinforcing black. However, prestressing of such a filled rubber can reduce the stress relaxation rate to a little more than the gum value. This superior elastic behaviour of

NR is a consequence of the high mobility of the molecules, which is also reflected in the relatively low glass transition temperature.

Stress relaxation rates are substantially independent of the type or amount of deformation, but creep rates depend on both the rate of stress relaxation and the load-deflection characteristics. In tension the creep rate may reach double the rate of stress relaxation, in shear it is about the same and in compression, it is lower. For unfilled rubbers having the same type of vulcanizing system the relaxation rate decreases with increasing hardness. Over the usual range of hardness possible with gum vulcanizates this will not alter the rate by more than about a third. In filled rubbers the relaxation rates increase with the amount of filler. The amount of creep is the largest during the first few weeks under load but should not exceed 20% (for 70 IRHD) of the initial deflection in this period. Thereafter, only a further 5-10% increase in deflection should occur over a period of many years.

Measurement of set under compression provides a practical evaluation of either the creep or the stress relaxation of rubber and has been very useful for those purposes where a high degree of precision is not expected. To get quick results the test conditions are made much more severe than the anticipated conditions of service by either increasing the temperature or the deformation, or both.

Creep and stress relaxation are particularly important in load-bearing applications such as springs. Partly because of its good creep characteristics, NR is the most widely used rubber in this field.

Resistance to abrasion

The terms wear and abrasion are very often used synonymously. Wear is a general term covering the loss of material by virtually any means. As wear usually occurs by the rubbing together of two surfaces, abrasion is usually used to mean wear. Abrasion resistance is the reciprocal of abrasion loss. The mechanisms by which abrasion occurs when a rubber is in moving contact with any surface, are somewhat complex, involving principally cutting of the rubber and its fatigue. These mechanisms have been extensively reviewed (James, 1967; Zhang, 1984; Gent and Pulford, 1983; Grosch and Schallamach, 1969; Mathew and De, 1983b). It is possible to categorise wear mechanisms of rubber in various ways and one convenient system is to differentiate between four main factors:

(a)   Abrasive wear, which is caused by hard asperities cutting the rubber.
(b)   Fatigue wear, which is caused by particles of rubber being detached as a result of dynamic stress on a localized scale.

(c) Wear through roll formation, which occurs with a relatively high coefficient of friction between the rubber and the abrading surface.

(d) Smearing resulting from degradation of rubber from either thermal or mechanical stress.

One of the most important applications of rubber where resistance to abrasion is of great importance is tyre tread. It is generally accepted that some of the synthetic rubbers such as SBR and BR are superior to NR in abrasion resistance. However, it may be pointed out that the relative wear rating of compounds depends on the nature of the track as well as load. On very smooth surfaces SBR is superior to NR, but with increasing sharpness the difference between the two is reduced and reversals in ranking can in fact be observed. These effects could be seen in tyre tests on actual road surfaces and on vehicles of varying load. In going from passenger car tyres to truck tyres and to aircraft tyres, loading increases and the proportion of NR increases. Car tyre treads are in general based on synthetic rubber. However, truck tyre treads contain 50-100% NR and aircraft tyres, in general, are made entirely of NR.

Tyre surface temperature is another important factor influencing the relative wear rating of NR and SBR (Grosch, 1967). At low tyre surface temperatures, as encountered during winter, NR is superior; at high surface temperatures the reverse is true, the reversal occurring at about 35°C.

Resistance to degradation

Natural rubber being an unsaturated polymer, is highly susceptible to degradation by oxygen, ozone, radiation, heat, chemicals etc. Hevea rubber contains natural antioxidants, proteins and complex phenols, which protect it from deterioration during coagulation of the latex and the subsequent processing and drying of the coagulum. These natural anti-degradants are lost or destroyed during further processing and hence additional protectants are needed to ensure adequate service life of end products.

The changes occurring during the degradation of rubbers could be described in three ways.

(a) Chain scission, resulting in a reduction in chain length and average molecular weight.

(b) Crosslinking resulting in a three-dimensional structure and higher molecular weight.

(c) Chemical alteration of the molecules by introduction of new chemical groups.

Natural and butyl rubbers degrade mostly by chain scission, resulting in a weak softened stock, often showing surface tackiness. Chemical analysis shows the presence of aldehyde, ketone, alcohol and other groups, resulting from oxidative attack at alpha hydrogens and double bonds. Synthetic rubbers such as SBR, polychloroprene and nitrile rubber degrade by cross-linking, resulting in brittle stocks.

Oxidative ageing

Oxygen is considered to be the most important degradant for NR. A small amount of one to two per cent of combined oxygen in rubber serves to render it useless for most applications. The oxidation of rubber is believed to take place through a free-radical chain reaction whose mechanism was first proposed by Bolland and Gee (1946). In order to prevent extensive deterioration of the rubber, it is necessary to interrupt the chain reaction and stop autocatalysis. This could be accomplished by either terminating the free radicals or by decomposing the peroxides into harmless products. Antioxidants, in fact, function this way. It is established that amine anti-oxidants act both by reacting with free radicals and by decomposing peroxides. Phenolic antioxidants, on the other hand, react primarily as free radical sinks or chain stoppers. Phosphites react readily with free peroxides such as ROOH to give ROH and a phosphate.

The attack by oxygen on raw rubber is different from that on vulcanized rubber. In the former case, an initial induction period is followed by rapid uptake of oxygen. With vulcanized rubber, there is no induction period and the oxygen uptake is essentially linear with time. The net result of oxygen attack on NR is an overall decrease in all properties. Tensile strength, elongation, flex life and abrasion resistance decrease progressively as oxidative ageing increases. Initially modulus and hardness increase slightly but then fall off. Antioxidants have great effect on the oxidation of rubber and as little as 0.001 per cent of a good amine antioxidant can protect rubber against oxidation for long periods.

Materials like heavy metal ions and peroxides catalyse oxidative ageing of rubbers and these are called pro-oxidants. Heavy metal ions such as copper, manganese and iron are pro-oxidants of NR. Some of the standard antioxidants, notably the aromatic diamines, are effective against metal catalysed oxidation of rubber, by forming stable coordination complexes with the ions.

The effect of heat and oxygen on rubber, in general practice, are never separated and the practical result of heat on rubber is a combination of crosslinking and an increase in the rate of oxidation.

Ozone

Ozone reacts readily with NR and the effect manifests itself in two ways.

(a)   Cracks appear on the surface of rubber perpendicular to the direction
of stress in rubber.

(b)   A silvery film appears on the surface in unstressed rubber. This
is usually called frosting.

The mechanism of ozone attack is thought to be the reaction of ozone
with the double bonds in rubber to form ozonides. These are easily
decomposed to break the double bond, and under strain, a crack appears.
As the reaction proceeds the cracks become deeper.

Two factors which greatly influence ozone cracking are ozone concentration and the strain in the rubber. Initiation of crack is favoured
by high strain and high ozone concentration.

Protection of NR stocks from ozone cracking can be accomplished
by using antiozonants. Under static conditions physical antiozonants like
wax which forms a bloom could be used. The bloom can act as a surface
barrier against ozone. Under dynamic conditions, waxes are unsuitable and
hence chemical antiozonants are employed. Blending of NR with more saturated
rubber such as EPDM or EPM has also been found to be effective in
protecting NR from ozone attack (Mathew, 1983; Mathew et al. 1988).

Light and weathering

Light promotes the action of oxygen at the surface of rubber,
producing a film of oxidized rubber, having physical properties different
from those of the original stock. The film then undergoes action by water
vapour and heat to produce crazing. The oxidized layer expands and
contracts on heating and drying. Finally, the oxidized layer washes away,
leaving the filler exposed. The greatest amount of light damage is done
by UV light. A familiar form of light ageing is the stiffening of the surface
of rubber products. Products may be protected against light and weathering
by the use of opaque pigments such as zinc oxide, titanium dioxide, carbon
black etc. and also by certain chemicals.

Atomic radiation

The effect of radiation damage is very similar to that of heat ageing
of vulcanizates. Loss of tensile strength and increase in modulus have been
observed. Radiation degradation of NR has been found to be accelerated
if the rubber is under strain (Alex et al. 1989). Antidegradants like

p-phenylene diamine derivatives and fillers like carbon black have been found to improve radiation resistance of NR.

Permeability

Permeability of a rubber film is a measure of the ease with which a liquid or gas passes through it. The process of permeation involves absorption or solution on one side of the rubber film followed by diffusion through the film to the opposite side where evaporation takes place. In an ideal case, the quantity of gas or vapour being transmitted builds up to a constant steady level after a period of time and then

$$q = \frac{PtpA}{b} \qquad (7)$$

where q = volume of gas transmitted
P = permeability coefficient
t = time
p = partial pressure difference across the test piece
A = test piece area, and
b = test piece thickness

In many cases P is a constant for a given gas and polymer combination. The permeation of a gas through a polymer takes place in two steps, the gas dissolving in the polymer and then the dissolved gas diffusing through the polymer. The solubility constant is the amount of a substance which will dissolve in unit quantity of the polymers under specified conditions, and the diffusion constant is the amount of substance passing through unit area of a given plane in the polymer in unit time for a unit concentration gradient of the substance across the plane. It can be shown that:

$$P = SD \qquad (8)$$

where S = the solubility constant, and
D = the diffusion constant

Gases differ considerably in permeability rate since this is affected by the size of the gas molecule and its solubility in rubber. By far the lowest gas permeability among the common rubbers is shown by butyl rubber. Air permeability of NR is almost twenty times that of butyl rubber. Nitrile and chloroprene rubbers are in between. Epoxidised NR, a chemically modified form of natural rubber, has much lower air permeability than NR itself (Gelling and Porter, 1989).

Electrical properties

In general rubbers are electrically insulating and this property is widely taken advantage of in cables and in various components in electrical appliances. They can also be made anti-static and even conducting by suitable compounding. In all cases, it is the combination of the electrical properties and the inherent flexibility of rubbers which make them attractive for electrical applications. The electrical properties most commonly considered are:

Resistance or resistivity,

Power factor, and

Dielectric strength.

Maintenance of electrical properties on exposure to water is especially important when the product is to be used in wet environments.

Resistance and resistivity: As the surface of rubbers may conduct electricity more easily than the bulk, it is usual to distinguish between volume resistivity and surface resistivity. Volume resistivity is defined as the electrical resistance between opposite faces of a unit cube, whereas surface resistivity is defined as the resistance between opposite sides of a square on the surface. Insulation resistance is the resistance measured between any two particular electrodes on or in the rubber and hence is a function of both surface and volume resistivities and of the test piece geometry. Conductance and conductivity are simply the reciprocals of resistance and resistivity respectively.

It is not easy to make a clear distinction among insulating, anti-static and conducting rubbers. The definitions should be made with respect to the resistance between two relevant points on a product rather than to the resistivity of the rubber. Generally, resistances of up to 10^4 ohms are considered conductive, between 10^4 and 10^8 ohms anti-static and above 10^8 ohms insulating.

Dielectric strength: The dielectric strength of an elastomer is the voltage required to puncture a sample of known thickness and is expressed as volts per mil of thickness. The rate of voltage application, the geometry of the electrodes and of the test specimen have profound influence on the results obtained.

Dielectric constant and power factor: The dielectric constant or specific conductive capacity is a measure of an insulation's ability to store electrical energy. The dielectric constant is the ratio of the electrical capacity of a condenser using the elastomer under test as the dielectric, to the capacity of a similar condenser using air as the dielectric.

The power factor of an insulating material indicates its tendency to generate heat in service. If a capacitor using an elastomer as the dielectric is charged by a direct current and then immediately discharged, there is an energy loss in the form of heat. If this capacitor is repeatedly charged and discharged by an alternating current, the electrical loss results in heating of the dielectric. The ratio of this loss to the energy required to charge the capacitor is known as power factor.

The best electrical properties are obtained with hydrocarbon rubbers such as NR, butyl, EPDM etc. Natural rubber vulcanizates, in particular can be made to give very high electrical resistance. A comparative assessment of electrical properties of NR and chloroprene rubber has been reported (Anon. 1963) and the data are given in Table 1.

TABLE 1

	Chloroprene rubber	Natural rubber
Insulation resistance of 1.15 mm cover on No. 12 AWG wire, megohms per 300 meters.	4	4000
D.C. resistivity, ohm-cm	10^{12}	10^{15}
Dielectric strength, V per mil	400-600	400-600
Dielectric constant	6.7	2.3
Power factor, %	2.5	0.5

Thermal properties

Properties like specific heat, thermal conductivity, thermal expansion and Joule effect in rubbers are of great practical importance to the designer of rubber products, but have not been properly recognised.

Specific heat: Specific heat is the quantity of heat required to raise unit mass of the material through 1°C. It is usually determined by supplying heat to a calorimeter containing the test piece and measuring the resulting temperature rise. Except when the highest precision is required, when

TABLE 2

Physical constants of natural rubber

Property	Unvulcanized	Pure gum vulcanizate	Vulcanizate with 50 phr carbon black	Hard rubber (Ebonite)
Density, Mg m^{-3}	0.913 (0.906-0.916)	0.970 (0.920-1.000)	1.120 (1.120-1.180)	1.170 (1.130-1.180)
Thermal coefficient of volume expansion, $\beta = (1/V)(\delta V/\delta T)$, K^{-1}	670x10^{-6}	660x10^{-6}	530x10^{-6} (450-550x10^{-6})	190x10^{-6}
Glass transition temperature, K	210 (199-204)	210 (201-212)	208	353
Specific heat, Cp, Cal g^{-1} (°C)$^{-1}$	0.449	0.437	0.357	0.331
Heat capacity, Cp, kJ kg^{-1} K^{-1} $\delta C_p/\delta T$, kJ kg^{-1} K^{-2}	1.905 3.54x10^{-3}	1.828	1.404	1.385
Thermal conductivity, W m^{-1} K^{-1}	0.134	0.153 (0.14-0.15)	0.280	0.163 (0.160-0.180)
Heat of combustion, MJ kg^{-1}	-45.2	-44.4		-33.0
Equilibrium melting temperature, K	301 (303-312)			
Heat of fusion of crystal, kJ kg^{-1}	64.0			
Optical: Refractive index, nD	1.5191	1.5264		1.6
dn$_D$/dT, K^{-1}	-37x10^{-5}	-37x10^{-5}		
Electrical: Dielectric constant (1 kHz)	2.37-2.45	2.68 (2.5-3.0)		2.82 (2.8-2.9)
Dissipation factor (1 kHz)	0.001-0.003	0.002-0.04		0.0043-0.009
Conductivity (60s), fS m^{-1}	2-57	2-100		2-3000

Property	Unvulcanized	Pre-gum vulcanizate	Vulcanizate with 50 phr carbon black	Hard rubber (Ebonite)
Mechanical: Compressibility, $B = -(1/V_0)(\delta V/\delta P)$,				
MPa⁻¹ ... MPa^{-1}	515×10^{-6}	514×10^{-6}	410×10^{-6}	240×10^{-6}
$\delta B/\delta P$, MPa^{-2}	-2.1×10^{-6}	-2.4×10^{-6}	-1.8×10^{-6}	-0.41×10^{-6}
$\delta B/\delta T$, $MPa^{-1}\ K^{-1}$	$+2.3 \times 10^{-6}$	$+2.1 \times 10^{-6}$		$+1.1 \times 10^{-6}$
Thermal pressure coefficient,				
$\gamma = B/\beta$, $MPa\ K^{-1}$		1.22		
$\delta\gamma/\delta T$, $MPa\ K^{-2}$		-0.0052		
Bulk modulus (isothermal), GPa	1.94	1.95	2.44	4.17
Bulk modulus (adiabatic), GPa	2.27	2.26		
Bulk wave velocity, Vb (longitudinal wave) km s⁻¹	1.58	1.58 (1.5-1.58)	1.49	
$\delta vb/\delta T$, $m\ s^{-1}\ K^{-1}$	-3	-3		
Strip (longitudinal wave) velocity, V_1 (1 kHz), $m\ s^{-1}$		45 (35-51)	141	1540
Ultimate elongation, %		750-850	550-650	6 (3-8)
Tensile strength, MPa		17-25	25-35	60-80
Initiation slope of stress-strain curve,				
Young's modulus, E(60s), MPa		1.3 (1.0-2.0)	3.0-8.0	3000
Shear modulus, G(60s), MPa		0.43 (0.3-0.7)		
Shear compliance, J(60s), MPa^{-1}		2.3 (1.5-3.5)	0.5-0.7	0.0017
Creep rate (1/J) ($\delta J/\delta \log t$), % / unit log t		2 (1-3)	8 (7-12)	
Dynamic properties				
Storage modulus, G', MPa	0.41(0.34-0.56)	0.41(0.31-0.60)	6.2(1.9-13)	1100
Loss modulus, G", MPa	0.029 (0.027-0.045)	0.0063 (0.0052-0.030)	0.68 (0.32-1.3)	45
Loss tangent, G"/G'	0.09(0.07-0.13)	0.016(0.01-0.05)	0.11(0.10-0.17)	0.040
Resilience (rebound) %	75-77	75-84	50 (45-55)	(63-67)

Values in parenthesis indicate the range.

an adiabatic calorimeter would be used, it is now usual to measure specific heat by a comparative method using differential scanning calorimetry (Richardson, 1976). It is also possible to determine the dry rubber content of raw rubber and rubber coagulum through specific heat measurements (Harris et al. 1985).

Thermal conductivity: Thermal conductivity is different for different rubber compounds, varying with the amount and conductivity of each constituent in the composition. Thermal conductivities of typical rubber compositions are (Btu/hr/Sq ft/in/°F): Chloroprene rubber 1.45, SBR 1.70, NBR 1.70 and NR 1.15. Conductivity is important to rubber technologists, because it affects the time required to heat the interior of a rubber product to the vulcanization temperature. It is also important to the designer of products in which heat is generated by vibration, flexing or friction. Care has to be taken to provide for heat dissipation in such products.

Coefficient of expansion: The coefficient of thermal expansion of rubber compositions vary with the kind and amount of filler used. Addition of fillers lowers the coefficient. The coefficient of volume expansion of rubbers is in the range of $4x10^{-4}$ to $7x10^{-4}$ per °C while that for steel is $0.3x10^{-4}$ per °C (Juve and Beatty, 1955). This results in shrinkage of moulded products, leading to difficulties in moulding rubber to close dimensional tolerance. In general, linear shrinkage figures fall within the range of 1.5-3.0%, depending on polymer type and filler loading.

Joule effect: When a rubber is heated under strain, it tries to contract. Its modulus of elasticity increases with rise in temperature. If the rubber is under constant load it will contract and if under constant strain it will exert a higher stress. This phenomenon is known as Joule effect and occurs only when the rubber is strained first and then heated.

Physical constants at a glance

The values of various physical properties of NR have been compiled by Wood (1966, 1976) and the same are reproduced in Table 2.

REFERENCES

Allen, P.W. and Bloomfield, G.F. 1963. Natural rubber hydrocarbon. In: L. Bateman (Ed.), The Chemistry and Physics of Rubber like substances. Maclaren, London, Chapter 1.
Alex, R., Mathew, N.M. and De, S.K. 1989. Effects of gamma radiation on natural rubber vulcanizates under tension. Radiat. Phys. Chem., 33(2): 91-95.
Andrews, E.H. and Gent, A.N. 1963. Crystallization in natural rubber. In: L. Bateman (Ed.), The Chemistry and Physics of Rubber like substances. Maclaren, London, Chapter 9.

Anonymous, 1963. The language of rubber. E.I. du Pont de Nemours & Co. (Inc), Wilmington, Delaware, 32pp.

Anonymous, 1981. Technical note - raw natural rubber; correlation between Mooney viscosity and rapid plasticity number. NR Technol., 12: 40pp.

Bloomfield, G.F. 1951. Studies in Hevea rubber. Part II. The gel content of rubber in freshly tapped latex. J. Rubb. Res. Inst. Malaya, 13: 18-24.

Bolland, J.L. and Gee, G. 1946. Kinetic studies in the chemistry of rubber and related materials. Part 3. Thermochemistry and mechanisms of olefin oxidation. Trans. Faraday Soc., 42: 244 pp.

Boyer, R.F. 1963. The relation of transition temperatures to chemical structure in high polymers. Rubber Chem. Technol., 36: 1303-1421.

Bristow, G.M. 1962. The Huggin's parameter for polyisoprenes. J. Polym. Sci., 62: 168 pp.

Bristow, G.M. and Place, M.R. 1962. Estimation of $\bar{M}n$ from osmotic pressure measurements. J. Polym. Sci., 60: 21.

Bristow, G.M. 1965. Relation between stress-strain behaviour and equilibrium volume swelling for peroxide vulcanizates of natural rubber and cis-1,4,polyisoprene. J. Appl. Polym. Sci., 9: 1571 pp.

Bristow, G.M. and Sears, A.G. 1982. The freezing and thawing of natural rubber. NR Technol., 13: 73 pp.

Bristow, G.M. and Sears, A.G. 1988. A new comparison of sheet and crumb rubber. Part I. Raw rubber composition and rheology. J. Nat. Rubb. Res., 3(4): 223 pp.

Bristow, G.M. and Sears, A.G. 1989. A new comparison of sheet and crumb rubber. Part II. Some aspects of processing. J. Nat. Rubb. Res., 4(1): 22-40.

Bristow, G.M. 1985. Capillary flow of natural rubber. NR Technol., 16: 6 pp.

Fielding, J.H. 1943. Flex life and crystallization of synthetic rubbers. Ind. Engg. Chem., 35: 1259 pp.

Flory, P.J. 1950. Statistical mechanics of swelling of network structures. J. Chem. Phys., 18: 108 pp.

Flory, P.J. and Rehner, J. 1943. Statistical mechanics of crosslinked polymer networks. J. Chem. Phys., 11: 521 pp.

Gelling, I.R. and Porter, M. 1989. Chemical modification of natural rubber. In: A.D. Roberts (Ed.). Natural Rubber Science and Technology. Oxford Science Publications, Chapter 10.

Gent, A.N. 1978. Strength of elastomers. In: F.R. Eirich (Ed.), Science and Technology of Rubber. Academic Press, New York, Chapter 10.

Gent, A.N. and Kim, H.J. 1978. Tear strength of stretched rubbers. Rubber Chem. Technol., 51: 35 pp.

Gent, A.N., Lindley, P.B. and Thomas, A.G. 1964. Cut growth and fatigue of rubbers. Part I. The relationship between cut growth and fatigue. J. Appl. Polym. Sci., 8: 455 pp.

Gent, A.N. and Pulford, C.T.R. 1983. Mechanisms of rubber abrasion. J. Appl. Polym. Sci., 28(3): 943 pp.

Greensmith, H.W., Mullins, L. and Thomas, A.G. 1963. In: L. Bateman (Ed.). The Chemistry and Physics of Rubber like substances. Maclaren, London,. p. 249.

Greogry, M.J. and Tan, A.S. 1975. Some observations on storage hardening of natural rubber. Proc. Int. Rubb. Conf., 1975, Kuala Lumpur, Vol. IV, 28 pp.

Grosch, K.A. 1967. The effect of tyre surface temperature on the wear rating of tread compounds. J. Inst. Rubb. Ind., 1: 35 pp.

Grosch, K.A. and Schallamach, A. 1969. The load dependence of laboratory abrasion and tyre wear. Kautschuk Gummi Kunststoffe, 22: 288 pp.

Gupta, B.R. 1989. Rheological and die swell behaviour of natural rubber melts. Indian J. Nat. Rubb. Res., 2(1): 38-46.

424

Hamed, G.R. 1981. Tack and green strength of NR, SBR and NR/SBR blends. Rubber Chem. Technol., 54: 403-414.

Harris, E.M., Aziz, Nor Aisah, A. and Amiruddin, Morani. 1985. Measurement of dry rubber content of natural rubber latex and crepe using specific heat. Proc. Int. Rubb. Conf., 1985, Kuala Lumpur.

Hofmann, W. 1967. Vulcanization and vulcanizing agents. Maclaren, London, 5 pp.

James, D.I. 1967. Abrasion of rubber. Maclaren, London.

Juve, A.E. 1944. Processing characteristics of synthetic rubbers and their use in the manufacture of extruded products. Rubber Chem. Technol., 17: 932-940.

Juve, A.E. and Beatty, J.R. 1955. The shrinkage of mould cured elastomer compositions. Rubber Chem. Technol., 28: 1141-1156.

Lake, G.J. and Lindley, P.B. 1964a. Cut growth and fatigue of rubbers. Part 2. Experiments on a noncrystallizing rubber. J. Appl. Polym. Sci., 8: 707 pp.

Lake, G.J. and Lindley, P.B. 1964b. Ozone cracking, flex cracking and fatigue life of rubber. Part 1. Cut growth mechanisms and how they result in fatigue failure. Part 2. Technological aspects. Rubber J., 146(10): 24 pp and 146(11): 30 pp.

Lake, G.J. and Thomas, A.G. 1967. The strength of highly elastic materials. Proc. Roy. Sco. A., 300: 108 pp.

Lindley, P.B. 1974. Non-relaxing crack growth and fatigue in a non-crystallizing rubber. Rubber Chem. Technol., 47: 1253-1264.

Mathew, N.M. 1983a. Scanning electron microscopic studies on ozone cracking of NR and a NR/EPDM blend. J. Polym. Sci: Polymer Letters edition, 22: 135-141.

Mathew, N.M. and De, S.K. 1983a. Scanning electron microscopy studies on flexing and tension fatigue failure of rubber. Int. J. Fatigue, January, 23-28.

Mathew, N.M. and De, S.K. 1983b. Scanning electron microscopy studies in abrasion of NR/BR blends under different test conditions. J. Mater. Sci., 18: 515-524.

Mathew, N.M., Thomas, K.T. and Philipose, E. 1988. Studies on ozone resistance of natural rubber - ethylene/propylene rubber blends. Indian J. Nat. Rubb. Res., 1(1): 3-17.

Montes, S. and White, J.L. 1982. A comparative rheological investigation of natural and synthetic cis-1,4-polyisoprene and their carbon black compounds. Rubber Chem. Technol., 55: 1354 pp.

Nair, S. 1970. Dependence of bulk viscosities (Mooney and Wallace) on molecular parameters of natural rubber. J. Rubb. Res. Inst. Malaya, 23: 76 pp.

Ong, E.L. and Lim, C.L. 1983. Rheological properties of raw and black filled natural rubber stocks. NR Technol., 14: 25 pp.

Ong, E.L. and Subramanian, A. 1975. Rheological studies of raw elastomers with the Mooney viscometer. Proc. Int. Rubb. Conf., 1975, Kuala Lumpur, Vol. IV, 39-60.

Onyon, P.F. 1959. In: P.W. Allen (Ed.) Techniques in polymer characterization. Butterworths, London.

Richardson, M.J. 1976. Quantitative interpretation of DSC results. Plastics and Rubber, Materials & Application, 1: 3-4.

Schulz, G.V., Altgelt, K. and Cantow, H.J. 1956. Determination of molecular size and structure of natural rubber by light scattering measurements. I. Method and measurement. Macromol. Chem., 21: 13-36.

Schulz, G.V. and Mula, A. 1960. New investigations on the size, form and flexibility of the rubber molecule. Proc. Nat. Rubb. Res. Conf., 1960, Kuala Lumpur, 602-610.

Sekhar, B.C. 1958. Aeration of natural rubber latex. II. Graft polymerization of vinyl monomers with aerated latex rubber. Rubber Chem. Technol., 31: 425-430.

Sekhar, B.C. 1962. Abnormal groups in rubber and microgel. Proc. Fourth Rubb. Tech. Conf., 1962, London, Paper No. 38.

Skewis, John, D. 1966. Self-diffusion coefficients and tack of some rubbery polymers. Rubber Chem. Technol., 39: 217-225.

Subramaniam, A. 1971. Effect of ethrel stimulation on raw rubber properties. RRIM Planters' Conf. (eds. Ng Siew Ku and J.C. Raja Rao), RRIM Kuala Lumpur, pp. 12-14.

Subramaniam, A. 1972. Gel permeation chromatography of NR. Rubber Chem. Technol., 45: 346-358.

Subramaniam, A. 1975. Molecular weight and other properties of NR: A study of clonal variations. Proc. Int. Rubb. Conf. Kuala Lumpur, Vol. IV, 3 pp.

Summers, W.R., Tewari, Y.B. and Sehreiber, H.P. 1972. Thermodynamic interaction in poly (dimethyl siloxane) - hydrocarbon systems from gas-liquid chromatography. Rubber Chem. Technol., 45: 1655-1664.

Tanaka, Y. 1985. Structural characterization of cis-polyisoprenes from sun flower, Hevea and guayule. Proc. Int. Rubb. Conf. Kuala Lumpur, 2, 73, Proc. Int. Rubber Conf., Kyoto, Japan, 141 pp.

Thomas, A.G. 1960. Rupture of rubber. VI. Further experiments on the tear criterion. J. Appl. Polym. Sci., 3: 168-174.

Trelor, L.R.G. 1975. The physics of rubber elasticity (3rd edition). Clarendon Press, Oxford.

Trick, G.S. 1977. Some basic physical properties. In: W.M. Saltman (Ed.). The Stereo Rubbers. John Wiley & Sons, New York, Chapter 10.

Wood, Lawrence, A. 1966. Tables of physical constants of rubbers. Rubber Chem. Technol., 39: 132-142.

Wood, L.A. 1976. Physical constants of different rubbers. Rubber Chem. Technol., 49: 189-199.

Wood, L.A., Bultman, G.M. and Decker, G.E. 1972. Modulus of natural rubber crosslinked by dicumyl peroxide. I. Experimental observations. Rubber Chem. Technol., 45: 1388-1402.

Wood, R.I. 1953. Mooney viscosity changes in freshly prepared raw natural rubber. Rubber Chem. Technol., 26: 1-16.

Zhang, S.W. 1984. Mechanisms of rubber abrasion in unsteady state. Rubber Chem. Technol., 57: 755-768.

CHAPTER 19

COMPOUNDING AND VULCANIZATION

C. METHERELL

Tun Abdul Razak Laboratory, Malaysian Rubber Producers' Research Association, Brickendonbury, Hertford SG13 8NL, England.

INTRODUCTION

Compounding is the term used to describe formulating rubber appropriate to its end use eg. tyres, beltings, engine or building mounts, seals or hot water bottles. Vulcanization is the most important aspect of compounding natural rubber, and this is generally achieved by reaction with sulphur and accelerators at an elevated temperature. Vulcanization transforms the linear polymer, obtained by coagulation and drying of latex, into a three-dimensional macromolecule by the insertion of a relatively few crosslinks between the polymer chains. Vulcanized natural rubber possesses high strength (without the need for reinforcing fillers), high resilience and rapid recovery from deformation. Vulcanization may be carried out using substances other than sulphur, but these are far less frequently used. Non-sulphur vulcanizing systems include organic peroxides, urethanes and certain phenol-formaldehyde resins.

Other aspects of compounding include incorporation of additives to protect rubber against degradation by heat, oxygen, ozone, etc., incorporation of fillers to increase modulus, tear strength, or to reduce cost, addition of oils in small amounts to improve processing or in larger levels to reduce modulus or cost. However, addition of fillers also decreases resilience. It is immediately apparent that there may be many ways of achieving a particular requirement eg. modulus or tensile strength. Furthermore, some of the requirements for a given product may even be conflicting eg. high tear strength and high resilience. Hence compromises frequently have to be made. In the early days of the industrial use of rubber, compounding was developed by an empirical, trial and error approach with little scope for application of science. However, over the last 50 years considerable progress has been made in understanding the science of compounding rubber. These developments have been described in two excellent books; the first edited by Bateman (1963) became as standard text in polymer research

427

and teaching institutions worldwide, and the second edited by Roberts (1988) is destined to do likewise. At a more practical level the technical publications, information sheets, and bulletins of Rubber Research Institute of Malaysia and Malaysian Rubber Producers' Research Association give advice and information on compounding NR for specific applications. Particular attention is drawn to the Formulary and Property Index (MRPRA, 1984). This, as its name implies, provides a property index enabling formulations to be selected possessing specific properties as well as giving starting point formulations for specific applications.

This chapter concentrates on aspects of vulcanization of NR and its protection against ageing.

VULCANIZATION

The chemistry of sulphur vulcanization

The generally accepted mechanism of reaction of sulphur with rubber is shown in a simiplified form in Fig. 1, where RH denotes the rubber hydrocarbon, H being an allylic hydrogen in a methyl or methylene group. The first step is the formation of the active sulphuration agent from sulphur, accelerators and activators. After abstracting a hydrogen atom from the rubber, a polysulphidic rubber-bound intermediate is formed which initially forms polysulphidic crosslinks.

The structural features present in the initial vulcanizate network, are summarized in Fig. 2. The polysulphides in the crosslinks and network-bound accelerator fragments shorten at a rate which depends on the ratio of sulphur to accelerator and the cure temperature. For example if there is a high accelerator to sulphur ratio, the final network contains a high proportion of monosulphide crosslinks and pendent groups, as shown in reaction A in Fig. 3. However, if the ratio of accelerator to sulphur is low then polysulphidic crosslinks will persist, together with some di- and mono-sulphidic crosslinks, as shown in reaction scheme B. The rubber is also modified by formation of cyclic sulphides and other reactions involving the main chain. It is evident that a considerable proportion of the sulphur may be combined with the rubber in forms which do not contribute to the crosslink structure. Estimates of the number of sulphur atoms chemically combined with the rubber per physically effective crosslink vary between 2 and 100. Vulcanization systems using a high ratio of accelerator to sulphur which give predominately monosulphidic crosslinks are called efficient vulcanization (EV) systems. Those using a low ratio of accelerator to sulphur are known as conventional systems, as these were well established in use

Figure 1 Overall course of accelerated sulphur vulcanization.

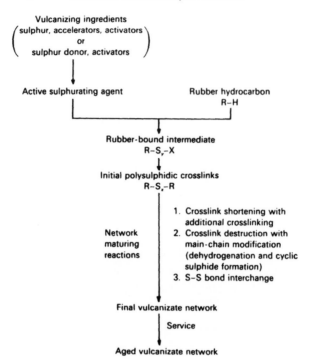

Vulcanizing ingredients
$\left(\begin{array}{c}\text{sulphur, accelerators, activators}\\ \text{or}\\ \text{sulphur donor, activators}\end{array}\right)$

↓

Active sulphurating agent   Rubber hydrocarbon
R–H

↓

Rubber-bound intermediate
R–S$_y$–X

↓

Initial polysulphidic crosslinks
R–S$_x$–R

Network
maturing
reactions

1. Crosslink shortening with
   additional crosslinking
2. Crosslink destruction with
   main-chain modification
   (dehydrogenation and cyclic
   sulphide formation)
3. S–S bond interchange

↓

Final vulcanizate network

Service

↓

Aged vulcanizate network

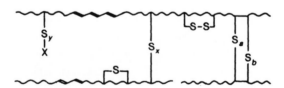

Figure 2 A diagrammatic representation of the network structure of a sulphur vulcanizate (x y, a and b = 1–9); X = accelerator fragment.

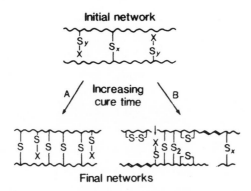

Figure 3 Dependence of network structure on vulcanizing system. A, high accelerator : sulphur ratio (or sulphur donor/accelerator system) and high soluble zinc concentration; B, low accelerator : sulphur ratio or low soluble zinc concentration. The symbols used are as in

before the more modern accelerators were developed. It should be recognised that the network-maturing reactions shown in Fig. 3 can also take place during service of a rubber product, if the temperature rises sufficiently.

The structure of the crosslinks and the extent and type of main chain modifications have important effects on the physical properties of a vulcanizate. High levels of polysulphidic crosslinks confer high tensile and tear strengths particularly in unfilled vulcanizates, which has been attributed to the ability of these crosslinks to break under stress. High resistance to fatigue on repeated stressing is also obtained together with high resilience, rapid recovery from deformation at ambient temperature. Resistance to heat and oxidation is limited, as also is resistance to set at elevated temperature. By contrast, monosulphidic crosslinks give rise to high thermal stability and resistance to oxidation and reversion (loss of modulus etc. on heating in the absence of oxygen). However, vulcanizates having high proportions of monosulphidic crosslinks possess marginally reduced strength characteristics compared to those possessing essentially polysulphidic crosslinks, but resistance to creep, stress relaxation and set at elevated temperature are substnatially improved. The resistance to crystallization at low temperature of EV vulcanizates is inferior to that of conventional vulcanizates. This latter effect is attributed to the low level of main chain modifications, particularly cyclic sulphides, in EV vulcanizates. Thus the range of temperatures over which EV vulcanizates may be used is shifted to a higher level.

Vulcanizing systems with roughly equal levels of sulphur and accelerator, the semi-EV systems, are frequently used when compromises between high and low temperature properties or between fatigue life and heat resistance are required.

For a more authoritative account on the chemistry of vulcanization and the effect of crosslink structure on physical properties the reader is referred to a review by Chapman and Porter (1988).

Practical vulcanization systems

Sulphur vulcanization systems contain activators such as zinc oxide and stearic acid and for optimum properties it is important to maintain sufficient concentration of each. Normal levels of zinc oxide are 3 to 5 parts per hundred rubber (pphr), and stearic acid at 1 to 3 pphr. Zinc soaps may be used to replace stearic acid. Zinc 2-ethylhexanoate (ZEH) is a rubber-soluble zinc soap and is used as a replacement for stearic acid where low creep or stress relaxation is required, particularly in soluble EV system (see later). However, it is important to ensure that an excess of free zinc oxide is maintained for all sulphur cure systems.

High levels of stearic acid, ie. 4-7 pphr, improve the reversion resistance of conventional and semi-EV systems. This has been exploited in NR truck retread formulations, where considerable improvement in abrasion resistance on overcure was obtained (Jones et al. 1988). A classification of sulphur vulcanization systems, giving approximate levels of sulphur and accelerator required for the types described in the previous section, is given in Table 1.

TABLE 1

Classification of sulphur vulcanization systems

	Conventional	Semi-EV	EV
Sulphur, pphr	2.0-3.5	2.0-1.0	1.0
Accelerator, pphr	1.0-0.5	1.0-2.0	3-6
Accelerator:sulphur ratio	0.5	0.5-2.0	3

Sulphenamides are the most widely used class of accelerator for natural rubber. They give fairly long delay period before the onset of vulcanization, ie. good processing safety, and a high rate of vulcanization. The most commonly used sulphenamides include N-cyclohexylbenzothiazole-2-sulphenamide (CBS), and the tert-butyl and morpholino derivatives (TBBS and MBS respectively). Processing safety generally improves in the order listed. The dicyclohexylbenzothiazole-2-sulphenamide (DCBS) is less widely used but gives substantially greater processing safety, although vulcanizate moduli are reduced. Thiazole accelerators, eg. mercaptobenzothiazole (MBT) or dibenzthiazyl disulphide (MBTS) are slow curing when used alone but when boosted with tetramethylthiuram monosulphide (TMTM) or N,n'-diphenyl-guanidine (DPG), are used in fast curing conventional vulcanization systems. Vulcanizates incorporating thiazoles possess particularly good resistance to light discolouration. However, processing safety is limited, particularly in vulcanizates containing reinforcing grades of carbon black.

Zinc dialkyldithiocarbamates are very fast-acting accelerators, most frequently used as secondary accelerators to reduce cure times. Most of them are insoluble in rubber and may form a white powdery bloom on the surface of a vulcanizate even at levels as low as 0.2 pphr. However, zinc dibutyldithiocarbamate (ZDBC) is soluble in rubber to a much greater extent and does not normally bloom.

Thiuram disulphides, eg. tetramethylthiuram disulphide (TMTD), are used as sulphur donors in EV systems where good resistance to heat is required, and as secondary accelerators to increase the rate of vulcanization in conventional and semi-EV systems. Thiuram disulphides may be used as the sole vulcanizing agent in the so-called 'sulphurless' systems which may also contain up to 0.5 pphr of a thiazole, for improved processing safety. Zinc dialkyldithiocarbamates are formed during vulcanization. Advantage is taken of the solubility of the zinc salt of tetrabutylthiuram disulphide in the soluble EV systems (see later).

Other accelerators in common use include N-oxydiethylene-thiocarbamyl-N'-oxydiethylene sulphenamide (OTOS) and zinc dibutyldithiophosphate (ZDBP). The former is a primary accelerator introduced quite recently and gives good processing safety in semi-EV and EV systems, with a fast rate of cure. Pronounced synergism is obtained with MBS. ZDBP gives a fast rate of cure with rather poor processing safety, but reversion resistance is quite high in some semi-EV system (Metherell, 1986).

The most widely used sulphur donor, apart from the thiuram di-sulphides, is 4,4'-dithiodimorpholine (DTDM) and this is commonly used in Semi-EV and EV systems. Triethoxysilylpropyl tetrasulphide (TESPT), although primarily a silica coupling agent, is also used as a sulphur donor in the so-called 'equilibrium cure' system.

Acidic materials such as phthalic anhydride or salicylic acid retard vulcanization and were widely used to improve processing safety. However, in recent years far more effective agents have been introduced, ie. n-cyclo-hexylthiophthalimide (CTP) and Vulcatard PRS. These two chemicals extend processing safety without increasing the overall cure time significantly. They both are most effective in conventional and semi-EV sulphenamide accelerated systems, but PRS is a little more effective than CTP in systems containing thiurams or dithiocarbamates.

Some typical conventional, semi-EV and EV systems are listed in Table 2, together with notes on their selection. The use of combinations of accelerators is widespread, (a) to adjust rates of cure and (b) to exploit the synergism that often occurs between accelerators of different structure eg. thiurams and sulphenamides.

There are a wide variety of cure systems classified as semi-EV and EV. For example, EV systems may comprise low sulphur/high sulphenamide (eg. 0.4 pphr sulphur/6.0 pphr CBS) or more commonly, sulphur with synergistic combinations of accelerators (eg. S/CBS/TMTD). The latter system gives a faster cure rate, and a higher modulus for a given total amount of accelerator. However, processing safety is less for this system. Other alternative EV systems use sulphur donors, sometimes used with no elemental sulphur. The sulphur donor DTDM is claimed to confer improved resistance to fatigue in semi-EV and EV systems. Improvements in processing safety may be obtained by replacing part of the elemental sulphur by DTDM.

The so-called 'soluble EV' system was developed by Elliott et al. (1970) for engineering applications requiring the lowest rates of creep and stress relaxation. In this system the rubber-soluble activator ZEH is used together with accelerators ie. TBTD and MBS, which are also rubber-soluble and which form rubber-soluble zinc salts. The modulus of soluble EV vulcanizates has been shown to be more reproducible than that of con-ventional cure systems. However, the resistance to high temperature ageing and set of soluble EV vulcanizates is marginally less than those of the EV systems of the synergistic combinations of thiuram and sulphenamide, low sulphur/high CBS, or sulphurless TMTD. Analysis of the network structure of soluble EV vulcanizates has shown that they contain more sulphur atoms per chemical crosslink than most EV systems.

For best resistance to oxidative ageing EV systems should be protected with a combination of a powerful antioxidant and a hydroperoxide decomposer (see later).

TABLE 2

Some typical vulcanization systems.

Amounts quoted are pphr.

(a) Conventional

2 to 3.5 sulphur + 0.5 to 1.0 sulphenamide	Good processing safety.
2 to 3.5 sulphur + 0.5 to 1.0 thiazole + 0.1 to 0.5 DPG or TMTM	Fast curing.
2 to 3.5 sulphur + 0.3 to 0.5 thiuram or dithiocarbamate	Very fast vulcanization but limited safety.

(b) Semi-EV range

2.0 sulphur + 1.0 sulphenamide	Increasing resistance to reversion and ageing.
1.5 sulphur + 1.5 sulphenamide	
1.0 sulphur + 2.0 sulphenamide	
1.5 sulphur + 0.6 DTDM + 0.6 sulphenamide	Improved fatigue resistance.
1.0 sulphur + 1.0 MBS + 2.0 ZDBP	Fast curing, moderate reversion resistance.
1.0 sulphur + 1.4 MBTS + 3.0 TESPT	"Equilibrium cure".

(c) Low sulphur EV systems

0.8 sulphur + 1.0 OTOS + 0.5 MBS	Fast cure, high processing safety
0.7 sulphur + 1.7 MBS + 0.7 TBTD	Soluble EV, Low creep, stress relaxation.
0.6 sulphur + 1.6 DTDM + 1.0 sulphenamide	Improved fatigue resistance.
0.4 sulphur + 6.0 CBS	High processing safety.
0.3 sulphur + 3.0 CBS + 2.0 TMTD	High modulus.
0.25 sulphur + 1.8 CBS + 1.2 TMTD	Lower modulus.

(d) Sulphurless EV systems

3 TMTD	Processing safety improved by addition of 0.5 MBT or MBTS.
1 DTDM + 1 CBS + 1 TMTD	Improved fatigue resistance.
3 TESPT + 2 DDTS	High abrasion resistance, low set with silica.
1.5 DTDM + 2 MBTS	For light colours, non-copper staining, low modulus.

The effect of sulphur cure systems on physical properties will be discussed in more detail after discussing the application of the most important non-sulphur vulcanization systems.

Non-sulphur vulcanization

Non-sulphur vulcanization systems are of far less industrial significance for natural rubber than sulphur systems. These systems were reviewed by Baker (1988) and Kempermann (1988). Of the many systems investigated, only two have achieved commercial significance: vulcanization with organic peroxides and urethanes.

Vulcanization by organic peroxides is a free radical process. The peroxide, POOP, decomposes into alkoxy radicals:

$$POOP \longrightarrow 2PO. \tag{1}$$

the alkoxy radicals abstract a hydrogen atom from the rubber, RH,

$$RH + PO. \longrightarrow R. + POH \tag{2}$$

crosslinks are formed by combination of two rubber radicals:

$$2R. \longrightarrow RR \tag{3}$$

Effective crosslinking can only take place in the complete absence of oxygen, otherwise peroxy radicals will be formed leading to oxidation of the rubber. Peroxide vulcanization is therefore limited in products made by press curing.

The structure of a peroxide vulcanizate is relatively simple compared with that cured with sulphur systems, as rubber chains are linked by simple carbon-carbon crosslinks which are highly thermally stable. Peroxide vulcanizates possess extremely good ageing characteristics, somewhat superior to those of EV systems, and also lower compression set at elevated temperature. However, they suffer from the disadvantages of EV systems, ie. low tension fatigue lives and poor resistance to low temperature crystallization.

The rate of vulcanization with organic peroxides is controlled essentially by the decomposition of the peroxide, which is a first-order chemical process, in which the time for 50% decomposition of the peroxide is a constant at a given temperature (the half-life). Almost complete decomposition of the peroxide is essential, otherwise unreacted peroxide

will cause oxidative degradation of the rubber in service at elevated temperature. The proportion of unreacted peroxide (p) after a given number
of half-lives, n, is given by

$$p = 0.5^n \qquad\qquad (4)$$

Hence, for 98% decomposition of peroxide the cure time must be about
5.5 half-lives. The most frequently used organic peroxide is dicumyl peroxide, which has a half-life of about 6 min at 160°C. Hence the time for
98% decomposition of the peroxide should be about 30 min at 160°C. The
use of cure times substantially less than this will result in failure to
achieve the advantages of peroxide cures over sulphur systems. For the
ultimate in resistance to heat ageing, cure times of 10 half-lives have
been recommended, ca.1 hour at 160°C for dicumyl peroxide.

A substantial disadvantage of simple peroxide systems is their lack
of processing safety as crosslinking commences as soon as the temperature
rises appreciably. For example, the half-life of dicumyl peroxide at 120°C
is approximately 8 h; thus about 1-2% decomposition and subsequent crosslinking of the rubber will occur within 10 min at 120°C, which corresponds
approximately to the processing safety in a Mooney scorch test.

Chow and Knight (1977) showed that the introduction of 0.1 to 0.3 pphr
of a radical scavenger such as a p-phenylenediamine, or a nitroso radical
capture agent, will result in reasonable processing safety. The level of
peroxide used should be increased to compensate for the amount of peroxide
consumed by the scavenger. Injection moulding of these systems is possible.
Acceleration of peroxide vulcanization is not possible: cure rate is adjusted
by changing to a more or less reactive peroxide.

The efficiency of utilization of the peroxide may be increased by
addition of a co-agent such as trimethylpropanetrimethacrylate. The
commercial product, Saret 500, incorporates a coagent and inhibitor and
has sufficient processing safety to allow injection moulding.

Best protection against oxidative ageing is achieved with high molecular weight quinoline derivatives eg. TMQ, preferably in combination with
a hydroperoxide decomposer eg. ZDMC and/or ZMBI (see later). Addition
of a base such as calcium hydroxide is beneficial. Owing to the interference
of p-phenylenediamines with peroxide vulcanization, substantially increased
levels of peroxide are required to achieve a given modulus when chemical
antiozonants of this type are present. Therefore waxes are preferred for
protection of peroxide vulcanizates against ozone. Peroxide vulcanization

is most effective under basic conditions. Zinc oxide and stearic acid are not necessary.

Urethane vulcanizing systems

Urethane vulcanizing agents available under the trade name Novor, developed by Baker (1989) are now being used increasingly in a variety of applications, in particular those demanding high temperature vulcanization or high temperature service. The general reaction scheme of urethane vulcanization is represented by Fig. 4. Novor 924 contains an addition

Figure 4

Mechanism of crosslinking with urethane reagents

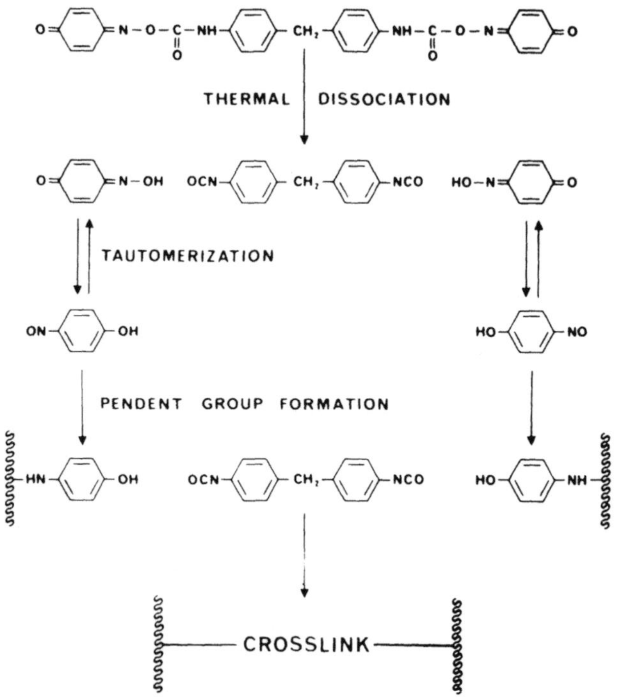

product of a nitrosophenol and a diisocyanate, and contains no free nitroso or isocyanate groups. At vulcanization temperatures the adduct dissociates and the free nitrosophenol adds on to the rubber chain to form pendent aminophenol groups, which react with the liberated diisocyanate to form crosslinks, which are mainly of the urea-urea type. The principal benefits Novor systems are outstanding reversion resistance and good ageing resistance coupled with good initial vulcanizate properties: a combination that cannot be achieved by established sulphur vulcanization systems. Another feature is that the urethane agents can be blended with sulphur vulcanizing systems to give a family of systems designed to vary in cost, cure behaviour and other properties. These mixed systems can also be devised to give vulcanizates with properties virtually unaffected by cure temperature - something which sulphur systems alone cannot provide.

All-Novor vulcanizing systems

The outstanding feature of the all-Novor 924 vulcanizing system is its ability to give vulcanizates having exceptional reversion resistance. Typical formulations are shown in Table 3.

TABLE 3
All-Novor vulcanizing system

Ingredient	Parts phr	Function
Novor 924	6.7	Crosslinking agent
ZDMC	2.0	Activator/catalyst
Calcium oxide	3.0	Drying agent

The best heat-ageing resistance is given by the incorporation of a blend of TMQ and zinc 2-mercaptobenzimidazole (ZMBI), each at 2 pphr. The best resistance to fatigue cracking is obtained with a p-phenylenediamine type antidegradant without a drying agent. Ageing resistance is inferior to that obtained with the TMQ/ZMBI system, although it is still better than that of a well-protected high sulphur vulcanizate.

Mixer Novor/sulphur systems

Adding a conventional sulphur vulcanizing system to Novor 924 reduces cost and gives a much faster cure rate, whilst retaining many of the desirable features of the urethane system. The two systems can be blended in any proportion, but formulations having less than 50% Novor are not generally recommended since they do not offer any advantage over semi-EV systems. Resistance to reversion decreases as the Novor/sulphur ratio is reduced but 90/10 and 80/20 Novor 924/sulphur systems are still able to match EV systems. With certain mixed systems the same modulus/strength/ resilience properties are obtained whether curing takes place at 140°C or 200°C. Many mixed systems, like the all-Novor system, give vulcanizates whose fatigue resistance can improve upon ageing.

A blend of a Novor and a sulphur system will give a higher crosslink density than that expected from the two components, and the overall levels must be adjusted to maintain modulus. In the formulations given in Table 4 the levels of Novor, catalyst, sulphur and accelerator have been reduced by 20% (ie. to 80% vulcanization potential).

TABLE 4

Mixed Novor/sulphur systems

	Novor/sulphur ratio				
	90/10	80/20	70/30	60/40	50/50
	Parts phr				
Novor 924	4.8	4.2	3.8	3.2	2.7
TMTM[a]	1.4	1.3	1.2	1.1	1.0
Zinc oxide	5.0	5.0	5.0	5.0	5.0
Stearic acid	1.0	1.0	1.0	1.0	1.0
Sulphur	0.2	0.4	0.6	0.8	1.0
TBBS	0.04	0.08	0.12	0.16	0.20

[a] Activator for Novor providing better scorch safety than ZDMC.

Mixed Novor/sulphur systems are particularly useful in high-temperature curing, eg. continuous vulcanization and injection moulding, where advantage can be taken of the combination of high reversion resistance, fast cure rate and insensitivity of properties to cure temperature. Optimum cures of 2-4 min at 180°C, 1-3 min at 190°C and 0.5-1 min at 200°C have been used in injection moulding with an 80/20 system. These systems are also useful in the high-temperature curing of conveyor belting and, because of their reversion resistance, give more uniform vulcanization of rubber-covered rollers.

Effect of vulcanization system on physical properties

Table 5 presents a comparison of the physical properties of

TABLE 5

Effect of vulcanization system on physical properties

Base formulation, SMR 10, 100, Light oil 10, zinc oxide 5, stearic acid 2.

	Conventional	Semi EV	Soluble[a] EV	EV	Novor	Peroxide[b]
N550 FEF black	45	45	50	50	45	50
TMQ	2	2	2	2	2	2
HPPD	2	2	2	2	-	-
ZMBI	-	-	-	-	2	2
Sulphur	2.5	1.5	0.6	0.25	0.4	-
MBTS	1.0	-	-	-	-	-
TBBS	-	1.2	-	2.1	0.1	-
MBS	-	-	1.5	-	-	-
DPG	0.1	-	-	-	-	-
TMTD	-	0.2	-	1.0	-	-
TBTD	-	-	0.6	-	-	-
TMTM	-	-	-	-	1.8	-
Novor 924	-	-	-	-	4.2	-
ZDMC	-	-	-	-	-	2.0
Dicumyl peroxide	-	-	-	-	-	2.5
Calcium hydroxide	-	-	-	-	-	2.0
Physical properties						
Cure time, min	5[c]	15[c]	30[c]	30[c]	30[c]	60[d]
Hardness, IRHD	60	64	58	59	68	59
Lupke resilience, %	75	74	73	67	69	72
M100, MPa	1.9	2.0	1.8	1.6	1.5	1.7
M300, MPa	10.0	13.5	12.0	9.5	9.0	12.5
TS, MPa	24.0	24.0	23.0	23.0	23.0	18.0
EB, %	540	475	520	540	555	395

Ring fatigue, 0-100% extension, kc to failure

	150	85	60	55	36	56

Compression set, %

1 day at -26°C	10	16	23	77	43	85
1 day at 70°C	27	26	14	17	25	12
1 day at 100°C	48	46	28	31	41	20

Air oven ageing for 3 days at 125°C, % change in

TS	-30	-25	-10	-15	-20	-5
EB	-40	-17	-10	-15	-25	-10

Air-oven ageing for 3 days at 100°C, % change in

TS	-	-	-65	-50	-50	-30
EB	-	-	-50	-45	-40	-10

[a] In the soluble EV system stearic acid was replaced by ZEH.

[b] In the peroxide system zinc oxide and stearic acid were omitted.

[c] Sulphur and Novor vulcanizates were cured at 150°C.

[d] Peroxide vulcanizates were cured for 60 min at 160°C.

vulcanizates cured with a fast-curing conventional system with semi-EV, soluble EV, EV, Novor and peroxide systems. Minor adjustments in filler loading were made in order to improve the match of hardness and moduli. It should be noted that it is not always possible to match both hardness and moduli when comparing vulcanization systems because of differences in the shapes of stress-strain curves. For example, the Novor system gives a particularly high hardness for a given modulus. However, with the other systems a better match could be achieved by altering the sulphur and accelerator levels.

The data demonstrate:

(1) the high resilience of the conventional vulcanizate compared with the most heat-resistant systems, ie. EV, Novor and peroxide

(2) in the presence of a moderately fine grade of carbon black, ie. FEF, differences in tensile strength are hardly significant, except for the peroxide system, which gives rise to a low strength

(3) there is a decrease in fatigue life from conventional through semi-EV to the EV and other systems

(4) compression set at -26°C is the lowest for the conventional and the highest for the EV and peroxide systems

(5) compression set at 70°C and 100°C is improved on changing from a conventional to the EV and peroxide systems

(6) improvements in ageing are found on changing to more efficient systems. Peroxide vulcanizates protected by the best combination of antioxidants give the best ageing resistance. The results for ageing for three days at 100°C indicates that the soluble EV system gives a little better ageing resistance than the other EV system, but past experience has shown that on ageing for longer times at 100°C, or for higher temperatures as shown in Table 5, the full EV system gives superior ageing to the soluble EV.

When the chemical structure of vulcanizates was first established, it was believed that the sulphur to accelerator ratio was the most important factor influencing the resistance to reversion and ageing. However, in the light of new developments in accelerators it is now recognized that other factors such as the type of accelerator have an influence on vulcanizate performance. Some of these effects were described by Metherell (1986). It must also be recognized that the state of cure is also important. For example, compression set may be reduced considerably by increasing the cure time from Rheometer $t_c 90$ to t_{max} and even beyond. This is due to the reduction in the proportion of polysulphidic crosslinks and increased proportion of more stable mono- and di-sulphide crosslinks and also the higher degree of crosslinking that occur as the cure time increases. Improvements in low temperature compression set performance are also obtained, but these are probably due to the increase in main chain modifications of the type able to inhibit crystallization.

Another approach to obtaining vulcanization systems with high resistance to reversion is the 'equilibrium cure system'. This was introduced by Wolff (1979) for vulcanizates containing carbon black and silica. In this type of vulcanization system the levels of accelerator are chosen to balance the loss of crosslinks normally associated with sulphur/sulphenamide, by the introduction of further crosslinks via the slow reaction of triethoxy-silylpropyltetrasulphide (TESPT). This type of system is claimed to almost eliminate the change in physical properties on extended cure and was intended primarily to aid the production of thick articles, such as off-the-road tyres, by obtaining a more uniform crosslinking throughout the thickness. It should be pointed out that the pseudo-reversion resistance obtained via an equilibrium between crosslink loss and further insertion could have undesirable effects if elevated temperatures should occur in subsequent testing or service. For example, high temperature compression set could be impaired by insertion of further crosslinks in the deformed state.

Another novel cure system based on TESPT was developed by Wolff (1981), for silica-filled vulcanizates, in which advantage is taken of the rubber-to-filler coupling action of TESPT. Vulcanization with the sulphur donor dimethyldiphenylthiuram disulphide (DDTS) is recommended for this system. Optimum performance is obtained at a 1:1 mole ratio of TESPT to DDTS with 3 pphr, or more, TESPT. Outstanding resistance to compression set for silica-filled vulcanizates is obtained with good abrasion resistance. A comparison of the performance of these systems and high sulphur systems is given in Table 6.

Another method of obtaining resistance to reversion via a hybrid crosslink was described by Lloyd (1988). The addition of 3 pphr sodium hexamethylene-1,6-bis(thiosulphate) dihydrate (Duralink HTS) to a high sulphur vulcanization system results in the interposition of a thermally stable alkyl chain terminated by monosulphide links to the rubber by reaction with a polysulphide crosslink. Thus resistance to extended cure is increased. The effect on physical properties is shown in Table 7. The outstanding feature of this system is that fatigue life is maintained, or improved as reversion resistance is improved. The material is becoming established in thick products where improved uniformity is required. Improvements in the environmental resistance of brass-bonded components are also observed. These developments are examples of a new concept in

TABLE 6

Rubber-filler crosslinks in silica-filled NR[a]

Ultrasil VN$_3$	40	40
TESPT [b]	3	5
Sulphur	3.5	–
CBS	1.0	–
DEG	1.5	–
DDTS	–	3.4
Cure time, min at 150°C	35	60
Hardness, IRHD	66	66
Tensile properties		
M100, MPa	2.6	3.0
M300, MPa	8.1	12.0
TS, MPa	26.0	25.5
EB, %	620	480
Compression set, %		
1 day at 70°C	28	18
Lupke resilience, %	66	69
DIN abrasion index [c]	80	105
Tear, ISO trouser, N/mm	23	12

[a] Base formulation: SMRCV 100, Calcium soap process aid 4, zinc oxide 5, stearic acid 2, Wingstay L 1, wax blend 5.

[b] SI69 (Degussa)

[c] Tested to ISO 4649 using control mix B2, volume loss 187 mm^3.

TABLE 7

Effect of duralink HTS on reversion resistance[a]

Sulphur	2.5	2.5	
TBBS	0.6	0.6	
Duralink HTS	–	3.0	
Rheometery 160°C			
tc95	9	12	
t5% reversion	19	39	
Vulcanizate properties cured 160°C to tc95	tc95	t5% reversion	
Hardness, IRHD	66	67	65
Lupke resilience, %	59	55	56
M100	2.6	2.7	2.4
M300	13.4	13.0	12.5
TS	26.0	25.0	25.5
EB	490	475	510
Compression set			
1 day at 70°C	43	33	26
Ring fatigue, kc to failure			
0-100% extension	228	329	234

[a] Base formulation SMR L 100, N330 HAF 50, Aromatic process oil 5, zinc oxide 2, HPPD 2.

vulcanization chemistry involving 'post-crosslink accelerator systems' or reversion stabilizers. The former term was introduced by Eholzer and Kempermann (1985). Further developments are discussed by Layer (1987) and Davis et al. (1987).

For further information on vulcanization systems reference may be made to reviews by Crowther et al. (1988), Skinner and Watson (1967), Elliott and Tidd (1973/4) and that by Kempermann (1988).

PROTECTION AGAINST OXYGEN AND OZONE

Oxidation of natural rubber vulcanizates occurs by an autocatalytic free radical chain mechanism which has certain features in common with the oxidation of most olefins. The chain reaction is initiated by the formation of radicals by loss of an α-methylenic hydrogen from the rubber hydrocarbon (RH):

$$RH \longrightarrow R. + H. \tag{5}$$

The chain reaction may be propagated by formation of peroxy radicals and reaction with rubber molecules generating a new radical:

$$R. + O_2 \longrightarrow RO_2. \tag{6}$$

$$RO_2. + RH \longrightarrow RO_sH + R. \tag{7}$$

The chain reaction may be terminated in a variety of reactions leading to non-radical products.

$$R. + R. \longrightarrow \qquad] \tag{8}$$

$$R. + RO_2. \longrightarrow \qquad] \text{ non-radical products} \tag{9}$$

$$RO_2. + RO_2. \longrightarrow \qquad] \tag{10}$$

Scission of the rubber chain occurs by decomposition of peroxy radicals, although the exact mechanism is still not well understood. It has been established that oxidation of purified peroxide vulcanizates follows the above sequence. However, oxidation of sulphur vulcanizates is considerably more complex and our understanding has been enhanced by studies of the oxidation of model sulphides.

Monosulphides are oxidized to sulphoxides, some of which are unstable and oxidized further leading to crosslink scission with formation of sulphenic acids. The latter are believed to be powerful antioxidants. Hence pronounced inhibition of oxidation may occur. Disulphides react with hydroperoxides leading to a variety of products, including sulphenic acids. Thus disulphide crosslinks also give rise to inhibition of oxidation (and crosslink scission). Little is known of the reactions of oxidation of polysulphides, but anti-oxidant behaviour by these structures is observed, complicated by an antagonism shown to conventional antioxidants.

The foregoing discussion indicates that one of the consequences of oxidation of vulcanizates would be softening due to scission of the hydro-carbon chain, or scission of crosslinks. However, it is common knowledge that oxidation of sulphur-vulcanized NR initially leads to hardening, before softening is observed on prolonged oxidation. This feature is not observed in the oxidation of peroxide vulcanizates. Hence the crosslinking observed during oxidation of sulphur vulcanizates is associated with oxidation of the sulphur crosslinks or other network-bound sulphur species. Combination of radicals formed from oxidation of cyclic mono- or disulphides (see Fig. 2) are believed to be responsible for hardening.

Most anitoxidants function as hydrogen donors (AH) and react with peroxy radicals:

$$AH + RO_2 \cdot \longrightarrow ROOH + A \cdot \qquad (11)$$

The radical produced is incapable of abstraction of hydrogen from the rubber molecule. Reaction (11) is in direct competition with the propagation reaction (7). Hence rubber peroxy radicals are destroyed and oxidation of the rubber is minimised. A second class of antioxidants is the hydroperoxide decomposers, which includes zinc dithiocarbamates, 2-mercaptobenzimidazole (MBI), its zinc salt (ZMBI) and the corresponding tolyl derivatives. Some other divalent sulphur compounds also act in this manner. Oxidation may be catalysed by the so-called 'redox' catalysts, particularly certain active forms of copper, manganese, iron etc. Many amines and some phenolic antioxidants are effective inhibitors of metal-catalysed oxidation; examples are p-phenylenediamines and 2-mercapto-benzimidazole and some of its derivatives.

SELECTION OF ANTIOXIDANTS, ANTI-FLEX CRACKING AGENTS AND WAXES

There are well over 100 antidegradants currently in commercial use; hence an indication of factors influencing their choice is appropriate. The most important of these are, degree of staining that can be tolerated for the application, volatility and/or ease of extraction and the extent of flexing involved, if any. Product thickness is also important, and special methods such as the use of rubber-bound antioxidants can be applied to protect thin products.

The most effective antioxidants and anti-flex cracking agents discolour on exposure to sunlight and stain other materials they contact in service. Hence the extent of staining is often used to classify antioxidants:

(1) Staining antidegradants giving protection against oxidation, flex cracking and ozone.

(2) Staining antioxidants with flex cracking protection.

(3) Other amine antioxidants.

(4) Non-staining antioxidants.

(5) Materials giving static protection against ozone.

Staining antidegradants giving protection against oxidation, ozone and flex cracking

The p-phenylenediamines are by far the most effective types of anti-degradant used in natural rubber. Trade examples are shown in Table 8. There are three main classes: N-alkyl-N'-phenyl-, N,N'-dialkyl- and N,N'-diaryl-p-phenylenediamines. Those having the lowest molecular weight are the most effective but are the most volatile, most easily extracted and worst staining.

Alkyl-phenyl- and dialkyl-p-phenylenediamines reduce the rate of growth of ozone-induced cracks under both static and dynamic conditions. The dialkyl compounds also increase the threshold strain for static ozone cracking; they may, therefore, be used alone. Alkyl-phenyl compounds need an additional antiozonant or wax for static protection; the synergistic effect obtained with such blends allows lower loadings of each constituent to be used.

6-Ethoxy-2,2,4-trimethyl-1,2-dihydroquinoline is less effective than p-phenylenediamines. It is highly staining and reduces only the crack growth rate. It does not protect against static cracking unless blended with waxes, when a synergistic effect is obtained.

Typical protective systems	Protection
2 pphr IPPD or HPPD + 2-3 pphr wax	static and dynamic
3-4 pphr DOPPD	static and dynamic
1 pphr IPPD or HPPD + 1 pphr	static and dynamic
diaryl PPD + 2-3 pphr wax	More resistance to extraction and volatilization

TABLE 8

Antidegradants

Antidegradant	Abbreviation	Comments	Trade examples
P-Phenylenediamines			
Alkyl-phenyl N-isopropyl-N'-phenyl	IPPD	high overall antidegradant activity	Flexzone 3C Permanax IPPD Santoflex IP Vulkanox 4010NA
N-(1,3-dimethylbutyl)-N'-phenyl	HPPD or 6PPD	as IPPD but more resistant to leaching less volatile.	Antiozite 67 Permanax 6PPD Santoflex 13 UOP 588 Vulkanox 4020
Dialkyl N,N'-bis-(1,4-dimethyl-pentyl)	7PPD	high antiozonant activity under static strain	Eastozone 33 Flexzone 41 Santoflex 77 Vulkanox 4030
N,N'-di(1-ethyl-3-methylpentyl)	DOPPD	high antiozonant activity under static strain	Antiozite 2 Flexzone 8L Santoflex 17 UOP 88
Diaryl mixed diaryl		lowest antiozonant activity but most resistant to aqueous leaching; least staining; low volatility	Akroflex AZ Antigene D Wingstay 100 Vulkanox 3100
Hydroquinoline 6-ethoxy-2,2,4-trimethyl-1,2-dihydroquinoline		less effective than IPPD or HPPD	Anox W Santoflex AW

The systems shown may be used without additional antioxidant, but dialkyl-p-phenylenediamines are weaker antioxidants than alkyl-phenyl compounds and their activity is less well retained after outdoor exposure: for the best oxidation resistance an antioxidant such as TMQ should also be included. Diaryl-p-phenylenediamines are most resistant to chemical and physical losses in activity, but are weaker antiozonants. Replacement of up to half the loading of an alkyl-phenyl compound with a diaryl compound will confer greater resistance to extraction and possibly more long-term protection under severe conditions.

It should be noted that the antiozonant system required to meet a specification is not always appropriate for long-term protection against ozone, particularly if the ozone concentration in the test is high and/or if the pre-conditioning time is low (Lews, 1972).

Staining antioxidants with anti-flex cracking protection

Diphenylamine-acetone condensates and naphthylamine derivatives are good anti-flex cracking agents, but they are not as effective as p-phenylenediamines. Table 9 lists some of the more common examples. It should be noted that phenyl-β-naphthylamine is not permitted to be used in some countries for health reasons.

TABLE 9

Staining antioxidants giving flex-cracking resistance

Type of antioxidant	Comments	Trade examples
Acetone/diphenylamine condensates	general-purpose including tyre components	Agerite Superflex BLE 25, Cyanoflex Permanax B
Napthylamine derivatives phenyl-β-napthylamine	not recommended for high temperatures	Agerite powder Antigene D Vulkanox PBN

Other amine antioxidants

These include polymerized 2,2,4-trimethyl-1-2-dihydroquinoline (TMQ), di-β-naphthyl-p-phenylenediamine and some substituted diphenylamines; see Table 10 for trade examples. They provide good protection against heat and oxygen, and the p-phenylenediamine derivative is also very effective as a metal deactivator. Substituted diphenylamines also confer some resistance to flex cracking.

Non-staining antioxidants

The least staining and discolouration is given by phenolic antioxidants; the more important types are shown in Table 11. The most effective in natural rubber are phenol-alkanes, in particular hindered methylene-bisphenols, but they are susceptible to pinking on exposure to light. Better resistance to pinking is given by some of the more complex phenol-alkanes.

448

TABLE 10

Antioxidants with no flex-cracking or ozone protection

Type of antioxidant	Comments	Trade examples
Polymerized 2,2,4-trimethyl-1,2-dihydro-quinoline	general purpose	Agerite Resin D Flectol Pastilles Permanax TQ Vulkanox HS
Di-β-naphthyl-p-phenylenediamine	metal deactivator	Agerite White Antigene F
Aklylated or aralkylated diphenylamines	minimal stain	Agerite Stalite Octamine Permanax OD Vulkanox DDA Wingstay 29

TABLE 11

Non-stainiing antioxidants

Type of antioxidant	Comments	Trade examples
Phenol-alkanes methylene-bis-phenols	good antioxidant protection, some pinking	Antioxidant 425 Antioxidant 2246 Naugawhite Permanax WSP Vulkanox BKF Vulkanox ZKF
Complex phenol-alkanes	more resistant to pinking	Lowinox 22IB46 Wingstay L
Hindered phenols styrenated phenols		Montaclere Permanax SP Singstay S
Other phenolics*		Permanax WSL Santowhite 54 Wingstay T
Thio-bis-phenols		Ethyl Antioxidant 736
Hydroquinones		Agerite Alba Santovar A

* eg. Alkylated and aralkylated phenols.

Most simple phenolic antioxidants are much cheaper than phenol-alkanes, and many are more resistant to discolouration. Some hindered thio-bis-phenols and hydroquinone derivatives are also available; the latter are often used in unvulcanized compositions and adhesives.

Static protection against ozone

Waxes increase the threshold strain and provide non-staining protection against ozone attack under static conditions. The type and level of wax used depend on the service and storage temperature and the product specification test temperatures. As a general guide, the melting temperature of the wax should not be less than 20°C above the maximum surface temperature of the product. To be effective, the wax must bloom to the surfce of the vulcanizate and form a coherent layer. The diffusion rate of a wax in rubber increases dramatically with temperature, but its solubility also rises. Since both of these factors govern the rate of blooming, differences will be found between various waxes. For year-round protection and retention of activity over a wide temperature range, proprietary blends containing the so-called 'microcrystalline' waxes are recommended.

Waxes are more soluble in oil-extended vulcanizates and higher levels are required to give the same level of protection to the vulcanizate.

Some types of wax are given in Table 12.

TABLE 12

Waxes for static ozone resistance

Temperature	Trade examples
below 20°C	Antilux, Antilux L Okerin 444, 587
15-40°C	Antilux 654 Sunolite 666
0-40°C	Antilux 600 Sunolite 127, 154, 240 Sunproof Improved

Several non-staining antiozonants have been introduced, but these are not very effective in natural rubber. Marginal protection to NR may be obtained using Vulkanox AFS (Bayer) in conjunction with a wax, or by incorporating Permanax OZNS (Rhone Poulenc). However, some waxes reduce the effectiveness of the latter.

450

REFERENCES

Baker, C.S.L. 1988. Non-sulphur vulcanization. In: Natural Rubber Science and Technology. (ed.) Roberts, A.D. Oxford University Press. pp 457-510.

Baker, C.S.L. 1989. Vulcanization with urethane reagents. NR Technical Bulletin, MRPRA.

Barnard, D. and Lewis, P.M. 1988. Oxidative ageing. In: Natural Rubber Science and Technology. (ed.) Roberts, A.D. Oxford University Press. pp. 621-678.

Bateman, L. 1963. The Chemistry and Physics of Rubber-like Substances. McClaren Press (now Applied Science Publishers).

Chapman, A.V. and Porter, M. 1988. Sulphur vulcanization chemistry. In: Natural Rubber Science and Technology. (ed.) Roberts, A.D. Oxford University Press. pp. 511-620.

Chow, Y.W. and Knight, G.T. 1977. Delayed action peroxide vulcanization systems. Rubbercon 1977, Brighton. paper 24.

Crowther, B.G., Lewis, P.M. and Metherell, C. 1988. Compound. In: Natural Rubber Science and Technology. (ed.) Roberts, A.D. Oxford University Press. pp. 177-234.

Davis, L.H., Sullivan, A.B. and Coran, A.Y. 1987. New curing system components. Rubb. Chem. Technol., 60: 125-139.

Eholzer, U. and Kempermann, Th. 1985. Improvements of reversion resistance with post crosslinking accelerator systems. Kaut und Gummi Kunst. 38(8): 710-720.

Elliott, D.J., Skinner, T.D. and Smith, J.F. 1970. Compounding natural rubber for engineering applications. NR Technol., 1(3), No. 13.

Elliott, D.J. and Tidd, B.K. 1973/4. Development in curing systems for natural rubber. Prog. Rubb. Technol., 37: 83.

Jones, K.P., Lewis, P.M., Swift, P.McL. and Wallace, I.R. 1988. Compounding for tyres. In: Natural Rubber Science and Technology. (ed.) Roberts, A.D. Oxford University Press. pp. 283-326.

Kempermann, Th. 1988. Sulphur-free vulcanization systems for diene rubber. Rubb. Chem. Technol. 61: 422-447.

Layer, W. 1987. A postcrosslinking accelerator system for natural rubber based on thiocarbamyl sulphenamides. Rubb. Chem. Technol., 60(1): 89.

Lewis, P.M. 1972. Protection of natural rubber against ozone cracking. NR Technol., 3(1): 1-36.

Lloyd, D. 1988. Long term stability. European Rubb. J., 170(1): 27-29.

Metherell, C. 1986. A comparison of low-sulphur vulcanization systems for natural rubber. NR Technol., 17(2): 27-32.

Malaysian Rubber Producers' Research Association, 1984. Natural Rubber Formulary and Property Index, MRPRA, Kuala Lumpur, Malaysia.

Roberts, A.D. 1988. Natural Rubber Science and Technology. Oxford University Press.

Skinner, T.D. and Watson, A.A. 1967. EV systems for NR. Rubb. Age, 99(11): 76 and 99(12): 69.

Wolff, S. 1979. A new development for reversion stable sulphur-cured NR compounds. Kaut & Gummi Kunst., 32(10: 760-765.

Wolff, S. 1981. Reinforcing and vulcanization effects of silane SI69 in silica-filled vulcanizates. Kaut & Gummi Kunst., 34(4): 280-284.

CHAPTER 20

CHEMICAL MODIFICATION OF NATURAL RUBBER

D.S. CAMPBELL

MRPRA, Tun Abdul Razak Laboratory, Brickendonbury, Hertford, England.

INTRODUCTION

For the first half of this century, natural rubber occupied a unique position as an industrial material. Its properties were very useful to a rapidly developing technological society and indeed played no small part in directing technological change through the development of a pneumatic tyre. Vulcanization was centrally important to this development and the processes used to achieve the vulcanized state must always be regarded as the most important form of chemical modification that can be made to natural rubber. Vulcanization chemistry has a very extensive literature and has been reviewed recently (Chapman and Porter, 1988; Baker, 1988a). It will receive only indirect attention in this chapter, as will the oxidative modifications which occur during ageing processes of raw rubber and of vulcanizates (Barnard and Lewis, 1988).

Commercial cultivation of the rubber tree was essential to meet the growing demand for raw rubber but the increased availability in turn increased the interest in rubber as a raw material for modification into non-rubbery products. Ebonite was a long-standing example of chemical modification to a level which transformed the raw rubber to a rigid solid (Scott, 1958) and a later development was hydrochlorination to give the film-forming material, 'Pliofilm', used for a time in food packaging (Le Bras and Delalande, 1950). Scientific curiosity also ensured that the unusual and poorly understood product from the rubber tree was subjected to a wide range of reactions familiar to conventional organic chemistry.

The Second World War imposed a heavy demand on the vulcanized but otherwise unmodified material and decreased its supply. The synthetic polymer industry emerged as a major market force and natural rubber's commercial position changed. Its usage is now dictated by a complex balance of technical quality and cost. Synthetic materials which compete in one or other of these directions have become available and at the same time the older markets for chemically modified derivatives of natural rubber have become swamped by a range of synthetic polymers.

This altered climate superimposes increasingly strong economic constraints on top of the older ones of chemical and technical feasibility when the value of work on chemical modification is being assessed. On the other hand, there is an increasing awareness of the value of renewable resources to the total world economy, and natural rubber is a renewable resource. Extension of its cultivation also bears similarities to reafforestation and has similar implications in contributing to global carbon dioxide balance. There are, therefore, long term reasons for continuing to regard natural rubber as a useful raw material for modification as well as short term reasons for being interested in more specific chemical modification processes which contribute to the material's current competitive position as an elastomer. These interests are reflected in the appearance of no fewer than three recent review articles on the chemical modification of natural rubber (Baker, 1988b; Gelling and Porter, 1988; Subramaniam, 1988).

The modification of natural rubber should also be viewed in the wider context of the modification of other elastomers (Schulz et al., 1982; Brydson, 1978; Pinazzi et al., 1975) and in the much more extensive field of reactions of polymers (Marechal, 1989; Sherrington, 1988; Soutiff and Brosse, 1990). The subject has many parallels in the history of the chemical modification of cellulose (Jett, 1989).

In this chapter, some consideration will initially be given to general economic aspects of chemical modification of pre-formed polymers and the position of natural rubber within this economic spectrum. The inherent chemical reactivity of the backbone will be discussed in relation to that of other unsaturated polymers, together with the importance of the physical reaction conditions under which chemical modification is attempted. Finally, a limited selection of more recent modification processes will be discussed.

ECONOMICS OF MODIFICATION

The chemical constitution of natural rubber is determined entirely by the rubber tree. There is, as yet, no procedure for the planned incorporation of co-monomer units into the rubber backbone during the bio-synthetic process. In this respect, natural rubber is at a disadvantage relative to its synthetic competitors, which are frequently the products of carefully manipulated copolymerization processes. Therefore, it must rely entirely on modification of the pre-formed polymer when modified materials are required.

The two major categories of modification are those at low levels which maximise the utility of the inherent elastic properties of the polyisoprene backbone (eg., vulcanization) and those at high levels which cause a major change in the fundamental physical behaviour (eg., the formation of ebonite). There can be a further subdivision where high levels of modification are confined to the surface of the rubber (eg., chlorination for modification of friction behaviour and bonding) or where very low levels of modification on the backbone act as attachment sites for substantial amounts of additional polymer or filler (eg., in graft copolymer formation and carbon black or silica reinforcement).

The present-day incentives for achieving different categories of modification can be identified to some extent from the data in Table 1, where elastomer prices and other polymer prices are listed relative to that of natural rubber. The raw material is comparable in price to the more expensive grades of polyethylene (PE) and polypropylene (PP) and to the emulsion grades of styrene-butadiene rubber (SBR). It is significantly cheaper than synthetic polyisoprene (IR), butyl rubber (IIR), ethylene-propylene terpolymer (EPDM) and polychloroprene (CR) and an order of magnitude cheaper than the very highly specialized elastomers.

Several of the polymers listed in Table 1 are derived from others by chemical modification processes. The relevant pairs are shown in Table 2, together with the ratio of the market values of the modified and unmodified polymers. Differentiation between modification costs and profit-ability is not possible but since all the types of modified polymer have been available from more than one manufacturer, it can be assumed that the profit margins are not excessively high using the current modification procedures. The price ratios should therefore represent measures of the costs of the modification processes. Chlorination of polyethylene and hydro-genation of SBS block copolymer are the least expensive of the modifications. Both are based on low cost reagents (chlorine and hydrogen) yet the products are two-and-a-half to three times as expensive as the base polymers. Levels of modification must affect the cost of the product to some extent, but with low cost reagents, the technical aspects of the modification may be of considerably greater importance. The two hydro-genations use the same reagent but different catalysts and different process arrangements. The original patent coverage for producing materials related to Kraton G (hydrogenated SBS) used a relatively inexpensive heterogeneous catalyst system with moderately high hydrogen pressure, but also specified a solution viscosity range for the polymer feed (Shell, 1965). The viability of the process may well be related to the fact that the viscosity limit is not exceeded at quite high polymer concentrations (10%) and that the polymer solution for hydrogenation can be prepared directly from the monomers with no intermediate polymer separation and re-dissolution. By

TABLE 1

Relative market value of Natural Rubber[a]

Vulcanizable Rubbers	Usage (%)	Relative Price	Plastics and Thermoplastic Rubbers
		0.79	PP
		0.86	LDPE, PVC (Pipe grade)
NR, SBR (Emulsion)		1.0	PP (Copolymer), HDPE
BR		1.3	
	65	1.4	PS
SBR (solution)		1.5	SBS
IR		1.6	PET
EPDM		2.0	
IIR		2.6	
EVM, NBR		2.7	
CM	32	3.0	
		3.5	Santoprene[b]
		3.8	Kraton G[c]
CR		4.2	
		4.3	Geolast[d]
CSM		5.0	
EAM		7.0	
MVQ		8.0	
ACM	3	9.5	
		10.1	PU
HNBR		28	
FPM		78	
FMQ		140	

a Based on data from Casper and Rhode (1988) and industrial press (Anon, 1990). Abbreviations conform to general technical usage (Pethric, 1985).
b Monsanto thermoplastic rubber.
c Shell Chemicals, hydrogenated SBS.
d Monsanto oil resistant thermoplastic rubber.

TABLE 2

Ratio of base polymer value to modified polymer value for post-polymerization modification processes.

Base Polymer	Modified Polymer	Price ratio	Chemical Process
PE	CM	3	Chlorination
PE	CSM	5	Chlorosulphonation
SBS	Kraton G	2.5	Hydrogenation
NBR	HNBR	10	Hydrogenation

a comparison, hydrogenated nitrile rubbers (HNBR) are currently made by re-dissolving pre-formed NBR prior to the hydrogenation step (Nippon Zeon, 1990). The costs of re-dissolution and a second polymer recovery must therefore be included in the overall product cost.

The general principle of chemical modification of a pre-formed polymer can be commercially acceptable in the current economic climate, and could assume increasing importance as the costs of producing new monomers for the synthetic polymer industry increases. The ultimate economic judgement on the viability of any modification process for natural rubber must rest on the cost of the product relative to the technical performance that it offers. Much prediction and scientific inference can be fed into the assessment of viable areas of exploration but there is an ultimate uncertainty which is not resolved until market forces have had their say and which necessitates the acceptance of development costs which can greatly outweigh the research costs for initial chemical investigation.

TECHNOLOGY OF CHEMICAL MODIFICATION

Technical aspects of chemical modification of natural rubber seldom receive much discussion when the subject of modification is under review. This may be because some of the basically important factors are regarded as self-evident. And yet, the very large literature on modifications that have been studied at research level compared with the number of chemically modified forms of natural rubber in the market place shows that the technical development of a chemically feasible process cannot be taken for granted. The situation is no different from that which prevails in any chemical industry, although natural rubber, in common with other polydiene rubbers, is amenable to a surprisingly large variety of modification reactions.

From the perspective of chemical engineering, there are three ways in which natural rubber reactions can be managed. These are as reactions in solution, in bulk rubber and in latex.

A large proportion of chemical modification work has, in the past, started with, and proceeded no further than the study of reactions in solution. This, in itself, has not prevented the development of commercial processes, eg., the chlorination procedure for making the base material for chlorinated rubber paints (Bloomfield, 1943; Kraus and Reynols, 1950; Baker, 1988b). But natural rubber from all Hevea brasiliensis sources has a high molecular weight and an appreciable (and widely variable) gel content (Subramaniam, 1975). The process of preparing a solution is time consuming and solution viscosities are high at even modest solid content

(2-3% w/w). Technical processes involving solvents therefore tend to be dealing with rubber doughs rather than the solutions familiar in laboratory experiments. The solubility characteristics of the polymer dictate the range of solvents that can be used (Gundert and Wolf, 1989), which in turn, has a considerable influence on the progress of the modification reaction. Poor reagent solubility can transform an eminently suitable reaction of a low molecular weight model olefin into a total failure because the reaction solvent will not dissolve rubber.

Reaction by direct mixing with dry rubber has advantages over solution processes for large scale operation, and is eminently practicable in the context of vulcanization. The process demands heavy machinery to accomplish the dispersion of reactants and the energy consumed in this dispersion is of considerable significance to the rubber industry (Sears, 1984; Fuller, 1988). Aside from the ubiquitous vulcanization process, there are several examples of the study of chemical modification in bulk rubber. The copolymerization of monomers is a case where specific advantage is taken of the fact that the high energy mixing causes rupture of the rubber chains to form carbon-centred radicals (Watson, 1958; Ceresa, 1973). Grafting reactions of pre-formed functional polymers can also occur in direct mixing operations (Campbell, 1985, 1988). The phenol-formaldehyde resin cure (Cuneen, 1943; Baker, 1988b) is broadly related to grafting processes, although reaction occurs mainly during the curing cycle. Examples of direct reaction of small molecules are the mastication (Le Bras and Compagnon, 1947) and high-temperature (Pinazzi et al. 1960) reactions of maleic anhydride, isomerization caused by sulphur dioxide (Cuneen et al., 1963) and reactions of azodicarboxylate derivatives (Barnard et al. 1975; Gelling and Porter, 1988). Reagent solubility becomes even more critical when there is no mediating solvent present and the reaction environment is the low polarity polymer matrix. The formation of rubber-soluble zinc-accelerator complexes is believed to be critically important in sulphur vulcanization (Chapman and Porter, 1988) and the reactivity of the inherently soluble azodicarboxylate group can be strongly influenced by attachment to polymers wich have varying degrees of compatibility with a polyisoprene backbone (Wong et al. 1987).

In mechanical terms, reaction in latex resembles reaction in a low viscosity solution, with the distinction that the rubber content can be much higher than for a true solution. However, latex is a dispersion of micron-sized particles of essentially dry rubber in an aqueous medium and mechanical mixing only achieves dispersion of reagents in this aqueous phase.

The greater part of the total weight of rubber in latex is contained in particles which are considerably larger than the mid-range size of about 0.5 microns (Blakley, 1966). Simple geometric calculations can demonstrate that even a 0.5 micron particle contains of the order of 10^5 rubber molecules and that the layer of isoprene units on the surface represents approximately 0.5% of the total number of units in the particle. Surface reactions of latex particles by reagents which are confined to the aqueous phase can therefore affect only a small proportion of the total isoprene units unless thermal motion of the rubber molecules replenishes the particle surface with fresh units at a rate which is at least comparable with the time scale of reaction. There are no experimental results which define thermal motion of the rubber molecules within latex particles but, because the majority of chemical modifications of rubber involve some increase in the polarity at the sites of modification, it seems intuitively probable that modified isoprene units will tend to remain at the polar interface. Penetration of reagents into the rubber particles must then rely on diffusion and once the reagents is in the rubber phase the reaction is subject to the same polarity constraints that apply to dry rubber reactions.

Reactions in latex have several attractions from an engineering point of view, but there are also some major limitations. Reagents must be chemically compatible with the aqueous serum and must not interfere with the stability of the latex. The permissible range of reaction temperatures is approximately 0 to 80°C, which means that rubber phase reactions which require high temperatures for acceptable rates in dry rubber cannot be forced to completion. There have, nevertheless, been several important latex modification processes. Latex prevulcanization is probably the most ubiquitous but perhaps the least well understood in terms of reaction mechanism. Not only does the process involve separate rubber and aqueous phases but reagents (sulphur, zinc oxide and accelerator) are introduced as three additional solid phases. Heveaplus MG, the graft copolymer of natural rubber and poly(methyl methacrylate), is prepared from latex by radical initiation in the presence of the methacrylate monomer (Allen, 1963; Pendle, 1973). The most recent commercial development of a latex modification process is epoxidation (vide infra).

CHEMICAL REACTIVITY OF NATURAL RUBBER

Natural rubber hydrocarbon is almost entirely <u>cis</u>-1,4-polyisoprene. The reactive chemical unit (<u>1</u>) has a double bond between carbons C1 and C2 and a set of seven allylic hydrogen atoms on carbon C1, C4 and Cm. All five carbon atoms are potential sites for modification. Those which become

458

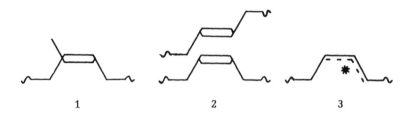

1   2   3

involved in any given reaction are determined by the details of the reaction mechanism and by the electronic and steric imbalance which is created by the presence of the methyl group. Electronic density (and hence chemical reactivity) at the double bond is influenced by hyperconjugative interaction with the methyl group (Cm) and the methylene groups (C1 and C4). The corresponding reactive unit of polybutadiene or butadiene-based copolymers (2, cis or trans) lacks the hyperconjugative interaction of the methyl group and consequently is less reactive towards electrophilic attack. The seven allylic hydrogen atoms in the polyisoprene repeat unit are activated towards abstraction by virtue of the resonance stabilization which can occur in the resultant carbanion, carbon radical or carbenium ion (eg. (3), where * represents a negative charge, an unpaired electron or a positive charge respectively).

The asymmetry caused by the methyl group increases the isomeric complexity of the products of chemical reaction compared with those obtained from the reactive unit of polybutadiene homopolymer. On the other hand, the absence of copolymer units, or indeed of isomeric methylpentenyl units, removes many of the complexities of identification which are associated with butadiene copolymers and with the less stereospecific synthetic poly-isoprenes.

The essential allylic hydrogen and double bond reactivity of natural rubber is displayed by the small molecule model, 2-methylpent-2-ene(4). However, the molecule cannot represent the stereospecific cis configuration of the polymer chain, nor can it model the interaction of adjacent repeat units during reaction. The two-unit model, 2,6-dimethylocta-2,6-diene (5), offers improved opportunity for characterizing interaction of adjacent units but neither unit is completely representative of the polymer. Progressive increase in the number of repeat units in the model gradually reduces the significance of the end groups, but this is achieved at the expense of increasing synthetic difficulties and greatly increased complications in identification of products by traditional methods of separation and analysis.

Squalene (6) has been used quite extensively to model polyisoprene reactions, mainly because of its easy availability, but the molecule has a trans configuration about all the double bonds and the head-to-head linkage in the centre largely negates the possible advantage of having six repeat units.

4   5

6

Progressive advances in the sophistication of instrumental analyses, particularly NMR, has led to the direct investigation of detailed chemical structure in the modified polymer chains themselves as an alternative to the study of multi-unit models (Bradbury and Perera, 1985; Bradbury et al., 1987; Koenig and Patterson, 1987; Perera et al., 1987; Estina et al., 1990). Estimates of the relative numbers of adjacent unit and alternate unit modifications can be made (Bradbury and Perera, 1985; Davey and Loadman, 1984) and compared with the expected frequency of occurrence from random reactions, leading to information on the degree of non-random modification for a process. Synthetic polyisoprenes with varying degrees of stereospecificity are frequently regarded as models for natural rubber. However, differences in microstructure must be continually borne in mind in such comparisons, as has been demonstrated in a study of the hydrochlorination of a synthetic polyisoprene having 93% 1,4-structure but only 71% cis-configuration (Tran and Prud'home, 1990). The natural rubber product has long been known to be crystalline as a result of the steric control which can occur in the course of the reaction (Gordon and Taylor, 1955) whereas this control is disrupted in the synthetic polymer by the presence of the trans double bonds.

Interaction between adjacent repeat units was recognized when a satisfactory interpretation of the acid-catalyzed resinification of natural rubber was given by D'Ianni (1946), based on earlier remarks by Bloomfield (1943) on the acid catalyzed cyclization of 2,6-dimethylocta-2,6-diene. The interaction takes the form of attack of an electron deficient carbenium ion - formed by initial protonation or Lewis acid attack - on the π-electron system of the adjacent double bond. Similar interaction can occur during hydrochlorination of natural rubber under polar conditions (Schultz et al., 1982) and an analogous free radical process has been invoked as a side reaction during chlorination of synthetic polyisoprene (Schultz et al. 1982). Rotation around the carbon-carbon single bonds along the polymer chain can allow the double bonds in two adjacent repeat units to become aligned. The alignment causes increased steric interaction between C1 and C4 of adjacent units. Simultaneous alignment of a third double bond introduces an additional C1 - C4 interaction, and also because of bond angle constraints, results in a bending of the overall chain configuration (7). The energetic penalty for successive alignment of the double bonds rapidly becomes very high.

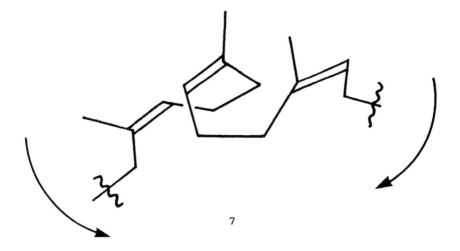

7

A further interesting feature of this alignment, as depicted by molecular models, is that the methyl groups have little influence on the steric interactions. These steric and conformational considerations offer a satisfying explanation for the fact that acid-catalyzed cyclization does not proceed in long unbroken runs along the polymer chain (Schultz, 1982) and the irrelevance of the methyl group in the steric interactions means that the higher electrophilic reactivity of the polyisoprene double bond relative to that of polybutadiene is allowed to manifest itself in a higher rate of cyclization (Shelton and Lee, 1958).

The effects of adjacent repeat units may not be as specific as in the case of acid catalyzed cyclization. The reactivity of one unit may simply be enhanced or reduced to some extent by the presence of modification on an adjacent unit, resulting in deviations from a totally random reaction along the chain. The ability to detect these deviations in some instances has encouraged attempts to predict the effects in butadiene polymers by mathematical modelling (Litmanovich, 1980; Martl and Hummel, 1990), but such modelling has not yet been applied to natural rubber.

Interaction along the chain can involve a hetero-atom as an intermediary, eg., oxygen in the course of autoxidation reactions (Barnard and Lewis, 1988) and sulphur in vulcanization chemistry (Chapman and Porter, 1988). Interaction can also arise in reactions of the products of initial modification, eg., the acid-catalyzed interaction of epoxide groups, which requires specific control in the preparation of epoxidized natural rubber (Gelling, 1985).

Although the cis-1,4-polyisoprene structure is the usual basis for discussion of the chemical modification of natural rubber, the raw material also contains numerous biological constituents which are not chemically bound to the rubber molecules (Archer et al., 1963) and there are reports of the presence of several types of attached chemical groups. Thus, the biosynthetic pathway is thought to involve an allylic pyrophosphate structure at the growing chain end (Audley and Archer, 1988) resulting in a phosphate, alcohol or carboxylic ester end group. Tanaka (1989) failed to detect this group in NR although the free alcohol was identified in isoprenoids from other sources. He has also identified the expected dimethylallyl group at the opposite end of the chain and, associated with it, a small number of specifically trans units. No significance has been attached to these groups in the context of chemical modification reactions. There is, on the other hand, ample evidence for a natural gelation or crosslinking process which occurs spontaneously on storage of rubber and which has variously been ascribed to the presence of a limited number of carbonyl groups or epoxide groups on the polymer backbone (Sekhar, 1961; Burfield, 1975; Gregory and Tan, 1975). Although the extent of chemistry involved in the gelation is limited, its physical effect is of some significance in relation to chemical modification processes when these are attempted in solution. For example, the relative ease of producing solutions of synthetic polyisoprenes, compared with solutions of natural rubber, has been a factor in diverting interest to the synthetic polymers as raw material for chlorination in the production of chlorinated rubber based paints.

The unbound biological residues also exert an influence in chemical modification, particularly when the modifications involve small numbers of isoprene units. The effects are clearly seen in the relative yields of grafting reactions on natural and synthetic polyisoprenes when specific chemical linking of pre-formed graft chains is attempted (Campbell, 1988), and have been invoked in the interpretation of quantitative crosslinking with peroxides (Bristow et al. 1965).

Table 3 illustrates the diversity of chemical reactivity that is known for the cis-1,4-polyisoprene chain. The majority of these reactions have been documented for natural rubber and are discussed in the reviews by Baker (1988b), Gelling and Porter (1988) and Subramaniam (1988). An increasing amount of work is being reported for synthetic cis-polyisoprene which, given the necessary attention to microstructural detail, the effect of gel content and the effect of non-rubbers, is closely relevant to modification of natural rubber and is extensively discussed in the review by Schultz et al. (1982). Cycloaddition reactions are defined as those which result in the incorporation of the two carbon atoms of the double bond (C2 and C3) into a 3-, 4-, 5- or 6-membered ring in the reaction product. They are classified as (x+2) in the table, where x is the number of atoms of the reagent which are incorporated into the ring system. No distinction is made as to whether or not these reactions are concerted multi-atom processes in terms of mechanistic organic chemistry.

Three areas of chemical modification will be selected for further discussion. These are: (a) epoxidation, (b) ene additions, with particular reference to azodicarbonyl addition and (c) graft copolymer formation and blending. Epoxidation is an area of current commercial interest, and represents the use of substantial levels of modification to alter the material properties of natural rubber. Azodicarbonyl addition, although currently of little economic interest, illustrates the use of a well defined reaction to achieve specific objectives. Graft copolymer formation and blending do not represent specific processes given in Table 3. These areas where chemical modification of the natural polymer chain is at a very low level but is centrally important to the properties of the product materials.

EPOXIDATION OF NATURAL RUBBER

Epoxidation with peroxyacids (Scheme 1) is a reaction which responds readily to electron availability at the double bond. The bicyclic transition state originally proposed by Bartlett (1950) is consistent with much of the original data for reaction in homogenous solution (Kahil and Prtzkow, 1973 Swern, 1971) and explains the total retention of the configuration - of the alkyl groups in the product -

TABLE 3

Classification of modification reactions of the cis-1,4-polyisoprene chain (based on Gelling and Porter, 1988).

Isomerization, rearrangement and molecular weight reduction

      Thiolacid isomerization
      Sulphur dioxide isomerization
      Photo- and radiation-induced isomerization
      Cyclization
      Mechanical chain scission
      Oxidative chain scission
      Photochemically initiated chain scission (nitrobenzene)

Simple addition to the double bond

      Hydrogenation
      Hydroformylation
      Halogen addition
      Halogen addition
      Halogen acid addition
      Halocarbon addition
      Sulphenyl halide addition
      Hydroboration
      Hydrosilylation

Substitution of allylic hydrogen

      Peroxide and accelerated sulphur crosslinking
      Autoxidation (primary steps)
      Maleic anhydride reaction (radical initiated)
      Quinoneimine and quinonediimine reaction

Cycloaddition

      Epoxidation (1+2)
      Carbene addition (1+2)
      Nitrene addition (1+2)
      Photolytic carbonyl addition (2+2)
      Chlorosulphonyl isocyanate addition (predominantly 2+2)
      Ozone reaction (primary step) (3+2)
      Nitrile oxide addition (3+2)
      Nitrone addition (3+2)
      Nitrilimmine addition (3+2)
      Sydnone addition (3+2)
      Azide addition (primary step) (3+2)
      Sulphur diimide addition (3+2)
      o-Quinonemethide addition (resin cure) (4+2)

Ene addition

      Singlet oxygen addition
      Activated carbonyl and thiocarbonyl addition
      Maleic anhydride addition (high temperature)
      Chlorosulphonyl isocyanate addition (minor reaction)
      C-Nitroso addition
      Azodicarbonyl addition

epoxide (Witnaur and Swern, 1950). Molecular orbital calculations support a transition state of this type for reaction of ethylene with peroxyformic acid, although they suggest that the interacting oxygen is not symmetrically placed between the two carbon atoms (Plesnicar et al., 1978). Dryuk (1976) proposes the formation of an association complex between the double bond and the interacting oxygen atom and also discusses the effects of interaction of the complex with the reaction medium.

Epoxides are susceptible to both nucleophilic and electrophilic attack (Grozynski-Smith, 1984), the balance being in favour of electrophilic attack for the trialkyl substituted epoxide structure of epoxidized natural rubber. This characteristic susceptibility to secondary reaction under acidic conditions was responsible for a considerable amount of confusion in the early identification of the products of reaction of natural rubber with peroxyacids and other epoxidizing reagents. The subject has been reviewed by Gelling and Porter (1988) and, in the general context of unsaturated rubbers, by Greenspan (1964) and by Schultz et al. (1982).

Current interest in epoxidized natural rubber (ENR) has been stimulated by the realization that the material can be obtained in quantity, at a commercially realistic cost, by epoxidation of the rubber as latex. Epoxidation of unsaturated rubber latices is not, in itself, novel (Colclough,1962; Hercules, 1962). The advances that have been made (Gelling, 1982; Gelling, 1985; Burfield et al., 1984a; Gan and Ng, 1986) are primarily related to the specification of conditions which will prevent, or minimize secondary reactions. Scheme 2 summarizes reactions of rubber epoxide which can be initiated by protic acid in the presence of the oxygenated species of an aqueous peroxyacid epoxidation mixture. The presentations are as ionic processes but it is unlikely that fully ionized species are involved in the majority of the reaction situations. Reactions a, b and c represent alternative consummation of the protonated oxirane ring by carboxylic acid, water and a second epoxide group respectively. The last of these is not commonly observed in the reactions of low-molecular-weight epoxides but there is evidence that it is responsible for gelation reactions in the polymer situation, where there may be more restricted access of water and other oxygenated species (Gelling and Morrison, 1985). Reaction d is one possible representation of an intramolecular analogue of reaction c. The furanoid structure is considered to be predominant cyclic ether formed (Gelling and Porter, 1988; Gan and Ng, 1981), although published evidence for differentiation of furan, pyran and ether crosslink structures is not unambiguous. The formation of cyclic ethers only occurs in samples which contain oxirane structures on adjacent repeat units of the polymer chain, and thus increases considerably in importance as the level of epoxidation increases. At high levels of epoxidation the cyclization can proceed along the chain until it is interrupted by either an unreacted double bond or the intervention of water or some other nucleophilic species.

Two reagents can be considered for the epoxidation process on a tonnage scale. These are commercially available peroxyacetic acid or a mixture of formic acid and hydrogen peroxide. Pre-formed peroxyacetic acid can give essentially quantitative conversion to epoxide relative to the amount of peroxyacid used (Gelling, 1982; Burfield et al., 1984a) but results in the formation of a molar equivalent of acetic acid in the product latex. This acid would have to be recovered in an additional processing step to prevent serious effluent problems with the reaction waste. A mixture of formic acid and hydrogen peroxide forms peroxyformic acid spontaneously. The epoxidation reaction is then more or less stoichiometric in hydrogen peroxide but can proceed with a deficiency of formic acid because the acid is regenerated in the epoxidation step. The over-all sequence is shown in Scheme 3, where a differential is made between formation of the peroxyacid in the aqueous phase of the reaction medium and interaction with the double bond in the rubber phase of the latex particles.

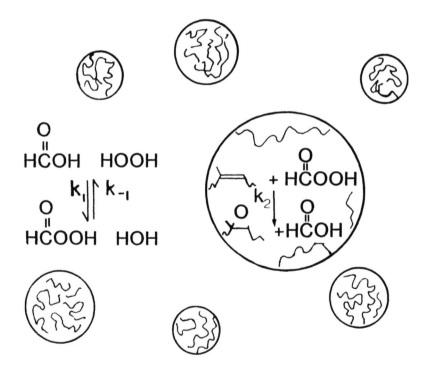

Chemical intuition provides strong justification for this segregation. Formic acid and hydrogen peroxide are highly water-soluble materials and their interaction is catalyzed by proton acid. Their interaction to form peroxyformic acid within the rubber phase of the latex particles therefore seems unlikely. Gan and Ng (1986) presented this reasoning and also offered a kinetic analysis which predicted a first order dependence on hydrogen peroxide concentration (and zero order dependence on formic acid concentration) if the formation of peroxyformic acid is the rate controlling step in the over-all reaction. Their experimental results were consistent with this interpretation. A similar conclusion had been reached by Dittmann and Hamann (1971) for the epoxidation of polybutadiene oligomers. A more detailed investigation (Campbell, 1989) confirmed that the reaction rate and activation parameters for the interaction of formic acid and hydrogen peroxide in aqueous solution are almost identical to the rates and parameters for the epoxidation of both natural rubber latex and synthetic polyisoprene latex (Table 4). A closer study does, however, show that the epoxidation reaction is not far removed from a situation where the rate of interaction of the peroxyformic acid with the double bonds in the rubber phase has a significant influence on the over-all reaction rate.

TABLE 4

Kinetic parameters for formic acid/hydrogen peroxide reactions (Campbell, 1989).

	Equilibrium[a] reaction	Epoxidation of[b]	
		HA Latex	IR Latex
$10^5 k_1$ at $25°C$ (dm^3 mol^{-1} s^{-1})	0.252	0.252	0.245
Ea (kJ mol^{-1})	54.3	56.3	53.0
ΔH^* (kJ mol^{-1}) at $35°C$	51.4	53.7	50.4
ΔG^* (kJ mol^{-1}) at $35°C$	106.5	106.5	106.8
ΔS^* (J mol^{-1} K^{-1}) at $35°C$	-179.8	-171.5	-182.9

[HCOOH]o : a 1.37 mol dm^{-3} [HOOH]o : a 2.88 mol dm^{-3}
           b 1.26 mol dm^{-3} b 3.07 mol dm^{-3} drc 19.1%

The material properties of ENRs are closely related to two character-istic features of the epoxide modification. The glass transition temperature (Tg) of the polymer increases linearly with the epoxide content. The proportionality constant for the relationship is variously given as 1 (Davey and Loadman, 1984), 0.93 (Gelling and Porter, 1988) and 0.85 (Burfield et al., 1984b), the uncertainties largely arising from the difficulties of precise estimation of epoxide content. Secondly, the epoxidized material retains an unexpectedly high ability to achieve strain-induced crystallization (Davies et al. 1983), resulting in some of the high tensile strength characteristics that are associated with vulcanizates of unmodified natural rubber. This crystallization ability is explained by the high retention of stereoregularity which accompanies the conversion of the double bonds to epoxide groups and to the fact that the oxygen atom occupies a sufficiently small physical space to allow formation of the crystalline lattice with only a minimum of alteration to the unit cell dimensions (Davies et al. 1983). However, the effect of the oxygen atom on crystallization is not totally absent, and can be detected as a reduction in the rate of low-temperature crystallization of the rubber (Lee, 1980; Lee and Porter, 1979).

Vulcanization procedures of ENR, vulcanizate properties and potential applications have been discussed by Baker et al. (1985), Gelling and Morrison (1985) and by Gelling and Porter (1988). Space does not allow repetition of these discussions in the present review but it is appropriate to emphasize that epoxidationoffers a range of new materials, depending on the level of modification that is produced. A full assessment of the compositions that are most appropriate for different uses must take time, and it is probable that only a limited number of compositions adequately covers all the areas of technological interest for the materials. Commercial production clearly cannot offer a large number of different compositional grades and for the foreseeable future it seems likely that production will centre around ENR-25 and ENR-50, with 25 and 50 mole per cent epoxide content respectively.

ENE ADDITIONS AND AZODICARBONYL COMPOUNDS

The net effect of the ene addition process is to form a bond at one of the carbon atoms of a carbon-carbon bond system and to transfer one of the allylic hydrogen atoms, with resultant migration of the double bond (Hoffmann, 1969). The generalized process is represented by Scheme 4 for one of the possible modes of hydrogen transfer from a polyisoprene chain.

These reactions require no catalysis and do not give high charge separation in the transition stage. They proceed smoothly and can be readily controlled in the polymer environment and usually result in stable products. Early examples of the reaction being specifically recognized in polyisoprenes were the high-temperature uncatalyzed reaction of maleic anhydride (Pinazzi et al. 1963) and the reactions of aromatic C-nitroso compounds (nitroso-phenols and nitrosophenylamines) (Knight and Pepper, 1971). The latter compounds formed the basis of urethane crosslinking systems for natural rubber (Baker, 1988a) and rubber-bound antioxidant systems (Cain et al. 1972) respectively but had a disadvantage that the initial reaction product (a hydroxylamine) was subject to decomposition reactions during the modi-fication procedure. The decompositions were sources of decreased crosslinking efficiency in the case of the crosslinking system and extensive discolouration in the bound antioxidant applications.

Considerable improvements were made in controlling the extent of these secondary reactions, and the improvements were incorporated into the NOVOR crosslinking system for natural rubber, but at the same time, alternative ene reagents were sought whose primary products from reaction with the rubber chain were more stable than the hydroxylamines. The family of reagents which received greatest attention was that of the azodicarbonyl compounds (8). Ene reaction with the polyisoprene chain results in attachment

$$-\overset{\overset{\displaystyle O}{\|}}{C}-N=N-\overset{\overset{\displaystyle O}{\|}}{C}- \qquad\qquad Ph-NH-\overset{\overset{\displaystyle O}{\|}}{C}-N=N-\overset{\overset{\displaystyle O}{\|}}{C}-OEt$$

8   9

of a carbon atom to one of the doubly bonded nitrogens and an allylic hydrogen to the other. The resultant 1-alkylhydrazine-1,2-dicarbonyl derivatives have considerable thermal stability and remain unchanged under the reaction conditions and under many of the conditions to which the modified rubber may subsequently be exposed.

Like epoxidation, the ene reaction of azodicarbonyl compounds is an electrophilic process which is favoured by electron availability at the allylic centre but it is also hindered by steric cluttering around the double bond (Hoffmann, 1969). The trialkyl substitution pattern of natural rubber together with the cis configuration of the bulky chain segments offers a favourable compromise of steric and electronic effects which results in high reaction rates relative to butadiene rubbers. The mode of substitution at either end of the azodicarbonyl system also has a considerable influence on the rate of reaction, as is shown in Table 5 for reaction with the small-molecule model, 2-methylpent-2-ene. Electronic and steric factors are both relevant to the reactivity. Some polarization of the nitrogen-nitrogen double bond system by asymmetric substitution is an advantage and a ring system constraining the nitrogen double bond to a cis con-figuration has a very profound activating effect (reaction of 4-phenyl-triazolinedione). The relative reactivities have been discussed at greater length by Gelling and Porter (1988). Several observations point to the importance of the free amide hydrogen atom in enhancing reactivity of the compounds containing the structure -NH-CO-N=N-CO-OR.

TABLE 5

Second order rate constants for the addition of azodicarbonyl compounds to 2-methylpent-2-ene in benzene at 100°C (Barnard et al. 1975).

Azo compound	k^2 $(dm^3 mol^{-1} s^{-1})$ x 10^4
PhCO.N = NCO.Ph	2.20
EtOCO.N = NCO.OEt	2.84
PhNHCO.N = NCO.NHPh	3.95
PhCO.N = NCO.OEt	5.53
BuNHCO.N = NCO.OEt	19.5
PhNHCO.N = NCO.OEt	109.5
N = N O N O Ph	ca 10^5

A combination of an asymmetric azodicarbonyl compound with a trialkyl olefin results in six possible isomeric structures for the ene products if attachment of the methyl substituted carbon to either of the nitrogen atoms is taken into consideration. This latter situation is not favoured (Hoffmann, 1969), but there remains four possible isomeric structures for the products. Model chemistry and spectroscopic observations of modified rubbers show that reaction preferentially occurs by transfer of hydrogen from the methyl group rather than a methylene group and that, where amide derivatives are involved, the nitrogen atom nearest to the amide carbonyl becomes attached to the carbon atom of the allylic double bond system.

The compound, ethyl N-phenylcarbamoylazoformate (ENPCAF, 9) is soluble in dry rubber. The rate of reaction is modest below 100°C but at 120°C reaction can be completed in a matter of minutes. There is there-fore a margin for uniform incorporation of the material into dry rubber by standard mixing procedures and also a sufficient reactivity to allow completion of the reaction as part of the same mixing cycle. The high reactivity of 4-phenyltriazoline-2,3-dione precludes homogeneous modification in this way and confines the utility of this compound to surface treatment. The triazolinedione has, however, been used for homogeneous treatment of butadiene rubbers, where the lower reactivity of the dialkyl substituted double bond reduces reaction rate.

Substitution in the benzene ring of ENPCAF offers the opportunity of introducing functional groups on to the rubber. The substitutions cause some variation in reactivity which can be described by a Hammet relation-ship having a value of = +0.6 in benzene solution at 39.6°C (Gelling and Porter, 1988). The sign of the Hammet constant is in agreement with the proposed electrophilic character of the reagents. This potential for the introduction of functional groups is summarized in Table 6, which also

TABLE 6

Azodicarbonyl reagents for chemical modification of Natural Rubber (Gelling and Porter, 1988).

Functional group Y in $Y.NH.CO.N:N.CO_2Et$	Intended purpose
$p—MeO—C_6H_4—$	
$p—O_2N—C_6H_4—$	Crystallization
$C_6H_{11}—$	resistance
$p—Me.CO—C_6H_4—$	Crosslinking by bis-hydrazides
$p—HO_2C—C_6H_4—$	Ionic crosslinking
$p—Bu^tO_2C—C_6H_4—$	
$p—EtO_2C.CH:CH—C_6H_4—$	Photocrosslinking
$EtO_2C.N:N.CO_2CH_2CH_2—$	Differential reactivity
$p—EtO_2C.N:N.CO.NH—$ $—C_6H_4.CH_2.C_6H_4—$	Crosslinking in latex
$EtO_2C.N:N.CO.NH.CH_2—$ $—CMe_2.CH_2.CHMe.(CH_2)_3—$	Crosslinking at room temperature
$p—(MeO)_3Si(CH_2)_2—C_6H_4—$	Adhesion to silica and silicates
$(EtO)_3Si(CH_2)_3—$	

Functional group Z in $Z.CO.N:N.CO_2Et$	
$Cl(CH_2)_3O—$	Thermolabile
$Br(CH_2)_3O—$	crosslinking by
$Br(CH_2)_{11}O—$	quaternization

472

includes functional azodicarbonyl reagents not derived from ENPCAF. Much of this work is not freely available in the open literature, apart from an extended summary given by Gelling and Porter (1988). Only two of the modifications will be discussed further here. These are the modification with ENPCAF itself and the behaviour of the silane coupling reagent ethyl N-[3-triethoxysilyl-propyl]carbamoylazoformate (10) which became known as SILCAF during the investigation work.

The effects of progressive modification with ENPCAF are summarised in Fig. 1 and Table 7 (from Gelling and Porter, 1988). Low temperature

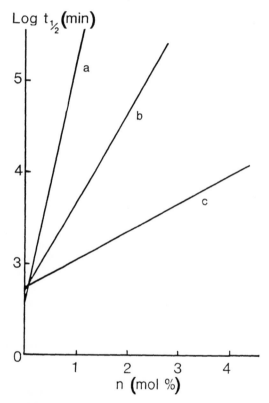

Fig. 1. Effect of concentration (n) of modifying groups on crystallization half life ($t_{1/2}$) at -26°C for Natural Rubber (Lee and Porter, 1979). a, C-C crosslinking; b, ENPCAF modification; c, cis,trans-isomeris-ation.

crystallization is retarded, and rebound resilience, nitrogen permeability, stress relaxation and linear swelling in light petroleum are all affected. For comparison on a molar basis, these effects are all considerably greater than the effects of epoxidation. There is, however, a factor of 13.8 difference between the atomic weight of oxygen and the molecular weight of ENPCAF, which results in a considerably smaller efficacy of ENPCAF on a weight for weight basis. This, together with the expense of the nitrogen chemistry necessary for its synthesis, prevented a commercial interest in the gross modification of rubber properties by this route.

TABLE 7

Selected properties of natural rubber modified with ENPCAF (vulcanized with dicumyl peroxide) (Gelling and Porter, 1988).

	mol % modification			
	0	4	9	12
Recovery (in 10 min after extension to 100% for 120 h)	99.0	97.8	–	29.0
Resilience (temperature of minimum rebound, 0°C)	–32	–18	0	+10
Permeability constant for nitrogen (cm^2 s^{-1} atm^{-1}) (x 10^8)	3.5	2.3	1.1	0.7
Stress relaxation (% per decade at 100% extension)	0.9	1.6	3.0	28
Linear swelling (l/lo) in light petroleum at 20°C	1.51	1.36	1.17	–

The molecular formula of SILCAF approaches the minimum molecular weight that can accommodate the azodicarbonyl function and the trialkoxysilane function necessary for interaction with silica surfaces. A synthesis is possible from the commercially available 2-aminopropyl triethoxysilane (Dawes and Rowley, 1978) and the material has a high efficiency of reinforcement in silica-filled vulcanizates compared with commercially available reagents (A189, Union Carbide, 11; Si69, Degussa, 12) which rely on interaction with sulphur vulcanization for their effect.

The enhancement of modulus in silica-filled natural rubber vulcanizates by the three reagents is compared in Fig. 2 on a weight for weight basis. On a molar basis, SILCAF is at least five times more effective than Si69. The increased efficiency, and the fact that the reinforcement mechanism does not confine the use of the material to sulphur vulcanization systems, might counterbalance the relatively high synthetic costs, but commercial development has been hindered by the occurrence of a slow, and as yet unidentified decomposition reaction which is not simply related to hydrolysis of the trialkoxysilane.

474

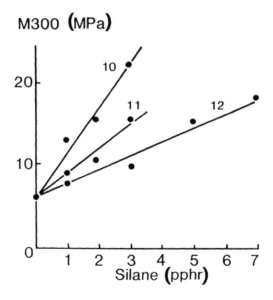

Fig. 2. Effect of SILCAF (10), mercaptosilane (11) and tetrathiodisilane (12) on modulus in silica-filled Natural Rubber (Dawes and Rowley, 1978).

GRAFT COPOLYMERS AND BLENDS

The formation of graft copolymers from natural rubber can be viewed as giving high or low levels of modification, depending on whether one is considering the overall weight composition of the product or the number of isoprene units of the backbone chain that become modified in the grafting process. In blends with other polymers - both plastics and rubbers - the amount of chemical interaction may be very small indeed. Blends of natural rubber with other rubbery polymers, where vulcanization is centrally important to the product properties, will not be considered here. The discussion will be confined to graft copolymers and blends with polymers which are above their glass transition temperatures or crystalline melting temperatures at ambient temperature.

Graft copolymerization on to natural rubber received a good deal of attention between 1950 and 1965, at a time when free radical polymerization and copolymerization processes were being extensively developed within the synthetic polymer industries. Allen (1963) has discussed the complexities of free radical grafting chemistry on polyisoprenes for initiation by conventional free radical initiators and by γ-radiation. Most of this work was centred around the graft copolymerization of methyl methacrylate, although attention was also given to styrene, alkyl acrylates, higher alkyl methacrylates, vinyl acetate, vinylidene chloride, and acrylonitrile (Bloomfield et al. 1954). The four possible modes of formation of a grafting site can be represented by the reactions in Scheme 5. Additional modes, involving combination of a growing polymer radical with a radical site on the rubber chain (Kobryner and Banderet, 1959) have been discounted by Allen et al. (1959).

$$I^{\bullet} + \quad -C(CH_3)=CH-CH_2- \longrightarrow IH + -C(CH_3)=CH-\overset{\bullet}{C}H- \qquad a$$

$$I^{\bullet} + \quad -C(CH_3)=CH-CH_2- \longrightarrow -\overset{\bullet}{C}(CH_3)-CHI-CH_2- \qquad b$$

$$(M)n^{\bullet} + -C(CH_3)=CH-CH_2- \longrightarrow (M)_nH + -C(CH_3)=CH-\overset{\bullet}{C}H- \qquad c$$

$$(M)n^{\bullet} + -C(CH_3)=CH-CH_2- \longrightarrow -\overset{\bullet}{C}(CH_3)-(M)_nCH-CH_2- \qquad d$$

For methyl methacrylate copolymerization, the two processes involving the growing polymer radicals (c and d) only make a significant contribution when the reaction medium has a high viscosity, eg., in monomer-swollen rubber phase at temperatures not far above ambient. Differentiation between radical addition of, and hydrogen atom abstraction by initiator radicals is not easy, but Allen et al. (1959) conclude that the two contributions are about equal for initiation by benzoyl peroxide.

Procedures for preparing graft copolymer in both monomer-swollen dry rubber and monomer-rubber latex mixtures were investigated. The latex processes became favoured for their practical simplicity and a good deal of effort was expended in assessing the utility of different initiation systems and reaction conditions. The result was the materialization of Heveaplus graft copolymer products, particularly the graft copolymer with methyl methacrylate which was commercialized as Heveaplus MG. Pendle (1973) has reviewed the properties and applications of these materials, with particular reference to latex preparation procedures and uses of the modified latex. Hourston and Roumaine (1989) have recently discussed composite materials prepared by polymerization of styrene in natural rubber latex.

In parallel with the work on conventionally initiated graft co-polymerization, Watson (1958) demonstrated that rubber radicals generated by mechanical scission of rubber chains in the course of mastication were capable of initiating the polymerization of vinylic monomers. The process initially produces a linear block copolymer radicals consisting of the rubber chain segment and the growing vinyl polymer chain. Termination by combination forms an A-B-A triblock copolymer where the A segments are rubber and the B segments are vinyl polymer. Alternatively, termination by disproportionation gives A-B diblock copolymer, or termination by abstraction from another rubber chain gives A-B diblock copolymer and a rubber radical which subsequently gives rise to graft copolymer. The over-all product from such a process is clearly a complex mixture of block and graft copolymers and, although the wide applicability of the process has been investigated, along with other mechanically related initiation processes such as vibro-milling and ultrasonic irradiation, well defined products are difficult to obtain in any quantity. The subject has been reviewed by Ceresa (1962, 1973) and, in a wider context, by Sohma (1989).

Even a superficial inspection of the kinetic relationships that are involved in the description of free radical graft copolymerization is sufficient to illustrate the difficulties of controlling the process to give well-defined products. The production of Heveaplus MG can be managed to give high yields of graft product with very little free poly(methyl methacrylate) but it is not possible to ensure that all the rubber chains become involved in the grafting reaction and it is also not possible to effectively control the number of graft sites and the molecular weight of the graft chains.

Improvements in the control of graft structure can be obtained by segregating the two steps of construction of the graft chain and assembly of the chains on the backbone molecule. The approach has been adopted for totally synthetic graft copolymers in the concept of 'macromer' copolymerization (Milkovich, 1980; Schultz and Milkovich, 1986; Yamashita et al. 1984), where a polymer chain is prepared with a terminal polymerizable group. The macromer is introduced into a polymerizing monomer system and graft copolymer is generated by copolymerization of the macromer end group and the monomer. The principle can be used for any polymerization mechanism but has been most commonly applied to free radical systems. The molecular weight of the graft chain is the molecular weight of the macromer, and is controlled by the polymerization conditions during macromer synthesis. The number of graft chains per backbone molecule is determined by the proportion of macromer to monomer in the final copolymerization and the reactivity ratios of the monomer and the macromer. The approach is clearly not applicable to the preparation of graft copolymers of natural rubber, where the backbone chain has already been formed by biosynthesis. A related approach was, however, tentatively put forward by Campbell et al. (1975) and was subsequently shown to have some range of applicability to natural rubber and other diene polymers (Campbell, 1988; Campbell et al. 1979).

The end group in the pre-formed polymer was chosen as one which would participate in molecular reaction with the unsaturated rubber backbone, rather than in copolymerization. Azodicarbonyl structures were favoured candidates and the azodicarboxylate group was initially chosen in preference to the carbamoylazoformate structure of ENPCAF and SILCAF for reasons of synthetic simplificity. The group can be synthesized hydroxyl functional polymer by either of the sequences summarised in Scheme 6 (Campbell et al. 1984a; Campbell, 1988).   Route be is experimentally more convenient and relies on the ability to prepare the highly reactive compound 2-ethyxycarbonylazocarbonyl chloride as a reagent solution (Campbell et al. 1983). The group was shown to have adequate reactivity for efficient grafting, given sufficient time, in solution (Campbell et al. 1984b). Within the limitations discussed below, it also allowed grafting on a time scale of minutes when solvent was omitted from the reaction system and the functional polymer and unsaturated rubber were mixed directly in an internal mixer at temperatures in the range 120 to 150°C (Campbell et al. 1981; Campbell, 1988).

The procedure works well for graft chains of molecular weights up to about 15,000. At higher molecular weights, the end group concentrations become sufficiently low to be sensitive to impurities in either the pre-polymer preparation or the grafting reaction. This is particularly so in reactions with whole natural rubber, where a number of the non-rubber constituents are specifically reactive towards the azodicarboxylate group (Coomarasami, 1980; Campbell, 1988).

For reaction in solution, the solvent acts as a mediator which helps to overcome incompatibility between the two polymer types. In the absence of solvent, no such mediation occurs, and it is not immediately obvious that the backbone and graft chain polymers will interact on a molecular level and allow the chemical grafting process to take place. Early work, with polystyrene as reactive polymer, demonstrated that the necessary inter-action could be induced under the high shear conditions of an internal mixer. Subsequent work with other functional polymers showed that the polystyrene process was close to a limiting case. The small increase in polymer-polymer incompatibility involved in changing from polystyrene reactive polymer to poly(methyl methacrylate) reactive polymer resulted in a requirement for a considerably higher mixing temperature to overcome the polymer-polymer interaction barrier (Campbell, 1988; Campbell and Seow, 1990). Further change, to a polyester functional polymer, gave a system which could not be induced to react at all by direct mixing but the same polyester reacted satisfactorily with epoxidized natural rubber (Tinker, 1981; Campbell, 1988). Correlation of these observations with solubility parameter differences for the various polymer pairs leads to the conclusion that grafting by direct mixing is only possible when the difference in solubility parameters is no greater that about 1.2.

The material properties of polystyrene grafts of natural rubber prepared by this procedure were not dissimilar to those of styrene-diene-styrene triblock copolymers. Elastic behaviour was obtained in the composition range from 20 to 50%, w/w, polystyrene, with a progressive increase in low extension modulus and yield behaviour as the polystyrene content increased. Grafting yields were, predictably, lower for natural rubber than for a synthetic polyisoprene with a negligible content of interfering non-rubbers (Cariflex IR305). However, the natural rubber products continued to exhibit extensive strain crystallization which resulted in higher modulus and tensile strength values at the same total polystyrene content (Fig. 3).

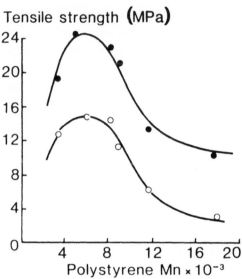

Fig. 3. Variation of tensile strength with polystyrene molecular weight for graft copolymers of polystyrene with Natural Rubber (o) and Cariflex IR305 polyisoprene (o); whole polymer from direct mixing reaction; 40% w/w polystyrene.

In comparison with the corresponding block copolymers, the graft copolymers had poorer stress relaxation and recovery properties. The deficiencies have been interpreted (Campbell, 1985) in terms of free backbone chain ends and the essentially random distribution of chain lengths between graft sites that are a necessary consequence of the grafting process. These considerations, together with the expense of the end-group chemistry prevented further development of the materials, although a limited technological compounding exercise was carried out. The economics of the polystyrene grafts had to be assessed against the relatively inexpensive styrene-diene block copolymers. However, the grafting procedure provided access to a range of materials which have never been fully assessed for useful properties.

480

The work on pre-polymer grafting was part of an effort to introduce natural rubber into the new and growing field of thermoplastic rubbers. An earlier concept of obtaining thermoplastic behaviour via networks with exchangeable crosslinks had been discarded (Campbell, 1973; Bain and McCall, 1976; Polysar, 1976), but during the investigation of the grafting process, a class of commercial thermoplastic rubbers emerged which were based on blends of rubbers with polyolefins. It was shown at an early stage (Campbell et al. 1978) that analogous materials could be obtained by blending natural rubber with polyethylene and polypropylene. This general class of thermoplastic rubbers is now commercially important in the rubber and plastics industries and natural rubber's position within the area has been continually under examination (Elliot and Tinker, 1988).

Blends of natural rubber with polypropylene (or polyethylene) vary considerably in modulus depending on their composition (Figure 4). It is

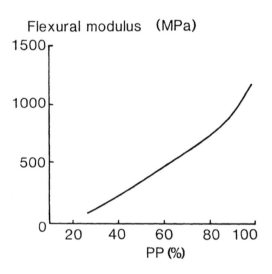

Fig. 4. The effect of polypropylene content on the flexural modulus of Natural Rubber - polypropylene blends (Elliott and Tinker, 1988).

possible to show (Elliott and Tinker, 1988; Tinker et al. 1989) that rubber and polyolefin form continuous phases at the lower polyolefin levels but at high polyolefin levels, the rubber phase becomes discrete. There is an arbitrary differentiation between soft blends, which exhibit something of the properties of rubbers, and hard blends which are more rigid materials with a greater ability to be self-supporting and one of whose main material characteristics is resistance to impact damage. Materials of

481

some interest can be prepared simply by direct mixing of polyolefin and rubber at temperatures above the melting temperature of the polyolefin. The importance of modification chemistry in such a process is difficult to establish, although interpolymers formed by combination of mechanically generated radicals may be formed and may make a contribution to reduction of interfacial energy and improvement of interfacial bonding. Chemistry attains greater significance when crosslinking or grafting is intentionally promoted during the blending process. Levels of crosslinking, rates of crosslinking relative to rates of dispersion and, in the case of polypropylene blends, rates of polyolefin degradation, become important in determining both the useful service properties and the processing behaviour of the final blend.

As work on these materials has progressed, there has been an increasing awareness that their successful use is as much dependent on the proper choice of processing conditions as on the preparation of the materials in the first instance (Tinker et al. 1989). The finer aspects of their properties are intimately related to the morphology of the blends and the dimensions of the phase structures which are initially established under the high shear conditions of their preparation. This morphology must be maintained by similar high shear conditions during processing. For this reason, there tends to be close interaction between producer and consumer. Consequently, there is only a limited amount of information about the commercially successful blends in the open literature. Natural rubber based materials have entered into this commercial relationship. Ranges of materials under the trade names TELCAR DVNR and VITACOM DVNR are available from Teknor Apex Company in the USA and Vitacom Ltd. in the UK, respectively.

SUMMARY

Chemical modification of natural rubber has had a long and varied history. The selection of topics for consideration in this chapter has been somewhat arbitrary, and predictably biased towards activities at the Malaysian Rubber Producers Research Association. Chemical treatments of rubber for control of bulk viscosity have been largely ignored, as has the process for reduction of nitrogen content by protein digestion. Both of these are important practical treatments of natural rubber which legitimately come under the heading of chemical modification in its fullest sense. Thermoplastic rubber blends have been included as a logical complement to some specific chemical work on graft copolymer formation, although physical processes are certainly of great importance during the preparation of these products.

Chemical ingenuity has, in the past, presented a wealth of reactions which can be more or less successful on a laboratory scale. There is a high probability that some of these reactions could benefit from further investigation either because advancing expertise has made us more able to understand and control the process - as has happened with epoxidation - or because changing circumstances make previously unattractive processes more interesting. Whether a 'green revolution' and a diminishing supply of petrochemicals can influence the latter situation remains to be seen.

REFERENCES

Allen, P.W., Ayrey, G., Moore, C.G. and Scanlan, J. 1959. Radiochemical studies of free-radical vinyl polymerizations. Part II. The polymerization of vinyl monomers in the presence of polyisoprenes: use of C^{14}-labeled initiators to determine the mechanism of graft interpolymer formation. J. Polym. Sci., 36: 55-67.

Allen, P.W. 1963. Graft copolymers from Natural Rubber. In: L. Bateman (Ed.). The Chemistry and Physics of Rubber-like Substanc es, Maclaren and Sons Ltd., London. pp. 97-134.

Archer, B.L., Barnard, D., Cockbain, E.G., Dickenson, P.B. and McMullen, A.I. 1963. Structure and composition and biochemistry of Hevea latex. In: L. Bateman (Ed.). The Chemistry and Physics of Rubber-like Substances, Maclaren and Sons Ltd., London. pp. 61-72.

Audley, B.G. and Archer, B.L. 1988. Biosynthesis of Rubber. In: A.D. Roberts (Ed.). Natural Rubber Science and Technology, Oxford University Press, Oxford. pp. 35-62.

Anon. 1990. Plastics and Rubber Weekly Issue 1323, p 6; Issue 1324, p 4.

Bain, P.J.S. and McCall, E.B. 1976. Polymer compositions. British Patent, 1 439 618.

Baker, C.S.L. 1988a. Non-sulphur vulcanization. In: a.d. Roberts (Ed.). Natural Rubber Science and Technology, Oxford University Press, Oxford. pp. 457-510.

Baker, C.S.L. 1988b. Modified Natural Rubber. In: A.K. Bhowmick and H.L. Stephens (Eds.). Handbook of Elastomers, New Developments and Technology, Marcel Dekker Inc., New York. pp 37-73.

Baker, C.S.L., Gelling, I.R. and Newall, R. 1985. Epoxidized Natural Rubber. Rubber Chem. and Technol., 58: 67-85.

Barnard, D., Dawes, K. and Mente, P.G. 1975. Chemical modification of natural rubber: past, present and future. In: Proc. Int. Rubb. Conf., 1975, Kuala Lumpur, Vol. IV, Rubber Research Institute of Malaysia, Kuala Lumpur. pp. 215-234.

Barnard, D. and Lewis, P.M. 1988. Oxidative ageing. In: A.D. Roberts (Ed.). Natural Rubber Science and Technology, Oxford University Press, Oxford. pp. 621-678.

Bartlett, P.D. 1950. Recent work on the mechanism of peroxide reactions. Record Chem. Prog., 11: 47-51.

Blackley, D.C. 1966. High Polymer latices, Vol. 1, Fundamental Principles. Maclaren and Sons Ltd., London.

Bloomfield, G.F. 1943. Rubber, polyisoprenes and allied compounds. Part 4, The relative tendencies towards additive and substitutive reactions during chlorination. J. Chem. Sco., 289-196.

Bloomfield, G.F., Merrett, F., Popham, F.J. and Swift, P.McL. 1954. Graft copolymers derived from Natural Rubber. In: T.H. Messenger (Ed.). Proceedings of the Third Rubber Technology Conference. W. Heffer and Sons Ltd., Cambridge. pp. 185-195.

Bradbury, J.H. and Perera, M.C.S. 1985. Epoxidation of Natural Rubber studied by NMR spectrosopy. J. Appl. Polym. Sci., 30: 3347-3364.

Bradbury, J.H., Elix, J.A. and Perera, M.C.S. 1987. Nuclear magnetic resonance study of the hydrobromination reaction and the microstructure of hydrobrominated Natural Rubber. Polymer, 28: 1098-1104.

Bristow, G.M., Moore, C.G. and Russel, R.M. 1965. Determination of degree of crosslinking in natural rubber vulcanizates. Part 7. Crosslinking efficiencies of di-t-butyl and dicumyl peroxides in the vulcanization of Natural Rubber and their dependence on the type of natural rubber. J. Polym. Sci. A 3: 3893-3904.

Brydson, J.A. 1978. Rubber Chemistry. Applied Science Publishers, London.

Burfield, D.R. and Gan, S.N. 1975. Nonoxidative crosslinking reactions in Natural Rubber. 1. Determination of crosslinking groups. J. Polym. Sci., Polym. Chem. Edn., 13: 2725-2734.

Burfield, D.R., Lim, K.L., Law, K.S. and Ng, S. 1984a. Analysis of epoxidized natural rubber. A comparative study of d.s.c., n.m.r., elemental analysis and direct titration methods. Polymer, 25: 995-998.

Burfield, D.R., Lim, K.L. and Law, K.S. 1984b. Epoxidation of Natural Rubber Latices; Methods of preparation and Properties of Modified Rubbers. J. Appl. Polym. Sci., 29: 1661-1673.

Cain, M.E., Gazely, K.F., Gelling, I.R. and Lewis, P.M. 1972. Rubber Chem. and Technol., 45: 204-221.

Campbell, D.S. 1973. Exchange reactions as a basis for thermoplastic behaviour in crosslinked polymers. Br. Polym. J., 5: 50-62.

Campbell, D.S., Loeber, D.E. and Tinker, A.J. 1975. New aspects of Natural Rubber graft copolymers. In: Proc. Int. Rubb. Conf. 1975, Kuala Lumpur, Vol. IV. Rubber Research Institute of Malaysia, Kuala Lumpur. pp. 149-162.

Campbell, D.S., Elliott, D.J. and Wheelans, M.A. 1978. Thermoplastic Natural Rubber blends. NR Technol., 9(2): 21-31.

Campbell, D.S., Loeber, D.E. and Tinker, A.J. 1979. A method of forming graft copolymers by attaching pre-polymerized side chains to a natural or unsaturated synthetic rubber backbone, and the resulting graft copolymers. Eur. Patent 0 000 976. Chem. Abs. 90: 169 988.

Campbell, D.S., Mente, P.G. and Tinker, A.J. 1981. Natural Rubber analogues of styrene-diene thermoplastic rubbers. Kaut. Gummi, 34: 636-640.

Campbell, D.S., Mente, P.G. and Tinker, A.J. 1983. Reagents for graft copolymers. European Patent 0 065 366. Chem. Abs. 98: 144415.

Campbell, D.S., Loeber, D.E. and Tinker, A.J. 1984a. Graft copolymers from azodicarboxylate-functional pre-polymers: 1. Synthesis of azodicarboxylate-functional polystyrene. Polymer, 25: 1141-1145.

Campbell, D.S. and Tinker, A.J. 1984b. Graft copolymers from azodicarboxylate-functional pre-polymers: 2. Preparation in solution of graft copolymers of polydiene with polystyrene. Polymer, 25: 1146-1150.

Campbell, D.S. 1985. Thermoplastic elastomeric graft copolymers. In: I. Goodman (Ed.). Developments in Block Copolymers - 2. Elsevier Applied Science Publishers, London. pp. 203-237.

Campbell, D.S. 1988. Graft copolymers from Natural Rubber. In: A.D. Roberts (Ed.). Natural Rubber Science and Technology. Oxford University Press, Oxford. pp. 679-730.

Campbell, D.S. 1989. The relevance of phase heterogeneity in epoxidations using formic acid and hydrogen peroxide. Polymer Latex III 17/1-17/9. The Plastics and Rubber Institute, London.

Campbell, D.S. and Seow, P.K. 1990. Graft copolymers of cis-1,4-polyisoprenes with poly(methyl methacrylate). J. Nat. Rubber Res. In press.

Ceresa, R.J. 1962. Block and Graft Copolymers, Butterworths, London.

Ceresa, R.J. 1973. Synthesis and characterization of Natural Rubber block and graft copolymers. In: R.J. Ceresa (Ed.). Block and Graft Copolymerization, Vol. 1. John Wiley and Sons Ltd., London. pp. 47-82.

484

Chapman, A.V. and Porter, M. 1988. Sulphur vulcanization chemistry. In: A.D. ROberts (Ed.). Natural Rubber Science and Technology. Oxford University Press, Oxford. pp. 511-620.

Colclough, T. 1962. New methods of crosslinking natural rubber. Part 2. Introduction of epoxide groups into natural rubber and their subsequent utilization for crosslinking. Trans. Inst. Rubber Ind., 38: 11-15.

Coomarasami. A. 1980. Unpublished work, MRPRA.

Cuneen, J.I., Farmer, E.H. and Koch, H.P. 1943. Rubber, polyisoprenes and allied compounds. Part V. The chemical linking of rubber and other olefins with phenol-formaldehyde resin. J. Chem. Soc., 472-476.

Cuneen, J.I. and Watson, W.F. 1963. Bulk isomerization of polyisoprene with sulphur dioxide. In: N.G. Gaylord (Ed.). Macromolecular synthesis, Volume 3. John Wiley and Sons, New York. pp. 34-37.

Davey, J.E. and Loadman, M.J.R.L. 1984. A chemical demonstration of the randomness of epoxidation of Natural Rubber. Br. Polym. J. 16: 134-138.

Davies, C.K.L., Wolfe, S.V., Gelling, I.R. and Thomas, A.G. 1983. Strain crystallization in random copolymers produced by epoxidation of cis-1,4-polyisoprene. Polymer, 24: 107-113.

Dawes, K. and Rowley, R.J. 1978. Chemical modification of Natural Rubber - A new silane coupling agent. Plastics and Rubber: Materials and Applications, 3: 23-26.

D'Ianni, J.D., Naples, F.N., Marsh, J.W. and Zarney, J.L. 1946. Chemical derivatives of synthetic polyisoprene rubbers. Ind. Eng. Chem., 38: 1171-1181.

Dittmann, V.W. and Hamann, K. 1971. Epoxydierung 1,2-dialkylsubstituierter Doppelbindung in Butadienoligomeren mit Perameisensaure 'in situ'. Chem. Zeit., 95: 857-863.

Dryuk, V.G. 1976. The Mechanism of Epoxidation of Olefins with Peracids. Tetrahedron, 32: 2855-2866.

Eskina, M.V., Khachaturov, A.S., Krentsel, L.B. and Litmanovich, A.D. 1990. On the structure of chlorinated Natural Rubber. ^{1}H and ^{13}C-NMR data. Eur. Polym. J. 26: 181-188.

Elliott, D.J. and Tinker, A.J. 1988. Blends of Natural Rubber with thermoplastics. In: A.D. Roberts (Ed.). Natural Rubber Science and Technology. Oxford University Press, Oxford. pp. 327-358.

Ferrnando, W.S.E. 1980. Unpublished work, MRPRA.

Fuller, K.N.G. 1988. Rheology of raw rubber. In: A.D. Roberts (Ed.). Natural Rubber Science and Technology. Oxford University Press, Oxford. pp. 141-176.

Gan, L.H. and Ng, S.C. 1981. Reactions of Natural Rubber latex with performic acid. Eur. Polym. J., 17: 1073-1077.

Gan, L.H. and Ng, S.C. 1986. Kinetic studies of the performic acid epoxidation of Natural Rubber Latex catalyzed by cationic surfactant. Eur. Polym. J., 22: 573-576.

Gelling, I.R. 1982. A method for making epoxidized cis-1,4-polyisoprene rubber. british Patent 2113692.

Gelling, I.R. 1985. Modification of Natural Rubber latex with peracetic acid. Rubber Chem. and Technol., 58: 86-96.

Gelling, I.R. and Morrison, N.J. 1985. Sulphur vulcanization and oxidative ageing of epoxidized natural rubber. Rubber Chem. and Technol., 58: 243-257.

Gelling, I.R. and Porter, M. 1988. Chemical modification of natural rubber. In: A.D. Roberts (Ed.). Natural Rubber Science and Technology. Oxford University Press, Oxford. pp. 359-456.

Gordon, M. and Raylor, J.S. 1955. Introduction to the mechanism of the hydrochlorination of rubber. Rubber Chem. and Technol., 28: 297-307.

Grozynski Smith, J. 1984. Synthetically useful reactions of epoxides. Synthesis: 629-708.

Gregory, M.J. and Tan, A.S. 1975. Some observations on the storage hardening of Natural Rubber. In: Proc. Int. Rubb. Conf. 1975, Kuala Lumpur, Vol. IV. Rubber Research Institute of Malaysia, Kuala Lumpur. pp. 28-38.

Gundert, F. and Wolf, B.A. 1989. Polymer-solvent interaction parameters. In: J. Brandrup and E.H. Immergut (Eds.). Polymer Handbook. John Wiley and Sons, New York. pp VII/174-VII/182.

Hercules Powder Co. 1962. Epoxidation of unsaturated polymers of conjugated dienes in latex form. British Patent 896 361. Chem. Abs. 57: 4877.

Hoffmann, H.M.R. 1969. The ene reaction. Angew. Chem. Int. Ed., 8: 556-577.

Hourston, D.J. and Roumaine, J. 1989. Modification of Natural Rubber latex 1. Natural rubber - polystyrene composite latices synthesized using amine-activated hydroperoxide. Eur. Polym. J., 25: 695-700.

Jett, A.C. Jr. 1989. Chemical modification of cellulose and its derivatives. In: G. Allen and J.C. Bevinton (Eds.). Comprehensive Polymer Science Vol. 6. Pergamon Press, Oxford. pp. 49-80.

Khalil, M.M. and Prizkow, W. 1973. Kinetische Untersuchungen uber die Reaktion offenkettiger aliphatischer Olefine mit Peressigsaure. J. prakt. Chem., 315: 58-64.

Knight, G.T. and Pepper, B. 1971. A rationalization of nitrosoarene-olefin reactions. Tetrahedron, 27: 6201-6208.

Kobryner, W. and Banderet, A. 1959. Sur le graffage du polymethacrylate de methyle sur le caoutchouc d'Hevea. J. Polym. Sci., 36: 381-396.

Koenig, J.L. and Patterson, D.J. 1987. A Fourier transform infrared and nuclear magnetic resonance study of cyclized natural rubber. Makromol. Chem., 138: 2325-2337.

Le Bras, J. and Delelande, A. 1950. Les Derives Chimiques du Caoutchouc Naturel. P. Piganiol (Ed.). Dunod, Paris.

Le Bras, J. and Compagnon, P. 1947. The chemistry of rubber. The inter-action of ethylenic compounds with rubber. Rubber Chem. and Technol., 29: 938-948.

Lee, T.K. 1980. Effect of Chemical Modification on the Low-Temperature crystallizationbehaviour of Natural Rubber. PhD Thesis, University of London.

Lee, T.K. and Porter, M. 1979. Effect of chemical modification on the cry-stallization of unvulcanized natural rubber at low temperatures. In: Proc. Int. Rubber Con. 1979, Venice. Airiel and Assogomma, Milan. p. 991.

Marechal, E. 1989. Chemical modification of synthetic polymers. In: G. Allen and J.C. Bevington (Eds.). Comprehensive Polymer Science Vol. 6. Pergamon Press, Oxford. pp. 1-47.

Milkovich, R. 1980. Synthesis of controlled polymer structures. Amer. Chem. Soc., Div. Polym. Chem., Polym. Preprints, 21(1): 40-41.

Litmanovich, A.D. 1980. Change in polymer reactivity in the course of macro-molecular reaction. Eur. Polym. J., 16: 269-275.

Nippon Zeon Co. Ltd. 1990. Zetpol Hydrogenated Nitrile Rubber. Nippon Zeon Co. Ltd., Rubber Division, Japan.

Martl, M.G. and Hummel, K. 1990. Simulation of coupled modification reactions of polymers, 4. Mono- and disubstitution of 1,4-polybutadiene, coupled with partiald ouble bond shift. Makromol. Chem., 191: 289-300.

Pendle, T.D. 1973. Properties and applications of block and graft copolymers of Natural Rubber. In: R.J. Ceresa (Ed.). Block and Graft Copolymeriz-ation, Vol. 1. John Wiley and Sons London. pp. 83-97.

Perera, M.C.S., Elix, J.A. and Bradbury, J.H. 1987. A ^{13}C NMR study of hydroxylated Natural Rubber. J. Appl. Polym. Sci., 33: 2731-2742.

Pethric, R.A. 1985. Polymer Yearbook. R.A. Pethric (Ed.). Harwood Academic Publishers, London.

Pinazzi, C., Danjard, J.C. and Pautrat, R. 1963. Addition of unsaturated monomers to rubber and similar polymers. Rubber Chem. and Tech., 36: 282-295.

486

Pinazzi, C., Brosse, J.C., Pleurdeau, A. and Reyx, D. 1975. Recent developments in chemical modification of polydienes. J. Appl. Polym. Sci., Appl. Polym. Symp., 26: 73-98.

Plesnicar, B., Tasevski,M. and Azman, A. 1978. The transition state for epoxidation of ethylene with peroxyformic acid. An ab initio molecular orbital study. J. Amer. Chem. Soc., 100: 743-746.

Polysar Limited, 1976. Amine modified remouldable rubbers. British Patent 1 455 939.

Schulz, D.N., Turner, S.R. and Golub, M.A. 1982. Recent advances in the chemical modification of unsaturated polymers. Rubber Chem. Technol., 55: 809-859.

Schulz, G.O. and Milkovich, R. 1986. Styrene/isoprene diblock macromer graft copolymer: synthesis and properties. Ind. Eng. Chem., Prod. Res. Dev., 25: 148-152.

Sears, A.G. 1984. Processing of SMRCV: effect of rubber viscosity level in single-pass mixing of stocks containing semi-reinforcing black. NR Technology, 15(3): 49-53.

Sekhar, B.C. 1961. Inhibition of hardening in Natural Rubber. In: Proc. Nat. Rubb. Res. Conf. 1960, Kuala Lumpur. Rubber Research Institute of Malaysia, Kuala Lumpur. pp. 512-514.

Shapilov, O.D. and Kostyukovskii,Ya.L. 1975. Reaction kinetics of formic acid with hydrogen peroxide in aqueous solution. Kinetics and Catalysis, 15: 947-948.

Shell Chemical Company. 1965. Synthetic elastomers. Neth. Appl. 6 404 532. Chem. Abs., 62, 6660.

Shelton, J.R. and Lee, L.H. 1985. The structure of cyclized polybutadiene. Rubber Chem. and Technol., 31: 415-423.

Sherrington, D.C. 1988. Reactions of Polymers. In: H.F. Mark and N.M. Bikales (Eds.). Encyclopedia of Polymer Science and Engineering, Vol. 14. 2nd ed., John Wiley and Sons, New York. pp. 101-169.

Sohma, J. 1989. Mechanochemistry of polymers. Prog. Polym. Sci., 14: 451-496.

Soutiff, J.C. and Brosse, J.C. 1990. Chemical Modification of Polymers 1. Applications and Synthetic Strategies. Reactive Polymers, 12: 3-29.

Subramaniam, A. 1988. Rubber Derivatives. In: H.F. Mark and N.M. BIkales. (Eds.). Encyclopedia of Polymer Science and Engineering, Vol. 14. 2nd ed., John Wiley and Sons, New York. pp. 762-786.

Swern, D. 1971. Organic Peroxyacids as Oxidizing Agents. In: D. Swern (Ed.). Organic Peroxides, Vol. 2. Wiley-Interscience, New YOrk. pp. 355-533.

Tanaka, Y. 1989. Structure and biosynthesis mechanism of natural polyisoprene. Prog. Polym. Sci., 14: 339-371.

Tinker, A.J., Iceogle, R.D. and Whittle, I. 1989. Natural rubber based TPEs. Rubber World, March: 25-9.

Tonnies, G. and Homiller, R. 1942. The oxidation of aminoacids by hydrogen peroxide and formic acid. J. Amer. Chem. Sco., 64: 3054-3046.

Tran, A. and Prud'homme, J. 1977. Microstructure and physical properties of hydrochlorinated 1,4-polyisoprene prepared by butyllithium in nonpolar solvent. Macromolecules, 10: 149-153.

Watson, W.F. 1958. Chemical reactions induced by polymer deformation. Trans. Inst. Rub. Ind., 34: 237-247.

Witnauer, L.P. and Swern, D. 1950. X-ray diffraction and melting point composition studies on 9,10-epoxy- and dihydroxystearic acids and 9,10-epoxyoctadecanols. J. Amer. Chem. Soc., 72: 3364-3368.

Wong, A.K., Campbell, D.S. and Tinker, A.J. 1987. Poly(isoprene-g-alkyl methacrylate) copolymers. 2. Graft copolymer formation from azodicarboxylate functional methacrylate prepolymers. Polymer, 28: 2161-2165.

Yamashita, Y. and Tsukahara, Y.W. 1984. Synthesis and application of tailored graft copolymers from polystyrene macromonomer. J. Macromol. Sci., Chem. A21: 997-1012.

CHAPTER 21

ENGINEERING PROPERTIES AND APPLICATIONS OF RUBBER

A. STEVENSON

Materials Engineering Research Laboratory, Tamworth Road,
Hertford SG13 7DG, England.

INTRODUCTION

The unique properties of rubber as an engineering material start from
its very low shear modulus which is only about one thousandth of its bulk
modulus, and its ability to deform elastically by several hundred per cent.
Rubber is thus highly extensible under modes of deformation which do not
constrain it hydrostatically, but much less so when hydrostatic compression
is involved (which causes direct straining of the molecular bonds). Thus
engineering components may be designed with very different stiffnesses in
different directions. Such components are frequently composites, exploiting
the properties of elastomers in combination with metals, plastics or fibres.
A wide range of engineering applications have been developed on the basis
of these properties.

An added advantage is the ability of rubber to function for many
years in aggressive environments - completely maintenance free. Rubber
is sometimes thought of by engineers as an unfamiliar or new material.
In fact, it has played an important part in many engineering developments
since the 1850's. There are examples of energy absorbing natural rubber
rings being used in rail buffers from 1850 and as load bearing pads
supporting a railway viaduct constructed in 1890 in Melbourne, Australia.
The use of rubber components often facilitates greater sophistication in
engineering design and enables compliance and flexibility to be introduced
in a controlled and generally failsafe manner. For example, the trend
towards longer bridge spans and closer tolerance designs has increased
the need for reliable and maintenance free bridge bearings. Elastomeric
bridge bearings are now preferred in most parts of the world. New developments in earthquake protection of buildings include a compliant foundation
system with elastomeric bearings isolating both the building and its contents
from seismic disturbances. Offshore oil platform design has evolved from
rigid towers of steel and concrete to compliant structures, capable of deep

water production, which rely on elastomeric flexjoints as an intrinsic part of the structural design. The ride comfort and safety of modern automobiles, trucks etc depend on advanced suspension systems which often use a large number of rubber components. Modern aircraft, helicopters and the space shuttle use rubber engineering components in key locations, where the consequences of failure may be catastrophic.

In this chapter, the properties of rubber and the principles for engineering design are reviewed and some applications from civil engineering and automotive engineering, selected from the author's personal work experience, are discussed. It is beyond the scope of the chapter to provide an exhaustive review of all rubber engineering applications. Finally there is a brief consideration of testing methods for engineering properties of rubber, where these go beyond current standard tests, and discussion of likely future requirements.

DESIGN PRINCIPLES AND MATERIAL PROPERTIES

LOAD/DEFLECTION CHARACTERISTICS

Most rubber engineering components contain rubber adhesively bonded to one or more layers of metal (or fabric) for reinforcing or fixing purposes. This prevents the rubber layer from slipping at load carrying surfaces and ensures reliable load/deflection characteristics.

The following equations may be used as the basis of engineering design and assume no slip at load carrying surfaces. Although the equations presented are not exact, they should give reasonable approximations (ie. within 20%) in most cases. This is normally adequate as an initial design calculation (Stevenson, 1986).

Hardness

Hardness is widely used as a simple measure to characterise vulcanized rubber. Readings are usually quoted either in International Rubber Hardness Degrees (IRHD) or in the Shore durometer A scale, which are approximately the same. The scales are non-linear in both cases, reading from 1 to 100. Hardness measurements are subject to considerable uncertainty of at least 1±1.5 degrees and often ±5 degrees in practice. Hardness cannot be used directly in design calculations and relates only approximately to shear modulus. Shear modulus measurements are intrinsically more accurate and reliable and will henceforth be used as the basis of the design principles described here.

Shear

When a shear force is applied to a flat bonded rubber layer, the principal effect is simple shear deformation (Fig. 1).

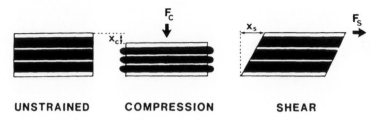

UNSTRAINED   COMPRESSION   SHEAR

Fig. 1. Shear and compression deformations.

Shear stiffness is given by:

$$K_s = F/x = \frac{G\ A}{t} \qquad (1)$$

where A is the loaded area and t the rubber layer thickness

F is the applied force and x shear deflection

Shear stress $q = F/A$ Shear strain $e_s = x/t$

For a laminated unit with n rubber layers separated by bonded rigid plates, this becomes:

$$K_s = \frac{G\ A}{nt} \qquad (2)$$

The incorporation of intermediate metal plates into a block of rubber does not affect the shear stiffness, which depends not on shape, but only on the total rubber thickness, nt.

The stress/strain curves of most types of rubber are approximately linear over a reasonably wide working range (upto 100% shear strain) (Fig. 2). The shear modulus, G is only approximately one third of the Young's modulus, E, the value expected from classicial elasticity theory. Figure 2 illustrates this for a filled Natural Rubber. The shear modulus is approximately constant between 10% and 100% strain.

Compression

The stiffness of rubber in compression depends on the shape factor, S (Fig. 3) defined by:

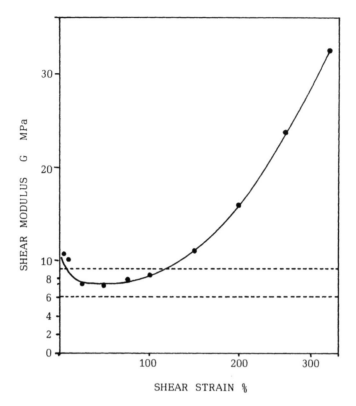

Fig. 2. Shear modulus as a function of shear strain for a black filled
natural rubber.

Shape Factor.   S = LOADED AREA / FORCE-FREE AREA

for a rectangular block: S = LB/[2t(L+B)]   (3)

Uniaxial compression of a rubber layer whose loaded areas are constrained laterally (eg. by bonding) produces a mixture of hydrostatic
compression (at high enough shape factors) towards the centre and shear
strain distributions towards the free edges. This causes the rubber to bulge
at the force free areas as shown in Figure 1. It has been shown (Gent
and Lindley, 1959) that to a first approximation, there is a parabolic
distribution of shear strains which are at a maximum at the bond edge
of the force free region. From this the following dependence of compression

$$\text{shape factor} \ = \ \frac{\text{loaded area}}{\text{force free area}}$$

$$\text{i.e.} \ S \ = \ \frac{LB}{2T(L+B)}$$

Fig. 3. Definition of Shape Factor.

modulus on shape factor can be derived:

$$E_c \ = \ G \ (3 + CS^2) \tag{4}$$

where G is the small strain shear modulus of the rubber and C a geometrical factor which depends on shape. Two extreme values for C are : for a plane circular disc, C = 6 and for a long thin strip or annulus, C = 4. This equation provides a good approximation for E_c for shape factors above about 0.5 and below about 10.

For most modes of deformation and low shape factors, rubber may be considered effectively incompressible. For very high shape factors, the effect of bulk compression needs to be included. This may be done by considering the strain in the elastomer arising from changes in molecular configuration and the strain arising from bulk compression to occur in parallel. Thus:

$$E_t \ = \ [\frac{1}{E_c} + \frac{1}{B}]^{-1} \tag{5}$$

where B is the rubber bulk modulus. As the shape factor increases, the total compression modulus E_t increases towards the bulk modulus of the material. This effect of shape factor is illustrated by Fig. 4 which shows data for a natural rubber engineering vulcanizate with carbon black content producing hardness range of 60-70 IRHD. It is difficult to determine accurate independent values for the bulk modulus B of rubber. Typical values however are B = 1500 - 3000 MPa, whereas the shear modulus will usually be between 0.5 and 10 MPa.

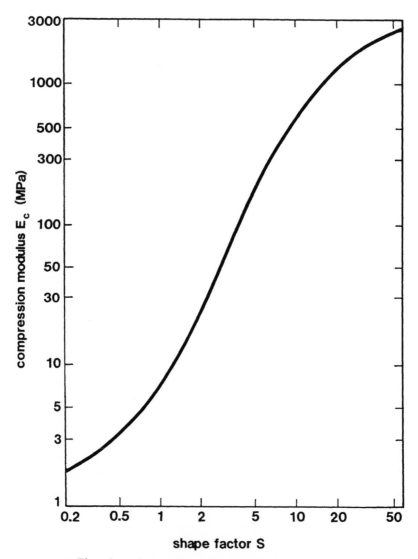

Fig. 4. Shape Factor vs Compression Modulus.

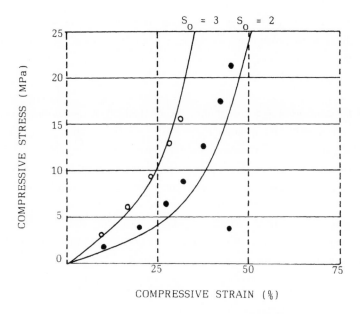

Fig. 5.   Typical stress/strain curves in Compression for two shape factors,
S = 2 and S = 3 ————o———— calculated – For a black filled
Natural Rubber ————•———— measured.

Non-linearity

The foregoing equations only calculate the stiffness of a component in compression for relatively small strains, up to about 10% (depending on shape factor). Force/deflection curves in compression are markedly non-linear over a wider range (Fig. 5) and other means may be needed to approximate the non-linear behaviour. There have been several attempts to calculate the non-linear behaviour of solid rubber in compression. There is no universally successful or generally approved method.

It is possible to consider changes in incremental modulus by considering large strain behaviour to consist of a series of small strain conditions. As the rubber is compressed, its thickness decreases, leading to an increased effective shape factor for small strains at that point (Stevenson, 1986).

$$S = S_0 \cdot t_0 / t \qquad (6)$$

where t = $t_0 (1 - e_c)$, the true thickness at strain e_c
        S_0 = initial shape factor
        t_0 = initial thickness
        e_c = compressive strain

At each successive stage the shape factor can be recalculated to obtain the incremental stiffness at strain e_c. This approach has become more practical with the advent of readily available microcomputers. Figure 5 shows typical results for natural rubber with shape factors at 2 and 3.

At very high compressions the shear strains in the rubber can become so great that the shear modulus G can no longer be considered constant. In that case, the appropriate non-linear shear data for the type of rubber concerned is needed as input. The design of building mounts usually only requires low compressive strains when approximations of linearity are quite adequate.

Combined shear and compression

A force F to a laminated unit at a mean angle a to the normal (or compressive) direction may be resolved into shear and compressive modes, so that:

$$F = F_c \cos a + F_s \sin a$$

$$x_c = x \cos a \qquad \text{\&} \qquad x_2 = x \sin a$$

then: $F = K_c x \cos^2 a + K_s x \sin^2 a$ (7)

where   K is the shear stiffness, and K_c the compression stiffness, as before.

The stiffness, K, in the direction of the force is thus:

$$K = K_c \cos^2 a + K_s \sin^2 a \qquad\qquad (8)$$

The use of one or more such inclined units enables elastomeric components to be designed with different stiffnesses in each of the rectangular co-ordinate directions. If the direction of the force shown above is in the y-direction, then the stiffness, K_z, perpendicular to the plane of the paper is simply the shear stiffness K_s.

Thus: $K_z = K_s$ (9)

There are two limiting cases of horizontal, or lateral stiffness, K_x, in the plane of the paper, depending on the degree of restraint against rotation:

(a) when a couple prevents rotation:

$$F_x = K_c \sin^2 a + K_s \cos^2 a$$ (10)

(b) when the unit is free to rotate (ie. no restraining couple):

$$K_x = K_c K_z / K_y = K_s K_c / (K_c \cos^2 a + K_s \sin^2 a)$$ (11)

For a unit consisting of a number, n of intermediate layers, bonded together,

$$K_c = E_c A / nt$$ (12)

The above design principles have been successfully used to develop a wide variety of rubber engineering components, from engine mounts to large multi-layer bearings for offshore platforms (Sedillot and Stevenson, 1983; Stevenson et al. 1986).

Creep and stress relaxation

Creep and stress relaxation are important properties to consider whenever the application requires the rubber component to support a dead load without settlement or maintain a stress. Creep is a time dependent increase in deformation under constant load, while stress relaxation is the time dependent reduction in stress under constant deformation. Both phenomena are referred to as 'relaxation'. Creep is an important consideration in the design of building mounts since uneven creep could lead to the equivalent of settlement problems with attendant structural damage. Theoretical relationships between creep and stress relaxation rates are mathematically complex, but in practice either rate can be deduced from the other if the shape of the force/deflection curve is known.

Both phenomena occur in all materials but are often more evident in elastomers due to the high initial deformations to which they may be subjected. There are two underlying relaxation mechanisms in elastomers - physical and chemical.

Physical relaxation involves relatively short term reconfiguration of the long molecular chains and is largely reversible in crosslinked polymers, when the load is removed. The amount of relaxation is approximately linear with the logarithm of time, so that physical rates are often expressed as 'per cent per decade' (ie. per cent increase in strain, or decrease in stress, for any factor of ten increase in time). This means that in practice, physical creep will slow up with time. The amount of creep in any tenfold increase in time (such as 1 week, 10 weeks or 100 weeks) will be approximately equal. Some values for typical engineering elastomers are as follows:

	NR	SBR	CR	BR	NBR
Unfilled (MPa)	1.6	2.5	3.1	1.5	3.5
Filled (MPa)	3.1	3.2	4.6	4.1	4.2

Chemical relaxation involves molecular chain scission - usually by oxidation - and is the dominant mechanism at longer times and/or elevated temperatures. Chemical relaxation rates are linear with time and vary with the chemical stability of the polymer and the extent to which it is protected by anti-degradants. A typical chemical creep rate for an engineering natural rubber is approximately 0.5% increase in initial deflection per year.

Creep which can never be recovered is called 'permanent set'.

The total creep between 1 day and t days will be:

$$\% \ \text{creep} = A \log t + Bt/365 \tag{13}$$

where A is the physical creep rate in % per decade and B is the chemical creep rate in % per year.

It was shown by Derham (1975) that the amount of creep for building mounts could be successfully predicted from laboratory experiments with good accuracy. Creep measurements on actual building mounts made over several years successfully correlated with laboratory predictions made several years earlier. Measurements made after 15 years have confirmed this.

DYNAMIC PROPERTIES

Rubber components are increasingly required to have precise dynamic properties for antivibration or energy absorbing applications. Although the force/deflection behaviour of rubber is in general non-linear, provided the main features of dynamic behaviour are well understood and the sources of non-linearity recognised, components can be successfully designed with precise and reproducible dynamic performance.

<u>Theoretical background</u>

A general approach is to represent the complex stiffness K* (spring rate) as:

$$K^* = K_1 + jK_2 \qquad (14)$$

In polar form, this is represented by a dynamic stiffness of magnitude $(K^*)^2$ and a phase angle δ, with:

$$(K^*)^2 = K_1^2 + K_2^2 \text{ and } \delta = \tan^{-1} (K_2/K_1) \qquad (15)$$

Where j signifies a component 90° out of phase, K_1 and K_2 are the truly elastic or in-phase and the viscuous and out-of-phase components respectively.

The dynamic properties of rubber are usually characterized in simple shear, where the dynamic shear modulus G* and phase angle δ are measured, and:

$$K^* = G^*A/t \qquad (16)$$

This representation assumes an underlying linear behaviour with an elliptical hystersis loop.

Transmissibility measures the amplification or attenuation of vibration from one point in the system (input) to another (response) and, assuming equilibrium, may be expressed as:

$$T = \left[\frac{x_o^*}{x_1} \right] = \frac{[(K_1^2 + K_2^2 - \omega^2 M \, K_1)^2 + \omega^4 M^2 K_2^2]^{1/2}}{(K_1 - \omega^2 M)^2 + K_2^2} \qquad (17)$$

This equation enables the transmissibility of the system to be calculated, provided the real and complex parts of the dynamic spring rate are known at the relevant frequency, temperature and strain amplitude.

Transmissibility is normally characterized in logarithmic units of dB.

The natural frequency, n_f, is determined by the stiffness K for a given period, or mass M applied to the spring:

$$n_f = 1/2\pi . \sqrt{(K/M)} \qquad (18)$$

<u>Effect of elastomer type and temperature</u>

Different types of unfilled rubber have different degrees of damping and also different degrees of sensitivity to frequency. Natural rubber (NR) is low damping and neither phase angle nor dynamic stiffness vary substantially with frequency over the range 1 to 200 Hz. Nitrile isoprene copolymer (NIR), on the other hand, is a higher damping material even at low frequencies and its dynamic stiffness and phase angle increase substantially between 1 and 200 Hz. Figure 6 shows results for some engineering elastomers. The data for phase angle and dynamic stiffness has been combined with the aid of the equation for T to present a series of transmissibility curves. This enables the overall effect in a spring mass system to be seen more clearly, and is probably the most useful format for engineering design. Decreasing the mass (or preload) or increasing the spring stiffness will shift any of these curves to higher frequencies.

Higher damping polymers inhibit the effect of resonance more successfully whenever there are input vibrations at resonance. However they are less successful at isolating the system from input vibrations at frequencies significantly above the natural frequency. In selecting a material for a specific design, the optimum will depend on the range of input vibrations of concern. High damping polymers also usually show high creep rates and more sensitivity to temperature. A rise in temperature in NIR from 23°C to 50°C is sufficient to increase the height of the peak transmissibility at resonance by 10 dB. At very low frequencies, the transmissibility tends to unity (0 dB) in each case.

Natural rubber has a good balance of properties for anti-vibration applications and so is still the preferred material for a wide category of applications.

Effect of fillers and strain amplitude

The addition of carbon black increases both the dynamic modulus and the phase angle (and hence damping) and these properties also become amplitude dependent. The magnitude of these effects depends on the type and quantity of carbon black. The resulting effect on transmissibility are illustrated in Fig. 7 for a natural rubber filled with 75 parts of HAF carbon black. At low strain amplitudes, the dynamic stiffness is unexpectedly high, and the phase angle low - resulting in a much higher natural frequency (curve a) and higher peak transmissibility than at lower amplitudes. This effect is important in the design of antivibration mounts that experience a range of input vibrations as in automotive suspensions. A mount designed on the basis of materials test data at a strain of 10% shear leads (curve c) to the expectation of an attenuation of 15 dB at an input frequency of 75 Hz. In fact, vibrations of 0.2% amplitude will be amplified by almost 20 dB.

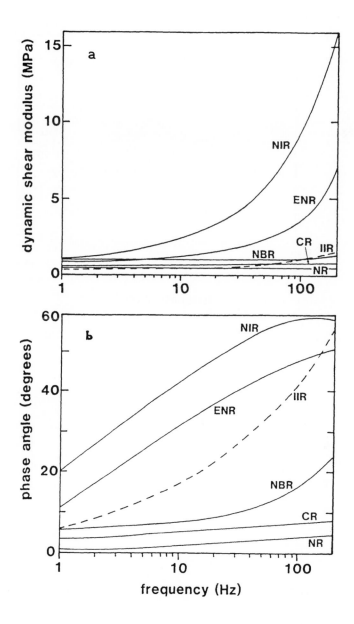

Fig. 6. Dynamic mechanical properties of different unfilled elastomers at 23°C. NR-Natural Rubber, CR-Polychloroprene, NBR-nitrile rubber, IIR-butyl rubber, NIR-nitrile isoprene copolymer.
(a) Dynamic shear modulus (b) phase angle

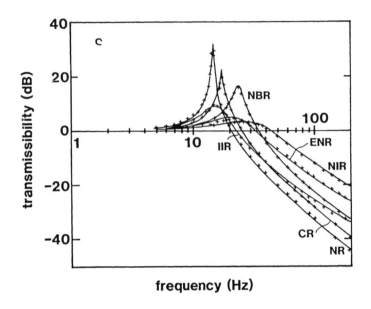

Fig. 6 (c) Transmissibility – as function of frequency strain amplitude 10% shear shear strain in all cases.

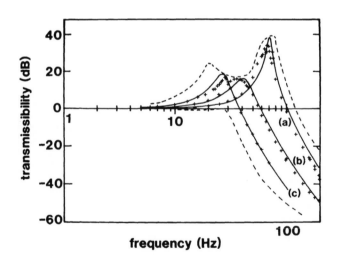

Fig. 7. Effect of strain amplitude on transmissibility for black filled NR at different shear strain amplitudes (a) 0.2% (b) 2% (c) 10%.
+ experimental ——— calculated - - - - overall response envelope
Effect of non-linearity on dynamic mechanical properties.

Effect of non-linearity from design

The geometrical design of a rubber spring can introduce non-linearities which can strongly affect its dynamic behaviour in addition to the effects outlined above. As discussed earlier, rubber is very non-linear in compression, resulting in shifts in dynamic modulus and phase angle with different preloads. This can cause shifts in the transmissibility behaviour, the nature of which will depend on the nature of the non-linearity. Thus for low shape factors over large compressive strains, there may be substantial changes in dynamic properties which will affect performance. This can be important for some applications, such as automotive suspensions, but non-linearity is not normally likely to be a factor in the design of building mounts where the strains are usually small and the range of input frequencies relatively limited. Figure 8 illustrates some effects of non-linearity on dynamic mechanical properties. Figure 8(a) shows the static force/deflection curve for a hollow rubber cylinder on which has been superimposed dynamic hysteresis loops at 5 selected points. The lower curve summarises the static modulus change with increasing deflection. At different points there will be a nested series of hysteresis loops with increasing deflection as shown in Figure 8(b). Figure 8(c) shows the resulting transmissibility curves at each of those 5 points. There are many novel properties achievable in this way and research programmes are under way at MERL to develop these ideas further.

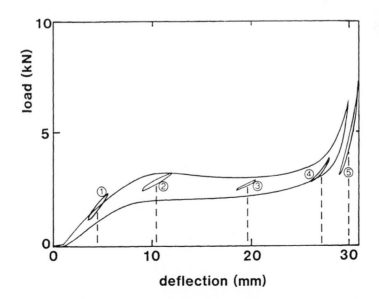

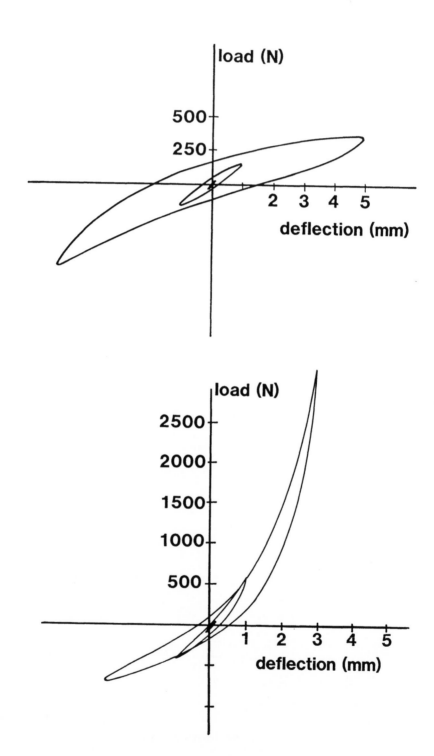

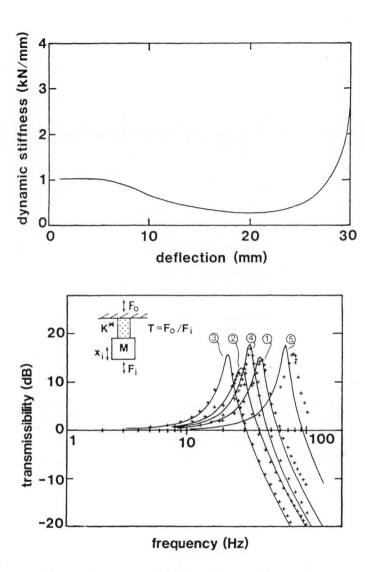

Fig. 8. (a) Upper - Static force/deflection curve for hollow rubber cylinder with dynamic hysteresis loops superimposed at points (1) to (5)

      Lower dynamic stiffness change with deflection corresponding to upper curve.

(b) Typical nested hysteresis loops at different points from 8(a).

(c) Transmissibility curves corresponding to the five points in Figure 8(a).

Further details on dynamic properties of rubber may be found in Payne (1964) and Harris and Stevenson (1986).

FATIGUE LIFE DETERMINATIONS

The tearing energy method

The application of a fracture mechanics approach to rubber has been very successful. This approach enables the resistance to fracture or fatigue to be characterised as a fundamental geometry independent characteristic of any particular type of rubber. Once this property is determined, then fatigue and fracture resistance of any geometry component can be calculated. This approach is normally discussed for rubber in terms of the tearing energy, which is defined as the energy required to cause unit area of new crack growth.

$$T = - \left[\frac{\delta U}{\delta A} \right]_l \qquad (19)$$

where U is the stored elastic energy;

      A is the area of new crack growth;

      l denotes that the partial differentiation is with respect to constant deformation l.

Calculation of tearing energy for different geometries

The tearing energy must be calculated for the different geometry units of interest for engineering components. Initial work in this field provided solutions only for different cases involving thin rubber strips. Subsequently, solutions have been derived for bonded rubber layers in simple shear and uniaxial compression. Derivation of new solutions requires a theoretical model for the energy balance between stored elastic energy and surface free energy for new crack area. Currently, solutions are available for several different geometry rubber strips (simple extension, pure shear, trouser etc), for simple shear and for uniaxial compression (high and low shape factor). This enables tearing energy to be estimated for many engineering components based on these or similar geometries.

The equations for tearing energy include an expression for the uniform (or average) stored energy density in the rubber layer(s), W, and for the appropriate multiplicative geometrical factor, which depends on the locus of crack growth and may contain explicit dependence on crack length, as in the case of simple extension of a rubber strip with a single central edge cut.

For uniaxial compression: $T = \frac{1}{2} W t$ (20)

For simple shear: $T = 0.4 W t$ short cracks   (21)

For simple extension: $T = 2 k W c$ (22)

Where W is the uniform stored energy density in the rubber layer.

In shear and in compression, the tearing energy is not a direct function of crack length, although in simple shear the initial numerical coefficient may vary between 0.2 and 1.0 depending on the size and configuration of the crack. In those cases the growth of a fatigue crack will not accelerate, as it will in simple extension - where the direct dependence on c causes T to increase as the crack grows. Thus rubber is not normally used in tension, but only in compression or shear. In many cases in compression, the strains are small enough for W to be given approximately by the linear expression:

$$W = \frac{1}{2} E_c e_c^2 \qquad\qquad\qquad (23)$$

The success of this approach in describing dynamic fatigue crack growth in shear, compression and tension is illustrated by Figure 9. This shows the results of a series of experiments performed with an engineering vulcanizate fatigued at a frequency of 2 Hz in shear, compression and simple extension. Although the data does show scatter, as is usual with fatigue tests, nevertheless, within the accuracy of the experiments, it was found that at a given calculated tearing energy the crack growth rates were the same for each of these geometries. This means that, provided the tearing energy can be calculated, fatigue lives in shear or compression may be determined by reference to dc/dN curves obtained on strip test pieces in simple extension. The large number of data points shown for the latter case reflects the ease of obtaining accurate measurements of crack growth rate with this geometry.

Characteristic relation dc/dN for different engineering elastomers

The relation between tearing energy and crack growth rate is a material characteristic, independent of component geometry. This relation differs for different rubbers, but once it has been derived, if the tearing energy can be calculated, the crack growth rate can be estimated from plots such as those shown in Figure 9. These relations follow a

506

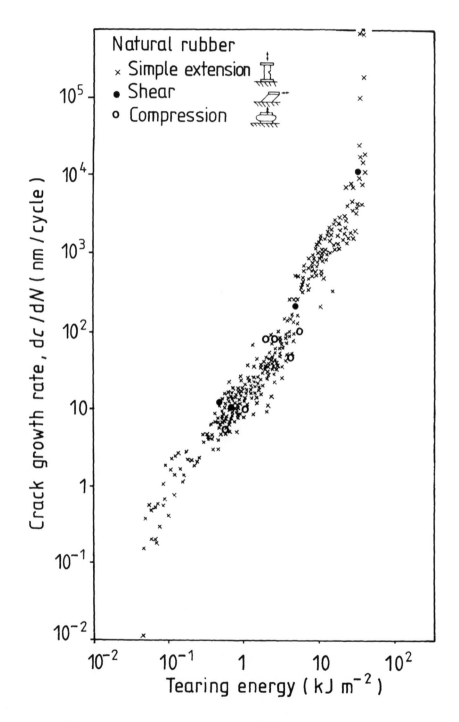

Fig. 9. Fatigue crack growth rate vs tearing energy for black filled Natural Rubber's different geometries - simple extension, shear and compression.

a characteristic form.

At very low tearing energies, only chemical failure mechanisms occur, and these proceed at an extremely low rate (0.1 mm/year). The minimum value for T below which no mechanical fatigue crack growth can occur is referred to as T_0. The value of this varies for different rubbers between 0.05 & 0.1 kJ m^{-2} and may be understood in terms of the molecular structure of the rubber.

As the tearing energy increases above T_0, the crack growth rate increases in accord with at least two different power law relations, depending on the value of T. At high enough values of T, failure will occur from a single cycle, and so dc/dN increases asymptotically towards T_c, the static critical tearing energy. Consideration of the shape of the T vs dc/dN curve immediately suggests the basis for good fatigue resistant design - namely to ensure that the maximum tearing energies of a component in service are in a stable region well away from the asymptote to T_c - where even approximate predictions of crack growth rate become very difficult.

Different rubber compounds have different forms of the T vs dc/dN relation, and the curves may even cross. The material which provides the longest fatigue life depends on which portion of the dc/dN curve predominates for the life of the component. For very large numbers of very low amplitude cycles CR will show less crack growth than NBR and NBR less than NR. Above T_0 there is a central region where the fatigue resistance of all three rubbers is very similar.

At high tearing energies, the situation is reversed, with NR sustaining only moderate crack growth rates, when NBR, in particular, would be expected to fail catastrophically, CR being an intermediate case. This leads to the general conclusion that NR or CR will be better suited to load bearing applications where high strains are possible while NBR will be better suited for components (like sealing and jointing systems) where primary load bearing capacity is not required and only small elastomer strains occur. This explains the generally good fatigue resistance of NR in mechanical engineering applications where strains (and energies) can be high

Application to the fatigue life determination of components

The factors considered so far indicate how crack growth rates may be calculated for rubber components subjected to known loads or motions. To apply this to a component, a load motion spectrum needs to be determined, either from past experience of similar applications, or from engineering analysis. This spectrum can then be converted to a spectrum of tearing energies by the methods discussed, and thus the cumulative crack growth

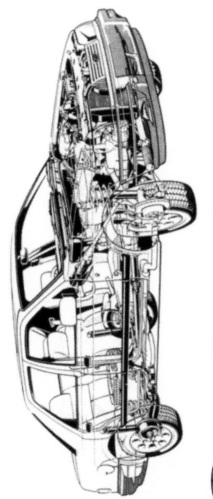

FORD SIERRA SAPPHIRE GHIA

Figure 10 Ford Sierra Sapphire GHIA

calculated. To interpret this in terms of component life requires this life to be defined in terms of crack growth. For any component, this definition requires some analysis of the effect that a crack will have on its performance.

If there is any requirement to perform in tension, a critical crack length can be defined for the onset of catastrophic failure. For components acting in compression, the situation is both safer and more complex. It has been shown that in a bonded rubber disc subjected to uniaxial compression cracks grow only in a region where the rubber has a free edge and is able to bulge. In the first instance, this restricts the effect of crack growth to the vicinity of an outer surface, where the volume of rubber involved may be small compared to the total. The effect on compression stiffness is then also small. Indeed, substantial crack growth in this region does not remove the load carrying capability in compression - so that a rubber bearing may be considered 'failsafe' in this mode. Resistance to crack growth can be enhanced by appropriate shaping of the force free surface of the rubber. A parabolic shape has been found optimum for some components in compression.

This approach has been successfully applied to the fatigue life analysis of rail car mounts, to automotive suspension mounts and to large flexible ball joints used in compliant offshore structures (Gent and Lindley, 1959; Stevenson et al. 1986). For further information on fatigue and fracture of elastomers, see Rivlin and Thomas (1953), Lake and Lindley (1965) and Stevenson (1983).

APPLICATIONS

There are an extremely large number of engineering applications of rubber in almost every industrial sector. In this chapter it is only possible to select some that the author has had personal experience and involving NR and discuss these in the context of the properties described in the previous section. In spite of the large number of applications it is still true that engineering design with rubber is often not founded very scientifically on precise material properties and principles. There is, however, now a strong move in that direction and this is likely to give design engineers more confidence that rubber can be relied on as a precise engineering material in critical applications. The applications selected for discussion are drawn from the fields of automotive engineering, civil engineering and offshore engineering.

Automotive engineering

Modern designs of automobile rely on a wide variety of elastomers for components as diverse as radiator hose, windscreen wipers, tyres, engine mounts and suspension mounts. In this chapter only the latter two will be considered. The design principles described earlier in the chapter apply more directly to these cases. A modern automobile, such as that shown in Figure 10 may contain more than ten different elastomeric bushes and mounts that together from the suspension system.

Engine mounts add another set of rubber engineering components with related anti-vibration functions. In this configuration, each rubber component will normally be required to have specified force/deflection behaviour in more than one direction. Compression of the rubber layers in the component may be combined with shear and/or torsion and/or rotation. Figure 11 shows an engine mount for the Ford Sierra. In this case the function is to accommodate static loads from the engine which create a combination of shear and compression and then superimposed dynamic deformations in a variety of directions. Full analysis in fundamental terms clearly becomes relatively complex. The equations in the earlier part of this chapter can be used to characterise the behaviour in each mode of deformation separately. Dynamic and static behaviour could be determined by a combination of theoretical analysis and careful testing.

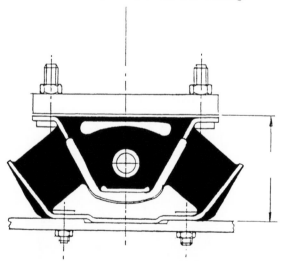

FORD SIERRA ENGINE MOUNT

Figure 11 Ford Sierra engine mount

bush example

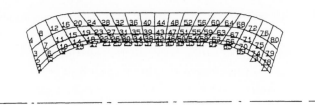

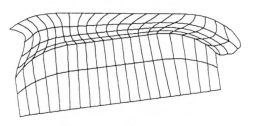

8 degrees rotation

Fig. 12. Finite element analysis of truck bush mount (Courtesy Dr. J. Harris)

For complex shaped components and perhaps many layers, computational tehniques can be of value. Conventional finite element programmes, such as ABAQUS have been adapted for use with elastomers and have had a degree of success for complex components, provided the material non-linearity is not too great. Figure 12 shows analysis of a truck bush mount with a single layer of rubber of complex shape, an inner metal cylindrical core and an outer metal casing. The effect of 8 degrees of rotation was modelled by Harris (1989) using the ABAQUS programme with the result shown in the lower half of Figure 12. This component was under engineering development by Metzeler (UK) Ltd. for a truck application. The effect of the rotation on this shaped rubber layer can be seen more clearly by FE analysis for this shape than would be easily derived from analytical analysis.

The limitation of all current finite element approaches is still however a difficulty in handling large strain non-linear behaviour with precision, and, in particular, at a crack tip. This restricts the value of such approaches for fatigue life predictions and means that at best qualitative answers are derived rather than quantitative life calculations. Scientists at MERL have developed specialist computer programmes to apply fracture mechanics analysis to calculate the fatigue life of complex rubber engineering components with considerable success.

Materials selection and property characterization are still often fairly basic for automotive applications. Materials classification is often based on the SAE J200 (or ASTM D2000) system, which classifies elastomers in terms of heat and oil resistance mainly using standard tensile test pieces. This approach does not, for example, clearly distinguish between NR, SBR or reclaimed rubber and ignores creep resistance, which is an important property for load bearing components. At present, the large majority of suspension and engine mounting components are made from natural rubber. However, synthetic elastomers can offer better resistance to heat and oil resistance and with the trend towards higher engine temperatures there is likely to be a greater requirement in future for improved natural and synthetic elastomers in this type of application, provided fatigue life is not downgraded to an unacceptable point. Fatigue life studies of many automotive components indicate however a very large margin with current components and materials, suggesting a degree of overdesign.

Civil Engineering

Laminated elastomeric bearings have been used to support bridge decks since about 1957 and plain rubber bearings since at least 1890. Previous

Fig. 13.   Sectioned bridge bearing - sectioned by the author to reveal internal structure and permit testing of rubber quality after 20 years service.

publications by the author have described detailed case histories of rubber from bridge bearings after 21 years and 100 years service (Stevenson, 1985, 1986). It has been found that elastomeric bearings are very much more reliable and long lasting than the metallic roller and knuckle bearings which they replace. During the last 10 years there has been a widespread trend towards using elastomeric bearings and major new bridges are nowadays rarely not fitted with rubber bearings in Europe and the USA. The type of rubber used for this application varies. In Great Britain, bridge bearings are almost always based on natural rubber. In the USA, France and West Germany they are more usually polychloroprene-based. For low temperatures natural rubber is to be preferred. For hot climates there can be an advantage in using polychloroprene which has better ozone and weathering resistance. In practice a bridge bearing is such a large bulky component that the bulk of rubber is an effective protection from oxidation. This arises because oxygen diffuses more slowly through oxidised rubber than fresh rubber so that a protective skin forms, even on relatively unprotected natural rubber. Thus studies have shown that unprotected NR after 100 years in a hot climate (Australia) has still functioned adequately in spite of the formation of a surface crust of harder oxidized material. Figure 13 shows a bridge bearing sectioned and studied by the author after 21 years service supporting the main M2 motorway in southern England. No deterioration could be discovered even in the surface layers of rubber. This bearing was fabricated from chemically protected natural rubber by BTR/Andre Rubber Ltd.

The calculation of bridge bearing stiffness for bearings of structure similar to that shown in Figure 13 is quite straightforward from equations 1-10. Normally, compression stiffness, shear stiffness and rotation stiffness should be calculated and then, the bearing designed and then prototypes tested before production and installation. One key function of bridge bearings is to allow thermal expansion and contraction of the reinforced concrete bridge deck without transmitting high forces into the supporting piers of the structure. Different considerations will apply depending on the bridge structure and loading requirements.

Another type of elastomeric bearing that does not accommodate thermal expansion but is effective in dealing with live load rotations is the pot bearing, illustrated typically in Figure 14. A PTFE slider plate then provides the mechanism for accommodating thermal expansions and contractions of the bridge deck.

Rotaflon bearings

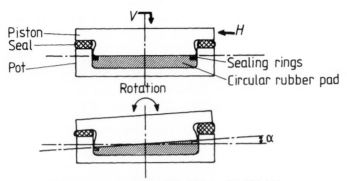

Vertical load capacity 'V'= 500 to 50 000 kN
Horizontal load capacity 'H'=30% of 'V'
Rotation α about any horizontal axis=1.5°

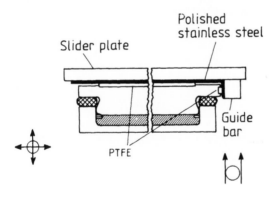

Fig. 14. Typical pot bearing structure (Courtesy BTR).

Offshore engineering

The last twelve years have seen many developments in the production of oil and gas from offshore fields, especially in the North Sea where relatively hostile conditions have stimulated many engineering advances. Great Britain has thus increasingly become a proving ground for the development of new advanced technology which may then be applied elsewhere. The foremost example of this is the Tension Leg Platform installed by Conoco (UK) Ltd for the Hutton Field in 1984. The TLP is a floating production platform held in place by 16 vertical tension legs which are attached to foundation templates piled to the seabed. The tension legs are held taut because the platform has more bouyancy than weight. The TLP is a compliant structure which responds dynamically to wave excitation. In a free floating condition, the platform would behave like semi-submersible and move in the 6 degrees of freedom associated with all floating bodies, namely surge, sway, heave, roll, pitch and yaw. The tension legs resist heave pitch and roll, but surge, sway and yaw still occur and depend on wave excitation and platform response. These mooring system assemblies provide flexible connections provided by rubber flexjoints. Each flexjoint contained 27 layers of nitrile rubber bonded to alternating layers of stainless steel. The flexjoints are primary structural components in the TLP mooring system, located in the crossload bearing assembly and the anchor connector and are required to have a design life equal to that of the platform structure. Extensive design analysis was performed to provide detailed information on flexjoint integrity. This included a fracture mechanics based analysis of fatigue life and extensive fatigue testing of full scale elements. The TLP is a success story by any standards. It was installed in 1984 and has now experienced several severe storms, which it has handled without incident. Recent detailed inspections have shown that the rubber flexjoints are still in excellent condition with no evidence of deterioration.

TESTING

For all of the engineering applications described in the previous section, simple standard tests have proven inadequate. The level of engineering performance required has made it essential that improved methods for the characterization of materials be developed. In automotive applications for bushes and mounts detailed knowledge of dynamic mechanical properties has become increasingly necessary for compound development and component evaluation. There is a growing use of sophisticated servo-hydraulic test equipment capable of providing this information in an efficient and precise manner for components and materials test pieces alike. An MTS servohydraulic test machine has been widely used for many tests between 1 and 400 Hz and temperatures between -50°C and +200°C. MERL staff have developed machine control software so that an automated sweep of frequency (at a constant amplitude) or amplitude (at a constant frequency) may be performed. This approach can then generate test data of the type shown in Figures 6-8.

Fatigue testing is essential for any critical engineering application. The most reliable method has generally proven to be based on a fracture mechanics approach, to derive tearing energy data in the format of Figure 9.

Materials fatigue tests may be performed very precisely on the servo-hydraulic test equipment, using a video camera with magnifying optics to provide crack tip measurements. This can, however, only test one sample at a time and so a multistation fatigue test equipment has been developed.

Environmental testing also needs special equipment. In an oil and gas environment, permeation of gas and diffusion of liquids are important material properties. A special permeation cell has been developed at MERL for elastomers under high pressures and at high temperatures. Data collection has been automated so that a computer output can provide a good measure of permeation properties at pressures of 10,000 psi and more.

In future there are likely to be increased needs for improved knowledge of rubber properties and improved techniques for their measurement. This will underpin the development of even more critical and sophisticated applications yet to be devised by engineers.

REFERENCES

Derham, C.J. 1975. Luxury without rumble. The Consulting Engineer, July 1975.

Gent, A.N. and Lindley, P.B. 1959. The compression of bonded rubber blocks. Proc. Inst. Mech. Eng., 173: 111.

Harris, J.A. 1989. Finite element study rubber truck mount using ABAQUS. Unpublished work. Material Engineering Research Laboratory.

Harris, J.A. and Stevenson, A. 1986. On the role of non-linearity in the dynamic behaviour of rubber components. Rubber Chem. Technol., 59: 740-764.

Lake, G.J. and Lindley, P.B. 1965. Cut growth and fatigue of rubbers. J. Appl. Polym. Sci., 9: 1233

McIntosh, W. and Stevenson, A. 1981. Operating experience with the Hutton tension leg platform. Proc. 3rd Int. Conf. Polymers in Offshore Engineering, Glenegles, Scotland, PRI, London.

Payne, A.R. 1964. Transmissibility through and wear effects in rubber. Rubber Chem. Technol., 37: 1190-1244.

Rivlin, R.S. and Thomas, A.G. 1953. Rupture of rubber. J. Polym. Sci., 10: 291.

Sedillot, F. and Stevenson, A. 1983. Laminated rubber activated joint for the deep gravity tower. A.S.M.E. Trans., J. Energy Res. Tech., 105, 480.

Stevenson, A. 1983. A fracture mechanics study of the fatigue of rubber in compression. Int. J. Fracture, 23: 47.

Stevenson, A. 1985. Longevity of natural rubber in structural bearings. Plast. and Rubb. Proc. and Applns., 5: 253.

Stevenson, A. 1986. Rubber in Engineering. Chapter G3 in Kemps Engineering Yearbook, Morgan Grampian Publishers.

Stevenson, A., Sedillot, F. and Monier, R. 1986. Proc. 18th Offshore Technology Conference, Houston, May 1986.

CHAPTER 22

THERMOPLASTIC NATURAL RUBBER

NAMITA ROY CHOUDHURY, ANIL K. BHOWMICK and S.K. De

Rubber Technology Centre, Indian Institute of Technology,
Kharagpur-721302, India.

INTRODUCTION

The technology of blending generically different polymers has developed into an important segment of polymer science in the last two decades. A large number of blends are possible: rubber-rubber blends (Roland, 1988; Corish, 1978; Marsh et al. 1967; 1968; Hess et al. 1967), rubber-plastic blends (Elliott, 1982; Newman, 1978; Morris, 1979) and plastic-plastic blends (Locke and Paul, 1973; Smith et al. 1981; Lipatov et al. 1981; Barlow and Paul, 1984; Ide and Hasegawa, 1974). Much attention is currently being devoted to rubber-plastic blends which combine the desirable properties of these polymers. The ratio of the main components of the blend can be varied to give different grades possessing properties ranging from the high elasticity of a soft rubber (thermoplastic elastomer) to the impact resistance of a toughened plastic (rubber modified thermo-plastics). Both thermoplastic elastomers and toughened plastics are multi-phase polymer systems composed of hard and soft domains. They may be copolymers or simple physical blends but in either case, it is their particular morphology which gives them their unique properties.

WHY THERMOPLASTIC ELASTOMER?

The economic attraction of thermoplastic processing compared to the traditional multistep thermoset rubber processing has led to the popularity of thermoplastic materials. Each step of rubber processing is capital, labour and energy intensive. Moreover, it leads to nonusable scrap. Although scrap losses resulting from mixing may be reduced through better process control, it is unlikely that this can be entirely eliminated. Many vulcanization processes such as lead press curing for industrial hose and cable jacketing or extended cure cycles for compression moulding are highly energy intensive. Thus, thermoplastic processing is more economical because of the reduced demand on capital, energy and labour which results from the simplicity of their processing. Apart from this, the key factors responsible for the continued acceptance and growth of thermoplastic rubbers are the raw material costs and the performance of the finished products over a wide range of applications.

The heat-fugitive linkage in thermoplastic elastomer allows it to soften and flow under shear at elevated temperature leading to easy processability as in the case of a true thermoplastic material. Thus to summarise, the advantages such as faster fabrication, lower material cost, scrap recycle and good performance contribute to the growth in demand for TPEs.

Theoretically, a large number of blends are possible, but only a few of them have attained technological importance (Coran, 1988; Coran et al. 1985). The difficulty in making a majority of blends technologically attractive is due mainly to the lack of specific interaction between the component phases and processing difficulties. Though most of the rubber-plastic blends are miscible at melting temperatures, they show segregation into respective phases upon cooling. The need for good adhesion (Kammer and Piglowski, 1984) between the component phases is thus of considerable importance. If it is not strong enough, voids would form at the interface and which would lead to crack initiation. Once such a crack is developed, it would propagate very quickly with little hindrance from the poorly anchored phases in its path. Thus the properties of a blend are a function of the characteristics of the blend components, their mutual wettability, the adhesion between them and phase morphology. Molecular interdiffusion may increase wetting of one phase by the other and reduce the effective interfacial tension. Increased adhesion would then be expected to confer improved ultimate properties on the blends.

THERMOPLASTIC NATURAL RUBBER (TPNR)

Thermoplastic natural rubber (TPNR) is usually a blend of natural rubber (NR) with polyolefins [polyethylene (PE) or polypropylene (PP)]. As for any blend system, the aim of developing TPNR is to combine the desirable properties of the two constituent polymers - eg., thermoplastic nature of the polyolefin and rubbery properties of NR. NR has many attributes to offer in TPE, but the challenge, as so often in blending, has been to make both polymers realise their full potential in the blend. Thus, TPNR is analogous to the olefinic TPEs, being based on physical blends of NR with polyoefins and is in a category of improved TPEs.

Another route to obtain TPNR is to graft onto NR some reactive thermoplastic in a precise, controlled and predictable manner. This overcomes the deficiencies of conventional latex graft methods used to make methyl methacrylate graft or styrene graft rubbers. This class of thermoplastic elastomers have properties and characteristics that are slightly different from the earlier described TPNR blends.

In this chapter emphasis will be on TPNR blends which are a family of materials derived from NR and polyolefins. With changes in composition of the blend, materials with a wide range of properties can be obtained. The family is considered to have two distinct types. At high rubber contents the blends behave as a thermoplastic elastomer (soft TPNR) whilst at low rubber contents, semi-rigid, rubber modified plastics are obtained (hard TPNR). Only the soft TPNR blends deserve detailed discussion here.

MIXING AND BLEND PREPARATION

Blending of polymers can be accomplished by melt mixing, solution blending or latex mixing (Gesner, 1969). Rubber-plastic blends, discussed here are prepared by melt mxing.

The basic factors to be considered for the preparation of thermoplastic natural rubber are:-

(1) High shear stress throughout the mass should be created for intensive mixing.

(2) Mixing time should be such that it permits the particles of the dispersed component to reach equilibrium size. However, it has been observed by Roy Choudhury and Bhowmick (1989a) that on prolonged mixing of a system with conventional curatives in rubber, the thermoplastic elastomer loses its processability and strength. The factors like time and temperature are important for large scale mixing.

(3) During blending local overheating should be avoided, because it causes degradation of the polymer (rubber or plastic) or it may greatly increase the fluidity of the plastic leading to poor dispersion of the rubber component.

Thus mixing can be carried out in:

(i) High speed, two rotor continuous mixers,

(ii) Twin screw extruders, or

(iii) Single screw extruders (L:D ratio > 30).

Simple rubber-plastic blends are prepared by mixing the components for about 3-4 mins, above the melting temperature of the plastic (150°C for PE, 180°C for PP). Higher temperatures are undesirable because some degradation of NR is likely to occur, especially above 200°C. Although this does not significantly affect the initial physical properties, it may influence the ageing behaviour of the blend. The blend is usually granulated or better, pelletised. For the purpose of granulation the blend, whilst still hot, is first sheeted by a single pass through a two roll mill. The sheets are then cut into strips before being fed into a rotary cutter for granulation. Grades of TPNR of all hardness have been produced in the form of pellets. The most suitable pelletising system for TPNR is an under-water die face cutter.

Gessler (1962) used the technique of dynamic vulcanization for 'semi-rigid' rubber plastic blends. Fischer (1973) used the process for the preparation of blends containing different amounts of partially vulcanized rubber. The use of HVA-2 (metaphenylene bismaleimide) has been recommended by Elliott (1986) for curing the NR phase and thus improving the properties of the blend. Recently Coran and Patel (1980), Kuriakose and De (1985b), Akhtar et al. (1985), Roy Choudhury and Bhowmick (1988b) prepared thermoplastic elastomeric compositions of rubber and plastics by dynamic vulcanization (in situ vulcanization). Rubber and plastics are mixed in the molten state as for simple blends. After homogeneous mixing vulcanizing agents are added. Vulcanization continues with the progress of mixing which is monitored as a function of mixing torque (Fig. 1). After discharge, it is processed in the same way as simple blends.

REUSE OF SCRAP AND RECYCLING

The main attraction of TPE is the possibility for recycling the products and reuse of scraps. In contrast to vulcanized rubbers, TPE scraps are reusable without much sacrifice in their processing characteristics. Incorporation of upto 30% of the recycled scrap into the virgin material does not register any change in the properties of the finished article.

After repeated recycling the processing characteristics of TPNR, which does not contain vulcanizing agents, do not change much, as indicated by the torque value. Only the stock temperature drops by a few degrees. However, tensile strength decreases after the first cycle. This may be attributed to thermooxidative degradation of rubber due to repeated mixing and moulding at elevated temperature. Akhtar and De (1987a) reported similar observations in connection with NR/HDPE blend (Fig. 2).

MORPHOLOGY OF THERMOPLASTIC NATURAL RUBBER

Generally, rubber-plastic blends exhibit structural heterogeneity on a molecular level as upon cooling one or more of the components separate from the mixture as a pure crystalline phase even though in the melt the components are miscible. An understanding of the properties thus requires

522

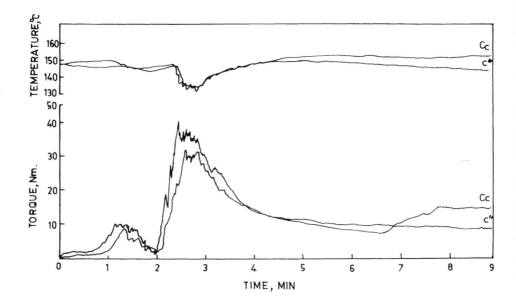

Fig. 1. Plastographs of blend C and C_c
C:NR/HDPE::70:30,
C_c:NR/HDPE/DCP::70:30:1.

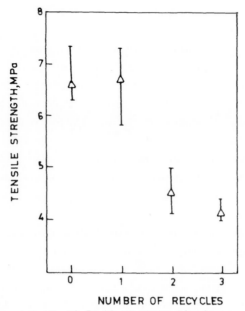

Fig. 2. Tensile strength of 70:30 NR/HDPE blends vs. number of recycles.

a study of the structure. Microscale morphology largely governs the properties of heterogeneous polymer blends. Again, the phase morphology of the thermoplastic elastomer can be controlled by the viscoelastic difference between the components. The fineness of dispersion of a particular component in the intensive mixing process depends on both the viscous and elastic behaviour of the component under the mixing conditions. It was reported (Kresge, 1978) that the phase morphology changes as a function of blend composition, viscosity of the components, shear rate and temperature during blending. Blends with a greater disparity in the components' melt viscosity show larger domains. Generally, the component having larger concentration forms the continuous phase when both components are of comparable viscosity at the conditions of mixing. At equal concentrations, the polymer with lower viscosity forms the continuous phase. In some cases the low viscosity of the plastic at the mixing temperature and the high proportion of rubber in the mixes favours both the components to form continuous phases (Kresge, 1978). Other characteristics of phase morphology are particle size and shape. Mixing time determines particle size and shape. The most interesting aspects of these crystallizable systems concern the influence of processing and crystallization conditions on the overall morphology, the rate of crystallization, the crystallinity, the melting behaviour and some structural parameters. The two components may thus influence each other in a number of ways.

The mixing torque and the mismatch in mixing torque also control the particle size. When the mixing torque is high and it is matching, the smallest domain size is obtained (Coran, 1988). Moreover, the particle size of the dispersed phase is largely affected by interaction between the phases which in turn is influenced by the difference between the surface energies of the two phases (Roy Choudhury and Bhowmick, 1988a). The large particles resulting from blending of thermodynamically incompatible polymers give blends of poor mechanical strength. However, the use of interaction promoters and compatibilizers greatly enhance the properties of such blends (Roy Choudhury and Bhowmick, 1989a).

The quality of the final TPNR product is greatly improved by dynamic vulcanization, which is the process of crosslinking the rubbery phase during mixing. It also offers a route to control the particle size by resisting agglomeration, unlike blends containing uncrosslinked rubber. It is reported (Coran et al. 1978; Coran and Patel, 1981) that increase in crosslink density in the case of thermoplastic rubbers gradually increases their elastic and strength properties and thermoplastic behaviour is still retained as the crosslinked rubber is dispersed as very small particles. Elliott (1982) studied the NR/PP system and reported that for a blend containing 65% of NR, partial crosslinking of the NR phase can improve some physical properties. Kresge (1978) reviewed the change in morphology resulting from processing a blend composition to finished part. Roy Choudhury and Bhowmick (1988a, 1988b) pointed out that by incorporation of a third component like EPDM or CPE, which possess some structural similarity with the plastic phase and at the same time, rubbery in nature, into NR/PE or NR/PP systems, the dispersed particle size could be greatly reduced (in some cases upto 33%) indicating better interaction between component phases through the third polymer (Fig. 3). For NR/EPDM/PE blend the average diameter of the dispersed domain was found to be 1.2 μm and both phases were continuous. The addition of modified rubber and modified plastic to NR/PE systems to improve their interaction has pronounced effect on their properties (Roy Choudhury and Bhowmick, 1989a). These systems offer interpenetrating network as observed from the respective photomicrographs. The morphological observations are in line with the properties of the blends. Thus the system which has improved mechanical strength also has better cocontinuous nature of the phases permitting direct load transfer between the components. The stress transfer area is also larger because of the existence of both components as continuous phases.

Tinker et al. (1989) reported a continuous PP phase at a PP content as low as 20%. The NR phase is dispersed at concentrations lower than 40%

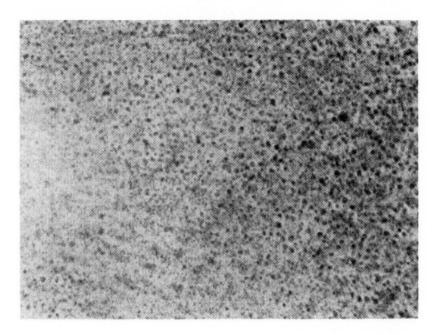

Fig. 3.

and becomes continuous at a composition between 40:60 and 50:50 NR/PP. The continuous nature of the PP phase was evident in the scanning electron micrograph of surfaces of blends extracted with chloroform. Spongy PP surface was left which retained its structural integrity.

Other microscopic techniques were also applied in the study of the morphology of dynamically vulcanized TPNR blends. Tinker et al. (1989) observed good contrast between the phases in optical microscopy of thin sections using either differential interference contrast or phase contrast. Phase morphology remains unaltered even after partial dynamic vulcanization.

DSC and DMA results of NR/PE and NR/PP are given in Tables 1-3. These blends are incompatible as observed from T_g values even with the use of modifiers (Roy Choudhury and Bhowmick, 1989b, 1990). The chemical structure of rubber and its proportion in the blend do not influence significantly the melting temperature (T_m) of plastics. The decrease in crystallinity (x_c) of plastics with the incorporation of rubber is related to incomplete crystallization observed from the lower value of ΔH (Table 1).

TABLE 1

Thermal properties of NR/PE blends

Parameters	PE	PEm	Sample reference					
			A 90/30 NR/PE	B 70/30 NR/PE	C 70/20/30 NR/EPDM/PE	H 50/50 NR/PE	K 50/20/50 NR/CPE/PE	N 70/20/3/27 NR/ENR/PEm/PE
Peak temperature - Tm (°C)	110	106.5	106.8	107.4	105.2	108.3	107.2	107.2
Heat of fusion - ΔH (J/g)	86	80	24.4	29.0	23.15	43.37	48.6	24.6
Percent crystallinity - Xc. (%)	30	25	8.4	10.0	8.3	15.0	17.0	8.8
Glass transition temperature - Tg (°C)	-120	-	-66	-65	-63	-66	-67	-63

TABLE 2

Thermoal properties of NR/PP blends.

Parameter	PP	Sample reference						NR
		A' 90/30 NR/PP	B' 70/30 NR/PP	C' 70/20/30 NR/EPDM/PP	D' 70/20/30 NR/CPE/PP	E' 50/50 NR/PP	F' 50/20/50 NR/EPDM/PP	
Peak temperature - Tm (°C)	165	160	161.2	160.5	160.9	163	162	-
Heat of fusion - ΔH (J/g)	84	18.8	22.15	18.01	21.84	44	36	-
Percent crystallinity (%)	61	13.6	16	13	15.8	32	26	-
Glass transition temperature - Tg (°C)	-65	-64	-64.4	-64	-63	-63.5	-	-70

TABLE 3

Dynamic mechanical properties of NR/PE blends

Sample reference	Tg_1* ($^\circ$C)	Tg_2** ($^\circ$C)	$\tan \delta_1$	$\tan \delta_2$	E''_1 (peak temp.) ($^\circ$C)	E''_2 (peak temp.) ($^\circ$C)
NR/PE/S-cure 70/30	-128	-47	0.031	0.564	-128	-53
NR/PE 90/30	-124.5	-53	0.030	0.885	-124.5	-58.7
NR/EPDM/PE 70/20/30	-119	-52.9	0.029	0.429	-119	-58.8
NR/ENR/PEm/PE 70/20/3/27	-113	-52.8 -26	0.031	0.420	-113	-55.9
EPDM/PE/S-cure 70/30	-136.7	-43	0.062	0.352	-136.7	-47
PE	-131	-10	0.058	0.086	-131	-20

* Tg_1 - Glass transition temperature of polyethylene phase.

** Tg_2 - Glass transition temperature of rubber phase.

The DMA spectra show separate T_g values for the blends. The storage modulus changes from a very high value to a much lower value with increase in temperature. On crosslinking, there is a shift in the transition region towards higher temperature (Kuriakose et al. 1986a). The loss modulus sharply increases in the transition zone till it attains maxima and then falls off with the rise in temperature.

X-ray diffraction patterns indicate decrease in crystallinity with increase in rubber content which also increases the interplanar distance due to migration of rubber into interchain space of polyethylene (Roy Choudhury and Bhowmick, 1989b).

EFFECT OF BLEND RATIO ON MECHANICAL PROPERTIES

Formulation and properties of the various blends are given in Table 4. Pure polyolefin showed the highest tensile strength. With the addition of rubber, tensile strength and modulus decrease, while elongation at break increases. The strength of TPNR depends on the strength of the hard phase, which in turn is dependent on the degree of crystallinity. Martuscelli (1984)

TABLE 4

Physical properties of NR-HDPE blends

Property	HDPE	Sample reference							NR
		A" 70/30 PE/NR	A_C 70/30/1 PE/NR/DCP	B" 50/50 PE/NR	B_C 50/50/1 PE/NR/DCP	C" 30/70 PE/NR	C_C 30/70/1 PE/NR/DCP	
Young's modulus - (MPa)	29.4	13.4	14.5	5.8	6.1	1.4	2.4	0.4
Modulus 100% - (MPa)	-	-	-	7.2	8.7	2.2	4.6	0.8
Tensile strength - (MPa)	32.1	13.5	14.6	10.9	11.3	9.5	13.0	2.3
Elongation at break - (%)	30	65	24	420	430	460	500	450
Tear strength - (kN/m)	118.0	72.5	74.4	47.2	55.7	28.1	40.8	4.35
Tension set - (%)	-	-	-	92	>100	72	64	>100

and Roy Choudhury and Bhowmick (1989b, 1990) have shown that the growth of polyolefin crystals is disturbed by the presence of the rubber phase. Hence the drop in strength is due to a drop in crystallinity. In the high plastic blends, yielding and plastic deformation are observed, whereas the thermo-plastic elastomeric blends exhibit rubbery behaviour.

On introduction of a small amount of crosslinks during mixing, there is uniform enhancement of tensile strength, modulus and tear resistance in 70/30 NR/polyolefin blends. Stress-strain curves of a few blends with and without crosslinking are shown in Fig. 4. However, this effect is not so

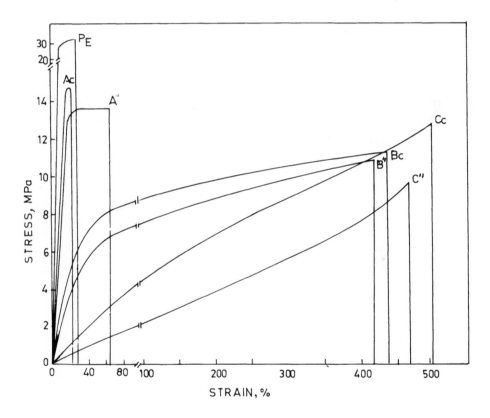

Fig. 4. Stress-strain curves of NR-HDPE blends.

prominent in plastic blends. Hardness decreased with an increase in the proportion of the elastomer phase in the blend. Dynamic crosslinking increases hardness. Flexural modulus also decreased with increasing rubber content in the blends. All the samples show very high set values as compared to rubber and the set increases with the plastic content. As usual the tensile impact energy increases with the increase in rubber content of the blends. Abrasion loss remains more or less constant upto a ratio of 50/50 and then increases with increasing proportion of rubber.

Kuriakose and De (1985b) reported that tear strength of thermoplastic polypropylene-natural rubber blends decreases with increase in rubber content in the blend but dynamic crosslinking of the elastomer phase gives higher tear strength values as compared with those of the uncrosslinked blends. Scanning electron microscopic examination of the tear fracture surfaces indicates that brittle fracture of polypropylene is changed to a ductile type with the addition of natural rubber and crosslinking of the elastomer phase changes the high deformation nature of the blends (plastic type flow) to restricted flow (elastic type), under tear fracture.

EFFECT OF MODIFIERS

Since natural rubber and polyoefins are incompatible, physical and chemical modifiers help in improving the properties of thermoplastic rubber (Roy Choudhury and Bowmick, 1988b, 1989a). It was observed that use of EPDM, CPE and CSPE at levels of 10 to 20 phr increases the strength of 70/30 NR/PE blends. This enhancement, which is almost double, is due to a reduction in the domain size as mentioned earlier. The cohesive strength of the third phase is also equally important. Further improvement could be achieved by chemical modification (use of maleic anhydride modified PE and ENR/sulphonated EPDM), in such a way that the blends form covalent linkages at the interface. The introduction of dynamic cross-linking to a physically modified system improves strength still further. However, the same is not achieved with a chemically modified system.

EFFECT OF ADDITIVES ON PHYSICAL PROPERTIES

Incorporation of fillers in polymers generally leads to modification of physical properties, optimisation of processing characteristics and reduction in cost. But practical modification of TPNR blends is almost limited to relatively small amounts of fillers for specific properties. In hard TPNR blends there is a general deterioration of physical properties with the addition of 40 phr of HAF black which causes a reduction in plastic-rubber interaction. With the increase in rubber content, the effect of carbon black becomes gradually favourable. According to Marsh et al. (1970) and Callan et al. (1971) carbon black has a greater tendency to disperse into polyisoprene than into polyethylene. Akhtar et al. (1985) observed 50% increase in tensile strength and modulus at 100% strain in high rubber blends of NR/HDPE on addition of HAF black. When both cross-linking agent and carbon black are present the increase in strength is less than that in either black or cured systems and there is a sharp fall in the elongation at break. The same trend, as observed by Elliott (1982), also reveals that general purpose blacks provide very little additional reinforcement in NR/PP blend to that given by the polyolefin when an optimum level of crosslinking agent is employed. However, at equal proportions of rubber and plastic, carbon black has no effect essentially. Moderate amount of cheap fillers may be used to reduce cost. Small amounts of clay have little effect on hardness, stiffness or strength though extensibility is reduced, and Young's modulus decreased. The effect of filler can be visualized in two ways (1) stiffening of the rubber, and (2) increasing the volume of its phase. Another effect of filler is that it drastically reduces thermoplasticity and hence, the expected ease of fabrication. In order to regain the full potential of fillers, plasticizers may be used.

Kuriakose and De (1986b) have reported that addition of silica filler improves tear strength of the 70:30 natural rubber:polypropylene thermoplastic elastomeric blend and incorporation of silane coupling agent enhances it further.

In order to protect against ultraviolet light 1-5% of carbon black or titanium dioxide may be added to TPNR blends. In some cases high levels of filler is useful. Abrasion resistance of light coloured TPNR could be improved by addition of precipitated silica. For better processability like improved extrusion or reduced mould shrinkage, use of whiting, china clay and talc was reported with little effect on most physical properties. Though the use of high levels of cheap fillers reduces the cost of TPNR, its chief attribute - low density, is lost which is particularly important in automotive applications for which light weight components are required in the interest of lower fuel consumption. Short fibre was also added to improve the properties of thermoplastic elastomeric blends (Akhtar et al. 1986a). Tensile and tear properties were substantially enhanced, but the ultimate elongation decreased sharply with increasing loading of short fibre in the blend. The effect of fibre orientation and the development of anisotropy in properties was also noted. Scanning electron microscopic (SEM) studies of the benzene extracted surfaces of NR/HDPE blends substantiated the theory of fibres providing anchoring between the rubber and the thermoplastic phases. The effect of fibre loading on the tear and tensile properties of the blends of NR/LDPE with varying blend ratios was studied. Improvement in physical properties on the addition of short fibres observed in the high rubber blends was higher than that in high plastic blends.

Akhtar et al (1986c) studied tensile rupture in short silk fibre-filled thermoplastic elastomer blends from low-density polyethylene and natural rubber, containing pre-cut fibre of different lengths. σ_b, the tensile strength of the blends, was found to decrease with increase in the cut length in accordance with the Griffith's theory of fracture. However, unlike in the case of vulcanized NR, no critical cut length (at which the strength drops abruptly) was found in the case of both blends and unvulcanized NR. The energy to fracture per unit volume, W_b, also varied inversely with the length of the pre-cut. Values of inherent flaw size, lo, of the composites, determined by extrapolation of W_b to the value obtained when no initial pre-cut was present, were found to increase with fibre loading. Blends with longitudinally oriented fibres showed higher lo values than those with tranversely oriented fibres.

ANTISTATIC COMPOUNDS

Antistatic compounds can be made with TPNR and conductive carbon blacks. By adding the black early in the mixing cycle, good dispersion of black is obtained, because of high shear force and if added late or in the second stage, poor dispersion and hence inferior strength results (Campbell et al. 1979).

EFFECT OF OIL AND PLASTICISERS

Mineral oil and plasticisers reduce hardness and melt viscosity of TPNR blends. Upto 15% of oil is recommended (Elliott, 1982). Oils of high viscosity have a tendency to bloom. They also improve processability but generally decrease tensile strength.

EFFECT OF CURATIVES

The key to gaining acceptable rubbery properties in TPNR is to partially crosslink the NR phase by a method which allows the thermo-plastic processability to be maintained. Crosslinking improves the properties of TPNR blends, which are high in NR content. Several cross-linking systems eg., sulphur, sulphur donors (Elliott and Tinker, 1985; Akhtar et al. 1985), peroxides, dimaleimides and combination of crosslinking systems (Kuriakose and De, 1985, 1986) have been studied. The inferior physical properties of TPNR based on NR/PP were reported to be due to the thermal instability of polysulphide crosslinks but favourable properties could be achieved with sulphur donor systems. DCP was found to increase hardness, modulus, tensile strength and elasticity at high rubber concentrations. DCP has some degradative action on PP (Elliott, 1986), but dimaleimide (NN' -metaphenylene dimaleimide -HVA 2) crosslinks NR. When initiated with a small quantity of peroxide, DCP primarily crosslinks the rubber phase of TPNR based on NR/LDPE, while the polyethylene phase remains largely unaffected. However, better strength was obtained with sulphur system (Roy Choudhury and Bhowmick, 1990). Dynamic crosslinking with Novor 924 (diurethane) and phenolic resin (SP 1045) has also been reported by Elliott (1989) for NR/PP system.

Kuriakose et al. (1985b) observed that tensile strength of thermoplastic PP-NR blends decreases with increase in the proportion of the elastomer, both in uncrosslinked and in crosslinked blends. Dynamic crosslinking of the elastomer phase increases the tensile strength of the blends and this effect is more prominent in the blends containing a higher proportion of the elastomer phase. Addition of rubber changes the brittle fracture of the polypropylene to a ductile type and crosslinking of the elastomer phase changes the deformation behaviour of the blends from plastic to elastic type. Kuriakose et al. (1986a) studied the dynamic mechanical properties of thermoplastic elastomers from polypropylene natural rubber blends with special reference to the effect of blend ratio and extent of dynamic cross-linking of the elastomer phase. The effects of HAF black and silica fillers have also been studied. It has been found that increasing the proportion of elastomer phase reduced the storage modulus and increased the loss tangent values of the blends. The effect of dynamic crosslinking was found to be more prominent in blends containing higher proportions of elastomer phase. The improvement in storage modulus and decrease in loss tangent values were quite remarkable with increase in the extent of crosslinking in these blends. The 70:30 NR:PP blend was found to exist as a two-phase system, both the components forming continuous phases of the blend.

RHEOLOGY

The rheological characteristics of molten polyolefins are important factors governing the supermolecular structure of their blends and hence controlling their processability and performance characteristics. The ratio of the rheological characteristics of the components determines the fineness of the dispersed phase. Thus a thorough knowledge of the rheological behaviour over a wide range of shear rate or stress is of paramount importance in order to predict the flow behaviour during processing. TPNR has high melt viscosity and hence it is necessary to use higher processing temperature and pressure during fabrication of these materials into useful products. Akhtar et al. (1987) studied the rheological characteristics of NR/HDPE blends by capillary rheometry, with reference to the effect of dynamic crosslinking, blend ratio and addition of carbon black. Since there is an intimate dependence of morphology on the melt rheology and hence the mixing process and shear, they also examined the morphology of these TPNR extrudates in order to understand how the rheological parameters influence morphology. Kuriakose and De (1985a) also made similar observations in the case of NR/PP blends. They found that viscosity of the blends decreases monotonically with increasing shear stress indicating their pseudoplastic behaviour. At low shear stresses, viscosity of the blend decreases with increasing proportion of rubber in the blend. One most important feature is that at most shear rates viscosity of the blend appears to be a nonadditive function of the viscosity of the homopolymers. Akhtar et al. (1987b) tested the Hashin's model using variational principles for NR/HDPE blend at a shear stress of 1.5×10^7 pa at $150^\circ C$ and found that the experimental values lie exactly on the lower limit of the model. Viscosity of the TPNR blends drops with increasing temperature. At low shear rates the high rubber blends have higher viscosities. At high shear rates the blends display the same viscosity values at all temperatures irrespective of the blend ratio.

Kuriakose and De (1985a) concluded that NR/PP blends could be processed like thermoplastic PP at high shear stress. In 70/30 NR/PP blends the observed difference in viscosity of blends is proportional to the degree of crosslinking of the rubber phase. The degradative effect of DCP on the blends was also observed. It was also seen that the effect of extent of cross-linking on viscosity was less pronounced for the blends in the 70/30 series at higher shear stresses unlike that observed for the 50/50 blends. At higher shear rates the effect of crosslinking of the rubber phase on viscosity is not prominent as the blend attained a sheath and core like structure and the plastic phase formed a lubricating layer at the capillary wall during extrusion. Increase in temperature reduces viscosity of the blends at all shear rates. The effect of blend ratio, dynamic crosslinking and shear rate on die swell for various blends are shown in Fig. 5. Die swell increased with increase in shear rate and their effect was more prominent in the uncrosslinked blends than in the crosslinked. Increasing rubber content in the blend reduces die swell of uncrosslinked blends but for blends containing crosslinked rubber phase die swell depends on shear rate and degree of crosslinking of the rubber phase. Fig. 6 shows that the deformation of extrudate increases with increase in shear rate for both uncrosslinked and crosslinked blends. Increasing rubber content of the blend beyond 30% increases melt fracture of extrudate. Extrudate deformation characteristics of the NR/PP blend is also shown in the same figure.

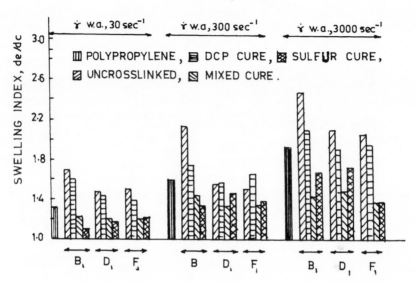

Fig. 5. Effect of blend ratio, dynamic crosslinking and shear rate on die
swell. B_1: 30:70 NR/PP blend (U,D,S,M) with 50 phr HAF black.
D_1: 50:50 NR/PP blend (U,D,S,M). F_1: 70:30 NR/PP blend (U,D,S,M).
U : Blends without curative. D : Blends containing 1.0 phr 40%
DCP based on rubber phase only. S : Blends with ZnO 5.0, St. acid
2.0, CBS 2.0, TMTD 2.5 and S 0.30 phr. M : Blends containing
1.0 phr 40% DCP, ZnO 5.0, St. acid 2.0, CBS 1.0, TMTD 1.25 and
S 0.15 phr.

These TPNR materials exhibit pronounced non-Newtonian flow and their
melt viscosities are highly sensitive to the rate of shear. But the melt
viscosities are less sensitive to temperature variations above the melting
point of the polyolefin. This helps in maintaining the properties despite
fluctuations in processing temperature.

Elliott (1987) observed that dynamic crosslinking increases melt
viscosity of TPNR, an effect most prominent at low shear rates. But at
high shear rates, in extruders and in injection moulding machines, the
material flows fast. Soft TPNR (based on NR/PP), processed by injection
moulding and extrusion, offers mat surface finish but these could be reduced
by increasing mould temperature to about 80°C. High injection speed and
pressure are needed for obtaining adequate flow in the mould. Since the
dispersed rubber phase has a higher viscosity than the continuous PP
phase, the continuous phase experiences the same deformation in all the
blends. The deformation undergone by the dispersed phase depends on its
melt viscosity (Danesi and Porter, 1978). Tinker et al. (1989) recommended

536

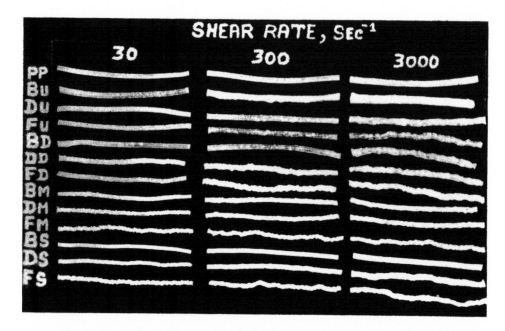

Fig. 6. Effect of shear rate and dynamic crosslinking on deformation of extrudates. B: 30:70 NR/PP blend (U,D,S,M) with 50 phr HAF black. D: 50:50 NR/PP blend (U,D,S,M). F: 70:30 NR/PP blend (U,D,S,M). U: Blends without curative. D: Blends containing 1.0 phr 40% DCP based on rubber phase only. S: Blends with ZnO 5.0, St. acid 2.0, CBS 2.0, TMTD 2.5 and S 0.30 phr. M: Blends containing 1.0 phr 40% DCP, ZnO 5.0 St. acid 2.0, CBS 1.0, TMTD 1.25 and S 0.15 phr.

L/D ratio of extruders of not less than 20 and a compression ratio of about 2.5 to 3.0.

A simple melt blend of NR/PP is not a power law fluid, the flow exponent is higher at low shear rates than at high shear rates. On dynamic vulcanization it becomes a power law fluid, over a wide range of shear rates with a flow exponent nearly equal to that of the simple blend at high shear rates. According to the analysis by Tinker et al. (1989) a substantial proportion of the dynamically vulcanized blend does not flow towards the centre of the duct relative to the surrounding material and though the blend is homogeneous (assumed) there is a tendency for PP to concentrate at the surface to form a thin layer during processing.

AGEING

Ageing of natural rubber and rubber vulcanizates has been extensively studied but there are only very limited number of reports on the ageing of thermoplastic natural rubber. To predict the service life of a material this study is essential. Elliott (1982) reported on the ageing behaviour of some natural rubber-polypropylene thermoplastic elastomeric composites. In comparison with vulcanized rubber, thermoplastic natural rubber blends are remarkably resistant to heat ageing, even at 100°C for seven days. Since interaction between the components of a blend governs its technical properties, it is quite natural that on ageing, interaction will change and hence, the technical properties. Roy Choudhury and Bhowmick (1989c) studied the ageing of natural rubber-polyethylene thermoplastic elastomeric composites with various levels of interaction in air at various temperatures and times of ageing. In general, at a particular ageing time the strength and modulus decrease with increase in temperature for all the blends (dynamically vulcanized and modified systems).

Hot air and nitric acid ageing behaviour of natural rubber-polyethylene blends have also been investigated (Akhtar et al. 1985). It has been observed that the high plastic blends show a much superior resistance to acid corrosion and hot air ageing as compared to the high rubber blends, which in turn, show much better resistance than pure natural rubber vulcanizates. Both crosslinking and HAF black have favourable effects on the hot air ageing resistance in the high plastic blends but acid resistance decreases slightly in the case of the crosslinked blends, and sharply when HAF black is present. However, when crosslinking and HAF black are present together, the high rubber blend shows extremely poor acid and hot air ageing resistance. The effect of ^{60}Co γ-radiation on the tensile properties of thermoplastic elastomer blends of natural rubber and high

538

density polyethylene has also been investigated (Akhtar et al. 1986d). The samples were irradiated to absorb doses ranging from 0.1 to 100 Mrad in air at room temperature (25°C) at a dose rate of 0.21 Mrad/h. The effect of blend ratio and addition of carbon black (N330) and dicumyl peroxide on the radiation resistance of the blends has also been studied. High energy radiation at a high dose rate was found to cause extensive crosslinking in the bulk, which in effect, caused a minimum in the ultimate tensile strength in the range of 10-25 Mrad and a continuous decline in the elongation at break in all the blends. Chain scission, on the other hand, was restricted to the surface and this was substantiated by the results of irradiation of the samples in nitrogen environment. Scanning electron microscopic (SEM) studies on the tensile fracture surface morphology of the blends have also been studied in order to gain insight into the failure mechanism.

APPLICATIONS

The major markets envisaged for soft TPNR are in moulded and extruded products eg. sports goods, domestic appliances, agricultural components, footwear, grommets, profiles for seals, hose and tubing.

The advantages of TPNR which influence its choice for these applications are:

(1)   good low temperature flexibility, elastic recovery, heat resistance and low compression set
(2)   relatively high softening point
(3)   low density
(4)   good affinity to common adhesives after surface pre-treatment
(5)   low processing cost with reusable scrap
(6)   low material cost

O'Connor and Fath (1981, 1982) have made a comparative evaluation of the properties of conventional rubber vulcanizates and TPE's. Soft TPNR, in fact, is making inroads into the market of vulcanized rubber and hence competitive in the market with vulcanized rubber, with other TPE's and with flexible plastics like plasticized PVC and EVA copolymer.

REFERENCES

Akhtar, S., De, P.P. and De, S.K. 1985. Effect of blend ratio, dynamic crosslinking and HAF black on failure properties and hot air and acid ageing resistance of thermoplastic elastomer from natural rubber-high density polyethylene blends. Mater. Chem. and Phys., 12: 235 pp.

Akhtar, S., De, P.P. and De, S.K. 1986a. Short fibre reinforced thermoplastic elastomers from blends of natural rubber and polyethylene. J. Appl. Polym. Sci., 32: 5123 pp.

Akhtar, S., De, P.P. and De, S.K. 1986b. Effect of blend ratio on the tear fracture topography of low-density polyethylene-natural rubber blends. J. Mat. Sci. Lett., 5: 399 pp.

Akhtar, S., Bhwomick, A.K., De, P.P. and De, S.K. 1986c. Tensile rupture of short fibre filled thermoplastic elastomer.   J. Mat. Sci., 21: 4179 pp.

Akhtar, S., De, P.P. and De, S.K. 1986d. Tensile failure of γ-ray irradiated blends of high density polyethylene and natural rubber. J. Appl. Polym. Sci., 32: 4169 pp.

Akhtar, S. and De, P.P. 1987. Processing characteristics and recyclability of thermoplastic elastomers from NR-HDPE blends. Kautsch. Gummi Kunst., 40(5): 464 pp.

Akhtar, S., Kuriakose, B., De, P.P. and De, S.K.1987. Rheological behaviour and extrudate morphology of thermoplastic elastomers from NR-HDPE blends. Plast. Rubb. Process. Appl., 7: 11 pp.

Barlow, J.W. and Paul, D.R. 1984. Mechanical compatibilization of immiscible blends. Polym. Eng. Sci., 24(8): 525 pp.

Callan, J.E., Hess, W.M. and Scott, C.E. 1971. Elastomer blends, compatibility and relative response to fillers. Rubb. Chem. Technol., 44: 814 pp.

Campbell, D.S., Elliott, D.J. and Wheelan, M.A. 1979. Thermoplastic natural rubber blends. NR Technol., 9(2): 21 pp.

Coran, A.Y. 1988. Thermoplastic Elastomeric rubber-plastic blends. In: A.K. Bhowmick and H.L. Stephens (Eds), Handbook of Elastomers-New Development and Technology. Marcel Dekker, New York, 288 pp.

Coran, A.Y., Patel, R. and Williams, Headd, D. 1985. Rubber thermoplastic composition, Part IX. Blends of dissimilar rubbers and plastics with technological compatibilization. Rubb. Chem. Technol., 58: 1014 pp.

Coran, A.Y. and Patel, R. 1980. Rubber thermoplastic composition: EPDM-Polypropylene thermoplastic vulcanizates. Rubb. Chem. Technol., 53: 141 pp.

Coran, A.Y., Das, B. and Patel, R. 1978. US Patent 4130535.

Coran, A.Y. and Patel, R. 1981. US Patent 4271049.

Corish, P.J. 1978. Elastomer Blends. In: F.R. Eirich (Ed.). Science and Technology of Rubbers. Academic Press Inc., New York, 489 pp.

Danesi, S. and Porter, R.S. 1978. Blends of isotactic polypropylene and ethylene propylene rubbers: rheology, morphology and mechanics. Polymer, 19: 448 pp.

Elliott, D.J. 1982. Natural Rubber Systems. In: A.J. Whelan and K.S. Lee (Eds.). Development in Rubber Technology-3. Applied Science Publ. London, 203 pp.

Elliott, D.J. 1986. Influence of crosslinking agents and interfacial adhesion promoters on the properties of natural rubber-polypropylene blends. Proc. Int. Conf. on Rubber and Rubber-like Materials. 6-8 Nov., 1986, Jamshedpur.

Elliott, D.J. and Tinker, A.J. 1986. Thermoplastic natural rubber blend. Proc. Int. Rubb. Conf. Kuala Lumpur. Vol. II. J.C. Rajarao and L.L. Amin (Eds.). Rubber Research Institute of Malaysia, Kuala Lumpur.

Elliott, D.J. 1987. Commercial prospects of thermoplastic natural rubber. Proc. Int. Conf. on Development in the Plastics and Rubber Product Industries. 5-6 July, 1987, Kuala Lumpur.

Fischer, W.K. 1973. US Patent 3758643.

Gesner, B.D. 1969. Polyblends. In: H.F. Mark and N.G. Gaylord (Eds.). Encyclopedia of Polym. Sci. and Technol. Vol. 10. Interscience Publishers, New York. 694 pp.

Gesler, A.M. 1962. US Patent 3037954.

Hess, W.M., Scott, C.E. and Callan, J.F. 1967. Carbon black distribution in elastomer blends. Rubb. Chem. Technol., 40: 371 pp.

Ide, F. and Hasegawa, A. 1974. Studies on polymer blend of nylon 6 and polypropylene and nylon 6 and polystyrene using the reaction of polymer. J. Appl. Polym. Sci., 18: 963 pp.

Kammer, H.W. and Piglowski, J. 1984. Adhesion between polymers. In: M. Kryszewski, A. Galeski and E. Martuscelli (Eds.). Polymer Blend Vol. 2, Plenum Press, New York, 19 pp.

Kresge, E.N. 1978. Rubber thermoplastic blends. In: D.R. Paul and S. Newman (Eds.). Polymer Blends-2. Academic Press Inc., New York, 293 pp.

Kuriakose, B. and De, S.K. 1985a. Studies on the melt flow behaviour of thermoplastic elastomers from polypropylene - natural rubber blends. Polym. Eng. Sci., 25(10): 630 pp.

Kuriakose, B., De, S.K., Bhagawan, S.S., Sivaramakrishnan, R. and Athithan, S.K. 1986a. Dynamic mechanical properties of thermoplastic elastomers from polypropylene-natural rubber blend. J. Appl. Polym. Sci., 32: 5509 pp.

Kuriakose, B. and De, S.K. 1986b. Tear and tear resistance of silica-filled thermoplastic polypropylene-natural rubber blend. Int. J. Polym. Mater., 11: 101 pp.

Kuriakose, B., Chakraborty, S.K. and De, S.K. 1985a. Scanning electron microscopy studies on tensile failure of thermoplastic elastomers from polypropylene-natural rubber blends. Mater. Chem. Phys., 12: 157 pp.

Kuriakose, B. and De, S.K. 1985b. Scanning electron microscopy studies on tear failure of thermoplastic elastomers from polypropylene-natural rubber blends. J. Mat. Sci. Lett., 4: 455 pp.

Lipatov, Y.S., Shumsky, V.F., Gorbatenko, A.N., Panov, Y.N. and Bolotnikova, L.S. 1981. Viscoelastic properties of polystyrene-polycarbonate blends in melt. J. Appl. Polym. Sci., 26: 499 pp.

Locke, C.E. and Paul, D.R. 1973. Chlorinated polyethylene modification of blends derived from waste plastics. Polym. Eng. Sci., 13(4): 308 pp.

Marsh, P.A., Voet, A. and Price, L.D. 1967. Electron microscopy of heterogeneous elastomer blends. Rubb. Chem. Technol., 40: 359 pp.

Marsh, P.A., Voet, A., Price, L.D. and Mullens, T.J. 1968. Fundamentals of electron microscopy of heterogeneous elastomer blends. II. Rubber Chem. Technol., 41: 344 pp.

Marsh, P.A., Mullens, T.J. and Price, L.D. 1970. Micrography of a triblend. Rubb. Chem. Technol., 43: 400 pp.

Martuscelli, E. 1984. Influence of composition, crystallization conditions and melt phase structure on solid morphology kinetics of crystallization and thermal behaviour of binary polymer/polymer blends. Polym. Eng. Sci., 24: 563 pp.

Morris, H.S. 1979. Polyolefin thermoplastic elastomers. In: B.M. Walker (Ed.). Handbook of Thermoplastic Elastomers. Van. Nostrand Reinhold, New York, 5 pp.

Newman, S.H.1978. Rubber modification of plastics. In: D.R. Paul and S. Newman (Eds.). Polymer Blends-2. Academic Press Inc., New York, 63 pp.

O'Connor, G.E. and Fath, M.A. 1981. Thermoplastic elastomers, Part I: Can TPEs compete against thermoset rubbers? Rubber World, 185(3): 25 pp.

O'Connor, G.E. and Fath, M.A. 1982. Part II. A new thermoplastic rubber. Rubber World, 185(4): 26 pp.

Roy Choudhury, N. and Bhowmick, A.K. 1988a. Influence of interaction promoter on the properties of thermoplastic elastomeric blends of natural rubber and polyethylene. J. Mater. Sci., 23: 2187 pp.

Roy Choudhury, N. and Bhowmick, A.K. 1988b. Adhesion between individual components and mechanical properties of natural rubber-polypropylene thermoplastic elastomeric blends. J. Adhesion Sci. Technol., 2: 167 pp.

Roy Choudhury, N. and Bhowmick, A.K. 1989a. Compatibilization of natural rubber-polyolefin thermoplastic elastomeric blends by phase modification. J. Appl. Polym. Sci., 38(6): 1089 pp.

Roy Choudhury, N. and Bhowmick, A.K. 1989b. Thermal, X-ray and dynamic mechanical properties of thermoplastic elastomeric natural rubber-polyethylene blends. Polymer, 30: 2047 pp.

Roy Choudhury, N. and Bhowmick, A.K. 1989c. Ageing of natural rubber-polyethylene thermoplastic elastomeric composites. Polym. Degrad. Stab., 25: 39 pp.

Roy Choudhury, N. and Bhowmick, A.K. 1990. Strength of thermoplastic elastomers from rubber-polyolefin blends. J. Mater. Sci., 25: 169 pp.

Roland, C.M. 1988. Rubber-Rubber Blends. In: A.K. Bhowmick and H.L. Stephens (Eds.). Handbook of Elastomers-New Developments and Technology. Marcel Dekker, New York. 183 pp.

Smith, W.A., Barlow, J.W. and Paul, D.R. 1981. Chemistry of miscible polycarbonate-copolyester blends. J. Appl. Polym. Sci., 26: 4233 pp.

Tinker, A.J., Icneogle, R.D. and Whittle, J. 1989. Natural rubber based thermoplastic elastomers. Rubber World, 199(6): 25 pp.

CHAPTER 23

TECHNICAL PROPERTIES AND UTILISATION OF RUBBER WOOD

A.C. SEKHAR
Sanjogtha, 146, N.E. Layout, Seethammadhara, Visakhapatnam-530013, India.

INTRODUCTION AND GENERAL ASPECTS

The most important product of the rubber tree is the latex and all efforts to improve the rubber tree have been from the point of obtaining higher yield of latex. After exploitation, the rubber tree is felled for replanting with high yielding clones. Till recently, most of the wood from the felled trees was used as fuel. With the depletion of forests in many parts of tropical regions, leading to shortage of wood for many industrial and engineering uses, attention has been given to rubber wood as an alternative source of timber. Research and development activities on the industrial applications of rubber wood are only of recent origin. New developments indicate the possibility of wider use of rubber wood for a variety of purposes. Table 1 lists the range of products which can be made of rubber wood.

Rubber trees grow to a height of 25 m and generally have straight trunks. Usually, at the time of felling, the girth varies between 100 to 110 cm at a height of 125 cm from the ground and gives 0.62 m^3 of stump wood and 0.4 m^3 of branch wood. At the time of felling, usually 180 to 185 trees will be available per hectare.

Methods of extraction, conversion and transport of rubber wood have been generally standardised in plantations. The usual implements for felling trees such as axes, saws, cant hooks, pickaroons, barking spades and cleavers, etc., are normally employed in the plantations. For crosscutting the stems and branches, power chain saws or bow saws are employed. The sites for conversion and storage are so chosen that they do not interfere with other plantation operations. The conversion of logs to smaller sizes or sections or chips depends upon the end destination of the material: whether it is for depots, saw mills or for specified industries (eg. plywood, pulping factories). For pulping factories, it would be more economical to convert small billets into chips before transport. The transport of the material within the plantation known as 'minor transporation' or yarding,

TABLE 1

Industrial articles from rubber wood which have been tried successfully in different countries.

Apron sets	Garden sets	Plywood
Bedroom sets	Gift boxes	Pulp
Benches	Hardboards	Restaurant furniture
Bread boards	Ice buckets	Railings
Building components	Irradiated timber	Rocking chairs
Black boards	Jewellery boxes	Rayon
Block boards	Kitchen cabinets	Salad bowls
Cabinets	Knife blocks	Screen partitions
Carving boards	Living room sets	Serving trays
Chairs	Lumber	Shelves
Chests	Magazine racks	Spice racks
Chopping boards	Moulded hardboards	Steak plates
Cement boards	Mouldings	Stools
Charcoal	Match splints	Suit cases
Chemicals	Match boxes	Tables
Dining sets	Packing cases	Tea trolleys
Doors	Pallets	Television cabinets
Drawing room sets	Panelling	Toilet gears
Drawer faces	Paper	Toys
Fibre boards	Particle boards	Treated lumber
Folding chairs	Patio	Vegetable boxes
Folding tables	Picture frames	Wine racks
Fruit boards	Pepper sets	Wood racks
Furniture	Parquet flooring	Wax trays
Furniture components	Plant stands	

TABLE 2

Properties of rubber tree bark wood elements

Property	Average value
Basic wood density (kg m^{-3})	543.7
Basic bark density (kg m^{-3})	620.8
Double bark thickness (mm)	11.6
Bark proportion (%)	7.7
Fibre length (mm)	1.189
Fibre proportion (%)	58.0
Vessel proportion (%)	8.5
Ray proportion (%)	22.0
Parenchyma (%)	11.5

is usually done by dragging or rolling manually using carts or tractors. The haulage of the material over long distances is done by trucks, train, or river craft. Storage of rubber wood for long periods either in the plantation or in transit depots or sales depots, or even in factories, requires prophylactic treatment and should not be exposed to direct sunlight. The storage area should also be clean without decomposed materials around.

ANATOMY OF RUBBER WOOD

Anatomy of rubber wood has been studied by several workers. The texture of the wood is fairly even with moderately straight and slightly interlocking grain. From whitish yellow when freshly cut, the wood turns to light brown as drying progresses. Latex vessels can be found with characteristic smell in some parts of the wood. The wood is soft to moderately hard with an average weight of 515 kg m^{-3} at 12% moisture content. Pores on the cross section are diffused and of medium to large size, mostly solitary but sometimes in short multiples of two to three, filled with tyloses. Vessel tissues are conspicuous in radial and tangential faces and are of the order of about 200 μ in diameter. Wood parenchyma are abundantly visible to the naked eye appearing as narrow, irregular and somewhat closely spaced bands forming a net like pattern with rays. The rays of the wood are moderately broad, rather few and fairly widespread. The pits found between the vessels and rays are half-bordered with narrow width. The length of the fibres is more than 1.0 mm on the average and the width is about 22 μ when dry. The cell wall thickness when dry is about 2.8 μ (Silva, 1963). The other characteristics of rubber wood are summarised in Table 2 (Bhatt et al. 1984). These authors further studied the variation in the properties of the wood and bark at different heights and came to the conclusion that there is no significant variation of bole quality between height levels of a tree or between trees in a plantation, in contrast with naturally grown trees in a forest.

There is insignificant heart wood formation and no transition appears between sapwood and heart wood, which is confined near the pith only. Growth rings or annual rings are not visible in rubber wood, unlike many other woods (ring porous woods). However, concentric false rings sometimes appear on the wood, depending on the presence of tension wood (gelatinous cells) which are fairly common in most of the clones. Maximum number of such rings are found in the basal portions with decreasing number towards the top. The tension wood may vary from 15 to 65%, and such erratic distribution tends to give a woolly appearance on the surface of wood. Such distribution and variation are supposed to be responsible for some of the commonly observed defects that may occur during drying and processing.

There are very few natural defects in rubber wood which may make it unworthy for general purpose applications. This is chiefly because unlike many forest based trees, rubber wood is from plantations where the trees are carefully nurtured. However, due to the presence of growth stresses and induced drying stresses, a few defects such as splits, cracks and checks are usually observed. These can be avoided or minimised by careful control measures during storage and drying. Decay or rot often occurs in rubber wood due to attack by fungi, which can be avoided by suitable chemical treatments. Similar defects due to other biological agencies like insects and birds or due to weather can also be suitably minimised by chemical treatments. Other defects like grain orientation, knots, woolly surfaces etc. can be rendered less effective by suitable machining and sorting. Logging defects like ruptured or crushed fibres can be eliminated by employing proper tools and observing necessary precautions while logging and transporting. Thus the defects that are commonly observed in rubber wood are not so serious as to render it useless.

PHYSICAL AND MECHANICAL PROPERTIES

Like all other wood species, rubber wood also exhibits orthotropicity in its properties, ie., its properties are different and independent in the three principal directions of growth: longitudinal, radial and tangential. Being nonhomogeneous in its structure, its density also varies from site to site inside the material. The variations in properties are attributable not only to the variations in density but also to the presence of latex particles in some locations and to the predominance of tension wood. Various laboratories in different countries have investigated strength properties of rubber wood, but systematic evaluation under any prescribed standard, or with necessary statistical design, as is usually done for most of the forest based wood species, is lacking. Comprehensive data on physical and mechanical properties of rubber wood are furnished by Sekhar (1989).

Edaphic, agrometeorological and plant factors such as elevation, air temperature, solar radiation, humidity, rainfall, soil characteristics, spacing, clonal difference and age of the tree can influence to a certain degree the properties of any species of wood. However, these changes may be significant with reference to the expected end use, and are generally taken care of in the system of evaluation itself by drawing samples representative of different growth conditions. However, it should be noted that strength in the green condition does not vary with moisture content. The strength in the dry condition (ie. below the fibre saturation point) varies according to the formula

$$\log S = a - bm$$

where S = strength at the moisture content m

m = moisture content of the piece of wood in question - average value

a&b = constants relating to the property

For most of the mechanical properties of rubber wood a = 2.5 and b = 0.015. When S is unknown and its density 'd' is known, the strength is given by the formula

$$S = Kd^n$$

where K and n are constants depending upon the strength property. In the case of most of the strength properties in both green and dry conditions n = 1, but otherwise it is found to vary between 0.75 and 2.5. Similarly K varies from 163 to 1665 in the green condition and from 140 to 1854 in the dry condition. These values for Indian species of wood were given by Sekhar and Rawat (1959). Most of the strength properties are generally determined either in the direction of grain ($0°$) or that perpendicular ($90°$) to the same. If strength S_θ is required in a direction at an angle θ to the grain direction, the same can be obtained by the formula

$$S_\theta = \frac{S_0 \times S_{90}}{S_0 \sin^2\theta + S_{90} \cos^2\theta}$$

where S_0 and S_{90} are strength values in the directions parallel and perpendicular to grain, respectively.

Like most of the wood species, the dynamic properties of rubber wood (ie. mechanical behaviour of rubber wood under dynamic forces) are higher than the static properties. In other words, under impact loads, rubber wood is capable of taking loads nearly twice that under slowly applied loads. However, it may be noted that the static properties of rubber wood in dry condition are higher than those in green condition, but in the case of dynamic properties, the reverse is the case for fibre stress at elastic limit and modulus of elasticity, and the increase is not significant in the case of maximum height of drop. This shows that in such cases where shocks come into play, presence of moisture in wood is helpful in taking up higher loads.

In some countries it is customary to explain the mechanical behaviour of any species for a specific function or end use, in terms of the mechanical behaviour of a popular species, widely used for a variety of purposes or for the same function and end use. In India teak is one such species and so the mechanical behaviour of all species is compared to that of teak as 100. The comparative figures are known as 'suitability figures' or 'suitability indices' and the same are indicated for rubber wood in Table 3.

TABLE 3

Comparative suitability figures of rubber wood, taking teak as 100

Parameters	Rating
Weight or heaviness	93
Strength as a beam	62
Stiffness as a beam	77
Suitability as a post	52
Shock resisting ability	75
Shear	92
Surface hardness	74
Splitting coefficient	75

From the table it may be seen that rubber wood is very near to the weight and shear properties of teak, and fairly comparable in other properties except in suitability as posts. However, rubber wood has to be used cautiously where compressive forces along the grain come into play. The suitability figures are derived by combining suitably the various properties in green and dry condition that become important for the particular function or end use. These figures serve only for comparison and not for any design or calculation of natural forces that come into action. Full details of how these figures are derived are discussed by Sekhar and Gulati (1972).

MACHINING AND FINISHING PROPERTIES

The first stage in the utilisation of any wood, involves some sort of machining. This includes sawing, wood working, fabricating, peeling, slicing, chipping and even defibrating using different kinds of machines and hand tools. In all these processes, the geometry of cutting tools, the speed of their cutting, the rate of feed and the manner of feed of the material play a prominent role in deciding the quality of the machined material and consumption of energy for the required operations. A suitable combination of these properties, determined quantitatively under standard conditions, is known as 'working quality'. In other words, economy, efficiency, and safety factors are governed by the machining properties and the working quality of the wood in question. Rubber wood is known to have been subjected to almost all types of the above machine operations and qualitative experience is available on its working. However, not much quantitative data are available on machining properties of rubber wood. Also, some of the traditional unscientific practices, as on any other species of wood, are still prevalent amongst carpenters and wood workers dealing with rubber wood.

'Wood finishing' is the effect of various types of surface finishes like paintings, polishing etc. on wood. In the case of rubber wood, quali- tative experience indicates that it can be worked to a good finish suitable for high class furniture. Shukla et al. (1984) determined the working quality and finish adaptability of rubber wood from India, and came to the conclusion that the working quality of rubber wood is 130 as compared to teak (100).

TABLE 4

Recommended wood working parameters for cutting green rubber wood.

1	Ripping rake angle	20°
2	Clearance angle	15°
3	Sharpness angle	55°
4	Top level angle when required	10°
5	Number of teeth on a circular saw	40 - 46
6	Thickness of a saw plate	15 - 17 BWG (=2 mm)
7	Diameter of a saw (d)	60 - 80 cm
8	Pitch (p)	0.068 cm
9	Depth of a gullet space	0.4 p cm
10	Tooth top	6.3 to 7.1 mm
11	Spring st (approx.)	0.4 mm
12	Material feed	8 - 10 cm/sec
13	Cutting speed	14 - 18 m/sec
14	Number of teeth in a band saw	20 - 25 / m

The finish adaptability is rated at 94% of that of teak under standard conditions. They found that rubber wood can be easily worked on lathe but is not that good in boring and mortising. They also found that ammonia treated rubber wood, free from blue stains, has exhibited better finish adaptability and water gloss than untreated or blue stained rubber wood. Colouring of rubber wood is reported to be giving very attractive appearance.

In Malaysia machining properties were determined by Lee and Lopez (1980). Some recommended working parameters for cutting of green rubber wood are given in Table 4 (Sekhar, 1988).

SEASONING BEHAVIOUR

The main purpose of seasoning is only to reduce and adjust the inherent moisture content of wood to a predetermined level as required for various end uses. Usually seasoning is done to a level of equilibrium moisture content of wood in the region where the material will be in use for a long time. Thus the absorption and desorption of moisture by wood will be very much minimised, and consequent swelling, shrinking, and warping in planks, and cut sizes of rubber wood are avoided. Other advantages of seasoning are reduced surface cracking and splitting, improved physical and mechanical properties, better working quality with different tools and easy finish uptake. Seasoned material, being comparatively lighter than unseasoned material (called green material ie. above the fibre saturation point), is easier and cheaper to transport.

There are several methods of seasoning but the most popular and economical methods are (1) kiln seasoning and (2) air seasoning or air drying. The former is done in an enclosed chamber in which temperature, humidity and circulation of air can be controlled to ensure a gradual removal of moisture. The required heat is provided either by steam or electricity, or by fuel gases or even by solar energy or a combination of any of them, depending upon local considerations and cost. Humidity is controlled by water sprays or steam jets. Suitable ducts and ventillators are provided for the same. Humidity is measured by suitable humidity meters. The circulation of air is done by placing reversible fans at appropriate places inside the chamber. Depending on the size and average moisture content of the wood suitably stacked inside the chamber, a proper schedule of seasoning is selected. Starting with low temperature and high humidity, the conditions inside the chamber are gradually changed to low humidity and high temperature and the process of seasoning is continued until the required moisture content of the material is attained. By careful observations of the various intermediate stages, all likely defects that may occur during seasoning are avoided. Table 5 gives typical seasoning schedules for rubber wood.

TABLE 5

Typical seasoning schedule for rubber wood

Moisture content of timber on the side of air inlet	Schedule A For stock upto 2.5 cm thick Time: Approx. 4 to 6 days			Schedule B For stock between 2.5 and 5 cm Time: Approx. 7 to 9 days			Schedule C For stock more than 5 cm Time; Approx. 9 to 12 days		
	Dry bulb temp. °C	Wet bulb temp. °C	R.H. %	Dry bulb temp. °C	Wet bulb temp. °C	R.H. %	Dry bulb temp. °C	Wet bulb temp. °C	R.H. %
Green	45	40	72	42	40	88	42	40	88
60	47	40	64	45	40	72	--	--	--
45	--	--	--	--	--	--	--	--	--
40	49	40	56	47	40	64	45	42	80
35	--	--	--	49	40	56	46	42	76
30	53	42	44	51	50	50	48	42	67
25	--	--	--	53	40	44	50	42	59
20	55	42	32.5	54	40	39	52	42	52
18	--	--	--	55	40	36	--	--	--
15	55	45	30	57	40	30	55	42	45

R.H. - Relative humidity.

Air drying of green timber is done in the open. The material is suitably stacked under cover such that free air passes through the stack. In air seasoning there are no control techniques as in kiln seasoning but several precautions have to be taken to avoid development of defects during storage in open air. All materials above 10 cm in thickness must necessarily be air dried first to bring down the moisture content at least to about the fibre saturation point. For a material of similar thickness and initial moisture content, air seasoning takes a comparatively longer time than kiln seasoning. While kiln seasoning requires skilled operations, careful supervision and higher cost, air seasoning does not require much skill and supervision and is comparatively cheaper in spite of blocked capital on wood, stacked for long periods.

Rubber wood is found to be amenable to extremely rapid movement of moisture and drying so that development of any severe drying stresses are unlikely to occur. Drying times are comparatively shorter than that in many other species under the same conditions. Planks about 25 mm thickness take about 55 to 60 days in air drying while only about 6 to 7 days in kiln drying. In solar kilns drying time is about 15 days for the same material. Table 7, which gives shrinkage figures of small, defect free specimens from green to oven dry conditions (Sharma and Kukreti, 1981) indicated significant longitudinal and volumetric shrinkage. The ratio of tangential

TABLE 6

Shrinkage of rubber wood

Description	Green to oven dry condition (%)		
	Minimum	Maximum	Average
Tangential (T)	5.7	6.5	6.1
Radial (R)	2.6	3.1	2.85
Longitudinal (L)	0.2	0.9	0.55
Volumetric (V)	10.1	12.0	11.05
Ratio (T/R)	2.19	2.09	2.14
Sum of T + R + L	8.5	10.5	9.5

to radial shrinkage, which is sometimes taken as an index of warping is also slightly high. Consequently the dimensional stability of rubber wood is to be viewed carefully as shrinkage defects may develop in some pieces of rubber wood. Greater attention is therefore required for selection of the material and proper seasoning before the material is used. Some of the unusual shrinkage characteristics are attributed to the presence of tension wood, cross grain and uneven growth of cells at different locations in the wood. However these defects may not assume objectionable proportions in such industries as furniture, packing cases and textile accessories where only short length, narrow width members are used.

DURABILITY AND PRESERVATION

Timbers are usually classified as non-durable or durable depending on whether their heartwood is liable to be attacked by fungi and insects or not. In some countries the durable woods are further divided into three groups as (1) highly durable (2) moderately durable and (3) less durable depending upon the degree of damage of a standard sized piece of wood on exposure directly to wood destroying agencies. Similarly the various species of wood are classified as easily treatable or not, depending upon the capacity of the species to take chemicals for preventing attack by fungi and insects. At present there are no adequate laboratory or field data on rubber wood to classify its durability or treatability, but there are various industrial and field experiences, from which it is known that rubber wood is definitely non-durable but fairly easily treatable by water soluble chemicals or even oil type preservatives.

There are several agencies which can damage rubber wood such as weather, fire, contact materials, fungi, and insects. Amongst all these the last two are the most important as against them the wood requires to be treated chemically. There are several types of fungi which are known to attack rubber wood. Some of the well known among them are Lenzites palisotti, Ganoderma applanatum, Tramates corrugata, Pollyporous zonzlis, Lentinus blapharods and Schizophyllum commune. Similarly some of the insects found to attack rubber wood are those belonging to families Cerambycidae, Bostrychidae, Lyctidae and Platypodidae.

In order to protect wood from the attack of the above agents, it has to be treated with chemical preservatives. The common types of preservatives are oil based and water based. In the oil based preservative, coal tar creosote, with or without admixture of fuel oil, is used mostly for exterior use of rubber wood. The discolouration and unpleasant odour restricts its use at many places. However this is toxic to many types of organisms and also prevents any further splitting in wood. Water based preservatives contain toxic inorganic salts dissolved/dispersed in water. The leachable types amongst these are zinc chloride, boric acid or borax and sodium pentachlorophenate. In the non-leachable type of water based preservatives, mention may be made of copper-chrome-arsenic (CCA) compositions, acid-cupric-chromate (ACC) compositions, and copper-chrome-boric (CCB) compositions. All of these can be used safely at most places with different concentrations, absorptions and penetrations to suit individual cases. These preservatives are particularly suitable for interior dry locations. Sodium pentachlorophenate is generally used for prophylactic treatment only and this may protect the surface of wood for three to four dry months only. The logs should preferably be end coated with antisplitting and preservative compositions.

Among the treating processes, usually two types of processes are employed: pressure treatments and non-pressure treatments. In the pressure treatment, the material is placed inside a closed vessel and the same is filled with the preservative in question. By means of vacuum pressure pumps, the required amount of preservative is so impregnated that adequate level of penetration is achieved. Usually a pressure of 3 to 7 kg cm^{-2} is built up and retained for about 20 to 30 min depending on the size, condition of wood and the preservatives employed. For most of the water soluble preservatives, an absorption of 5 to 10 kg m^{-3} and a penetration of at least 20 mm all around is obtained under the above conditions.

In the non-pressure type of treatments, the most common are (1) surface application by brushing, spraying, or dipping and (2) open tank process also known as hot and cold process. The former are comparatively cheaper and do not require any elaborate equipment or skills but the protection realised could be only for short duration. However, dipping methods have been fairly popular in the case of rubber wood. Some recent studies on rubber wood preservation indicated that introduction of boron compounds in rubber wood by diffusion process with the addition of a small percentage of NaPCP (Sodium pentachlorophenate) is fairly effective and cheap. Around 4 kg m^{-3} of dry salt absorption could be obtained by this process. In the open tank process, the timber is initially placed in tanks containing hot oil type preservative or hot water at temperatures of about 90-95°C so that moisture and air are thus partially expelled first. Then the timber is allowed to cool either in the same tank if it is oil type or in another tank with cold water-soluble preservative solution. During the cooling period the preservative gets inside the material to be treated. The heating and cooling times are suitably adjusted to get the required absorption and penetration of chemicals inside the timber.

RUBBER WOOD IN JOINERY AND FURNITURE

Timber joints are important in buildings and furniture as they occur at various places such as frames, stressed members, doors, windows, built-in and moving furniture of various sorts and designs. The joints are required for different purposes such as lengthening, cross jointing, bearing, heels, corners etc. Thus the exact design of a joint is based not only on architectural and constructional needs, but also on the strength required at the joint. The joints are classified as (1) end-to-end jonts (2) end-to-edge joints (3) edge-to-edge joints and (4) angle joints. In most of the joints, gluing of all mating surfaces is necessary for which proper control should be exercised on the quantity of glue, temperature and pressure depending on the glue and quality of mating surfaces.

The finishing of surfaces of jointed frames and surfaces as in buildings and furniture requires special attention. Sunken or misfit joints do not present a good appearance. It would be necessary to remove any undue discolouration that may have come at the joint due to chemical action between the glue and tannin content or other constituents of wood. All crevices such as those due to surface cracks, holes, and open grain etc. are required to be filled up with fine wood flour (known as patti in India). All rough surfaces require to be appropriately sanded so as to take up the required finish such as varnishes, paints, and polishes. The final finish on the exposed faces particularly in furniture or specially chosen decorative faces of panels and doors, should be such as to give proper protection without marring the appearance.

Rubber wood possesses all the requisite properties for meeting the above requirements of joinery and finish, provided some care is taken to process and prepare the material suitably. In fact it has been reported that rubber wood is quite suitable for large scale manufacture of prototype furniture on a properly designed production line system. Inter-changeable components of rubber wood furniture, popularly known as knock-down furniture, have found favour in export markets also. Sufficient data exist for preparing bent articles for fanciful drawing room furniture and curved shapes of other building components. A minimum radius of 7.5 cm on 1.3 cm thick pieces was reported by Sharma et al. (1982). Thus rubber wood furniture has been found to be quite comparable with furniture made from other types of wood such as Aini (Artocarpus hirsuta), Benteak (Lagerstroemia lanceolata), Bijasal (Pterocarpus marsupium), Kathal (Artocarpus heterophyllus), Kokko (Albizzia lebbeck), Pali (Palaquium ellipticum), Poon (Callophylum tomentosum), Pyinma (Lagerstroemia hypoleuca). Because of its generally white colour and medium density, rubber wood has the possibility of being adapted to produce 'improved woods' and imitation products. In Sri Lanka plastic coated rubber wood has been produced with highly improved properties for use in high class furniture. In India some attempts have been made to irradiate rubber wood impregnated with styrene, using gamma rays. Excellent surface finish was obtained for high class furniture and panelling. Compressed rubber wood has been found to successfully replace species like Cornel (Cornus sp), and Maple (Acer sp), in textile accessories.

RUBBER WOOD IN PLYWOOD INDUSTRIES

With increasing demand for wide panels of wood in buildings and some types of furniture, plywood has virtually replaced old type wide solid planks. Further, plywood has introduced uniform strength in all directions and greater dimensional stability in wood based panels. Plywoods are made up of odd number of thin veneers, sliced or rotary cut on machines with the grain of alternate veneers placed at right angles to each other and equidistant from a central core. Thus a balanced construction of plywood panel is obtained which prevents any warping in the panel and prevents delamination due to uneven stresses with varying moisture content and varying thickness of layers of veneers. The process of making such plywood involves preparation of veneers on a rotary lathe or a slicer, cutting them to size, drying, gluing and finally pressing them under high pressure and temperature.

Rubber wood has been found to be generally quite suitable for making plywood but a few special precautions may be necessary. The logs that are brought to the mills should first be given a prophylactic treatment against possible decay and insect attack. The peeling of logs may be done between 50 to 250 cm length of veneer per second depending upon the roughness of the grain of the log chosen for peeling. After the veneers are made and cut to proper size, they are again required to be immersed in borax solution as a precaution against borer and fungal attack. They should then be carefully dried to be able to take up adhesives for assembly and pressing. In small cottage industries in India, the veneers are generally dried under open sun in spite of some undue losses by such uncontrolled drying. Then well prepared adhesives are applied on the veneers, which along with other veneers are made into one pack for pressing at about 12 kg sqcm^{-1} either in a cold or hot press depending upon the type of adhesive. After making the plywood, it has to be conditioned to about 12% moisture at room temperature. The plywood panels are then cut to the required sizes and sanded to obtain necessary smoothness.

Several sizes of rubber wood plywood are produced commercially. Panels of size 8'x4' are reported to have been used for a number of purposes such as wall panels, built-in furniture and interior wood work. Plywood of small sizes of 2'x2' are normally used for tea chests, suit cases, back support of mirrors, and a variety of small articles and furniture. There is no evidence as yet for the use of rubber wood plywood in specialised fields like marine, or fire-proof constructions, shutters or even for decorative surfaces. However, if appropriate precautions are taken and properly processed the possibility of using plywood even for special purposes need not be eliminated. With the medium density and other favourable qualities of rubber wood there seem to be good possibilities for preparing other types of composite woods such as compreg, irradiated wood, sandwich boards, core boards, flush doors etc. Rubber wood is reported to have been successfully used for particle boards in Malaysia (Wong and Ong, 1979).

RUBBER WOOD IN MATCH INDUSTRY

Lighting matches form one of the most essential needs of every home, and the most important raw material for the match industry is wood for boxes and for splints. Several alternative materials have been tried for the above but by and large wood remained the most acceptable raw material. Of all wood species in the world, aspen and similar white, soft, straight grained woods were the most preferred ones, even though several other species of wood and bamboo have met the prescribed standards and requirements for the purpose. Although rubber wood as such does not seem to have found any special place in the prescribed specifications so far, it has been in wide use in the match wood industries, particularly the cottage type, in India. Although rubber wood exhibited some defects in the manufacture of splints and boxes, because of its easy availability and good working qualities, it has always found favour in most of the cottage industries in Kerala and Tamil Nadu states of India.

Various technical problems have been reported with rubber wood for match industry. The warping of splints in storage, lower absorption of wax, lack of adhesion of head composition etc are some of the commonly described problems and these are popularly attributed to the presence of latex particles or gelatinous cells of tension wood in the material. These have been overcome by treatment of match splints with caustic soda or reducing the viscosity of wax by addition of diesel oil etc. Treatment of veneers with sulphur vapours reduced the incidence of fungal attack in storage. Careful drying, levelling, and screening of match splints helped in reducing the internal sticking and quick assembly of the splints into frames for further dipping. Thus in spite of a few drawbacks, rubber wood is fairly well established in cottage scale manufacture and is also said to be gaining ground in the mechanised sector of the match industry.

RUBBER WOOD FOR FIBRE PRODUCTS LIKE PULP AND PAPER

For making any type of pulp and pulp products like paper and boards, the wood material requires to be debarked, chipped screened, digested, cleaned, washed, bleached, beaten, sheet formed and finally the product so obtained is properly cut and packed. In the various processes indicated above, several mechanical and chemical factors come into play for obtaining the required end product from the given wood species or any other ligno-cellulosic raw materials like grasses, agricultural wastes, rags, waste paper etc. At present there are a variety of papers and paper products and as such the individual parameters for any raw material vary considerably depending upon the process and the end product. The most common of the above processes are the chemical processes. There are two types, (1) alkali process, (2) acid process. In the former, the sulphate process is the one most commonly employed. In this process a solution of sodium hydroxide and sodium sulphide is used to cook the wood chips in a 'digester' under pressure and heat so as to separate the fibres for further operations. In the acid process the raw materials are digested with calcium bisulphite in the presence of sulphur dioxide and the process is known as sulphite process. Almost all processes have been tried with rubber wood with varying results.

Several workers tried to get acceptable pulp from rubber wood. Chipping, cooking and screening operations with rubber wood compare well with other commercial hardwoods. However, in storage, rubber wood chips are susceptible to fungal attack. Unbleached kraft pulp was found to be grey in colour in contrast to normal brown stock from other woods. Guha and Negi (1969) carried on the digestion for a period of four hours in a three litre capacity stainless steel autoclave at a maximum temperature of 170°C in the presence of chemicals in the ratio of $NaOH:Na_2S$ as 3:1. They worked separately with branches and stem wood. They removed latex by simple scraping of logs. The brightness of the pulp was 65%. The pulps were beaten to about 250 ml freeness and standard sheets of about 60 gsm were obtained. Good strength and good yield were obtained as indicated in Table 7. Nair (1971) also attempted mixing defibrated and bleached

TABLE 7

Properties of bleached pulps from rubber wood-stems and branches.

Sl.No. S-Stem B-Branch	Chlorine Consump- ption	Yield of bleached pulp	Burst factor	Tear factor	Breaking length in kilometers	Folding endurance double fold
1.S	14.1	41.0	33.9	86.0	6.76	112
1.B	9.0	40.8	46.6	82.7	5.74	266
2.S	12.8	39.0	26.9	64.2	5.60	22
2.B	10.0	39.7	30.0	72.4	5.90	24
3.S	9.6	29.3	28.0	69.0	5.21	36
3.B	8.5	38.0	20.0	62.1	4.20	12
4.S	10.0	38.4	18.9	57.3	4.70	14
4.B	8.0	28.9	14.2	41.0	3.46	4
5.S	8.8	37.6	14.3	40.0	3.13	2
5.B	7.6	25.0	10.5	38.0	3.26	4

rubber wood with chemical pulp from bamboo in the ratio of 1:1. He had also removed about 2.5 cm material under the bark to remove problems of residual latex in the wood. Allauddin (1971) prepared pulp from Indonesian rubber wood by the soda process (using NaOH only) and found that it contained soft, sticky, latex accumulations. They found that pulping

TABLE 8

Properties of fibre boards from rubber wood

Sl. No.	Amount of alkali %	Average moisture content %	Density g m^{-3}	Untempered		Heat tempered		Oil tempered	
				M of R kg cm^{-2}	Water absorption % in 24 h	M of R kg cm^{-2}	Water absorption % in 24 h	M of R kg cm^{-2}	Water absorption % in 24 h
1	0.0	9.5	0.96	180	183.7	184	67.8	391	31.7
2	1.0	10.6	1.04	188	146.5	269	65.7	441	49.4
3	2.0	7.6	1.01	398	130.5	457	6.6	617	43.5
4	3.0	9.0	1.03	366	130.2	384	60.4	597	42.4
5	4.0	4.3	1.03	309	128.3	346	71.5	562	63.4
6	5.0	5.7	1.18	384	147.2	417	65.2	515	45.7

actually hardens the latex particles and so the same could be removed to a large extent by screening. They also blended rubber wood kraft pulp with long fibred rice straw and bamboo for producing paper of acceptable quality. There is also evidence in literature that even commercially acceptable viscose grade pulp can be produced from rubber wood. In Japan, mill trials have been successful in obtaining 80 to 85% brightness (Nakayama et al. 1972). The kraft paper from rubber wood was found to be quite suitable for the manufacture of wrapping paper and corrugated boards of moderate quality (Hussain et al. 1977).

Some attempts were also made to make small sheets of fibre boards of 4 to 5 mm thickness from rubber wood (Jain, 1965). The wood was subjected to mild alkali treatment as given in Table 8. After thorough washing, the material was passed through a conduit mill, then felted and pressed at 56.34 kg cm^{-2} for about 20 min. The yield of the board varied from 50-70% for the solid to liquid ratio of 1:9 and cooking time of 2 h. The boards were also subjected to heat treatment and also tempered in cashew nut shell liquid. The strength, density and water absorption are indicated in Table 8. It may be seen that water absorption and modulus of rupture improved considerably on heat and oil treatments.

REFERENCES

Allauddin. 1971. Experiments on Indonesian rubber wood as raw material for pulp and paper. In: Proc. UNIDO Experts Group Meeting Pulp and Paper, Sept. 13-17.

Bhat, K.M., Bhat, K.V., Damodaran, T.K., Rugmani. 1984. Some wood and bark properties of Hevea brasiliensis. J. Tree Sci., 3(1&2).

Guha, S.R.D., Negi, J.S. 1969. Pulping of rubber wood. Indian Pulp and Paper, 24: 1-3.

Hussain, M., Siddique, A., Das, P. 1977. Investigations on the possibility of sulphate pulp from rubber wood (H. brasiliensis). Bano Bigyan Patrika, 6: 54-57.

Jain, N.C. 1965. Hardboards from rubber wood. Indian Pulp and Paper. XIX, 8.

Lee, Y.H., Lopez, D.T. 1980. The machining properties of some Malaysian timbers. Malaysian Forest Service Trade Leaflet No. 35.

Nair, V.S.K. 1971. Certain aspects on utilization of ever green hardwood for pulp and paper making in India. In: Proc. Conf. on utilization of hardwoods for pulp and paper. For. Res. Inst. Dehra Dun.

Nakayama, K., Usuri, T., Hairashi, S.I. 1972. Pulping of para rubber tree. Japan TAPPI, 20, 6.

Sekhar, A.C. 1989. Rubber Wood, Production and Utilization. Rubber Research Institute of India, Kottayam.

Sekhar, A.C., Gulati, A.S. 1972. Suitability indices of Indian timbers for industrial and engineering uses. Indian Forest Records (N.S). T.M. Vol. 2, No. 1, New Delhi.

Sekhar, A.C., Rawat, B.S. 1959. Studies on effect of specific gravity on strength considerations of Indian timbers. J. Inst. Eng. (India), Vol. 1, No. 8, Ft. I.

Sharma, S.N., Gandhi, B.L., Kukreti, D.P., Gaur, B.K. 1982. Shrinkage, hygroscopicity and steam bending properties of rubber wood (H. brasiliensis). J. Timb. Dev. Assn. India, 28(1): 19-24.

Sharma, S.N., Kukreti, D.P. 1981. Seasoning behaviour of rubber wood - an underutilized nonconventional timber resource. J. Timb. Dev. Assn. (India), 28: 2.

Shukla, K.S., Bhatnagar, R.C., Pant, B.C. 1984. A note on the working quality and finish adaptability of rubber wood. Ind. Forester, 110: 5.

Silva, S.S. 1963. Industrial utilization of rubber wood for wood based panel products FAO/PPP/Conf. Paper 2.5: 1-19.

Wong, W.C., Ong, C.L. 1979. The production of particle boards from rubber wood. Maly. Forester, 42(1).

CHAPTER 24

ANCILLARY INCOME FROM RUBBER PLANTATIONS

V. HARIDASAN
Rubber Research Institute of India, Kottayam-686009, Kerala, India.

Although the main source of income for the small grower is the rubber produced from the holding, certain ancillary products also fetch supplementary income to him. The most important ancillary product is of course the rubber wood. This is discussed separately (Chapter 23). Rubber seed is also a supplementary source of income for the grower though on a moderate scale. The other source is the honey produced from the rubber plantations.

RUBBER SEED, SEED OIL AND OIL CAKE

Among the rubber producing countries, perhaps it was in India that rubber seed began to be used commercially for extracting oil. Around 1965, the Khadi and Village Industries Commission of India took the initiative to introduce the processing of rubber seed. It has now assumed the proportion of a small scale industry. The industry is concentrated in the state of Tamil Nadu (Haridasan, 1976). Rubber seed is a minor source of non-edible oil in India.

Rubber seed production

Rubber seed production is not stable every year. Even within one year inter-clonal and intra-clonal variation in production is noted. Apart from the climatic factors, the incidence of diseases affect the availability of useful seeds for extracting oil. In India the attack of Phytophthora can restrict the number of seeds available for commercial use. The occurrence of severe rainfall during the seed fall season can also affect the availability of good quality seeds (Haridasan, 1976).

Large variation in the rubber seed production is noted all over the world. In a study published from Nigeria, seed production per hectare was found to vary from 73 kg (PB 86) per hectare to 424 kg (PB 5/51) per hectare (Adindu and Aghaka, 1985). Based on the studies undertaken in India, around 150 kg of useful seed can be obtained from one hectare of rubber plantation. The seed production can be considerably higher, if the incidence of disease is insignificant.

Studies show considerable variation in the seed weight. A study conducted in Nigeria showed the mean whole seed weight as 3.25 g for GT 1 and 4.57 g for RRIM 600. In India the weight of seeds (as on received basis) was found to vary from 5.2 to 5.7 g. However, on moisture free basis the weight is found to be around 4 g (Adindu and Aghaka, 1985). In a study conducted in India the percentage of shell and kernel was found to be at 42 and 58 respectively on moisture free basis (Azeemoddin and Thirumala Rao, 1962). From published literature, the figures are found to vary from 41% to 48% for kernel and 34% to 55% for shell. Fresh seed may contain moisture even upto 25% by weight.

Reports from Nigeria show the oil recovery from the kernel in the laboratory at 43%. In the study carried out in India mentioned above, oil content of kernel on moisture free basis varied from 38% to 46% with an average of 42%. This becomes 37% and 40% on 10% and 5% moisture basis respectively (Adindu and Aghaka, 1985). Under the commercial conditions in India, oil recovery is around 35%.

There are some problems in the collection and storing of rubber seeds. If the seeds are not collected, dried and stored in disease and insect free atmosphere, the seeds will become useless for extracting oil. Studies show that fresh seeds are more susceptible to the attack of fungi than dried seeds. Fungicidal treatment and drying of rubber seeds have to be carried out before storage, to ensure good quality seeds.

A certain percentage of seeds will be used in the plantations for raising stock material and only the balance will be available for extracting oil. In India, around 10% of the collected seeds is thus used in the plantations as stock materials (Haridasan, 1976).

The mobilisation of the collection of seeds is also a problem. In India most of the plantations allow the dependents of the workers to collect the seeds, while rubber dealers form the link between the oil millers and the collectors of seeds.

Method of processing

Three methods of extracting rubber seed oil are reported, viz., (1) solvent extraction, (2) by expeller, and (3) by rotary machine. To get the maximum yield from the seeds, solvent extraction process is the ideal method. But this process requires relatively higher investment and expertise. For optimum efficiency under this method large quantity of seeds will have to be mobilised. In India where the commercial production of rubber seed oil has been in vogue over two decades, the rotary and expeller machines are used for the purpose; the expeller being limited in number. These machines have been in use for extracting oil from other seeds as well.

In India rubber seed oil extraction is carried out by oil millers so as to utilise the mills fully. Although 90% of the rubber producing areas are located in Kerala on the western part of the Western Ghats, the oil mills are located in the rain shadow side of the Western Ghats, in Tamil Nadu. These mills were set up mainly for extracting groundnut oil. The groundnut crop reaches the mills after December, while the rubber seed fall season is between July and September and this enables the mills to fully utilise their capacity. In the recent past, a few mills have been operating exclusively for rubber seed throughout the year.

The rotary form of extraction is the simplest one, and is suited to the people of the developing countries where rubber is mainly grown. It is relatively more labour intensive. The machinery can be operated by a semi-skilled worker. A pair of rotary machines is necessary for giving full employment to one person. As a result, the number of rotary machines to be installed will have to be in multiples of two. A reasonably efficient worker can crush around 250 kg of dry rubber seed kernel during an eight hour working shift.

The main equipment required for processing are the rotary machine and an electric motor. A platform balance is also used in most mills. For removing shells, machines are available, but in the state of Tamil Nadu manual removal is practised widely. There is no additional expenditure for drying the seeds in Tamil Nadu, as the state is endowed with plenty of sunshine during the seed fall season. In other parts of the world where the above favourable climatic conditions are not available, machinery may be required for decortication and a kiln, for drying.

For processing rubber seed kernel under the above method, a certain amount of molasses is required. Usually for every 100 kg of dry rubber kernel, 20 to 25 kg of molasses is necessary. In the course of extraction of oil, around 10% to 15% of the total weight of the kernel is lost (Haridasan, 1976). Under normal conditions, the oil recovery would be around 35%.

Uses of rubber seed oil

In India rubber seed oil is mainly used in the manufacture of washing soap. Small quantities are used in the paint industry as a substitute for linseed oil (Haridasan, 1976). Studies conducted in India and other countries have shown that rubber seed oil suitably treated with sulphur produces factice which finds use in rubber compounds. Epoxidation of rubber seed oil with hydrogen peroxide and acetic acid has also been reported. Epoxidised oil is used in the formulation of anti-corrosive coatings, adhesives and alkyd resin casting (Vijayagopalan, 1971; Vijayagopalan and Gopalakrishnan, 1971). Rubber seed oil has been used for the production of fat liquor for the leather industry (Vijayalakshmi et al. 1988). The properties of rubber seed oil are given in Table 1.

TABLE 1

Acid value	4-40
Saponification value	190-195
Iodine value	132-141
Hydroxyl value	12-32
Unsaponification (%)	0.5-1.0
Refractive index, 40°C	1.466-1.469
Specific gravity, 15/15°C	0.924-0.930
Titra (°C)	28-32
Fatty acid composition (%):	
Palmitic acid	11
Stearic acid	12
Archidic acid	1
Oleic acid	17
Linoleic acid	35
Linolenic acid	24

	100

Rubber seed oil cake

Research carried out in India has shown that rubber seed oil cake is a good ingredient in cattle and poultry feed. In the cattle feed, upto 20% of the weight can be rubber seed oil cake and in India important cattle feed manufacturers use rubber seed oil cake for the purpose. The Kerala Agricultural University, Trichur, India conducted a 12 year long study utilising rubber seed cake in the ration of cattle, pigs and poultry. The University obtained the following results (Ananthasubramaniam, 1980):

The rubber seed cake had a crude protein content of 25% with a DCP of 15 and a TDN of 66 for cattle.

The DCP and TDN of the cake in respect of pigs were 16 and 78 respectively.

The cake contained nearly 9 mg/100 g of hydrocyanic acid which had no deleterious effect on feeding at the recommended levels.

Rubber seed oil cake can be fed up to 30% of the concentrate mixture for cattle and at 10% level in the ration for pigs and chicken.

RUBBER HONEY

Beekeeping as a vocation in the rubber plantations in India was introduced by European Missionaries in the twenties of this century. Since 1950, the Khadi and Village Industries Commission of the Central Government of India and the Khadi and Village Industries Boards of the State Governments have taken up the promotion of beekeeping by offering financial, technical and training facilities to the beekeepers (Haridasan et al. 1987). Since 1988, the Rubber Board of India also gives help in taking up beekeeping in rubber plantations.

Beekeeping in the present context means the rearing of certain varieties of domesticated bees. Although there are four such bees, Apis cerana indica is the one reared in the bee hives in the rubber plantations in India. Attempts are also made in rubber growing countries, particularly in Malaysia, to develop beekeeping in rubber plantations.

The rubber tree is a prolific producer of honey. Honey in the rubber tree is found in the extrafloral nectary glands at the tip of the petiole where the leaflets join (Jayarathnam, 1970). It is collected in India from January to March, when the rubber tree sheds its old leaves and produces new ones. About 45% of Indian honey originates from rubber plantations.

Studies made by the Rubber Research Institute of India indicate that an optimum number of 15 to 20 hives can be placed in a hectare of rubber plantation. Under experimental conditions about 20 kg of honey per hive per year has been produced in rubber plantations. Under commercial conditions it is reasonable to assume 10 kg per hive per year. Assuming a production potential of 10 kg of honey per hive and 15 hives per hectare, a minimum of 150 kg of honey can be produced from one hectare in a normal year (Haridasan et al. 1987). It can give a gross income of Rs. 3000 ha^{-1} (USS 120). An average of 30 hives (2 hectares) can be managed by a beekeeper.

Qualitatively, the most important drawback of rubber honey is the higher moisture content in it. The internationally accepted standards of quality tolerate moisture content only upto 19% (Haridasan et al. 1987). The higher moisture content creates problems during storage. Yeast is attracted to honey when the moisture content is higher and this leads to fermentation. Therefore, there is the need of reducing the moisture content immediately after collection to maintain the quality. The vacuum concentration process could be introduced for upgrading the quality of rubber honey. Granulation of honey, though not a symptom of bad quality, is suspected by the consumer, due to adulteration. To prevent granulation and fermentation, honey should be heated at around 63 degree celsius under controlled conditions.

The honey gathering activity in the rubber plantations lasts until the end of March. Afterwards the beekeeper has to see that the bee colonies are sustained till December. Honey provides the carbohydrates and pollen, the protein, required by the bees. An assured source of honey and pollen should be available in the vicinity of the plantations if beekeeping is to be carried out throughout the year. The need for raising plants which will flower in a cycle through out the year is therefore all the more important.

The Rubber Research Institute of India has identified five promising bee forage plants along with twenty one major and minor sources of nectar and pollen for off-season bee management (Nehru et al. 1983). These plants provide a source of nectar and pollen during the long dearth period from April to December every year. These can be raised on the hedges, boundaries, bunds or vacant spaces in the plantations.

Nehru et al. (1983) have reported the technical properties of rubber-honey (Table 2).

TABLE 2

Properties of honey from rubber estates

		Range	Average
1.	Viscosity (in centipoise) at 27°C	550-3800	1358
2.	Specific gravity at 27°C	1.3985-1.3400	1.379
3.	Moisture (%)	21.50-25.50	22.00
4.	Reducing sugars:	69.08-74.80	72.80
	(a) Levulose (%)	34.80-40.70	37.14
	(b) Dextrose (%)	33.57-37.97	35.98
5.	Non-reducing sugars (%)	0.78-3.14	1.71
6.	Acidity (%)	0.06-0.20	0.127
7.	Ash (%)	0.09-0.39	0.216
8.	Protein (%)	0.054-0.249	0.138
9.	Yeast (Million/g)	103.9-159.0	139.39

REFERENCES

Adindu, C. and Aghaka. 1985. Quantitative and commercial implications of seed size variation in rubber (Hevea brasiliensis) clones. Proceedings of the National Conference, Nigeria, January 22-24.

Ananthasubramaniam, C.R. 1980. Rubber seed cake, Technical Bulletin No. 5, Kerala Agricultural University, Vellanikkara, Trichur, p. 23.

Azeemoddin, G. and Thirumala Rao, S.D. 1962. Rubber seed and oil. Rubber Board Bulletin, 6(2): 59-68.

Haridasan, V. 1976. Utilisation of rubber seeds in India. Rubber Board Bulletin, 14(1&2): 19-24.

Haridasan, V., Jayarathnam, K. and Nehru, C.R. 1987. Honey from rubber plantation - A study of its potential. Rubber Board Bulletin, 23(1): 18-20.

Jayarathnam, K. 1970. Hevea brasiliensis as a source of honey. Journal of Palynology, 6: 101.

Nehru, C.R., Thankamony, S., Jayarathnam, K. and Levi Joseph, P.M. 1983. Studies on off-seasonal bee forage in rubber plantations. Indian Bee Journal, 45(4).

Radhakrishna Pillai, P.N. 1980. Handbook of Natural Rubber Production in India, Rubber Research Institute of India, Kottayam, pp.505-514.

Vijayagopalan, K. 1971. Factice from rubber seed oil. Rubber Board Bulletin, 11: 48-51.

Vijayagopalan, K. and Gopalakrishnan, K.S. 1971. Epoxidation of rubber seed oil. Rubber Board Bulletin, 11: 52-54.

Vijayalakshmi, K., Geetha Baskar, Parthasarathy, K., Rao, V.V.M. and Rajadurai, S. 1988. Prospects of rubber seed oil in leather industry. Chemical Weekly, Bombay, October 25. p.1-3.

CHAPTER 25

GUAYULE AS AN ALTERNATIVE SOURCE OF NATURAL RUBBER

F.S. NAKAYAMA

U.S. Water Conservation Laboratory, Phoenix, Arizona, USA.

HISTORICAL

Natural rubber was once produced commercially in the United States and Mexico from the guayule plant (Parthenium argentatum). Historical reviews on the start and demise of the guayule industry are presented by Hammond and Polhamus (1965), and McGinnies and Mills (1980). Production of guayule rubber (GR) began in the early 1900's when the source of natural rubber (NR) from Hevea brasiliensis (HR) was still limited, and the economics of GR still favourable. American companies and their Mexican counterparts produced GR on an intermittent basis up to the mid 1930's. Approximately 30 tonnes of GR per year were made in 1905, and production peaked to 9750 tonnes in 1910, declining to 5400 tonnes in 1927, and then to 4000 tonnes in 1934. There were also times when no production was recorded within this 1900 to 1930 interval. According to McGinnies (1983a), 1400 tonnes of GR were produced in the United States between 1931 and 1941 at Salinas, California in an area of approximately 1800 ha. The primary source of natural rubber has been concentrated in the Southeast Asian countries, but it could come full circle by getting re-established back to the American continents with guayule commercialization.

During the World War II period of 1941 to 1944, HR supply was essentially unavailable to the United States and the government formed the Emergency Rubber Project (ERP) in 1942 to make up the loss of NR. To get the programme started as soon as possible, the government purchased existing GR companies and acquired their fields, processing facilities and expertise. The U.S. Congress in 1942 had authorized the planting of 30,000 ha which was later increased to 200,000 ha and back to a realistic 81,000 ha, but only about 13,000 ha were eventually planted (McGinnies, 1983a). In the three and a half years of existence, the ERP had produced 1100 tonnes of rubber from the cultivated fields plus another 230 tonnes from native shrubs growing in Texas. This project involved about 5000 personnel including professional, technical and field support, and since the project's objective was primarily that of producing the maximum rubber in the minimum time, the basic aspects of rubber synthesis, chemistry, and plant genetics were not thoroughly explored. Instead, efforts were concentrated toward improving agronomic and process management. A new source of shrubs had to be established through cultivation to provide sufficient material needed at the levels of natural rubber consumed in the war effort since the native guayule shrub supply would have been quickly and completely depleted.

Other sources of natural rubber were investigated during the 1941 to 1945 period. The Russian dandelion (Taraxacum kom-saghyz) came closest to guayule's capability, but its yield was much lower and cost higher than guayule so that it was not considered a reliable source of rubber even under war-time conditions (Whaley, 1948). A desert shrub, rubber rabbitbrush (Chrysothamnus nauseous), native to the western United States was promoted as an alternative, but like guayule, interest of this plant waned with the close of the ERP programme (Weber et al. 1985).

Soon after the end of World War II, the ERP was terminated. Nurseries, laboratories, fields, and processing facilities were abandoned and the stockpiles of shrubs, seeds, and rubber destroyed. The liquidation involved the loss of about 9500 tonnes of rubber in the field plots (Hammond and Polhamus, 1965). The termination occurred rather abruptly so that many experiments still underway were never completed, and much of the genetic pool was lost, and only a small fraction of the technical information gathered, was formally published. The incident serves as an example of how quickly policies can change, and unfortunately, such events have occurred several times with guayule. Wolek (1985), in referring to past guayule project terminations, noted that "... the temporality of political support would continue to haunt supporters of the modern programme".

This report will attempt to cover the latest efforts at commercialization of guayule that began in the mid-1970's. In this endeavour, we need to go over some of the early research, however, to learn from past experiences, and to avoid making the same errors.

Natural rubber was designated as a strategic material and provisions were made by the United States government to stockpile NR in support of the country's defence programme. New research was started in 1947 to investigate the extraction and processing of GR. A few years later in 1950, an emergency programme was set up to stockpile seeds and seedlings, but this was stopped two years later. Some low-keyed studies on guayule culture and breeding, which were being conducted, were terminated in 1959. The petroleum crisis of the 1970's again regenerated strong interest in guayule rubber. The National Academy of Science's report of 1977 on guayule added credibility to the potential benefits of establishing a guayule rubber industry. Besdies the vulnerability of the United States to a cutoff from its external sources of NR, the deficit balance of payment in foreign trade markets prompted discussions into the use of guayule to help relieve these problems, and also to alleviate some of its surplus crop supplies by substituting guayule as an alternative crop. Presently, the United States imports approximately $1 billion worth of NR annually, and producing its own GR would aid in cutting down its trade deficit. Projections being made at that time (Foster et al. 1980) indicated that a shortage of NR would occur in the 1990's. This prediction, combined with a steady increase in the price of HR, and the abrupt price rise during the oil embargo, made the cultivation of guayule in the United States very attractive to the growers, processers and investors who would all eventually participate in commercializing the 'new' industry.

To promote the establishment of the guayule rubber industry, the United States Congress mandated the Department of Defence to acquire GR as part of the NR stockpile. The price of NR in the 1980's has been less than the 1970's peak, but more recently a noticeable price increase has occurred. This increase is probably due to another unanticipated crisis attributable to the extensive need for natural rubber-based hygienic products to minimise the spread of the human immunodeficiency disease more commonly termed AIDS.

Other countries have investigated the possibility of having a guayule industry within their own boundaries essentially for the same reasons as the U.S. Latest economic projections by Rawlins (1986) show that Australia could be self-sufficient in NR, and furthermore, could possibly be an exporter of GR. Mexico had taken an active role in the 1970's (Campos-Lopez et al. 1978), but has reduced its research efforts in the 1980's. Feasibility studies, planting trials and research in guayule culture have been conducted at a low priority level in other countries such as France (Serier, 1983), India (Muthana, 1983; Behl, 1984; Srivastava et al. 1986), South Africa (Patternson-Jones, 1983, 1986), New Zealand (Taylor, 1986), Argentina (Ayerza and Bengtson, 1986), and Israel (Forgacs et al. 1986). In addition, rubber biosynthesis studies have been conducted at various international institutions to determine the exact nature of rubber formation with the possibility of producing rubber in the laboratory.

PROPERTIES OF GUAYULE RUBBER

Unlike Hevea, the rubber polymer in guayule is distributed within the parenchymous cell of the stem and root tissues and the rubber does not flow out as a milky latex sap when the stem tissue is cut. Instead, GR plant must be extracted by disrupting the cells and separating the polymer from the plant material by special extraction procedures. Also unlike Hevea, the guayule hydrocarbon extract contains a resinous material that must be separated from the rubber component to get good quality rubber. The separated GR has similar chemical and physical properties as HR (Table 1). GR has less quantities of heavy metals than HR, but their

TABLE 1

Non-rubber constituents and properties of natural rubber (National Academy of Sciences, 1977).

Non-rubber constituent	Guayule	Hevea
Dirt , %	0.007	0.05
Nitrogen, %	0.16	0.70
Ash, %	0.79	0.50
Copper, ppm	trace	8
Manganese, ppm	0	10
Rubber properties		
Mooney viscosity (ML 1+4, 100 C)	105	85
Plasticity Retention Index, %	41	60
Tack		
rubber to rubber, psi	9.5	8.5
rubber to metal, psi	4.25	5.0
Green Strength, psi at 100% elongation	20	20

physical properties are similar. Winkler et al. (1978) studied the effects of various formulations on the vulcanization and mechanical properties of GR and concluded that with appropriate modifications, GR could be vulcanized to get properties similar to Hevea vulcanites so that GR could be a direct substitute for HR.

Chemically, GR is made up of cis-1,4-isoprene polymers and has identical analytical optical properties as HR (Campos-Lopez et al. 1978). However, the polymer size of GR based on molecular weight analysis is slightly less than that of HR (Swanson and Buchanan, 1979). Guayule rubber does not contain any natural anti-oxidant like Hevea. In fact, some of the unsaturated fatty acids in the guayule resin component accelerate the degradation of GR alter harvesting. Degradation can be controlled by acetone extraction. With HR, however, acetone extraction removes the naturally occurring antioxidants (Hammon and Polhamus, 1965). Antioxidants must be added during the rubber extraction process and prior to the drying and storage of the guayule shrub to control degradation. Guayule rubber does not harden under storage as HR, and maintains constant Mooney viscosity, which Angulo-Sanchez et al. (1981) attributed to the absence of 'abnormal' functional groups in GR.

In spite of certain observable differences between GR and HR, both have been used to fabricate commercial products such as tires, which behave similarly under testing. Vulcanized rubber properties of GR and HR are compared in Table 2. In road tests, tires made from GR performed as well

TABLE 2

Vulcanized properties of Guayule and Hevea Rubber (Winkler et al. 1978).

Properties	Guayule	Hevea
Cure Time, $t_c(90)$, min	16	12
Stress at 500% elongation, MPa	1.72	5.00
Tensile strength, MPa	11.38	21.21
Elongation, %	720	680
Shore A hardness	32	37
Bashore Rebound, %	60	64
Molecular weight between crosslink, M_c	15,000	11,000

as the HR tires (National Academy of Sciences, 1977). Berger and Fontanoz (1983) reported that aircraft tires made from GR appeared to be an acceptable substitute for HR tires when tested under operational conditions of takeoffs and landings.

The most significant difference between the two rubber sources is their plant characteristics. Guayule is a very drought tolerant, semi-arid, non-laticiferous, shrub, whereas, Hevea is a laticiferous tree that grows best in tropical climates. Both plants, however, originated in the American continents. Irrigated guayule shrubs can be harvested two to three years after planting, whereas, Hevea trees require six to seven years before the trees can be opened for tapping. Rubber yields are presently of the order of 500 and 3400 kg ha^{-1} y^{-1} for guayule and Hevea, respectively. When Hevea cultivation began, yields were of the order of 225 kg ha^{-1} y^{-1}, but through proper selection, breeding and improved management practices, yields have increased at least tenfold. We can expect similar improvement in GR production, but not necessarily the same magnitude of increase because of limitations in the rubber carying potential of the guayule plant.

PLANT DESCRIPTION

Rubber was identified by the early Spanish explorers of the American continent as the main ingredient of the bouncing ball used in a game by the native Indians. The botanical nomenclature and description of guayule were made by Asa Gray, noted botanist, in the 1850's from plant samples collected in Texas (Hammond and Polhamus, 1965).

Guayule is a perennial plant of one meter in height and is part of the Compositae family that includes the sunflower (Helianthus annuus). It is a native of the Chihuahuan desert of northeastern Mexico and the adjoining Big Bend area of the state of Texas. The native habitat is predominantly in Mexico at elevations of 750 to 2000 meters, with temperatures of 3 to 34°C, and low rainfall of 130 to 350 mm, and in shallow calcareous soils. Muller (1946) attributed guayule's survival over other competing plants in such an environment to its unique wide ranging rooting pattern that allows the plant to exploit soil moisture very effectively. Guayule grows well under irrigation, but its relatively slow growth rate makes the plant non-competitive against weeds, especially in its seedling stage. Guayule is limited in its cold tolerance and can be damaged when exposed to a temperature of -10°C for only a few hours unless it is hardened before exposure (Mitchell, 1944). Low temperatures may not necessarily be bad, as Bonner (1943) observed that low night temperatures of 7°C promoted rubber synthesis. Winter conditions apparently are favourable to rubber production in guayule.

Availability of native shrubs for rubber extraction is limited and scattered widely over 3,400,000 ha. Naqvi and Hanson (1983) noted in later surveys that the present distribution is definitely smaller than those provided in earlier investigations. There is a slow regeneration of native stands and progenies are not added every year, but must wait for ideal weather conditions. Potential cultivation areas in the United States are limited to the southwestern region with 2,000,000 ha within the states of California, Arizona, New Mexico and Texas (Fig. 1). Fort Stockton, Texas is at the northern edge of guayule's native habitat. Guayule has

Fig. 1. Potential guayule cultivation regions in the United States (Bullard, 1946).

been successfully cultivated in Salinas, California, which was the centre of research and development during the ERP era of the 1940's. Bullard (1946) has delineated the potential growing areas based primarily on temperature since guayule cannot tolerate freezing temperatures below -10°C. In addition, guayule requires good soil aeration, and this was also considered in the selection of the growing sites.

Extensive reviews on rubber synthesis have primarily covered the Hevea plant (Bonner and Galston, 1947; Archer and Audley, 1967). More recently, articles relating to guayule have become available (Backhaus, 1985; Benedict, 1982). Rubber synthesis in guayule occurs in the parenchyma cells, whereas it occurs in the cells of the latex vessels in Hevea. Both plants synthesize rubber of high molecular weight (MW) that can be used for making commercial products. Their molecular weight distributions (MWD) are similar (Angulo-Sanchez et al. 1978; Hager et al. 1979). Both exhibit bimodal MWD (Backhaus and Nakayama, 1988) and in guayule, the low MW rubber predominates during the active vegetative growth period and is absent during the semi-dormant winter period, possibly when the high molecular weight polymer is being formed. This is characterized by the seasonal cyclic rubber accumulation observed in field measurements of rubber content, with the high values occurring during the winter months and the low values in the late spring and throughout the summer months (Benedict et al. 1986). Rubber production in guayule can also be enhanced by imposing stress on the plant. Benedict et al. (1947) found that the plant could be subjected to alternate low and high moisture stresses so that the storage tissue production could be promoted during the vegetative, non-stress period and rubber accumulation could be enhanced during the stress period. Geographic location does not affect MW or MWD, although it can affect the rubber content (Angulo-Sanchez et al. 1978).

CULTIVATION

Research on guayule culture in the 1980's has been directed toward re-addressing problems that were not completely solved during the ERP era and utilizing new tools for improving rubber yields and collecting and analyzing data. In most instances, guayule cultivation essentially follows standard farming practices similar to row crops with machinery modifications made to fit the growth behavior of the plant. Cultivation in the United States is either under dryland (non-irrigated) or irrigated conditions. In dryland agriculture, plants are grown from three to five years before harvesting, whereas under irrigation, the shrubs could be harvested within two years. Agricultural land considered marginal for crop production because of water shortage caused by over-use could be put into guayule production. Other non-agricultural land or overgrazed rangeland could be planted with guayule for wind and water erosion control. They can serve as future rubber reserve for emergencies and also act as anti-desertification buffers.

The estimated water requirement for guayule cultivation is listed in Table 3 for the various climatic regions. Since guayule is extremely drought tolerant, short-term water shortages in the irrigated areas would be only a temporary setback and not a crop disaster as with other commercial crops. Similarly, dryland agriculture can suffer periods of drought, but again yield decrease only would occur without the complete loss of the crop.

Two major shortcomings, existing today as they did in the ERP program, are poor plant establishment with direct seeding and low rubber yields. Other problems such as diseases, insects and weeds are also present, but these are site specific and more readily handled with today's technology.

TABLE 3

Water quantity and quality requirements for guayule production in different climatic regions (McGinnies and Mills, 1980).

Water Quantity

Sutability class	Arid Zone ha cm	Intermediate Zone ha cm	Coastal Zone ha cm
1. Good	91.4	76.2	51.8
2. Permissible	76.2-91.4	61.0-76.2	39.6-51.8
3. Doubtful	61.0-73.2	45.7-57.9	30.5-36.6
4. Unsuitable	0.0-57.9	0.0-42.7	0.0-27.4

Note: Intermediate zone has late spring rains and Coastal zone has fog and cooler weather compared to the Arid zone.

Water Quality

Suitability class	Total dissolved Solids ppm or (dS/m)	Na %	B ppm	Cl ppm	SO_4 ppm
1. Good	< 525 (< 0.82)	<40	< 2.0	< 248	< 336
2. Permissible	525-1400 (0.82-2.19)	40-60	2.0-3.0	248-426	336-576
3. Doubtful	1401-2100	60-80	3.1-3.7	421-710	577-960
4. Unsuitable	>2100 (>3.28)	> 80	> 3.7	> 710	>960

Plant establishment

Direct seeding for plant establishment was not reliable during the 1940's, so special nurseries were set up to provide enough seedlings for field planting. The rearing of the seedlings involved tremendous manpower and equipment (Hammond and Polhamus, 1965). Cost estimates were between $0.02 to 0.05 per seedling. Planting rates were in the order of 35,000 plants ha^{-1}. Transplanting also involved both equipment and manpower and adequate management was required after the transplanting to control soil moisture. The guayule plant regardless of age is susceptible to excess moisture. Young seedlings are also readily damaged by soil salinity (Miyamoto et al. 1989). Once the plants are established, they can withstand salinities much higher than those by other commercial crops (Francois, 1986; Maas et al. 1988).

Significant advances have been made in the 1980's to make direct seeding a reliable method for plant establishment. Naqvi and Hanson (1980, 1982) have discussed improved seed germination techniques and the existence of inhibitors that affect the dormancy of the seeds. In the ERP years and until the early 1980's, seeds were being treated with hypochlorite solutions to break their dormancy. Presently, seeds are being osmo-conditioned with polyethylene glycols, and additionally treated with gibberellic and abscisic acid (Chandra and Bucks, 1986; Chandra ans Svrjcek, 1988). Earlier ERP reports recommended seed placement at the top or just below the soil surface (Hammond, 1959). Conditioned seeds could be placed at 10 mm or deeper, which provides a better moisture and temperature regime for germination and seedling growth (Fink et al. 1987). Mulching with vermiculite also provides protection from wind and water erosion (Foster et al. 1986, 1987). Fungicide applications also help in promoting growth and establishment (Mihail and Alcorn, 1988). Shading of the seed bed improves establishment of the directly sown seed (Allen et al. 1988; Bucks et al. 1987).

A better understanding of seedling mortality is being developed. Miyamoto et al. (1984b) found that the hypocotyl is very sensitive to salinity and salt build-up at the soil surface is detrimental to the young plants. Mortality is further increased by fungal infection under saline conditions (Mihail et al. 1987).

Soil and fertility management

Guayule grows best in light to medium textured calcareous soils with good drainage. Water logging can cause death or stunted growth. Disease problems are created or aggravated due to poor soil aeration. Detailed fertilizer recommendations for guayule have not been developed. The plant does not seem to have especially high fertility requirements, and thus, would eliminate part of the cultivation cost. Nutrient deficiency symptoms are neither prominent nor typical of other field crops. Therefore, plant appearance and elemental nutrient deficiency interrelations are difficult to establish. In greenhouse studies, Bonner (1944) recorded the highest rubber yield in plants grown in high nitrogen and phosphorus levels.

However, Tingey (1952) observed little increase in rubber yield with added fertilizers, and McGinnies (1983b) indicated that yields may be actually suppressed under high nutritional conditions. Plants with high growth rates under high nutrition had the lowest rubber contents. Hammond and Polhamus (1965) noted from an unpublished 1945 study that growth was better with the nitrate rather than the ammonium nitrogen source. Bonner (1944) reported that high ammonium to nitrate ratios decreased growth and rubber content. In field trials, Rubis (1983) observed that guayule responded better to calcium nitrate than either urea or ammonium phosphate nitrogen sources. Bucks et al. (1985a) showed that maximum rubber yields had to be accomplished with a favourable combination of high water and nitrogen applications.

Guayule does not respond to sulfate deficiency (Bonner, 1944); however, Mitchell et al. (1944) found that boron deficiency, which caused reduced plant growth and rubber content, could be treated with boron application. Low phosphorus conditions were imposed on guayule grown under gravel culture by Thomas (1986), but plants did not exhibit the anthocyanin pigmentation or purple coloration typical of phosphorus deficiency. Although the soil phosphorus content is high in the native habitat (Ostler et al. 1983), phosphorus does not appear to be a nutrition problem for guayule. Apparently, the growth rate of guayule is relatively slow so that the rate of soil nutrient availability can keep up with root mineral absorption. In the rapidly growing seedling stage, however, Bloss and Pfeiffer (1981, 1984) found that soil mycorrhiza aided in the nutrient uptake by roots and improved plant growth and rubber content.

Because nutrient deficiency appears to be only a minor problem with limited plant response to high fertilizer application rates, research emphasis on guayule nutrition and soil fertility is very low.

Water management

Guayule is a drought tolerant plant, and thus, it has the ability to survive and compete effectively against other plants in the semi-arid region. However, for commercialization, cultural practices must be adopted to increase plant growth and rubber accumulation rates above its native state. Kelley (1975) pointed out that where temperature and soil conditions are suitable for guayule, water is probably the most important factor for growth and rubber production. Water requirements of guayule in the various climatic zones and the quality of irrigation water are listed in Table 3. Dryland cultivation would prevail in south Texas where irrigation water is limited, but where rainfall is still sufficient to maintain plant growth.

Supplementary irrigation is definitely needed in the arid zone, and water presently being used for other crops could be redirected for this purpose.

Production of existing crops is being handicapped because water supply has become limited or erratic. Most crops require water at a critical point in their development, whereas guayule does not, so that several irrigations can be skipped without major impact on production. In fact, Benedict et al. (1947) found that imposing alternate periods of water stress on guayule could increase rubber yields. Admittedly, guayule requires water during plant establishment, but planting can be delayed or postponed in an emergency.

Since water stress has been noted to be an important part of rubber production, Nakayama and Bucks (1983, 1984) investigated the application of the crop water stress indexing (CWSI) technique to guayule. The water stress could be readily followed by measuring the canopy temperature, and when it was incorporated into the CWSI concept the rubber yields could be predicted. Ray et al. (1986) used CWSI for establishing the various irrigation treatments. Interrelations between stress and rubber production have been observed by Cornish and Backhaus (1988), and Reddy and Das (1988) who found that the activity of the rubber transferase enzyme increased as the plant was stressed. There is a limit to how much continuous stress can be imposed on the plant before rubber accumulation is curtailed since the plant requires energy for respiration and synthesis of tissues where the rubber polymer is stored.

Proper water management is important in plant establishment by direct seeding or transplanting. Salt accumulation on the soil surface can be controlled by frequent, light water applications. Miyamotor and Bucks (1985) have summarized the amounts and methods of water application by various investigators. Success in plant establishment involves keeping the soil surface moist, but not wet. As noted earlier in the stand establishment section, the availability of conditioned seeds has improved seed germination and seedling growth.

Several sets of irrigation experiments were conducted by Bucks and co-workers (Bucks et al. 1985a, 1985b, 1985c) and their results can be summarized in Fig. 2. Guayule under cultivation has a seasonal pattern of growth and rubber accumulation, with the biomass production the highest in the spring and fall months, and rubber accumulation in the winter. Hunter and Kelley (1946) reported little gain in guayule shrub weight from November through March. The optimal times for plant harvest are indicated by the dashed lines in Fig. 2 and occur during the January to March period. As expected, irrigation increases rubber yield. The dry treatment I_6 comes

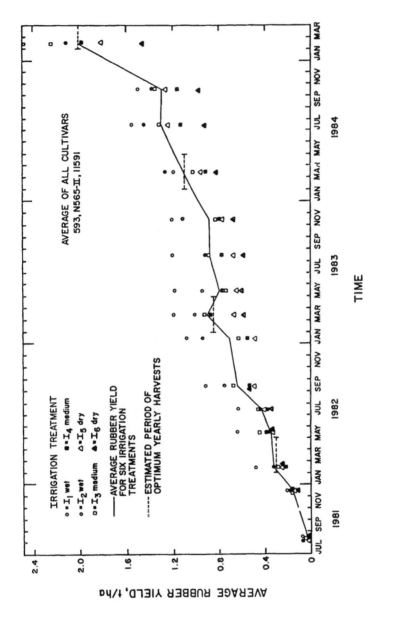

Fig. 2. Seasonal guayule rubber production for various irrigation treatments, Mesa, Arizona (Bucks et al. 1986).

closest to the yields being obtained in dryland culture (Gonzalez, 1988), which depends primarily on natural rainfall. The dry treatment received 740 mm of water per year and is in the Permissible-Doubtful suitability designation of Table 3. The wettest treatment with 1840 mm water per year exceeds the Good suitability class and would not be acceptable for economic production. Instead, irrigation application four to six times per year as exemplified by treatments I_5 (840 mm y^{-1}) and I_4 (1040 mm y^{-1}), could be used for rubber production in the 150 mm y^{-1} rainfall region.

Investigators in other climatic zones and under different irrigation levels have also obtained increased rubber yields with increasing water application rates (Fangmeier et al. 1985; Miyamoto et al. 1984a; Tingey, 1952). The highlight of these studies points to the fact that although guayule is drought resistant, it responds well to water application. Also, its water use efficiency is much less than other commercial crops. Based on biomass production, its water use efficiency is in the order of 0.8 kg/m^3 compared to that of other crops such as alfalfa and corn with values of 1.2 to 2.8 kg/m^3, respectively.

Where irrigation is practiced, water management also involves salinity management. As noted earlier, transplants and seedlings are very susceptible to salt damage. The severity of salinity damage varies with the stage of plant growth (Miyamoto et al. 1989), and aggravates disease problems (Mihail et al. 1987). According to the studies of Hoffman et al. (1988) and Maas et al. (1988), the threshold salinity of the saturated soil extract for guayule is about 7.5 dS m^{-1} with yield decrements of 11% with each increment in the salinity of the extract. In terms of irrigation water salinity, a 10% decrease in yield occurs when the salinity is increased from 1.2 dS m^{-1} (salinity of Colorado River water at Brawley, California) to 3.2 dS m^{-1}. At salinity of 24 dS m^{-1}, yield is down to zero.

Pest management

Plant pests are considered in terms of weed, insect and disease. Guayule is a poor competitor against weeds except in its native location. The plant is slow growing relative to other plants, so that good moisture conditions created by irrigation promote the proliferation of weeds. In the 1940's when herbicides were not available, weeds were controlled by hand weeding and diesel application (Benedict and Krofchek, 1946). More recently, the availability of various herbicides has helped in controlling weed problems in guayule culture. Elder et al. (1983) observed that Trifluralin was the best for preplant weed control and Simazine and Surflan for post-transplant control. Preplant applications combining Goal and Surflan appeared to be effective for transplant establishment (Ferraris, 1986). The herbicides Dacthal, Prowl and Prefar showed fair weed control and adequate selectivity in direct-seeded guayule, and for transplants the same three herbicides and Brake were effective (Whitworth, 1983). It was also noted that the cultivars had varying degrees of tolerance to a specific herbicide. Gonzalez and Latigo (1986) got best results with Prefar as a preplant herbicide for seed germination, whereas Surflan reduced seed germination. In young stands, Clark and Whitworth (1981) found three herbicides, viz., Diuron, Simazine and Fluridone, gave the best weed control. In mature four year-old stands, Foster et al. (1985) got good weed control with Glyphosate applied during dormancy.

Weed control with the advent of modern selective herbicides is not as great a problem as that encountered in the early history of guayule culti-vation. Some herbicide related chemicals are being tested as potential plant biological regulators for increasing rubber yields (Yokoyama et al. 1987; Benedict et al. 1987), but field results have been mixed (Allen et al. 1988; Bucks et al. 1985d).

Guayule is most susceptible to insect damage during its early growth stages. Leaf hoppers, crickets, leaf-feeding beetles, and caterpillars can cause considerable damage to the seedlings and young transplants. Control includes insecticides such as Diazinon, Malathion or other materials normally used in other domesticated crops. Animals such as rabbits have also been known to feed on the young shrubs and cause extensive damage. In the greenhouse, red spider mite, cricket and white fly are the predominant insects and can be controlled by Malathion, Maverick or Thiodan. In mature plants, Liygus can damage the young seeds and reduce their weight and viability (Romney et al. 1945).

Thomas and Goddard (1986) found that guayule was an atypical in respect to plant-parasite nematode relationships. Two common parasites of annual crops, Meloidogyne incognita and Pratylenchus scribneri, did not infect guayule, and they recommended the possibility of using guayule for alternate cropping. They suggested that the rubber in the roots prevented parasite movement and feeding. The nematode Criconemella xenoplax, however, infests guayule and could pose potential problems.

The general perception at the beginning of guayule culture was that the plant was relatively free of diseases. The first problems were observed in the nursery with both pre- and post-emergence damping off particularly on the heavy, poorly drained soils (Campbell and Presley, 1946; Presley, 1975). The fungal diseases Pythium root rot and Sclerotinia were observed in warm climates and Verticillium in cooler climates where its soil population was already high. Verticillium damage was less in thickly planted than sparsely planted fields. Plant damage by Rhizoctonia bataticola and Fusarium moniliform could make the plant susceptible to nematode infestation (Sidhu and Behl,1987).

Most of the recent work has been directed towards the control of pathogens either through the development of disease resistant cultivars (Mihail and Alcorn, 1986; Orum and Alcorn, 1986) and through the use of chemicals.

Mihail and Alcorn (1988) found that the combination of one of the three fungicides (Terrachlor/Super-X, Ridomil, Ridomil/PCNB) with solar screening enhanced seed germination and seedling growth. Thiram at low concentrations is being used by Chandra et al. (1987) to control fungi as part of the seed conditioning procedure.

Harvest management

Three aspects of harvesting are involved in guayule: seed harvesting, shrub harvesting and postharvest baling to prepare the shrub for the processing facility. Clipping or pollarding is also practiced, where the stems are cut at or above the soil surface and the remaining sucker allowed to regrow for another clipping or whole-plant harvest. Clipping equipment has been developed and is functional (Coates, 1986b). Whole-plant harvesting equipment has also been designed, built and tested and is presently operational (Coates and Wodrich, 1988). Similarly, guayule shrub balers have been perfected (Coates and Lorenzen, 1988). The equipment was essentially built from existing farm machinery with the necessary modifications to fit the growth pattern and geometry of the guayule shrub.

Although guayule is noted for its prolific seed production, collection still remains a problem. The seed head shatters readily and seeds can easily be lost by strong wind or rain at maturity. Under natural conditions, the exact time for seed collection depends upon rainfall distribution since flowering is highly related to soil moisture availability. Under natural or dryland culture two to three seed collections are possible, and under irrigation four to six or more collections can be made depending upon the irrigation frequency. Guayule is a long-day plant and flower initiation does not start until day lengths of 11 h are present (Higgins and Backhaus, 1983). Mid-summer seed harvest, however, gives poor quality seeds (McGinnies and Mills, 1980).

Equipment improvements in seed harvesting have been made by modifying a cotton harvester with the installation of brushes, beaters and a vacuum system to collect the seeds (Coates, 1985, 1986a; Lorenzen and Coates, 1986). Plant damage is minimal and an adequate rate of land coverage can be made.

Cleaning and threshing of guayule seeds have also been made by modfying existing seed cleaning equipment (Hammond and Polhamus, 1965). The equipment includes the clipper cleaner with vibrating screens and air flow control (Ranne et al. 1986). Tipton et al. (1981) made a cleaning apparatus for cleaning small quantities of seeds. Guayule seeds need to be cleaned to remove the loral parts, stems, and other foreign materials, which are included in the harvesting operation. Threshing was found to increase seed viability (McGinnies and Mills, 1980).

CROP IMPROVEMENT

Development of a new crop involves several important steps before commercialization is possible. These include (a) germplasm collection and evaluation, (b) chemical utilization and evaluation, (c) agronomic and horti-cultural evaluation, (d) breeding programs, (e) production and processing scale-up, and finally, (f) commercialization. Guayule can be considered as a 'new' old crop and has gone through several intermediate steps, but never attained the ultimate last step. At present, several different steps are being conducted simultaneously and independently to hasten development with varying intensity and research support. In the ERP programme, essentially all emphasis was directed toward rapid production of rubber and other aspects such as germplasm improvement were not fully exploited. Another programme was started in the 1950's to meet this objective, but it was unfortunately phased out before improved varieties could be developed.

Economic analyses indicate that rubber yields must be doubled to make guayule commercialization profitable. Other domesticated crops faced similar problems in their developmental stage and owe much of their success to germplasm improvements. This should also hold true for guayule. Besides increasing rubber yields, other genetic related factors could be improved such as disease resistance, germination and seedling vigor for reliable plant establishment, heat and cold tolerance, and regrowth after clipping.

A breakthrough in guayule breeding came about by the discovery of apomixis by Powers and Rollins (1945) where it was found that seed development could occur without pollination. Thus, the 'true' breeding programme started from this point, but unfortunately, the research work ended at the F_1 hybrid level before significant progress could be realized (Rollins, 1975). Since the early breeders were not aware of apomixis, selection for uniformity led to the prevailing apomictic lines and the selections made within these lines did not lead to germplasm improvement. Fortunately, guayule is not totally apomictic and sexual reproduction does occur. Thompson and Ray (1988) have recently reviewed apomixis and its consequence to the guayule breeding programme. Importantly, the genetic control of apomixis is not clearly understood.

Tysdal (1983) presented a flowchart for a 10-year guayule breeding programme. His approach was based on experience gained during the aborted 1950's breeding programme. His recommendation included selection within diploid lines, interspecific hybridization and backcrossing. Other important findings in guayule genetics were that the 36-chromosome diploid plants (2N = 36) were the sexual plants (Rollins, 1975), and that various levels of ploidy (3N = 54, 4N = 72, ...8N = 144) existed. Only a few sexually reproducible diploid lines (2N = 36) are presently available. Tysdal et al. (1983) and Estalai (1985, 1986) have registered diploid germplasms for use in the breeding research programme.

Detailed characterization of the 26 USDA lines from the ERP era have been made by Ray and Thompson (1986). Chromosome numbers ranged from 2N = 26 to 8N = 144. As expected most of the available germplasm collections were apomictic polyploids with the 3N and 4N characteristics. Field trials by Naqvi et al. (1986) showed a wide range of rubber content for these lines and indicated that they were not uniform.

Interspecific hybridization has not produced progenies with high rubber yielding qualities so far. Crosses have been made between the diploid P. argentatum and a tree-like species P. schottii. The offspring has high biomass, but low rubber content and quality, so that the hybrid yields were not much better than the existing lines (Naqvi et al. 1987). Also large biomass-low rubber content combinations create additional shrub material for handling by the processing unit. Further screening and back-crossing are needed to improve the best characteristics of the new lines produced.

Current breeding programme are following along the lines of (a) single plant selections from existing apomictic polyploid lines, (b) recurrent selections among sexually reproducing diploid plants, and (c) interspecific hybridization with desirable characteristics (Thompson and Ray, 1988). The doubling of yield has not been achieved yet.

PROCESSING

Two distinctly different approaches have been used for extracting rubber from the guayule shrub (Gartside, 1986). The classical water flotation method involves a preliminary step of par-boiling the plant to coagulate the cell rubber followed by grinding the plant in water. The rubber-resin particles with the plant wood float to the surface and the worm-like material is skimmed off or centrifuge-separated from the rest of the suspended material. The 'worm' is treated with solvents to extract the resin (acetone) and rubber (hexane or cyclohexane). The rubber quality by this process even up to the 1940's was poor because the resin constituent was not extensively removed from the rubber. The rubber was tacky, but could be used in blends with HR and later with synthetic rubber.

With improvements in the solvent extraction process, the flotation method could be entirely by-passed, so that large quantities of water are not required. A single stage, dual solvent (acetone-hexane) rubber extraction pilot processing facility was constructed by the Firestone Tire and Rubber Company in Sacaton, Arizona in 1988 and under test operation at present. In this continuous process, ground shrub material is mixed with the acetone-hexane mixture under pressure and elevated temperature to remove the resin and rubber constituents from the plant cell. The dissolved rubber is then selectively separated from the resin by adjustment of the acetone-to-hexane ratio. The rubber is then recovered by desolventization. The resin fraction is also concentrated by desolventization. Spent acetone and hexane solvents are recovered by fractional distillation and recycled back into the processing system. Other rubber and resin extraction techniques have been tried using existing soybean and cottonseed oil extracting equipment and with different solvent extracting combinations (Hamerstrand and Montgomery, 1984; Wagner et al. 1986).

The completion and reliable operation of the pilot processing facility would solve one of the critical problems leading to the commercialization of GR. Besides demonstrating that a reliable and economical rubber extraction process is attainable, sufficient guayule rubber and resin can be made available to industry for fabricating rubber and resin products for testing purposes.

BY-PRODUCTS

The resin output would be approximately the same as the rubber yields. Investigations are being conducted on the use and marketability of the resin component to help defray the rubber extraction cost. Optimistic viewpoints are held by the investigators presently involved in by-product or co-product research who believe that the resin may in time become more valuable than the rubber. Identification of resin components is important for developing their potential uses. Schloman et al. (1983) made composition profiles of the acetone-extractable resin and found the major components to be made up of sesquiterpene esters (10-15%), triterpenoids (27%) and fatty acid triglycerides (7-19%).

The hydrocarbon resin fraction is made up of many organic compounds. The monoterpenes such as α- and β-pinene, cadine, camphene, limonene, β-myrcene, ocimene, phellandene, sabine and terpenolene are part of the resin (Scora and Kumamoto, 1979; Kumamoto et al. 1985). Some of these compounds plus isoprene are present in vapor emissions from the plant

(Nakayama, 1984) and contribute to the typical odour of a pine forest notice-able around a guayule field planting.

Banigan et al. (1982) isolated a guayule wax, docosanyl ercosanate, from the leaves which is similar to the jojoba wax. Other compounds of interest present in the resin are the fatty palmitic, stearic and linoleic acids (Banigan and Meeks, 1953), which are responsible for the oxidation of the rubber.

Besides attempting to identify the numerous chemical compounds of the resin, studies have been conducted to determine its usefulness for pest control. Bultman et al. (1986) observed that guayule resin could protect wood in marine and terrestrial environments against crustacean borer and termites of the genera Coptotermes and Heterotermes. The protective action of guayule resin was also noted by Bultman and Bailey (1988) when it was incorporated into polyvinyl chloride (PVC) electrical insulation. They also found that fungal colonies were absent from the treated (PVC) material. Bultman et al. (1987) had earlier reported that pine and birch wood impregnated with the resin resisted the destructive activity of brown and white rot fungi.

Winkler and Stephens (1978) reported that guayule resin promoted reduction in the gel content of HR, styrene, and butadiene rubbers and had the advantage of not having the toxicity associated with synthetic peptizers. Guayule resin has also been tested as extender/plasticizer. Materials formulated with the resin performed equally well on metal surfaces as other epoxy-resin systems (Thames and Kaleem, 1988). Belmares and Jimenez (1978) indicated the possibility of using the resin in varnishes, adhesives, and pigment dispersers and tackifiers for rubber.

A large quantity of spent desolventized plant material or bagasse will be produced in the rubber processing. One possible use includes the burning of the bagasse with a thermal energy of 6.40×10^8 J kg^{-1} for the generation of steam and electricity. In the earlier rubber processing days, the bagasse was used as fuel for the steam boilers. Wagner et al. (1988) are investigating the possible use of bagasse for making fireplace logs. The possibility exists for utilizing the bagasse for the synthesis of furfural (Schloman, 1986). Kuester and Wang (1986) converted the bagasse to trans-portation grade diesel fuel, which is compatible with existing engines and fuel distribution systems. Thus, the fuel can be used by the farm equipment located close to the processing site. On-site consumption of bagasse would reduce cost and problems related to storage and disposal.

ECONOMICS

Opposing views have developed regarding the economics of guayule commercialization. Weihe and Nivert (1983) in some earlier analyses projected that guayule rubber production would be profitable. Based on 40,000 ha with a harvest of 10,000 ha per year on a 4-year cycle and 3,400 kg ha^{-1} rubber per year, GR at $0.66 kg^{-1} could be produced. This estimate includes a processing cost of $1.26 kg^{-1} rubber less by-product credit of $0.62 kg^{-1} would give a net processing cost of $0.64 kg^{-1}. A net positive return would be realized based on the rubber price of $1.54 kg^{-1} rubber. In an Australian analysis of rubber processing, Gartside (1986) arrived at a cost of $732 t^{-1} rubber based on a 15.2% rubber content and 90% extraction efficiency. The plant processing capacity would be 139,000 t shrub per year with an output of 18,700 t of rubber per year with capital costs of $27,900,000 for the plant.

In contrast, Wright and Connell (1983) presented analyses which showed that guayule rubber production would not be profitable in the United States or elsewhere at the present state of development. They indicated that GR yields or the price of HR must double before profitability can be attained, assuming that the by-products can cover the cost of processing. They estimated a total cost of $7,910 ha^{-1} based on a 5-year crop production period and includes all farming operations from land preparation, plant establishment, crop management to harvesting. Gross revenues of $4,050 ha^{-1} can be realized from a rubber yield of 3.81 t ha^{-1} y^{-1} assuming an 8% rubber content and NR price of $1.32 kg^{-1}.

Before all the objectives can be reached for increasing yields, improving rubber extraction process, and developing by-product applications to achieve profitability, some kind of temporary support programme must be provided for the various parties involved in the undertaking. The Council for Agricultural Sciences and Technology (1984) suggested developmental incentives which would reduce the risks for the grower or producer and processor, who would all participate in a well coordinated programme. For the producer, guayule could be grown in 'set aside' land withheld from the production of surplus crops so that penalties would not be imposed on the grower. Furthermore, loan guarantees could be made to the producer and processor based on a reasonable expectation of yield, quality and price. Such guarantees, however, should be made to meet economic needs so that future surplus production or the support of an unprofitable industry is not perpetuated. The financing could be similar to the surplus crop support programme, but in this case the multiple benefits would be to help create a new agriculturally based industry with an alternative crop for the farmers. It would also provide an internal source of a critical strategic material and help solve, in part, the balance of payment problem.

SUMMARY

Guayule rubber production even at the magnitude envisioned for the United States would still be a small fraction of the total natural and synthetic rubber production. However, it could develop into a very important competitor and its economic and political impact on the rubber industry would be immeasurable. The potentials of guayule resin have not been fully established and it may be a key raw material which could help the economics of guayule cultivation.

Guayule culture is adaptable to high mechanization that is appropriately suited for U.S. farming technology. Cultivation could be located on land that is not competitive with other crops. Guayule commercialization on a permanent basis with positive support from the private, industry and government sectors has now come closest to realization than any other time in its history. All aspects concerning the cultivation, processing and marketing are in place and ready for commercial acceptance in the next five to ten years.

A large burden is placed on the plant geneticists to at least double the rubber yield in the near future. Hevea rubber production faced similar problems, but progress in cultivation and breeding helped to increase yields from approximately 900 kg ha^{-1} y^{-1} to 4500 kg ha^{-1} y^{-1} in a span of 25 years. Tysdal (1975) alluded to the plant breeding miracles achieved with wheat, corn, soybeans and many other crops, and that guayule also needs a continuing, fruitful breeding programme to attain similar success.

Even if guayule rubber production does not become a viable entity because of insurmountable political, social and economic barriers, the potential always exists for guayule to become an additional source of natural rubber, and this implication alone could have a bearing on the pricing structure and supply of natural and synthetic rubber. Much will depend also on how rapidly the world's supply of easily extractable hydrocarbon resources are depeleted.

REFERENCES

Allen, S.G. and Nakayama, F.S. 1987. DCPTA bioregulation of growth and rubber production of several Parthenium species. In: Seventh Ann. Guayule Rubber Soc. Conf., Annapolis, Maryland. 16 pp.
Allen, S.G., Bucks, D.A., Powers, D.E., Nakayama, F.S. and Alexander, W.L. 1988. Irrigation and row cover effects on direct seeding of guayule. Eight Ann. Guayule Rubber Soc. Conf., Mesa, Arizona. 14 pp.

Angulo-Sanchez, J.L., Campos-Lopez, E. and Gonzalez-Serna, R. 1978. Geographic influence on guayule rubber. In: Campos-Lopez, E. and McGinnies, W.G. (Eds.). Guayule: Reencuentro en el Desierto. Publ. 371, CIQA, Saltillo, Coahuila, Mexico. pp. 177-190.

Angulo-Zanchez, J.L., Jimenez-Valdez, L.L. and Campos-Lopez, E. 1981. Storage hardening and 'abnormal' groups in guayule rubber. J. Appl. Polymer Sci., 26: 1511-1517.

Archer, B.L. and Audley, B.G. 1967. Biosynthesis of rubber. In: Nord, F.F. (Ed.). Advances in Enzymology, 29: 221-257.

Ayerza, R. and Bengtson, D. 1986. The guayule program in Argentina. In: Fangmeier, D.D. and Alcorn, S.M. (Eds.). Proc. Fourth Intern. Guayule Res. and Develop. Conf. Tucson, Arizona. pp. 11-19.

Backhaus, R.A. 1985. Rubber formation in plants-a mini-review. Israel J. Bot., 34: 283-293.

Backhaus, R.A. and Nakayama, F.S. 1988. Variation in the molecular weight distribution of rubber from cultivated guayule. Rubber Chem. Technol., 61: 78-85.

Banigan, T.F. and Meeks, I.W. 1953. The isolation of palmitic, stearic and linoleic acids from guayule resin. J. Amer. Chem. Soc., 75: 3829-3830.

Banigan, T.F., Verbiscar, A.J. and Weber, C.W. 1982. Composition of guayule leaves, seed and wood. J. Agrc. Food Chem. 30: 427-432.

Behl, H.M. 1984. Feasibility of commercial development of guayule crops in India. Fifth Ann. Guayule Rubber Soc. Conf. Washington, D.C. 82 pp.

Belmares, H. and Jimenez, L.L. 1978. Resina de guayule: caracterizaciony posibles usos industriales. In: Campos-Lopez, E. and McGinnies, W.G. (Eds.). Guayule: Reencuentro en el Desierto. PUbl. 371, CIQA, Saltiollo, Coahuila, Mexico. 315-329 pp.

Benedict, C.R. 1982. Biosynthesis of rubber. Porter, J.W. and Spurgeon, S.L. (Eds.). Biosynthesis of Isoprenoid Compounds. John Wiley and Sons, New York, 355-369 pp.

Benedict, C.R., Rosenfield, C.L., Gale, M.A. and Poster, M.A. 1986. The biochemistry and physiology of isopentenylpyrophosphate incorporation into cis-polyisoprene in guayule plants. In: Benedict, C.R. (Ed.). Biochemistry and Regulation of cis-Polyisoprene in Plants. Texas A&M University, College Station, Texas. pp. 85-105.

Benedict, C.R., Madhavan, S. and Foster, M.A. 1987. The enzymatic incorporation of isopentyl pyrophosphate into polyisoprene in Parthenium argentatum treated with DCPTA. In Seventh Ann. Guayule Rubber Soc. Conf. Annapolis, Maryland. 18 pp.

Benedict, H.M. and Krofchek, A.W. 1946. The effect of petroleum oil herbicides on the growth of guayule and weed seedlings. J. Amer. Soc. Agron., 38: 882-895.

Benedict, H.M., Mcrary, W.L. and Slattery, M.C. 1947. Response to guayule to alternating periods of low and high moisture stresses. Bot. Gaz. 108: 535-549.

Berger, M.S. and Fontanoz, Jr. M. 1983. Evaluation of guayule rubber in military applications. In: Gregg, E.C., Tipton, J.L. and Huang, H.T. (Eds.). Proc. Third Intern. Guayule Conf. Pasadena, California. pp. 203-208.

Bloss, H.E. and Pfeiffer, A.W. 1981. Growth and nutrition of mycorrhizal guayule plants. Ann. Appl. Biol., 99: 267-274.

Bloss, H.E. and Pfeiffer, A.W. 1984. Latex content and biomass increase in mycorrhizal guayule (Parthenium argentatum) under field conditions. Ann. Appl. Biol., 104: 175-183.

Bonner, J. 1943. Effects of temperature on rubber accumulation by the guayule plant. Bot. Gaz., 105: 233-243.

Bonner, J. 1944. Effect of varying nutritional treatments on growth and rubber accumulation in guayule. Bot. Gaz., 105: 352-364.

Bonner, J. and Galston, A. 1947. The physiology and biochemistry of rubber formation in plants. Bot. REv., 13: 543-596.

Bucks, D.A., Nakayama, F.S., French, O.F., Legard, W.W. and Alexander, W.L. 1965a. Irrigated guayule-evapotranspiration and plant water stress. Agric. Water Manage., 10: 61-79.

Bucks, D.A., Nakayama, F.S., French, O.F., Rasnick, B.A. and Alexander, W.L. 1985b. Irrigated guayule-plant growth and production. Agric. Water Manage., 10: 81-93.

Bucks, D.A., Nakayama, F.S., French, O.F., Legard, W.W. and Alexander, W.L. 1985c. Irrigated guayule-production and water use relationships. Agric. Water Manage., 10: 96-102.

Bucks, D.A., Roth, R.L., Nakayama, F.S. and Gardner, B.R. 1985d. Irrigation water, nitrogen, and bioregulation for guayule production. Trans. Amer. Soc. AGric. Eng., 28: 1196-1205.

Bucks, D.A., Nakayama, F.S. and Allen, S.G. 1986. Regulation of guayule rubber content and biomass by water stress. In: Benedict, C.R. (Ed.). Biochemistry and Regulation of cis-Polyisoprene in Plants. Texas A&M University, College Station, Texas. OO. 161-173.

Bucks, D.A., Powers, D.E., Chandra, G.R., Allen, S.G. and Fink, D.H. 1987. Shading techniques for the direct seeding of guayule. In: Seventh Ann. Guayule Rubber Soc. Conf. Annapolis, Maryland, 70 pp.

Bullard, W.E., Jr. 1946. Climate and Guayule Culture. Emergency rubber project report. Forest Services, U.S. Department of Agriculture. Washington, D.C. 49 pp.

Bultman, J.D. and Bailey, C.A. 1988. Guayule resin as an antitermatic additive for polyvinyl chloride electrical insulation - early field results. In: Eight Ann. Guayule Rubber Soc. Conf. Mesa, Arizona. 49 pp.

Bultman, J.D., Beal, R.H., Schloman, W,W. and Bailey, C.A. 1986. The evaluation of guayule resin as a pesticide. Proc. In: Fangmeier, D.D. and Alcorn, S.M. (Eds.). Proc. Fourth Intern. Conf. on Guayule Res. and Develop. Tucson, Arizona. pp. 353-356.

Bultman, J.D., Gilbertson, R.L., Adaskaveg, J.E., Amburgey, T.L., Parikh, S.V. and Bailey, C.A. 1987. Guayule resin as a potential fungicide. In: Seventh Ann. Guayule Rubber Soc. Conf. Annapolis, Maryland. 28 pp.

Campbell, W.A. and Presley, J.T. 1946. Diseases of Cultivated Guayule and Their Control. (USDA Circular No. 749). 42 pp.

Campos-Lopez, E., Neavez-Camacho, N. and Maldanado-Garcia, M. 1978. Guayule: Present state of knowledge. In: Campos-Lopez, E. and Mc-Ginnies, W.F. Guayule: Reencuentro en el Desierto. Publ. 371, CIQA, Saltillo, Coahuila, Mexico. pp. 375-410.

Chandra, G.R. and Bucks, D.A. 1986. Improved quality chemically treated guayule (Parthenium argentatum Gray) seeds. In: Fangmeier, D.D. and Alcorn, S.M. (Eds.). Proc. Fourth Intern. Conf. Guayule Res. and Develop. Tucson, Arizona. pp. 59-69.

Chandra, G.R. and Svrjcek, R. 1988. Improved quality of osmo-conditioned guayule seed. Eight Ann. Guayule Rubber Soc. Conf. Mesa, Arizona. 12 pp.

Chandra, G.R., Svrjcek, R. and Bucks, D.A. 1987. Effect of Thiram on guayule seed germination. In: Seventh Ann. Guayule Soc. Conf. Annapolis, Maryland. 73 pp.

Clark, D.C. and Whitworth, J.W. 1981. Effects of herbicides on guayule (Parthenium argentatum). Proc. Western Soc. Weed Sci., 34: 69-70.

Coates, W.E. 1985. Development of a guayule seed harvester. Trans. Amer. Soc. Agric. Eng., 28: 687-690.

Coates, W. 1986a. A guayule seed harvesting device. Appl. Eng. Agric. 2: 70-72.

Coates, W. 1986b. Evaluation of cutting devices for guayule harvesting. Appl. Eng. AGric., 2: 73-75.

592

Coates, W.E. and Lorenzen, B.R. 1988. Baling as a means of guayule shrub densification. Eight Ann. Guayule Rubber Soc. Conf. Mesa, Arizona. 19 pp.

Coates, W.E. and Wodrich, T.D. 1988. An implement for digging guayule. Eight Ann. Guayule Rubber Soc. Conf. Mesa, Arizona. 18 pp.

Cornish, K. and Backhaus, R.A. 1988. Rubber transferase activity in field-grown guayule. In: Eight Ann. Guayule Rubber Soc. Conf. Mesa, Arizona. 46 pp.

Council for Agricultural Science and Technology. 1984. Development of New Crops: Needs, Procedures, Strategies and Options. (Rept. No. 102), CAST. Ames, Iowa. pp. 30.

Elder, N.G., Elmore, C.L. and Beaupre, C. 1983. Effect of preplant and pre-emergence herbicides on weed control, growth, and rubber content of transplant guayule (Parthenium argentatum). In: Gregg, E.C., Tipton, J.L. and Huang, H.T. (Eds.). Proc. Third Intern. Guayule Con. Pasadena, California. p. 582.

Estalai, A. 1985. Registration of Cal-5 guayule germplasm. Crop Sic., 25: 369-370.

Estalai, A. 1986. Registration of Cal-6 and Cal-7 guayule germplasm. Crop Sci., 26: 1261-1262.

Estalai, A. and Waines, J.G. 1987. Variation in regrowth and its implication for multiple harvest of guayule. Crop Sci., 27: 100-103.

Fangmeier, D.D., Samani, Z., Garrot, D., Jr., and Ray, D.T. 1985. Water effects on guayule rubber production. Trans. Amer. Soc. Agric. Eng., 28: 1947-1950.

Ferraris, R. 1986. Agronomic practices for the production of guayule. In: Stewart, G.A. and Lucas, S.M. (Eds.). Guayule in Australia. Potential Production of Natural Rubber from Guayule (Parthenium argentatum) in Australia. CSIRO, Australia. pp. 97-110.

Fink, D.H., Allen, S.G., Bucks, D.A., Nakayama, F.S., Powers, D. and Patterson, K. 1987. Guayule seedling emergence as related to planting depth. Proc. Seventh Guayule Rubber Soc. Conf. Annapolis, Maryland. 71 pp.

Forgacs, D., Schecter, J., Wisnick, J. and Forti, M. 1986. Guayule as an industrial crop in semiarid zones. In: Fangmeier, D.D. and Alcron, S.M. (Eds.). Proc. Fourth Intern. Guayule Res. and Develop. Conf. Tucson, Arizona. pp. 285-290.

Foster, K.E., McGinnies, W.G., Taylor, J.G., Mills, J.L., Wilkinson, R.R. Hopkins, F.C., Lawless, E.L., Maloney, J. and Wyatt, R.C. 1980. A Technology Assessment of Guayule Rubber Commercialization. Final Report for National Science Foundation. MRI Rept. Office of Arid Lands Studies. University of Arizona. Tucson, Arizona. 267 pp.

Foster, M.A., Carrillo, T.D. and Moore, J. 1985. Weed control in established guayule stands. In: Fangmeier, D.D. and Alcorn, S.M. (Eds.). Proc. Fourth Intern. Conf. Guayule Res. and Develop. Tucson, Arizona. pp. 317-320.

Foster, M.A., Carrillo, T.D. and Moore, J. 1986. Direct seeding of guayule in west Texas. In: Fangmeier, D.D. and Alcorn, S.M. (Eds.). Proc. Fourth Intern. Conf. Guayule Res. and Develop. Tucson, Arizona. pp. 71-75.

Foster, M.A., Ranne, D.W., Gomez, V.J., Jr. and Moore, J. 1987. Effect of seed conditioning, planting depth, and mulch on direct seeding of guayule. In: Seventh Ann. Guayule Rubber Soc. Conf. Annapolis, Maryland. 72 pp.

Francois, L.E. 1986. Salinity effects on four arid-zone plants. J. Arid En-viron., 11: 103-109.

Gartside, G. 1986. Guayule processing, rubber quality and possible by-products. In: Stewart, G.A. and Lucas, S.M. (Eds.). Potential Pro-duction of Natural Rubber from Guayule in Australia. CSIRO, Melbourne, Asutralia. pp. 111-132.

Gonzalez, C.L. 1988. Effect of plant density and harvesting methods of guayule on rubber and resin production under dryland conditions. J. Rio Grande Valley Hort. Soc., 41: 51-57.

Gonzalez, C.L. and Latigo, G. 1986. Effects of two herbicides on germination of guayule seed. In: Sixth Ann. Guayule Rubber Conf. College Station, Texas. 26 pp.

Hager, T., MacArthur, A., Mcintyre, D. and Seeger, R. 1979. Chemistry and structure of natural rubbers. Rubber Chem. Technol., 52: 693-709.

Hamerstrand, G.E. and Montgomery, R.R. 1984. Pilot scale guayule processing using counter-current solvent extraction equipment. Rubber Chem. Technol., 57: 344-350.

Hammond, B.L. 1959. Effect of gibberellin, sodium hypochlorite, light, and plant depth on germination of guayule seed. Amer. Soc. Agron. J., 51: 621-623.

Hammond, B.L. and Polhamus, L.R. 1965. Research on Guayule (Parthenium argentatum): 1942-1959. USDA Tech. Bull. No. 1327. pp. 157.

Higgins, R.R. and Backhaus, R.A. 1983. Long-day flower induction in guayule. In: Gregg, E.C., Tipton, T.L. and Huang, H.T. Proc. Third Intern. Guayule Conf. Pasadena, California. pp. 225-233.

Hoffman, G.J., Shannon, M.C., Maas, E.V. and Grass, L. 1988. Rubber Production of salt-stressed guayule at various plant population. Irrig. Sci., 9: 213-226.

Hunter, A.S. and Kelley, O.J. 1946. The growth and rubber content of guayule as affected by variations in soils moisture stresses. Amer. Soc. Agron. J., 38: 118-134.

Kelley, O.J. 1975. Soil-plant relationships - guayule. In: McGinnies, W.G. and Haase, E.F. (Eds.) Proc. Intern. Conf. on the Utilization Guayule. Tucson, Arizona.

Kuester, J.L. and Wang, T. 1986. Conversion of guayule residues to diesel fuel. In: Fangmeier, D.D. and Alcorn, S.M. (Eds.) Proc. Fourth Intern Conf. Guayule Res. and Develop. Tucson, Arizona. pp. 385-391.

Kumamoto, J., Scora, R.W. and Clerx, W.A. 1985. Composition of leaf oils of the genus Parthenium L. (Compositae). J. Agric. Food Chem., 33: 650-652.

Lorenzen, B. and Coates, W. 1986. Guayule seed harvesting equipment. In: Fangmeier, D.D. and Alcron, S.M. (Eds.). Proc. Fourth Intern. Conf. Guayule Res. and Develop. Tucson, Arizona. pp. 51-58.

Maas, E.V., Donovan, T.J. and Francois, L.E. 1988. Salt tolerance of irrigated guayule. Irrig. Sci., 9: 199-211.

McGinnies, W.G. 1983a. History of guayule rubber production. In: Gregg, F.C., Tipton, J.L. and Huang, H.T. (Eds.) Proc. Third Intern. Guayule Conf. Pasadena, California, pp. 35-43.

McGinnies, W.G. 1983b. Idiosyncrasies of guayule (Parthenium argentatum) under cultivation. In: Gregg, E.C., Tipton, J.L. and Huang, H.T. (Eds.). Proc. Third Intern. Guayule Con. Pasadena, California. pp. 99-104.

McGinnies, W.G. and Mills, J.L. 1980. Guayule Rubber Production. The World War II Emergency Rubber Project: A Guide to Future Development. Office of Arid Land Studies. University of Arizona, Tucson, Arizona. 200 pp.

Mihail, J.D. and Alcorn, S.M. 1986. Greenhouse evaluation of fifteen tetraploid guayule lines for differential susceptibility to charcoal rot (Microphomena phaseolina). In: Sixth Ann. Guayule Rubber Soc. Conf. College Station, Texas. 25 pp.

Mihail, J.D. and Alcorn, S.M. 1988. Improving survival of guayule seedlings from directly sown seed. Eight Ann. Guayule Rubber Soc. conf. Mesa, Arizona. 16 pp.

Mihail, J.D., Alcorn, S.M. and Ray, D.T. 1987. The effects of saline irrigation water on the development of charcoal rot of guayule. In: Seventh Ann. Guayule Rubber Soc. Conf. Washington, D.C. 43 pp.

Mitchell, J.W. 1944. Winter hardiness in guayule. Bot. Gaz., 106: 95-102.

Mitchell, J.W., Benedict, H.M. and Whiting, A.G. 1944. Growth, rubber storage, and seed production by guayule as affected by boron supply. Bot. Gaz., 106: 148-157.

Miyamotor, S. and Bucks, D.A. 1985. Water quantity and quality requirements of guayule: Current assessment. Agric. Water Manage., 10: 205-219.

Miyamotor, S., Davis, J. and Piela, K. 1984a. Water use, growth and rubber yields of four guayule selections as related to irrigation regimes. Irrig. Sci., 5: 95-103.

Miyamotor, S., Piela, K., Davis, J. and Fenn, L.G. 1984b. Salt effects on emergence and seedling mortality of guayule. Agron. J., 76: 295-300.

Miyamotor, S., Davis, J. and Madrid, L. 1989. Salt tolerance of guayule (Parthenium argentatum). Texas Agric. Expt. Sta. Bull., El Paso, Texas (In press).

Muller, C.H. 1946. Root Development and Ecological Relations of Guayule. (USDA Tech. Bull. No. 923). 114 pp.

Muthana, K.D. 1983. Guayule: Its prospects in INdian and semiarid regions. In: Fourth Ann. Guayule Rubber Soc. Conf. Riverside, California. 13 p.

Nakayama, F.S. 1984. Hydrocarbon emission and carbon balance of guayule. J. Arid Environ., 7: 353-357.

Nakayama, F.S. and Bucks, D.A. 1983. Application of a foliage temperature based crop water stress index to guayule. J. Arid Environ., 6: 269-276.

Nakayama, F.S. and Bucks, D.A. 1984. Crop water stress index and rubber yield relations for the guayule plant. Agron. J., 76: 791-794.

Naqvi, H.H. and Hanson, G.P. 1980. Recent advances in guayule seed germination procedures. Crop. Sci., 20: 501-504.

Naqvi, H.H. and Hanson, G.P. 1982. Germination and growth inhibitors in guayule (Parthenium argentatum Gray) chaff and their possible influence in seed dormancy. Amer. J. Bot., 69: 985-989.

Naqvi, H.H. and Hanson, G.P. 1983. Observations on the distribution and ecology of native guayule stands in Mexico. In: Gregg, E.C., Tipton, J.L. and Huang, H.T. (Eds.). Proc. Third Intern. Guayule Conf. Pasadena, California. pp. 145-154.

Naqvi, H.H., Burch, T.J. and Waines, J.G. 1986. Rubber biomass, and genetics evaluation of newly developed guayule germplasm. In: Sixth Ann. Guayule Rubber Soc. Conf. College Station, Texas, 15 p.

Naqvi, H.H., Hashemi, A. Davey, J.R. and Waines, J.G. 1987. Morphological, chemical and cytogenetic characters of F1 hybrids between Parthenium argentatum (guayule) + P. fructucosum var fructucosum (Asteraceae) and their potential in rubber improvement. Econ. Bot., 41: 66-77.

National Academy of Sciences, 1977. Guayule: An Alternative Source of Natural Rubber. NAS, Washington, D.C. 80 pp.

Orum, T.V. and Alcorn, S.M. 1986. Greenhouse evaluation of tolerance of triploid and tetraploid guayule lines to Verticillium dahliae. In: Fangmeier, D.D. and Alcron, S.M. (Eds.) Proc. Fourth Intern. Conf. Guayule Res. and Develop. Tucson, Arizona. pp. 329-336.

Ostler, W.K., Alder, G.M. and Garza, R. 1983. The community ecology of guayule. In: Gregg, E.C., Tipton, J.L. and Huang, H.T. (Eds.). Third Intern. Guayule Con. Pasadena, California. pp. 155-169.

Patterson-Jones, J.C. 1983. The South African guayule program. In: Fourth Ann. Guayule Rubber Soc. Conf. Riverside, California. 7 p.

Patterson-Jones, J.C. 1986. An update on the South African guayule programme. In: Fangmeier, D.D. and Alcorn, S.M. (Eds.). Fourth Intern. Guayule Res. and Develop. Conf. Tucson, Arizona. pp. 273-284.

Powers, L. and Rollins, R.C. 1945. Reproduction and pollination studies on guayule, Parthenium argentatum Gray, and P. incanum. H.B.K. J. Amer. Soc. Agron., 37: 96-112.

Presley, J.T. 1975. Diseases of guayule. In: Mcginnies, W.F. and Haase, E.F. (Eds.). An International Conference on the Utilization of Guayule. Tucson, Arizona. pp. 38-41.

Ranne, D.W., Foster, M.A., Harbour, J.M. and Moore, J. 1986. Seed cleaning techniques for guayule. In: Sixth Ann. Guayule Rubber Soc. Conf. College Station, Texas. 24 p.

Rawlins, W.H.M. 1986. Guayule economics. In: Stewart, G.A. and Lucas, S.M. (Eds.). Potential Production of Natural Rubber from Guayule in Australia. CSIRO, Melbourne, Australia. pp. 133-150.

Ray, D.T., Garrot, D.J., Jr., Fangmeier, D.D. and Coates, W. 1986. Clipping as an agronomic practice in guayule. In: Proc. Fourth INtern. Conf. Guayule Res. and Develop. Tucson, Arizona. pp. 185-191.

Ray, D.T. and Thompson, A.E. 1986. Chemical and cytological characterization of the original 26 USDA lines. In: Sixth Ann. Guayule Rubber Soc. Conf., College Station, Texas. 20 pp.

Reddy, R.A. and Das, V.S.R. 1988. Enhanced rubber accumulation and rubber transferase activity in guayule under stress. J. Plant Physiol., 133: 152-155.

Rollins, R. 1975. Genetics and plant breeding. In: McGinnies, W.G. and Haase, E.F. (Eds.). Proc. Intern. Conf. on the Utilization of Guayule. University of Arizona. pp. 111-117.

Romney, V.E., York, G.T. and Cassidy, T.P. 1945. Effect of lygus spp on seed production and growth of guayule in California. J. Econ. Ent. 38: 45-50.

Rubis, D.D. 1983. Influence of irrigation and fertilizer treatments on guayule. In: Fourth Ann. Guayule Rubber Soc. Conf. Riverside, California. 47 pp.

Schloman, W.W., Jr. 1986. The utilization and economic impact of by-products derived from guayule. In: Sixth Ann. Guayule Rubber Soc. Conf. College Station, Texas. 73 pp.

Schloman, W.W., Jr., Hively, R.A., Krishen, A. and Andrews, A.M. 1983. Guayule by-product evaluation: Extract characterization. J. Agric. Food Chem., 31: 973-975.

Scora, R.W. and Kumamoto, J. 1979. Essential leaf oils of Parthenium argentatum A Gray. J. Agric. Food Chem., 27: 642-643.

Serier, J.B. 1983. IRCA's activities in guayule research. In: Fourth Ann. Guayule Rubber Soc. Conf. Riverside, California. 8 pp.

Sidhu, D.D. and Behl, H.M. 1987. Fungal and nematode pathogenicity of guayule in India. In: Seventh Ann. Guayule Rubber Soc. Conf. Annapolis, Maryland. 44 pp.

Srivastava, G.S., Ramadas, V.S., Shah, J.J., Gnanam, A., Iyengar, E.R.R., Behl, H.M., Bhatia, K. and Sethuraj, M.R. 1986. Guayule research in India-An overview. In: Proc. Fourth Intern. Conf. on Guayule Res. and Develop. Tucson, Arizona. pp. 359-404.

Swanson, C.L. and Buchanan, R.A. 1979. Molecular weights of natural rubbers from selected temperate zone plants. J. Appl Polymer Sci., 23: 743-748.

Taylor, J.O. 1986. Guayule-The New Zealand experience. In: Fangmeier, D.D. and Alcorn, S.M. (Eds.). Proc. Fourth Intern. Guayule Res. and Develop. Conf. Tucson, Arizona. pp. 23-29.

Thames, S.F. and Kaleem, K. 1988. Guayule coproduct development. In: Eight Ann. Guayule Rubber Soc. Conf. Mesa, Arizona. 52 pp.

Thomas, J.R. 1986. Comparative phosphorus requirements of guayule cultivars. In: FAngmeier, D.D. and Alcorn, S.M. (Eds.). Proc. Fourth Intern. Conf. Guayule Res. and Develop. Tucson, Arizona. pp. 229-235.

Thomas, S.H. and Goddard, C.J. 1986. Susceptibility of guayule transplants to plant-parasitic nematode injury. In: Fangmeier, D.D. and Alcron, S.M. (Eds.). Proc. Fourth Intern. Conf. Guayule Res. and Develop. Tucson, Arizona. pp. 341-346.

Thompson, A.E. and Ray, D.T. 1988. Breeding guayule. Janick, J. (Ed.). Plant Breeding Review, 6: 93-165.

Tingey, D.C. 1952. Effect of spacing, irrigation, and fertilization on rubber production in guayule sown directly in the field. Agron. J., 44: 298-302.

Tipton, J.L., Craver, J.L. and Blackwell, J. 1981. A method for harvesting, cleaning and treating achenes of guayule (Parthenium argentatum Gray). Hort. Sci., 16: 296-298.

Tysdal, H.M. 1975. Genetics and agronomic research. In: McGinnies, W.G. and Haase, E.F. (Eds.). Proc. Intern. Con. on the Utilization Guayule. Tucson, Arizona.

Tysdal, H.M. 1983. Blueprint for guayule improvement. In: Gregg, E.C., Tipton, J.L. and Huang, H.T. (Eds.). Proc. Third Intern. Guayule Conf. Pasadena, California. pp. 297-307.

Tysdal, H.M., Estilai, A., Siddiqui, I.A., Knowles, P.F. and Madison, P.F. 1983. Registrationof four guayule germplasms (Parthenium argentatum). Crop Sci., 23: 189.

Wagner, J.P., Engler, C.R., Parma, D.G. and Lusas, E.W. 1986. Development of a process for solvent extraction of natural rubber from guayule. In: Fangmeier, D.D. and Alcorn, S.M. (Eds.). Proc. Fourth Intern. Guayule Res. and Develop. Conf. Tucson, Arizona. pp. 357-369.

Wagner, J.P., Soderman, K.L. and Parma, D.G. 1988. Toxic species evolution from guayule fireplace logs. In: Eight Ann. Guayule Rubber Soc. Conf. Mesa, Arizona. 51 pp.

Weber, D.J., Davis, T.D., MacArthur, E.D. and Sankhla, N. 1985. Chrysothamnus nauseous (rubber rabbitbrush): Multiple-use shrub of the desert. Desert Plants, 7: 172-180, 208-210.

Weihe, D.L. and Nivert, J.J. 1983. Assessing guayule's potential. An update of Firestone's development program and economic studies. In: Gregg, E.C., Tipton, J.L. and Huang, H.T. (Eds.). Proc. Third Intern. Guayule Conf. Pasadena, California. pp. 115-125.

Whaley, W.G. 1948. Rubber - The primary sources for American production. Econ. Bot., 2: 198-216.

Whitworth, J.W. 1983. Guayule research in New Mexico. In: Gregg, E.C., Tipton, J.L. and Huang, H.T. (Eds.). Proc. Third Intern. Guayule Conf. Pasadena, California. pp. 105-110.

Winkler, D.S., Schostarez, H.J. and Stephens, H.L. 1978. Gum properties and filled stocks in guayule rubber. In: Campos-Lopez, E. and McGinnies, W.G. (Eds.). Guayule: Reencuentro en el Desierto. Publ. 371. CIQA, Saltiollo, Coahuila, Mexico. pp. 265-280.

Winkler, D.S. and Stephens, H.L. 1978. Plastificaion effect on guayule resin in raw rubber. In: Campos-Lopez, E. and McGinnies, W.G. (Eds.). Guayule: Encuentro en al Desierto. Publ. 371, CIQA, Saltillo, Coahuila, Mexico. pp. 303-314.

Wolek, F.W. 1985. Guayule: A case study in civilian technology. Technol. Soc., 7: 11-23.

Wright, N.G. and Connell, J. 1983. Economics and commercial guayule production in the southwest. In: Fourth Ann. Guayule Rubber Soc. Conf. Riverside, California. 5 pp.

Yokoyama, H., Keithly, J.H. and Hayman, E.P. 1987. Recent studies on regulation of biological responses in guayule. In: Seventh Ann. Guayule Rubber Soc. Conf. Annapolis, Maryland. 15 pp.

Subject Index

602

Printed and bound by CPI Group (UK) Ltd, Croydon, CR0 4YY

13/10/2024

01773610-0001